T0402055

SpringerWienNewYork

Inna Shingareva
Carlos Lizárraga-Celaya

Solving Nonlinear Partial Differential Equations with Maple and Mathematica

SpringerWienNewYork

Prof. Dr. Inna Shingareva
Department of Mathematics, University of Sonora, Sonora, Mexico
inna@gauss.mat.uson.mx

Dr. Carlos Lizárraga-Celaya
Department of Physics, University of Sonora, Sonora, Mexico
carlos@raramuri.fisica.uson.mx

SpringerWienNewYork is part of
Springer Science+Business Media
springer.at

Cover Design: WMX Design, 69126 Heidelberg, Germany
Typesetting: Camera ready by the authors

With 20 Figures

Printed on acid-free and chlorine-free bleached paper
SPIN: 80021221

Library of Congress Control Number: 2011929420

ISBN 978-3-7091-0516-0 e-ISBN 978-3-7091-0517-7
DOI 10.1007/978-3-7091-0517-7
SpringerWienNewYork

Preface

The study of partial differential equations (PDEs) goes back to the 18th century, as a result of analytical investigations of a large set of physical models (works by Euler, Cauchy, d'Alembert, Hamilton, Jacobi, Lagrange, Laplace, Monge, and many others). Since the mid 19th century (works by Riemann, Poincarè, Hilbert, and others), PDEs became an essential tool for studying other branches of mathematics.

The most important results in determining explicit solutions of nonlinear partial differential equations have been obtained by S. Lie [91]. Many analytical methods rely on the Lie symmetries (or symmetry continuous transformation groups). Nowadays these transformations can be performed using computer algebra systems (e.g., *Maple* and *Mathematica*).

Currently PDE theory plays a central role within the general advancement of mathematics, since they help us to describe the evolution of many phenomena in various fields of science, engineering, and numerous other applications.

Since the 20th century, the investigation of nonlinear PDEs has become an independent field expanding in many research directions. One of these directions is, symbolic and numerical computations of solutions of nonlinear PDEs, which is considered in this book.

It should be noted that the main ideas on practical computations of solutions of PDEs were first indicated by H. Poincarè in 1890 [121]. However the solution techniques of such problems required such technology that was not available or was limited at that time. In modern day mathematics there exist computers, supercomputers, and computer algebra systems (such as *Maple* and *Mathematica*) that can aid to perform various mathematical operations for which humans have limited capacity, and where symbolic and numerical computations play a central role in scientific progress.

It is known that there exist various analytic solution methods for special nonlinear PDEs, however in the general case there is no central theory for nonlinear PDEs. There is no unified method that can be applied for all types of nonlinear PDEs. Although the "nonlinearity" makes each equation or each problem unique, we have to discover new methods for solving at least a class of nonlinear PDEs. Moreover, the functions and data in nonlinear PDE problems are frequently defined in discrete points. Therefore we have to study numerical approximation methods for nonlinear PDEs.

Scientists usually apply different approaches for studying nonlinear partial differential equations.

In the present book, we follow different approaches to solve nonlinear partial differential equations and nonlinear systems with the aid of computer algebra systems (CAS), *Maple* and *Mathematica*. We distinguish such approaches, in which it is very useful to apply computer algebra for solving nonlinear PDEs and their systems (e.g., algebraic, geometric-qualitative, general analytical, approximate analytical, numerical, and analytical-numerical approaches).

Within each approach we choose the most important and recently developed methods which allow us to construct solutions of nonlinear PDEs or nonlinear systems (e.g., transformations methods, traveling-wave and self-similarity methods, ansatz methods, method of separation of variables and its generalizations, group analysis methods, method of characteristics and its generalization, qualitative methods, Painlevè test methods, truncated expansion methods, Hirota method and its generalizations, Adomian decomposition method and its generalizations, perturbation methods, finite difference methods, method of lines, spectral collocation methods).

The book addresses a wide set of nonlinear PDEs of various types (e.g., parabolic, hyperbolic, elliptic, mixed) and orders (from the first-order up to n-th order). These methods have been recently applied in numerous research works, and our goal in this work will be the development of new computer algebra procedures, the generalization, modification, and implementation of most important methods in *Maple* and *Mathematica* to handle nonlinear partial differential equations and nonlinear systems.

The emphasis of the book is given in how to construct different types of solutions (exact, approximate analytical, numerical, graphical) of numerous nonlinear PDEs correctly, easily, and quickly with the aid of CAS. With this book the reader can learn to understand and solve numerous nonlinear PDEs included into the book and many other differential equations, simplifying and transforming the equations and solutions, arbitrary functions and parameters, presented in the book.

This book contains many comparisons and relationships between various types of solutions, different methods and approaches, the results obtained in *Maple* and *Mathematica*, which provide a more deep understanding of the subject.

Among the large number of CAS available, we choose two systems, *Maple* and *Mathematica*, that are used by students, research mathematicians, scientists, and engineers worldwide. As in the our other books, we propose the idea to use in parallel both systems, *Maple* and *Mathematica*, since in many research problems frequently it is required to compare independent results obtained by using different computer algebra systems, *Maple* and/or *Mathematica*, at all stages of the solution process.

One of the main points (related to CAS) is based on the implementation of a whole solution method, e.g., starting from an analytical derivation of exact governing equations, constructing discretizations and analytical formulas of a numerical method, performing numerical procedure, obtaining various visualizations, and comparing the numerical solution obtained with other types of solutions (considered in the book, e.g., with asymptotic solution).

This book is appropriate for graduate students, scientists, engineers, and other people interested in application of CAS (*Maple* and/or *Mathematica*) for solving various nonlinear partial differential equations and systems that arise in science and engineering. It is assumed that the areas of mathematics (specifically concerning differential equations) considered in the book have meaning for the reader and that the reader has some knowledge of at least one of these popular computer algebra systems (*Maple* or *Mathematica*). We believe that the book can be accessible to students and researchers with diverse backgrounds.

The core of the present book is a large number of nonlinear PDEs and their solutions that have been obtained with *Maple* and *Mathematica*. The book consists of 7 Chapters, where different approaches for solving nonlinear PDEs are discussed: introduction and analytical approach via predefined functions, algebraic approach, geometric-qualitative approach, general analytical approach and integrability for nonlinear PDEs and systems (Chapters 1–4), approximate analytical approach for nonlinear PDEs and systems (Chapter 5), numerical approach and analytical-numerical approach (Chapters 6, 7). There are two Appendices. In Appendix A and B, respectively, the computer algebra systems *Maple* and *Mathematica* are briefly discussed (basic concepts and programming language). An updated Bibliography and expanded Index are included to stimulate and facilitate further investigation and interest in future study.

In this book, following the most important ideas and methods, we propose and develop new computer algebra ideas and methods to obtain analytical, numerical, and graphical solutions for studying nonlinear partial differential equations and systems. We compute analytical and numerical solutions via predefined functions (that are an implementation of known methods for solving PDEs) and develop new procedures for constructing new solutions using *Maple* and *Mathematica*. We show a very helpful role of computer algebra systems for analytical derivation of numerical methods, calculation of numerical solutions, and comparison of numerical and analytical solutions.

This book does not serve as an automatic translation the codes, since one of the ideas of this book is to give the reader a possibility to develop problem-solving skills using both systems, to solve various nonlinear PDEs in both systems. To achieve equal results in both systems, it is not sufficient simply "to translate" one code to another code. There are numerous examples, where there exists some predefined function in one system and does not exist in another. Therefore, to get equal results in both systems, it is necessary to define new functions knowing the method or algorithm of calculation. In this book the reader can find several definitions of new functions. However, if it is sufficiently long and complicated to define new functions, we do not present the corresponding solution (in most cases, this is *Mathematica* solutions). Moreover, definitions of many predefined functions in both systems are different, but the reader expects to achieve the same results in both systems. There are other "thin" differences in results obtained via predefined functions (e.g., between predefined functions `pdsolve` and `DSolve`), etc.

The programs in this book are sufficiently simple, compact and at the same time detailed programs, in which we tried to make each one to be understandable without any need of the author's comment. Only in some more or less difficult cases we put some notes about technical details. The reader can obtain an amount of serious analytical, numerical, and graphical solutions by means of a sufficient compact computer code (that it is easy to modify for another problem).

We believe that the best strategy in understanding something, consists in the possibility to modify and simplify the programs by the reader (having the correct results). Each reader may prefer another style of programming and that is fine. Therefore the authors give to the reader a possibility to modify, simplify, experiment with the programs, apply it for solving other nonlinear partial differential equations and systems, and to generalize them. The only thing necessary, is to understand the given solution. Moreover, in this book the authors try to show different styles of programming to the reader, so each reader can choose a more suitable style of programming.

When we wrote this book, the idea was to write a concise practical book that can be a valuable resource for advanced-undergraduate and graduate students, professors, scientists and research engineers in the fields of mathematics, the life sciences, etc., and in general people interested in application of CAS (*Maple* and/or *Mathematica*) for constructing various types of solutions (exact, approximate analytical, numerical, graphical) of numerous nonlinear PDEs and systems that arise in science and engineering. Moreover, another idea was not to depend on a specific version of *Maple* or *Mathematica*, we tried to write programs that allow the reader to solve a nonlinear PDE in *Maple* and *Mathematica* for any sufficiently recent version (although the dominant versions for *Maple* and *Mathematica* are 14 and 8).

We would be grateful for any suggestions and comments related to this book. Please send your e-mail to `inna@gauss.mat.uson.mx` or `carlos.lizarraga@correo.fisica.uson.mx`.

We would like to express our gratitude to the Mexican Department of Public Education (SEP) and the National Council for Science and Technology (CONACYT), for supporting this work under grant no. 55463. Also we would like to express our sincere gratitude to Prof. Andrei Dmitrievich Polyanin, for his helpful ideas, commentaries, and inspiration that we have got in the process of writing the three chapters for his "Handbook of Nonlinear Partial Differential Equations" (second edition). Finally, we wish to express our special thanks to Mr. Stephen Soehnlen and Mag. Wolfgang Dollhäubl from Springer Vienna for their invaluable and continuous support.

May 2011

Inna Shingareva
Carlos Lizárraga-Celaya

Contents

Chapter 1
Introduction

This chapter deals with basic concepts and a set of important nonlinear partial differential equations arising in a wide variety of problems in applied sciences. Various types of nonlinear PDEs, nonlinear systems, and their solutions are discussed. Applying various predefined functions embedded in *Maple* and *Mathematica*, we construct and visualize various types of analytical solutions of nonlinear PDEs and nonlinear systems. Moreover, applying the *Maple* predefined function `pdsolve`, we construct exact solutions of nonlinear PDEs and their systems subject to initial and/or boundary conditions.

1.1 Basic Concepts

A partial differential equation for an unknown function $u(x_1,\ldots,x_n)$ or *dependent variable* is a relationship between u and its partial derivatives and can be represented in the *general form*:

$$\mathcal{F}\left(x_1,x_2,\ldots,u,u_{x_1},u_{x_2},\ldots,u_{x_1x_1},u_{x_1x_2},\ldots,u_{x_ix_j},\ldots\right)=0, \quad (1.1)$$

where $\mathcal{F}$ is a given function, $u=u(x_1,\ldots,x_n)$ is an unknown function of the *independent variables* $(x_1,\ldots,x_n)$. We denote the partial derivatives $u_{x_1}=\partial u/\partial x_1$, etc. This equation is defined in a domain $\mathcal{D}$, where $\mathbf{x}=(x_1,\ldots,x_n)\in\mathcal{D}\subset\mathbb{R}^n$. The partial differential equation (1.1) can be written in the *operator form*:

$$D_{\mathbf{x}}u(\mathbf{x})=\mathcal{G}(\mathbf{x}), \quad (1.2)$$

where $D_{\mathbf{x}}$ is a partial differential operator and $\mathcal{G}(\mathbf{x})$ is a given function of *independent variables* $\mathbf{x}=(x_1,\ldots,x_n)$.

Definition 1.1 The operator $D_{\mathbf{x}}$ is called a *linear operator* if the property $D_{\mathbf{x}}(au+bv)=aD_{\mathbf{x}}u+bD_{\mathbf{x}}v$ is valid for any functions, u,v, and any constants, a,b.

1.1.1 Types of Partial Differential Equations

Definition 1.2 Partial differential equation (1.2) is called *linear* if $D_{\mathbf{x}}$ is a linear partial differential operator and *nonlinear* if $D_{\mathbf{x}}$ is not a linear partial differential operator.

Definition 1.3 Partial differential equation (1.2) is called *inhomogeneous* (or *nonhomogeneous*) if $\mathcal{G}(\mathbf{x}) \neq 0$ and *homogeneous* if $\mathcal{G}(\mathbf{x}) = 0$.

For example, the nonlinear first-order and the second-order partial differential equations, e.g., in two independent variables $\mathbf{x} = (x_1, x_2) = (x, y)$, can be represented, respectively, as follows:

$$\mathcal{F}(x, y, u, u_x, u_y) = 0, \quad \mathcal{F}(x, y, u, u_x, u_y, u_{xx}, u_{xy}, u_{yy}) = 0. \quad (1.3)$$

These equations are defined in a domain $\mathcal{D}$, where $(x, y) \in \mathcal{D} \subset \mathbb{R}^2$, $\mathcal{F}$ is a given function, $u = u(x, y)$ is an unknown function (or dependent variable) of the independent variables $(x,\ y)$. These equations can be written in terms of *standard notation*:

$$\mathcal{F}(x, y, u, p, q) = 0, \quad \mathcal{F}(x, y, u, u_x, u_y, p, q, r) = 0, \qquad (1.4)$$

where $p = u_x$, $q = u_y$ (for the first-order PDE), and $p = u_{xx}$, $q = u_{xy}$, $r = u_{yy}$ (for the second-order PDE).

Definition 1.4 Partial differential equations (1.3) are called *quasilinear* if they are linear in first/second-partial derivatives of the unknown function $u(x, y)$.

Definition 1.5 Partial differential equations (1.3) are called *semilinear* if their coefficients in first/second-partial derivatives are independent of u.

NOTATION. In this book we will use the following conventions in

Maple:

`_Cn` `(n=1,2,...)`, for arbitrary constants; `_Fn`, for arbitrary functions;
`_c[n]`, for arbitrary constants while separating the variables;
`_s`, for a parameter in the characteristic system;
`&where`, for a solution structure, $\_\varepsilon$, for a Lie group parameter,

and

Mathematica:

`C[n]` `(n=1,2,...)`, for arbitrary constants or arbitrary functions.*

*In general, arbitrary parameters can be specified, e.g., F_1, F_2, ..., by applying the option `GeneratedParameters->(Subscript[F,#]&)` of the predefined function `DSolve`.

Also we introduce the following notation for the solutions in

Maple and Mathematica:
`Eqn` and `eqn`*, for equations (`n=1,2,...`);
`PDEn/ODEn` and `pden/oden`, for PDEs/ODEs;
`trn`, for transformations; `Sysn` and `sysn`, for systems;
`IC`, `BC`, `IBC` and `ic`, `bc`, `ibc`, for initial and/or boundary conditions;
`Ln` and `ln`, for lists of expressions; `Gn` and `gn`, for graphs of solutions.

Problem 1.1 *Linear, semilinear, quasilinear, and nonlinear equations. Standard notation.* We consider the following linear, semilinear, quasilinear, and nonlinear PDEs:

$$u_{xx}+u_{yy}=0, \quad xu_x+yu_y=x^2+y^2, \quad v_t+vv_x=0, \quad u_x^2+u_y^2=n^2(x,y).$$

Verify that these equations, written in the standard notation (1.4), have the form: $p+q=0$, $xp+yq=x^2+y^2$, $q+vp=0$, and $p^2+q^2=n^2(x,y)$.

Maple:

```
with(PDEtools): declare(u(x,y),v(x,t));
U,V:=diff_table(u(x,y)),diff_table(v(x,t));
Eq1:=U[x,x]+U[y,y]=0;     Eq2:=x*U[x]+y*U[y]=x^2+y^2;
Eq3:=V[t]+v(x,t)*V[x]=0; Eq4:=U[x]^2+U[y]^2=n(x,y)^2;
tr1:=(x,y,U)->{U[x,x]=p,U[y,y]=q}; tr2:=(x,y,U)->{U[x]=p,U[y]=q};
F1:=(p,q)->subs(tr1(x,y,U),Eq1); F2:=(p,q)->subs(tr2(x,y,U),Eq2);
F3:=(p,q)->subs(tr2(x,t,V),Eq3); F4:=(p,q)->subs(tr2(x,y,U),Eq4);
F1(p,q); F2(p,q); F3(p,q); F4(p,q);
```

Mathematica:

```
{eq1=D[u[x,y],{x,2}]+D[u[x,y],{y,2}]==0, eq2=x*D[u[x,y],x]
 +y*D[u[x,y],y]==x^2+y^2, eq3=D[v[x,t],t]+v[x,t]*D[v[x,t],x]==0,
 eq4=D[u[x,y],x]^2+D[u[x,y],y]^2==n[x,y]^2}
tr1[x_,y_,u_]:={D[u[x,y],{x,2}]->p,D[u[x,y],{y,2}]->q};
tr2[x_,y_,u_]:={D[u[x,y],x]->p,D[u[x,y],y]->q};
f1[p_,q_]:=eq1/.tr1[x,y,u]; f2[p_,q_]:=eq2/.tr2[x,y,u];
f3[p_,q_]:=eq3/.tr2[x,t,v]; f4[p_,q_]:=eq4/.tr2[x,y,u];
{f1[p,q], f2[p,q], f3[p,q], f4[p,q]}
```
□

*Since all *Mathematica* functions begin with a capital letter, it is best to begin with a lower-case letter for all user-defined symbols.

Now let us consider the most important classes of the second-order PDEs, i.e., semilinear, quasilinear, and nonlinear equations.

For the semilinear second-order PDEs, we consider the classification of equations (that does not depend on their solutions and it is determined by the coefficients of the highest derivatives) and the reduction of a given equation to appropriate canonical and normal forms.

Let us introduce the new variables $a=\mathfrak{F}_p$, $b=\frac{1}{2}\mathfrak{F}_q$, $c=\mathfrak{F}_r$, and calculate the discriminant $\delta=b^2-ac$ at some point. Depending on the sign of the discriminant δ, the type of equation at a specific point can be *parabolic* (if $\delta=0$), *hyperbolic* (if $\delta>0$), and *elliptic* (if $\delta<0$). Let us call the following equations

$$u_{y_1y_2}=f_1(y_1,y_2,u,u_{y_1},u_{y_2}), \quad u_{z_1z_1}-u_{z_2z_2}=f_2(z_1,z_2,u,u_{z_1},u_{z_2}),$$

respectively, the *first canonical form* (or *normal form*) and the *second canonical form* for hyperbolic PDEs.

Problem 1.2 *Semilinear second-order equation. Classification, normal and canonical forms.* Let us consider the semilinear second-order PDE

$$-2y^2u_{xx}+\tfrac{1}{2}x^2u_{yy}=0.$$

Verify that this equation is *hyperbolic* everywhere except at the point $x=0$, $y=0$, find a change of variables that transforms the PDE to the *normal form*, and determine the *canonical form*.

1. Classification. In the standard notation (1.4), this semilinear equation takes the form $F_1=-2y^2p+\frac{1}{2}x^2r=0$, the new variables $a=-2y^2$, $b=0$, $c=\frac{1}{2}x^2$ (`tr2(F1)`), and the discriminant $\delta=b^2-ac=x^2y^2$ (`delta1`) is positive except the point $x=0$, $y=0$.

Maple:

```
with(PDEtools): declare(u(x,y),F1(p,r,q)); U:=diff_table(u(x,y));
PDE1:=-2*y^2*U[x,x]+x^2*U[y,y]/2=0; show;
tr1:=(x,y,U)->{U[x,x]=p,U[y,y]=r,U[x,y]=q};
tr2:=F->{a=diff(lhs(F(p,q,r)),p),b=1/2*diff(lhs(F(p,q,r)),q),
         c=diff(lhs(F(p,q,r)),r)}; delta:=b^2-a*c;
F1:=(p,r,q)->subs(tr1(x,y,U),PDE1); F1(p,r,q); tr2(F1);
delta1:=subs(tr2(F1),delta)-rhs(F1(p,r,q));
is(delta1,'positive'); coulditbe(delta1,'positive');
```

Mathematica:

```
pde1=-2*y^2*D[u[x,y],{x,2}]+x^2*D[u[x,y],{y,2}]/2==0
tr1[x_,y_,u_]:={D[u[x,y],{x,2}]->p,D[u[x,y],{y,2}]->r,
 D[u[x,y],{x,y}]->q}; tr2[f_]:={a->D[f[p,q,r][[1]],p],
 b->1/2*D[f[p,q,r][[1]],q],c->D[f[p,q,r][[1]],r]};
f1[p_,r_,q_]:=pde1/.tr1[x,y,u]; delta=b^2-a*c
{f1[p,r,q], tr2[f1], delta1=delta/.tr2[f1]-f1[p,r,q][[2]]}
{Reduce[delta1>0], FindInstance[delta1>0,{x,y}]}
```

The same result can be obtained, in both systems with the principal part coefficient matrix as follows:

Maple:

```
interface(showassumed=0): assume(x<0 or x>0, y<0 or y>0);
with(LinearAlgebra): A1:=Matrix([[-2*y^2,0],[0,x^2/2]]);
D1:=Determinant(A1); is(D1,'negative'); coulditbe(D1,'negative');
```

Mathematica:

```
{a1={{-2*y^2,0},{0,x^2/2}},d1=Det[a1],Reduce[d1<0],
 FindInstance[d1<0,{x,y}]}
```

Here we calculate the determinant `D1` of the matrix `A1`. The PDEs can be classified according to the eigenvalues of the matrix `A1`, i.e., depending on the sign of `D1`: if `D1=0`, parabolic, if `D1<0`, hyperbolic, and `D1>0`, elliptic equations.

2. Normal and canonical forms. Let us find a change of variables that transforms the PDE to the normal form $v_{\eta\xi}+\dfrac{v_\xi\eta-v_\eta\xi}{2(\eta^2-\xi^2)}=0$, and determine the canonical form $v_{\lambda\lambda}-v_{\mu\mu}+\frac{1}{2}\Big(\frac{v_\lambda}{\lambda}-\frac{v_\mu}{\mu}\Big)=0$:

Maple:

```
with(LinearAlgebra): with(VectorCalculus): with(PDEtools):
declare(v(xi,eta)); interface(showassumed=0):
vars:=x,y; varsN:=xi,eta; assume(x<0 or x>0,y<0 or y>0);
Op1:=Expr->subs(y=y(x),Expr); Op2:=Expr->subs(y(x)=y,Expr);
A1:=Matrix([[-2*y^2,0],[0,x^2/2]]); D1:=Determinant(A1);
is(D1,'negative'); coulditbe(D1,'negative');
m1:=simplify((-A1[1,2]+sqrt(-D1))/A1[1,1],radical,symbolic);
m2:=simplify((-A1[1,2]-sqrt(-D1))/A1[1,1],radical,symbolic);
```

```
Eq1:=dsolve(diff(y(x),x)=-Op1(m1),y(x));
Eq11:=lhs(Eq1[1])^2=rhs(Eq1[1])^2; Eq12:=solve(Eq11,_C1);
g1:=Op2(Eq12); Eq2:=dsolve(diff(y(x),x)=-Op1(m2),y(x));
Eq21:=lhs(Eq2[1])^2=rhs(Eq2[1])^2; Eq22:=solve(Eq21,_C1);
g2:=Op2(Eq22); Jg:=Jacobian(Vector(2,[g1,g2]),[vars]);
dv:=Gradient(v(varsN),[varsN]); ddv:=Hessian(v(varsN),[varsN]);
ddu:=Jg^%T.ddv.Jg+add(dv[i]*Hessian(g||i,[vars]),i=1..2);
Eq3:=simplify(Trace(A1.ddu))=0;
tr1:={isolate(subs(isolate(g1=xi,x^2),g2=eta),y^2),
      isolate(subs(isolate(g1=xi,y^2),g2=eta),x^2)};
NormalForm:=collect(expand(subs(tr1,Eq3)),diff(v(varsN),varsN));
c1:=coeff(lhs(NormalForm),diff(v(varsN),varsN));
NormalFormF:=collect(NormalForm/c1,diff(v(varsN),varsN));
CanonicalForm:=expand(expand(dchange(
 {xi=lambda+mu,eta=mu-lambda},NormalFormF))*(-4));
```

Mathematica:

```
jacobianM[f_List?VectorQ,x_List]:=Outer[D,f,x]/;Equal@@(
 Dimensions/@{f,x}); hessianH[f_,x_List?VectorQ]:=D[f,{x,2}];
gradF[f_,x_List?VectorQ]:=D[f,{x}]; op1[expr_]:=expr/.y->y[x];
op2[expr_]:=expr/.y[x]->y; {vars=Sequence[x,y],
 varsN=Sequence[xi,eta], a1={{-2*y^2,0},{0,x^2/2}}, d1=Det[a1],
 Reduce[d1<0],FindInstance[d1<0,{x,y}], m1=Assuming[{x>0,y>0},
 Simplify[(-a1[[1,2]]+Sqrt[-d1])/a1[[1,1]]]], m2=Assuming[
 {x>0,y>0},Simplify[(-a1[[1,2]]-Sqrt[-d1])/a1[[1,1]]]]}
{eq1=DSolve[D[y[x],x]==-op1[m1],y[x],x], eq11=eq1[[1,1,1]]^2==
 eq1[[1,1,2]]^2,eq12=Solve[eq11,C[1]][[1,1,2]], g[1]=Expand[
 op2[eq12]*2], eq2=DSolve[D[y[x],x]==-op1[m2],y[x],x], eq21=
 eq2[[1,1,1]]^2==eq2[[1,1,2]]^2,eq22=Solve[eq21,C[1]][[1,1,2]],
 g[2]=Expand[op2[eq22]*2]}
{jg=jacobianM[{g[1],g[2]},{vars}], dv=gradF[v[varsN],{varsN}],
 ddv=hessianH[v[varsN],{varsN}]}
{ddu=Transpose[jg].ddv.jg+Sum[dv[[i]]*hessianH[g[i],{vars}],
 {i,1,2}], eq3=Simplify[Tr[a1.ddu]]==0, tr0={y^2->Y,x^2->X},
 tr01={Y->y^2,X->x^2}, tr1=Flatten[{Expand[Solve[First[g[2]==
 eta/.{Solve[g[1]==xi/.tr0,X]/.tr01}/.tr0],Y]/.tr01],Expand[
 Solve[First[g[1]==xi/.{Solve[g[2]==eta/.tr0,Y]/.tr01}/.tr0],
 X]/.tr01]}], nForm=Collect[Expand[eq3/.tr1],D[v[varsN],varsN]]}
c1=Coefficient[nForm[[1]],D[v[varsN],varsN]]
normalFormF=Collect[Thread[nForm/c1,Equal],D[v[varsN],varsN]]
nF[x_,t_]:=D[D[u[x,t],x],t]+(2*t*D[u[x,t],x]
          -2*x*D[u[x,t],t])/(4*t^2-4*x^2)==0; nF[xi,eta]
```

```
tr2={xi->lambda+mu,eta->mu-lambda}; nFT[v_]:=((Simplify[
 nF[xi,eta]/.u->Function[{xi,eta},u[(xi-eta)/2,(xi+eta)/2]]])
 /.tr2//ExpandAll)/.{u->v}; canonicalForm=nFT[v]
```

□

Problem 1.3 *Semilinear second-order equation. Classification, normal and canonical forms.* Let us consider the semilinear second-order PDE

$$x^2u_{xx}+2xyu_{xy}+y^2u_{yy}=0.$$

Verify that the PDE is *parabolic* everywhere and that the normal and canonical forms of the PDE, respectively, are $x^2v_{\xi\xi}=0$, $v_{\xi\xi}=0$.

Maple:

```
with(LinearAlgebra): with(VectorCalculus): with(PDEtools):
declare(v(xi,eta)); Op1:=Expr->subs(y=y(x),Expr);
Op2:=Expr->subs(y(x)=y,Expr); vars:=x,y; varsN:=xi,eta;
A1:=Matrix([[x^2,x*y],[x*y,y^2]]); D1:=Determinant(A1);
m1:=simplify((-A1[1,2]+sqrt(-D1))/A1[1,1],radical,symbolic);
Eq1:=dsolve(diff(y(x),x)=-Op1(m1),y(x)); Eq11:=solve(Eq1,_C1);
g1:=Op2(Eq11); g2:=x; Jg:=Jacobian(Vector(2,[g1,g2]),[vars]);
dv:=Gradient(v(varsN),[varsN]); ddv:=Hessian(v(varsN),[varsN]);
ddu:=Jg^%T.ddv.Jg+add(dv[i]*Hessian(g||i,[vars]),i=1..2);
NorF:=simplify(Trace(A1.ddu))=0; CanF:=expand(NorF/x^2);
```

Mathematica:

```
jacobianM[f_List?VectorQ, x_List]:=Outer[D,f,x]/;Equal@@(
 Dimensions/@{f,x}); hessianH[f_,x_List?VectorQ]:=D[f,{x,2}];
gradF[f_,x_List?VectorQ]:=D[f,{x}]; op1[expr_]:=expr/.y->y[x];
op2[expr_]:=expr/.y[x]->y; {vars=Sequence[x,y],varsN=Sequence[
 xi,eta], a1={{x^2,x*y},{x*y,y^2}}, d1=Det[a1]}
m1=Assuming[{x>0,y>0},Simplify[(-a1[[1,2]]+Sqrt[-d1])/a1[[1,1]]]]
eq1=DSolve[D[y[x],x]==-op1[m1],y[x],x]/.Rule->Equal//First
{eq11=Solve[eq1,C[1]][[1,1,2]], g[1]=op2[Eq11], g[2]=x}
{jg=jacobianM[{g[1],g[2]},{vars}], dv=gradF[v[varsN],{varsN}],
 ddv=hessianH[v[varsN],{varsN}], ddu=Transpose[jg].ddv.jg
 +Sum[dv[[i]]*hessianH[g[i],{vars}],{i,1,2}]}
{norF=Simplify[Tr[a1.ddu]]==0,
 canF=Thread[norF/x^2,Equal]//Expand}
```

□

Nonlinear second-order partial differential equations can be classified as one of the three types, hyperbolic, parabolic, and elliptic and reduced to appropriate canonical and normal forms. For the nonlinear second-order PDEs, we consider the classification of equations (that, in general, can depend on the selection of the point and the specific solution).

Problem 1.4 *Nonlinear second-order equations. Classification.* Let us consider the nonhomogeneous Monge–Ampère equation [124] and the nonlinear wave equation:

$$(u_{xy})^2-u_{xx}u_{yy}=F(x,y), \quad v_{tt}-(G(v)v_x)_x=0.$$

Verify that the type of the nonhomogeneous Monge–Ampère equation at a point (x,y) depends on the sign of the given function $F(x,y)$ and is independent of the selection of a specific solution, while the type of the nonlinear wave equation depends on a specific point (x,t) and on the sign of a specific solution $v(x,t)$.

1. In the standard notation (1.4), these nonlinear equations, respectively, take the form: $F_1=q^2-pr=F(x,y)$ and $F_2=r-G(v)p-G_v v_x^2=0$. In these two cases, we select a special solution $u=u(x,y)$, $v=v(x,t)$, and calculate the discriminant $\delta=b^2-ac$ at some point (x,y), (x,t), where $a=\mathcal{F}_p$, $b=\frac{1}{2}\mathcal{F}_q$, $c=\mathcal{F}_r$.*

2. Let us verify that the type of the nonhomogeneous Monge–Ampère equation at a point (x,y) depends on the sign of the given function $F(x,y)$ and is independent of the selection of a specific solution. Therefore, at the points where $F(x,y)=0$, the equation is of *parabolic type*, at the points where $F(x,y)>0$, the equation is of *hyperbolic type*, and at the points where $F(x,y)<0$, the equation is of *elliptic type*. We verify that the type of the nonlinear wave equation at a point (x,t) depends on a specific point (x,t) and on the sign of a specific solution $v(x,t)$, i.e., it is impossible to determine the sign of δ for the unknown solution $v(x,t)$.

Maple:

```
with(PDEtools): declare((u,v)(x,y),(F1,F2)(p,r,q),G(u(x,t)));
U,V,GV:=diff_table(u(x,y)),diff_table(v(x,t)),
  diff_table(G(v(x,t))); PDE1:=U[x,y]^2-U[x,x]*U[y,y]=F(x,y);
tr1:=(x,y,U)->{U[x,x]=p,U[y,y]=r,U[x,y]=q};
tr2:=H->{a=diff(lhs(H(p,q,r)),p),b=1/2*diff(lhs(H(p,q,r)),q),
        c=diff(lhs(H(p,q,r)),r)}; delta:=b^2-a*c;
```

*In general, the coefficients a, b, and c can depend not only on the selection of a specific point, but also on the selection of a specific solution.

```
F1:=(p,r,q)->subs(tr1(x,y,U),PDE1); F1(p,r,q); tr2(F1);
delta1:=subs(tr2(F1),delta)=rhs(F1(p,r,q));
PDE2:=V[t,t]-G(v)*V[x,x]-GV[x]*V[x]=0;
F2:=(p,r,q)->subs(tr1(x,t,V),PDE2); F2(p,r,q); tr2(F2);
delta2:=subs(tr2(F2),delta)=rhs(F2(p,r,q));
```

Mathematica:

```
pde1=D[u[x,y],{x,y}]^2-D[u[x,y],{x,2}]*D[u[x,y],{y,2}]==f[x,y]
tr1[x_,y_,u_]:={D[u[x,y],{x,2}]->p, D[u[x,y],{y,2}]->r,
                D[u[x,y],{x,y}]->q};
f1[p_,r_,q_]:=pde1/.tr1[x,y,u]; f1[p,r,q]
tr2[f_]:={a->D[f[p,q,r][[1]],p], b->1/2*D[f[p,q,r][[1]],q],
          c->D[f[p,q,r][[1]],r]}; delta=b^2-a*c; tr2[f1]
delta1=(delta/.tr2[f1])==f1[p,r,q][[2]]
dgDv=D[g[v[x,t]],x]*D[v[x,t],x];
pde2=D[v[x,t],{t,2}]-g[v[x,t]]*D[v[x,t],{x,2}]-dgDv==0
f2[p_,r_,q_]:=pde2/.tr1[x,t,v]; {f2[p,r,q], tr2[f2]}
delta2=(delta/.tr2[f2])==f2[p,r,q][[2]]
```

□

Let us consider hyperbolic systems of nonlinear first-order PDEs:

$$\mathbf{u}_t+\sum_{i=1}^{n}\mathbf{B}_i(\mathbf{x},t,\mathbf{u})\mathbf{u}_{x_i}=\mathbf{f}, \tag{1.5}$$

subject to the initial condition $\mathbf{u}=\mathbf{g}$ on $\mathbb{R}^n\times\{t=0\}$. Here the unknown function is $\mathbf{u}=(u_1,\ldots,u_m)$, the functions $\mathbf{B}_i(\mathbf{x},t,\mathbf{u})$, $\mathbf{f}$, $\mathbf{g}$ are given, and $\mathbf{x}=(x_1,\ldots,x_n)\in\mathbb{R}^n$, $t\geq 0$.

Definition 1.6 The nonlinear system of PDEs (1.5) is called hyperbolic if the $m\times m$ matrix $\mathbf{B}(\mathbf{x},t,\mathbf{u},\beta)=\sum_{i=1}^{n}\beta_i\mathbf{B}_i(\mathbf{x},t,\mathbf{u})$ (where $\beta\in\mathbb{R}^n$, $x\in\mathbb{R}^n$, $t\geq 0$) is diagonalizable for each $\mathbf{x}\in\mathbb{R}^n$, $t\geq 0$, i.e., the matrix $\mathbf{B}(\mathbf{x},t,\mathbf{u},\beta)$ has m real eigenvalues and corresponding eigenvectors that form a basis in $\mathbb{R}^m$.

There are two important special cases:
(1) The nonlinear system (1.5) is a *symmetric hyperbolic system* if $\mathbf{B}_i(\mathbf{x},t,\mathbf{u})$ is a symmetric $m\times m$ matrix for each $\mathbf{x}\in\mathbb{R}^n$, $t\geq 0$ $(i=1,\ldots,m)$.
(2) The nonlinear system (1.5) is *strictly hyperbolic system* if for each $\mathbf{x}\in\mathbb{R}^n$, $t\geq 0$, the matrix $\mathbf{B}(\mathbf{x},t,\mathbf{u},\beta)$ has m distinct real eigenvalues.

Problem 1.5 *Nonlinear hyperbolic systems. Classification.* Let us consider the nonlinear system [38]:

$$u_t+\big(uF_1(u,v)\big)_x+\big(uF_2(u,v)\big)_y=0, \quad v_t+\big(vF_1(u,v)\big)_x+\big(vF_2(u,v)\big)_y=0,$$

where $(u,v)\big|_{t=0}=\big(u_0(x,y),v_0(x,y)\big)$. Verify that this system is a nonstrictly hyperbolic system (also considered in Problem 2.5) and this system is symmetric if $u(F_i(u,v))_v=v(F_i(u,v))_u$, $i=1,2$.

We rewrite the above system in the matrix form $\mathbf{u}_t+B_1\mathbf{u}_x+B_2\mathbf{u}_y=0$, where

$$\mathbf{u}=\begin{pmatrix} u \\ v \end{pmatrix}, \; B_1=\begin{pmatrix} F_1+u\,(F_1)_u & u\,(F_1)_v \\ v\,(F_1)_u & F_1+v\,(F_1)_v \end{pmatrix}, \; B_2=\begin{pmatrix} F_2+u\,(F_2)_u & u\,(F_2)_v \\ v\,(F_2)_u & F_2+v\,(F_2)_v \end{pmatrix}.$$

The eigenvalues of the this system are (L1, L2): $\lambda_1=\beta_1F_1+\beta_2F_2$ and $\lambda_2=\beta_1F_1+\beta_2F_2+\beta_1v(F_1)_v+\beta_1u(F_1)_u+\beta_2v(F_2)_v+\beta_2u(F_2)_u$. The eigenvalues are equal, $\lambda_1=\lambda_2$, if $[\beta_1(F_1)_u+\beta_2(F_2)_u]u+[\beta_1(F_1)_v+\beta_2(F_2)_v]v=0$ (L12). Therefore, the system is nonstrictly hyperbolic.

Maple:

```
with(PDETools): with(LinearAlgebra): declare((F1,F2)(u,v));
B1:=<<F1(u,v)+u*diff(F1(u,v),u),v*diff(F1(u,v),u)>|
    <u*diff(F1(u,v),v),F1(u,v)+v*diff(F1(u,v),v)>>;
B2:=<<F2(u,v)+u*diff(F2(u,v),u),v*diff(F2(u,v),u)>|
    <u*diff(F2(u,v),v),F2(u,v)+v*diff(F2(u,v),v)>>;
Eq1:=beta1*B1+beta2*B2-lambda*Matrix(2,2,shape=identity)=0;
Eq2:=Determinant(lhs(Eq1))=0; Eq3:=factor(Eq2);
L1:=solve(op(1,lhs(Eq3)),lambda);
L2:=solve(op(2,lhs(Eq3)),lambda); L12:=L2-L1;
A1:=subs(u*diff(F1(u,v),v)=v*diff(F1(u,v),u),B1);
A2:=subs(u*diff(F2(u,v),v)=v*diff(F2(u,v),u),B2);
```

Mathematica:

```
b1={{f1[u,v]+u*D[f1[u,v],u],u*D[f1[u,v],v]},{v*D[f1[u,v],u],
 f1[u,v]+v*D[f1[u,v],v]}}; b2={{f2[u,v]+u*D[f2[u,v],u],
 u*D[f2[u,v],v]},{v*D[f2[u,v],u],f2[u,v]+v*D[f2[u,v],v]}};
Map[MatrixForm,{b1,b2}]
{eq1=beta1*b1+beta2*b2-lambda*IdentityMatrix[2]==0,
 eq2=Det[eq1[[1]]]==0, eq3=Factor[eq2]}
{l1=Solve[eq3[[1,1]]==0,lambda][[1,1,2]],
 l2=Solve[eq3[[1,2]]==0,lambda][[1,1,2]], l12=l2-l1}
a1=b1/.u*D[f1[u,v],v]->v*D[f1[u,v],u]
a2=b2/.u*D[f2[u,v],v]->v*D[f2[u,v],u]
```

□

1.1.2 Nonlinear PDEs and Systems Arising in Applied Sciences

Nonlinear partial differential equations arise in a variety of physical problems (e.g., in problems of solid mechanics, fluid dynamics, acoustics, nonlinear optics, plasma physics, quantum field theory, etc.), chemical and biological problems, in formulating fundamental laws of nature, and numerous applications.

There exists an important class of nonlinear PDEs, called the *soliton equations*, which admit many physically interesting solutions, called *solitons*. These nonlinear equations have introduced remarkable achievements in the field of applied sciences. A collection of the most important nonlinear equations (considered in the book) is represented in Tab. 1.1.

The *eikonal equation*[*] arises in nonlinear optics and describes the propagation of wave fronts and discontinuities for acoustic wave equations, Maxwell's equations, and equations of elastic wave propagation. The eikonal equation can be derived from Maxwell's equations, and it is a special case of the Hamilton–Jacobi equation (see Sect. 3.2.1). This equation also is of general interest in such fields as geometric optics, seismology, electromagnetics, computational geometry, multiphase flow.

The *nonlinear heat (or diffusion) equation* describes the flow of heat or a concentration of particles, the diffusion of thermal energy in a homogeneous medium, the unsteady boundary-layer flow in the Stokes and Rayleigh problems.

The *Burgers equation* has been introduced by J. M. Burgers in 1948 for studying the turbulence phenomenon described by the interaction of the two physical transport phenomena convection and diffusion. It is the important nonlinear model equation representing phenomena described by a balance between time evolution, nonlinearity, and diffusion. It is one of the fundamental model equations in fluid mechanics. The Burgers equation arises in many physical problems (e.g., one-dimensional turbulence, traffic flow, sound and shock waves in a viscous medium, magnetohydrodynamic waves). The Burgers equation is completely integrable (see Chap. 4). The wave solutions of the Burgers equation are single-front and multiple-front solutions.

The *kinematic wave equation* (or the nonlinear first-order wave equation is a special case of the Burgers equation (if the viscosity $\nu = 0$) and describes the propagation of nonlinear waves (e.g., waves in traffic flow on highways, shock waves, flood waves, waves in plasmas, sediment transport in rivers, chemical exchange processes in chromatography, etc.).

[*] *Eikonal* is a German word, which is from *eikon*, a Greek word for image or figure.

Table 1.1. Selected nonlinear equations considered in the book

Nonlinear PDE	Equation name	Problem
$(u_x)^2+(u_y)^2=n^2$	Eikonal eq.	1.18, 3.12, 3.13
$u_t-\left(F(u)u_x\right)_x=0$	Nonlinear heat eq.	2.3, 2.17, 2.27 2.39, 2.45, 2.49
$u_t+uu_x=\nu u_{xx}$	Burgers eq.	1.16, 2.10, 4.1 5.5, 6.1, 7.1
$u_t+c(u)u_x=0$	Kinematic wave eq.	3.6, 3.7
$u_t+uu_x=0$	Inviscid Burgers eq.	3.3, 5.2, 6.4
$u_t+G(u)u_x=H(u)$	Generalized inviscid Burgers eq.	3.8, 3.9
$u_t-u_{xx}=au(1-u)$	Fisher eq.	3.20, 5.7
$u_t+auu_x-\nu u_{xx}=bu(1-u)(u-c)$	Burgers–Huxley eq.	1.10
$u_t+auu_x+bu_{xxx}=0$	Korteweg–deVries eq.	1.12, 2.18, 4.5 4.10, 5.9, 6.13
$u_t+(2au-3bu^2)u_x+u_{xxx}=0$	Gardner eq.	1.12, 2.13, 4.6
$u_t+6u^2u_x+u_{xxx}=0$	Modified KdV eq.	2.41, 4.13
$u_{tt}-\left(F(u)u_x\right)_x=0$	Nonlinear wave eq.	1.4, 1.11, 2.25 2.28, 2.46, 6.2
$u_{tt}+(uu_x)_x+u_{xxxx}=0$	Boussinesq eq.	1.6
$(u_t+auu_x+u_{xxx})_x+bu_{yy}=0$	Kadomtsev–Petviashvili eq.	1.15
$u_t+au_x+buu_x-cu_{xxt}=0$	Benjamin–Bona–Mahony eq.	1.14
$u_t+u_x+u^2u_x+au_{xxx}+bu_{xxxxx}=0$	Generalized Kawahara eq.	4.2, 4.4
$iu_t+u_{xx}+\gamma\|u\|^2u=0$	Nonlinear Schrödinger eq.	1.7, 2.42, 4.12
$u_t-au_{xx}-bu+c\|u\|^2u=0$	Ginzburg–Landau eq.	1.13
$u_{tt}-u_{xx}=F(u)$	Klein–Gordon eq.	2.21, 4.14, 5.4
$u_{tt}-u_{xx}=\sin u$	sine–Gordon eq.	2.11, 2.20, 2.32 3.21, 4.8, 6.14
$u_{xx}+u_{yy}=F(u)$	Nonlinear Poisson eq.	2.43, 2.44, 6.15
$(u_{xy})^2-u_{xx}u_{yy}=F(x,y)$	Monge–Ampère eq.	1.4, 2.19

The *inviscid Burgers equation* (or the Hopf equation) is a special case of the kinematic wave equation ($c(u) = u$). The Burgers equation is parabolic, whereas the inviscid Burgers equation is hyperbolic. The properties of the solution of the parabolic equation are significantly different than those of the hyperbolic equation.

The *generalized inviscid Burgers equation* appears in several physical problems, in particular it describes a population model [98].

The *Fisher equation* has been introduced by R. A. Fisher in 1936 for studying wave propagation phenomena of a gene in a population and logistic growth-diffusion phenomena. This equation describes wave propagation phenomena in various biological and chemical systems, in the theory of combustion, diffusion and mass transfer, nonlinear diffusion, chemical kinetics, ecology, chemical wave propagation, neutron population in a nuclear reactor, etc.

The *Burgers–Huxley equation* describes nonlinear wave processes in physics, mathematical biology, economics, ecology [107].

The *Korteweg–de Vries equation* has been introduced by D. Korteweg and G. de Vries in 1895 for a mathematical explanation of the solitary wave phenomenon discovered by S. Russell in 1844. This equation describes long time evolution of dispersive waves and in particular, the propagation of long waves of small or moderate amplitude, traveling in nearly one direction without dissipation in water of uniform shallow depth (this case is relevant to *tsunami waves*). The KdV equation admits a special form of the exact solution, the *soliton*, which arises in many physical processes, e.g., water waves, internal gravity waves in a stratified fluid, ion-acoustic waves in a plasma, etc.

The *Gardner equation*, introduced by R. M. Miura, C. S. Gardner, and M. D. Kruskal in 1968 [103] as a generalization of the KdV equation, appears in various branches of physics (e.g., fluid mechanics, plasma physics, quantum field theory). The Gardner equation can be used to model several nonlinear phenomena, e.g., internal waves in the ocean.

The *modified KdV equation* (mKdV), the *KdV-type equation*, and the *modified KdV-type equation* are the nonlinear evolution equations that describe approximately the evolution of long waves of small or moderate amplitude in shallow water of uniform depth, nonlinear acoustic waves in an inharmonic lattice, Alfvén waves in a collisionless plasma, and many other important physical phenomena.

The *nonlinear wave equation* describes the propagation of waves, which arises in a wide variety of physical problems.

The mathematical theory of water waves goes back to G. G. Stokes in 1847, who was first to derive the equations of motion of an incompressible, inviscid heavy fluid bounded below by a rigid bottom and above by a free surface. These equations are still hard to solve in a general case because of the moving boundary whose location can be determined by solving two nonlinear PDEs. Therefore, most advances in the theory of water waves can be obtained through approximations (e.g., see Problem 5.10, where we construct approximate analytical solutions describing nonlinear standing waves on the free surface of a fluid).

However, Korteweg and de Vries in 1895, instead of solving the equations of motion approximately, considered a limit case, which is relevant to *tsunami waves*. This limit case describes long waves of small or moderate amplitude, traveling in nearly one direction without dissipation in water of uniform shallow depth. There are alternative equations to the KdV equation that belong to the family of the KdV-type equations, e.g., the Boussinesq equations (1872), the Kadomtsev–Petviashvili equation (KP, 1970), the Benjamin–Bona–Mahony equation (1972), the Camassa–Holm equation (1993), the Kawahara equation (1972).

The *Boussinesq equation*, introduced by J. V. Boussinesq in 1872 [21], appears in many scientific applications and physical phenomena (e.g., the propagation of long waves in shallow water, nonlinear lattice waves, iron sound waves in a plasma, vibrations in a nonlinear string). The main properties are: the Boussinesq equation is completely integrable (see Sect. 4.2), admits an infinite number of conservation laws, N-soliton solutions, and inverse scattering formalism.

The *Kadomtsev–Petviashvili equation* is a generalization of the KdV equation, it is a completely integrable equation by the inverse scattering transform method. In 1970, B. B. Kadomtsev and V. I. Petviashvili [77] generalized the KdV equation from (1+1) to (2+1) dimensions. The KP equation describes shallow-water waves (with weakly non-linear restoring forces), waves in ferromagnetic media, shallow long waves in the x-direction with some mild dispersion in the y-direction.

The *Benjamin–Bona–Mahony equation* has been introduced by T. B. Benjamin, J. L. Bona, and J. J. Mahony in 1972 [16] for studying propagation of long waves (where nonlinear dispersion is incorporated). The BBM equation belongs to the family of KdV-type equations. As we stated above, the KdV equation is a model for propagation of one-dimensional small amplitude, weakly dispersive waves. Both BBM and KdV equations are applicable for studying shallow water waves, surface waves of long wavelength in liquids, acoustic-gravity waves in compressible fluids, hydromagnetic waves in cold plasma, acoustic waves in anharmonic crystals, etc.

The *Kawahara equation*, introduced by T. Kawahara in 1972, is a generalization of the KdV equation (it belongs to the family of KdV-type equations). The Kawahara equation arises in a wide range of physical problems (e.g., capillary-gravity water waves, shallow water waves with surface tension, plasma waves, magneto-acoustic waves in a cold collision free plasma, etc).

The *nonlinear Schrödinger equation* (NLS), introduced by the physicist E. Schrödinger in 1926, describes the *evolution of water waves* and other *nonlinear waves* arising in different physical systems, e.g., nonlinear optical waves, hydromagnetic and plasma waves, nonlinear waves in fluid-filled viscoelastic tubes, solitary waves in piezoelectric semiconductors, and also many important physical phenomena, e.g., nonlinear instability, heat pulse in a solid, etc. V. E. Zakharov and A. B. Shabat in 1972 have developed the inverse scattering method to prove that the NLS equation is completely integrable.

The *Ginzburg–Landau theory*, developed by V. L. Ginzburg and L. Landau in 1950, is a mathematical theory for studying superconductivity. The *Ginzburg–Landau equations* are based on several key concepts developed in the framework of this theory. Real Ginzburg–Landau equations were first derived as long-wave amplitude equations by A. C. Newell and J. A. Whitehead and by L. A. Segel in 1969; complex Ginzburg–Landau equations were first derived by K. Stewartson and J. T. Stuart in 1971 and by G. B. Ermentrout in 1981. The nonlinear equations describe the evolution of amplitudes of unstable modes for any process exhibiting a Hopf bifurcation. The Ginzburg–Landau equations arise in many applications (e.g., nonlinear waves, hydrodynamical stability problems, nonlinear optics, reaction-diffusion systems, second-order phase transitions, Rayleigh–Bénard convection, superconductivity, chemical turbulence, etc).

The *Klein–Gordon equation* (or Klein–Gordon–Fock equation), introduced by the physicists O. Klein and W. Gordon in 1927, describes relativistic electrons. The Klein–Gordon equation was first considered as a quantum wave equation by Schrödinger. In 1926 (after the Schrödinger equation was introduced), V. Fock wrote an article about its generalization for the case of magnetic fields and independently derived this equation. The Klein–Gordon equations play a significant role in many scientific applications (e.g., nonlinear dispersion, solid state physical problems, nonlinear optics, quantum field theory, nonlinear meson theory).

The *sine–Gordon equation** has a long history that begins in the 19th century in the course of study of surfaces of constant negative curvature. This equation attracted a lot of attention since 1962 [120] due to discovering of soliton solutions and now is one of the basic nonlinear evolution equations that describes various important nonlinear physical

*The name "sine–Gordon equation" is a wordplay on the Klein–Gordon equation.

phenomena. The sine–Gordon equation has a wide range of applications in mathematics and physics (e.g., in differential geometry, relativistic field theory, solid-state physics, nonlinear optics, etc.).

The *nonlinear Poisson equation* describes a variety of steady-state phenomena in the presence of external sources (e.g., velocity potential for an incompressible fluid flow, temperature in a steady-state heat conduction problem).

The *Monge–Ampère equations*, introduced by G. Monge in 1784 and A. M. Ampère in 1820, appear in differential geometry, gas dynamics, meteorology.

Systems of nonlinear partial differential equations arise in various applications (e.g., chemical, biological). A selection of the most important systems of nonlinear PDEs (considered in the book) is represented in the following table:

Table 1.2. Selected nonlinear systems considered in the book

Nonlinear Systems	**Name of Nonlin. System**	**Problem**
$u_t+\big(uF_1(u,v)\big)_x+\big(uF_2(u,v)\big)_y=0$ $v_t+\big(vF_1(u,v)\big)_x+\big(vF_2(u,v)\big)_y=0$	Nonstrictly hyperbolic system	1.5, 2.5
$u_x=F(u,v),\ \ v_t=G(u,v)$	First–order system	2.53, 2.55
$u_t=vu_x+u+1,\ \ v_t=-uv_x-v+1$	First–order system	1.21, 5.6, 6.7
$z_x=F(x,y,z),\ \ z_y=G(x,y,z)$	Overdetermined system	4.15–4.17
$u_t-v_x=0,\ \ v_t-F\big(u(x,t)\big)u_x-G\big(u(x,t)\big)=0$	Nonlinear telegraph system	1.19, 2.3
$v_x-2u=0,\ \ v_t-2u_x+u^2=0$	Burgers system	1.20, 2.6
$u_t=a_1u_{xx}+F(u,v),\ \ v_t=a_2v_{xx}+G(u,v)$	Second–order system	2.54
$u_t=u_{xx}+a_1uF(u-v)+a_2G_1(u-v)$ $v_t=v_{xx}+a_1vF(u-v)+a_2G_2(u-v)$	Second-order system	2.56
$u_t=u_{xx}+u(\alpha-u)(u-1)-v,\ \ v_t=\beta u$	FitzHugh–Nagumo system	6.8
$\phi_{xx}+\phi_{yy}=0$ $\phi_t+\frac{1}{2}[(\phi_x)^2+(\phi_y)^2]+gy+p/\rho=0$	Equations for inviscid fluid (Eulerian framework)	7.3
$x_{tt}x_a+(y_{tt}+g)y_a+p_a/\rho=0$ $x_{tt}x_b+(y_{tt}+g)y_b+p_b/\rho=0,\ \ x_ay_b-x_by_a=1$	Equations for inviscid fluid (Lagrangian framework)	5.10, 3.22

The *nonstrictly hyperbolic systems* arise in various problems of elastic theory, magnetohydrodynamics.

The *nonlinear telegraph systems* arise in the study of propagation of electrical signals.

The *Burgers system* can be considered for modeling shock waves (e.g., shock reflection, see [105]), it also arises in nonlinear acoustics and nonlinear geometrical optics.

The *nonlinear first-order systems* of the form $u_x=F(u,v)$, $v_t=G(u,v)$ can describe various physical, chemical, and biological processes, e.g., convective mass transfer, suspension transport in porous media, migration of bacteria or virus, industrial filtering, etc.

The *nonlinear second-order systems* of the form $u_t=a_1u_{xx}+F(u,v)$, $v_t=a_2v_{xx}+G(u,v)$ can describe reaction-diffusion phenomena.

The *FitzHugh–Nagumo equations*, introduced by R. FitzHugh [43] and J. S. Nagumo [108] in 1961, arise in mathematical biology and can be considered for modeling the nerve impulse propagation along an axon (see [107]).

Euler's equations of motion and the *continuity equation* are fundamental equations for studying water wave motions that are of great importance, e.g., for studying nonlinear standing wave motion on the free surface of a fluid, surface waves generated by wind, flood waves in rivers, ship waves in channels, tsunami waves, tidal waves, solitary waves in channels, waves generated by underwater explosions, etc. There are two ways for representing the fluid motion: the *Eulerian framework* (in which the coordinates are fixed in the reference frame of the observer) and the *Lagrangian framework* (in which the coordinates are fixed in the reference frame of the moving fluid).

1.1.3 Types of Solutions of Nonlinear PDEs

There are many types of solutions of nonlinear partial differential equations. Now let us mention only some of these types which we will consider in the book.

A *classical solution* of a nonlinear equation (1.1) is a function $u = u(x_1,\dots,x_n)$, defined in a domain $\mathcal{D}$, which is continuously differentiable such that all its partial derivatives involved in the equation exist and satisfy Eq. (1.1) identically.

The concept of classical solution can be extended by introducing the notion of *weak solution* (also known as generalized solution) in order to include discontinuous and nondifferentiable functions [37]. For example, for the nonlinear equation $u_t+F(u)_x=0$ (where $F(u)$ is a nonlinear convex function of $u(x,t)$, and $\{x \in \mathbb{R}, t \geq 0\}$), with bounded measurable initial data $u(x,0)$, we say that the bounded measurable function $u{=}u(x,t)$ is a *weak solution* (or generalized solution) if

$$\int_0^{\infty}\int_{-\infty}^{\infty}[\phi_t u{+}\phi_x F(u)]\,dxdt{=}-\int_{-\infty}^{\infty}\phi(x,0)u(x,0)\,dx$$

holds for all *test functions* $\phi(x,t)\in C_0^1(\mathbb{R}\times\mathbb{R})$.*

On *general solution* of a nonlinear PDE we understand an explicit closed form expression which may contain movable critical singularities. In particular, a general solution (or general integral) of a first-order PDE is an equation of the form $f(\phi,\psi){=}0$, where f is an arbitrary function of the known functions $\phi{=}\phi(x,y)$, $\psi{=}\psi(x,y)$ and provides a solution of this partial differential equation.

Exact solution of a nonlinear PDE is a solution, defined in the whole domain of definition of the PDE, which can be represented in closed form, i.e., as a finite expression (infinite functional series and products are not included). The exact solution of a nonlinear PDE is *not new* if it is possible to reduce it to known exact solution. The exact solution of a nonlinear PDE is *redundant* if there exist more general solutions such that this redundant solution can be considered as a particular case of more general solutions. A *vacuum solution* of a given nonlinear PDE is the constant solution.

A *traveling wave* is a wave of permanent form moving with a constant velocity. Applying the ansatz $u(x,t){=}u(\xi)$, $\xi{=}x{-}ct$ (where c is the wave velocity), it is possible to transform the PDE (in x,t) into an ODE (in ξ), which can be solved by appropriate methods. In other words, a traveling wave solution of a given nonlinear PDE is a solution of the reduction $\xi{=}x{-}ct$ (see Definition 2.8) if it exists.

Periodic solutions are traveling wave solutions that are periodic, e.g., $\cos(x{-}t)$.

Kink solutions are traveling wave solutions which rise or descend from one asymptotic state to another, the kink solutions approach a constant at infinity.

Standing wave solutions are two superposed traveling wave solutions of equal amplitude and speed (but in opposite direction). The wave amplitude of standing waves varies with time but it does not move spatially.

A *solitary wave* of a given nonlinear PDE is a traveling wave such that the solution or its derivative obeys some decreasing conditions as $\xi\to\pm\infty$ ($\xi{=}x{-}ct$), i.e., solitary waves are localized traveling waves, asymptotically zero at large distances.

* C_0^1 is the space of functions that are continuously differentiable with compact support.

Solitons are special kinds of solitary waves. The soliton solution is a spatially localized solution, i.e., $u'(\xi) \to 0$, $u''(\xi) \to 0$, $u'''(\xi) \to 0$ as $\xi \to \pm\infty$ ($\xi = x - ct$). The main property of solitons: in the process of interaction with other solitons, it keeps its identity.

A *one-soliton solution* is the first iterate of the vacuum solution via the *Bäcklund transformation* (see Definition 2.6).

An *N-soliton solution*, where N is an arbitrary positive integer, is the N-th iterate of the vacuum solution via the Bäcklund transformation.

A *soliton equation* is a nonlinear PDE admitting an N-soliton solution, where N is arbitrary positive integer.

A *similarity* or *invariant solution* is a solution of a PDE arising from invariance under a one-parameter Lie group of transformations.

A *self-similar* or *automodel solution* is an invariant solution arising from invariance under a one-parameter Lie group of scalings.

A *symmetry* of a differential equation (or system) is a transformation that maps any solution to another solution of the equation (or system).

1.2 Embedded Analytical Methods

Computer algebra systems *Maple* and *Mathematica* have various embedded analytical methods and the corresponding predefined functions based on symbolic algorithms for constructing analytical solutions of linear and nonlinear PDEs (see more detailed description in [28]).

Although the predefined functions are an implementation of known methods for solving PDEs, it allows us to solve nonlinear equations and obtain solutions automatically (via predefined functions) and develop new methods and procedures for constructing new solutions.

1.2.1 Nonlinear PDEs

In this section, we will consider the most important functions for finding all possible analytical solutions of a given nonlinear PDE.

Maple:

```
pdsolve(PDE);              pdetest(sol,PDE); pdsolve(PDE,build);
pdsolve(PDE,funcs,HINT=val,INTEGRATE,build,singsol=val);
infolevel[procname]:=val;   with(PDEtools);     declare(funcs);
dchange(rules,PDE);        casesplit(PDEs);  diff_table(funcs);
separability(PDE,DepVar);          SimilaritySolutions(PDE,ops);
Infinitesimals(PDE);          F:=TWSolutions(functions_allowed);
TWSolutions(PDE,parameters=val,singsol=false,functions=F);
TWSolutions(PDE,output=ODE);           TWSolutions(PDE,extended);
```

NOTE. `HINT=val` are some hints, e.g., with ``HINT=`+` `` or ``HINT=`*` `` a solution can be constructed by separation of variables (in the form of sum or product, respectively), with `HINT='TWS'` or `HINT='TWS(MathFuncName)'` a traveling wave solution can be constructed as power series in $\tanh(\xi)$ or several mathematical functions (including special functions), where ξ represents a linear combination of the independent variables; with `build` an explicit expression can be constructed for the indeterminate function `func`, etc.

`pdsolve`, finding analytical solutions for a given partial differential equation `PDE` and systems of `PDEs`,

`PDEtools`, a collection of functions for finding analytical solutions for `PDEs`, e.g.,

`declare`, declaring functions and derivatives on the screen for a simple, compact display,

`separability`, determining under what conditions it is possible to obtain a complete solution through separation of variables,

`SimilaritySolutions`, determining the group invariant (point symmetry) solutions for a given PDE or system of PDEs,

`TWSolutions`, constructing traveling wave solutions for autonomous PDEs or systems of them, etc.

In the computer algebra system Mathematica, analytical solutions of a given nonlinear PDE can be found with the aid of the predefined function `DSolve`:

Mathematica:

```
DSolve[pde,u,{x1,..,xn}]  DSolve[pde,u[x1,...,xn],{x1,...,xn}]
DSolve[pde, u[x1,...,xn], {x1,...,xn}, GeneratedParameters->C]
```

`DSolve`, finding analytical solutions of a PDE for the function `u`, with independent variables `x1,...xn` ("pure function" solution),

`DSolve`, finding analytical solutions of a PDE for the function `u`, with independent variables `x1,...xn`,

`DSolve`, `GeneratedParameters`, finding analytical solutions of a PDE for the function `u`, with independent variables `x1,...xn` and specifying the arbitrary constants.

Problem 1.6 *Boussinesq equations. Exact solutions.* We consider the Boussinesq equation and its modifications,

$u_{tt}+(uu_x)_x+u_{xxxx}=0,\ u_{tt}-a(uu_x)_x-bu_{xxxx},\ u_{tt}-u_{xx}-(3u^2)_{xx}-u_{xxxx},$

where $\{x \in \mathbb{R}, t \geq 0\}$. Verify that the following solutions of these equations [124]

$$u(x,t) = -\frac{3\lambda^2}{\cos\left(\frac{1}{2}\lambda(x \pm \lambda t) + C\right)^2}, \quad u(x,t) = \frac{3\lambda^2}{a\cosh\left(\frac{1}{2}\lambda(x \pm \lambda t)/\sqrt{b} + C\right)^2},$$

$$u(x,t) = -\frac{2Ak^2 \exp\left(k(x + t\sqrt{1+k^2})\right)}{\left(-1 + A\exp\left(k(x + t\sqrt{1+k^2})\right)\right)^2},$$

are exact solutions of the given nonlinear PDEs. Here a, b, A, C, k, λ are arbitrary real constants.

Maple:

```
with(PDEtools); declare(u(x,t)); PDE1:=diff(u(x,t),t$2)+
 diff(u(x,t)*diff(u(x,t),x),x)+diff(u(x,t),x$4)=0; PDE2:=expand(
 diff(u(x,t),t$2)-a*diff(u(x,t)*diff(u(x,t),x),x)-b*diff(u(x,t),
 x$4)); PDE3:=diff(u(x,t),t$2)-diff(u(x,t),x$2)-diff(3*u(x,t)^2,
 x$2)-diff(u(x,t),x$4); Sol1:=S->u(x,t)=-3*lambda^2*cos(lambda*
 (x+S*lambda*t)/2+C1)^(-2); Test11:=pdetest(Sol1(1),PDE1);
Test12:=pdetest(Sol1(-1),PDE1); Sol2:=S->u(x,t)=3*lambda^2/a*
 cosh(lambda*(x+S*lambda*t)/2/sqrt(b)+C1)^(-2);
Test21:=pdetest(Sol2(1),PDE2); Test22:=pdetest(Sol2(-1),PDE2);
f3:=S->1-A*exp(k*x+S*k*t*sqrt(1+k^2)); Sol3:=S->u(x,t)=2*diff(
 log(f3(S)),x$2); factor(Sol3(1)); factor(Sol3(-1));
Test31:=pdetest(Sol3(1),PDE3); Test32:=pdetest(Sol3(-1),PDE3);
```

Mathematica:

```
{pde1=D[u[x,t],{t,2}]+D[u[x,t]*D[u[x,t],x],x]+D[u[x,t],{x,4}]==0,
 pde2=Expand[D[u[x,t],{t,2}]-a*D[u[x,t]*D[u[x,t],x],x]
 -b*D[u[x,t],{x,4}]]==0, pde3=D[u[x,t],{t,2}]-D[u[x,t],{x,2}]
 -D[3*u[x,t]^2,{x,2}]-D[u[x,t],{x,4}]==0}
sol1[s_]:=u->Function[{x,t},-3*lambda^2*Cos[lambda*(x
 +s*lambda*t)/2+c1]^(-2)]; Map[FullSimplify,
 {test11=pde1/.sol1[1],test12=pde1/.sol1[-1]}]
sol2[s_]:=u->Function[{x,t},3*lambda^2/a*Cosh[lambda*(x
 +s*lambda*t)/2/Sqrt[b]+c1]^(-2)]; Map[FullSimplify,
 {test21=pde2/.sol2[1],test22=pde2/.sol2[-1]}]
f3[s_]:=1-a*Exp[k*x+s*k*t*Sqrt[1+k^2]]; sol3[s_]:=u->Function[
 {x,t},2*D[Log[f3[s]],{x,2}]]; Map[Factor,{sol3[1],sol3[-1]}]
Map[FullSimplify,{test31=pde3/.sol3[1],test32=pde3/.sol3[-1]}]
```

□

Problem 1.7 *Nonlinear Schrödinger equation. Exact solutions.* Lets us consider the nonlinear Schrödinger equation (NLS)

$$iu_t + u_{xx} + \gamma|u|^2 u = 0,$$

where u is a complex function of real variables x and t: $\{x \in \mathbb{R}, t \geq 0\}$, and $\gamma \in \mathbb{R}$ is a constant. Verify that the following solutions [124]

$$u(x,t) = C_1 \exp\left[i(C_2 x + (\gamma C_1^2 - C_2^2)t + C_3)\right],$$
$$u(x,t) = A\sqrt{\frac{2}{\gamma}}\frac{\exp\left[iBx + i(A^2 - B^2)t + iC_1\right]}{\cosh(Ax - 2ABt + C_2)}$$

are exact solutions of the NLS equation.

Maple:

```
with(PDEtools);  declare(u(x,t)); U:=diff_table(u(x,t));
interface(showassumed=0); assume(gamm>0);
PDE1:=I*U[t]+U[x,x]+gamm*abs(U[])^2*U[]=0;
Sol1:=u(x,t)=C1*exp(I*(C2*x+(gamm*C1^2-C2^2)*t+C3));
Test1:=pdetest(Sol1,PDE1); Test11:=simplify(evalc(Test1));
Sol2:=u(x,t)=A*sqrt(2/gamm)*(exp(I*B*x+I*(A^2-B^2)*t+I*C1))
      /(cosh(A*x-2*A*B*t+C2));
Test2:=pdetest(Sol2,PDE1); Test21:=simplify(evalc(Test2));
```

Here A, B, C_1, C_2, and C_3 are arbitrary real constants, and the second solution is valid for $\gamma > 0$.

Mathematica:

```
pde1=I*D[u[x,t],t]+D[u[x,t],{x,2}]+gamm*Abs[u[x,t]]^2*u[x,t]==0
sol1=u->Function[{x,t},c1*Exp[I*(c2*x+(gamm*c1^2-c2^2)*t+c3)]]
test1=pde1/.sol1
test11=Assuming[{c1>0},test1//ComplexExpand//FullSimplify]
sol2=u->Function[{x,t},a*Sqrt[2/gamm]*(Exp[I*b*x+I*(a^2-b^2)*t
       +I*c1])/(Cosh[a*x-2*a*b*t+c2])]
test2=pde1/.sol2
test21=Assuming[{gamm>0,{a,b,c,c1,c2,x,t}\[Element] Reals},
       FullSimplify[test2]]
```

□

Problem 1.8 *Nonlinear first-order equation. General solution.* Let us consider the nonlinear first-order PDE,

$$yu_y - xu_x - f(x)u^{-n} = 0,$$

where $\{x \in \mathbb{R}, y \in \mathbb{R}\}$ and $f(x)$ is an arbitrary real function. Verify that the general solution of the given nonlinear PDE has the form:

$$u=\Big(-I_1(1+n)+F_1(xy)\Big)^{1/(1+n)}, \quad \text{where} \quad I_1=\int \frac{f(x)}{x}\,dx.$$

Maple:

```
with(PDEtools); declare(u(x,y));
PDE1:=y*diff(u(x,y),y)-x*diff(u(x,y),x)-f(x)/u(x,y)^n=0;
Sol1:=pdsolve(PDE1);
Test1:=pdetest(subs(n=9,Sol1),subs(n=9,PDE1));
Test2:=simplify(factor(expand(simplify(pdetest(Sol1,PDE1))
       assuming n>1)));
```

Mathematica:

```
Off[Solve::"ifun"];
pde1=y*D[u[x,y],y]-x*D[u[x,y],x]-f[x]/u[x,y]^n==0
sol1=DSolve[pde1,u,{x,y}]//FullSimplify//First
Print["sol1=",sol1]; HoldForm[sol1]==sol1
{test1=pde1/.sol1//FullSimplify, sol2=u[x,y]/.sol1}
Print["sol2=",sol2]; tr1={C[1][var_]:>f[var],K[1]->s}
sol3=(u[x,y]/.sol1)/.tr1//FullSimplify
```

Solving this nonlinear PDE, *Mathematica* generates a warning message, which can be ignored or suppressed with the `Off` function. According to the *Mathematica* notation, `sol1` is the "pure function" solution for $u(x, y)$ (where `C[1]` is an arbitrary function, `K[1]` is the integration variable), `sol2` represents the solution $u(x, y)$, and `sol3` represents the solution $u(x, y)$ in more convenient form, with arbitrary function f and integration variable s. □

Problem 1.9 *Nonlinear first-order equations. Method of characteristics.* Let us consider the nonlinear first-order PDEs:

$$uu_x=u_y, \qquad (u_x)^2+u_y=0,$$

where $\{x \in \mathbb{R}, y \in \mathbb{R}\}$. Applying the *Maple* predefined functions `pdsolve` with option `HINT=strip`, `dsolve`, and `charstrip`, verify that the solutions (`Sol1`, `Sol2`), obtained by the method of characteristics read in the *Maple* notation:

$Sol1 := \{u(\_s) = \_C2, x(\_s) = \_C2\,\_s + \_C1, y(\_s) = -\_s + \_C3\}$

$Sol2 := (u_x{}^2 + u_y = 0)$ &where $[\{\{u(\_s) = -\_C3\,\_s + \_C2, x(\_s) = 2\,\_C4\,\_s + \_C1, y(\_s) = \_s + \_C5, \_p_1(\_s) = \_C4, \_p_2(\_s) = \_C3\}\}, \{\_p_1 = u_x, \_p_2 = u_y\}]$

Maple:

```
with(PDEtools);  declare(u(x,y));
PDE1:=u(x,y)*diff(u(x,y),x)=diff(u(x,y),y);
PDE2:=diff(u(x,y),x)^2+diff(u(x,y),y)=0;
sysCh:=charstrip(PDE1,u(x,y)); funcs:=indets(sysCh,Function);
Sol1:=dsolve(sysCh,funcs,explicit);
Sol2:=pdsolve(PDE2,HINT=strip);
```

For the given equation `PDE1`, we first obtain the characteristic system (depending on a parameter `_s`) via `charstrip`, and solve this system via `dsolve` to obtain the solution `Sol1` in the parametric form. We solve `PDE2` by the characteristic strip method directly via `pdsolve` with the option `HINT=strip`. □

Problem 1.10 *Burgers–Huxley equation. Traveling wave solutions.* Let us consider the Burgers–Huxley equation,

$$u_t + a u u_x - \nu u_{xx} = b u(1-u)(u-c),$$

where $\{x \in \mathbb{R}, t \geq 0\}$ and a, b, c are arbitrary real constants, and $\nu \in \mathbb{R}$ is the kinematic viscosity. Applying predefined functions, construct various forms of analytical solutions of the Burgers–Huxley equation.

Maple:

```
infolevel[pdsolve]:=5; with(PDEtools); declare(u(x,t));
U:=diff_table(u(x,t));
PDE1:=U[t]-nu*U[x,x]+a*U[]*U[x]=b*U[]*(1-U[])*(U[]-c);
casesplit(PDE1); separability(PDE1,u(x,t));
separability(PDE1,u(x,t),`*`); params1:={nu=1,a=1,b=1,c=1};
params2:={nu=1,a=1,b=1}; params3:={nu=1,a=-1,b=1};
Sol1:=pdsolve(subs(params1,PDE1),u(x,t));
Sol2:=pdsolve(subs(params2,PDE1),HINT='TWS');
Sol3:=pdsolve(subs(params3,PDE1),HINT='TWS(coth)');
for i from 1 to 3 do
      Test||i:=pdetest(Sol||i,subs(params||i,PDE1)) od;
pdsolve(subs(params1,PDE1),HINT=f(x)*g(t));
```

First, we can split the nonlinear problem into all the related regular cases (the general case and all the possible singular cases). For our case, we have the general case. Then we check the separability conditions (additive, by default, and multiplicative): if there is a separable solution, the result is 0. For our problem, we have, respectively, the expressions $-\left[\left((6\nu b+a^2)u-2\nu b(1+c)\right)u_x-(-u_t+bu(u-1)(-u+c))a\right]u_t\nu$ and $-[u_{xx}a-bu_x(1-4u+c)]u_t\nu$ that must vanish for the PDE to become separable. Finally, choosing some particular values for the arbitrary parameters (`params1`, `params2`, `params3`), we construct exact solutions, i.e., traveling wave solutions (`Sol1`, `Sol2`, `Sol3`), including options `HINT='TWS'` and `HINT='TWS(coth)'`:

$$Sol1 := u = 1/2 - 1/2\tanh(-\_C1 - 1/4\,x + 3/8\,t)$$

$$Sol2 := u = 1/2 - 1/2\tanh(-\_C1 - 1/4\,x - (-1/2\,c + 1/8)\,t)$$

$$Sol3 := u = 1/2 - 1/2\coth(-\_C1 - 1/2\,x - (-1/2\,c + 1/2)\,t)$$

If we try to construct separable solutions with the functional `HINT`, for example `HINT=f(x)*g(t)`, *Maple* cannot find the solution of this form and, according to the general strategy, described by Cheb-Terrab and von Bulow [28], applies another method and gives the result that coincides to one of the known solutions, i.e., the traveling wave solution (`Sol1`). We note that adding the function `infolevel` with the values (0–5), we can obtain more and more detailed information on each of the steps of the solving process.

Mathematica:

```
Off[DSolve::"nlpde"];
pde1=D[u[x,t],t]-nu*D[u[x,t],{x,2}]+a*u[x,t]*D[u[x,t],x]==
 b*u[x,t]*(1-u[x,t])*(u[x,t]-c);
{params1={nu->1,a->1,b->1,c->1},params2={nu->1,a->1,b->1}}
sol1=DSolve[pde1/.params1,u,{x,t}]
sol2=DSolve[pde1/.params2,u,{x,t}]
{n1,n2}=Map[Length,{sol1,sol2}]
test1=Table[pde1/.params1/.sol1[[i]]//FullSimplify,{i,1,n1}]
test2=Table[pde1/.params2/.sol2[[i]]//FullSimplify,{i,1,n2}]
```

Solving this nonlinear PDE, Mathematica generates a warning message, which can be ignored or suppressed with the `Off` function. □

Problem 1.11 *Nonlinear wave equation. Separable and self-similar solutions.* Let us consider the nonlinear wave equation of the form

$$u_{tt} = ae^{\lambda u}u_{xx},$$

where $\{x \in \mathbb{R}, t \geq 0\}$ and a $(a > 0)$, λ are arbitrary real constants. Applying *Maple* predefined functions to the nonlinear wave equation, find additive separable and self-similar solutions and verify that these solutions are exact solutions.

Maple:

```
with(PDEtools); declare(u(x,t));
PDE1:=diff(u(x,t),t$2)=a*exp(lambda*u(x,t))*diff(u(x,t),x$2);
params1:={a=1,lambda=1}; Sol1:=pdsolve(subs(params1,PDE1));
Sol2:=pdsolve(subs(params1,PDE1),build);
Sol3:=pdsolve(subs(params1,PDE1),HINT=`+`,build);
Sol4:=[SimilaritySolutions(subs(params1,PDE1))];
for i from 1 to 4 do pdetest(Sol||i,subs(params1,PDE1)); od;
```

Choosing some particular values for the arbitrary parameters (`params1`), we begin with the function `pdsolve(PDE1)` (without options) and we obtain a structure of the exact solution (additive separable solution) `Sol1`, for which we build an explicit expression in `Sol2`. Then we find the additive separable solution `Sol3` via option `HINT`. The solutions `Sol2` and `Sol3` are equivalent. Finally, we construct similarity solutions in `Sol4` and verify that the solutions obtained (`Sol1`–`Sol4`) are exact solutions of the given nonlinear wave equation. □

Problem 1.12 *Korteweg–de Vries equations (KdV). Gardner equation. Traveling wave solutions.* Let us consider the family of nonlinear equations,

$$u_t+G(u)u_x+u_{xxx}=0,$$

where $\{x\in\mathbb{R}, t\geq 0\}$ and the function $G(u)$ takes, for example, the following forms: $G(u)=\mp 6u$ (`tr1`, `tr2`), $G(u)=2au-3bu^2$ (`tr3`). This family is called the *third-order KdV equations.* Applying *Maple* and *Mathematica* predefined functions to the third-order KdV equations, find traveling wave solutions and verify that these solutions are exact solutions of the given PDEs.

Maple:

```
with(PDEtools); declare(u(x,t),G(u(x,t)));
KdVF:=diff(u(x,t),t)+G(u)*diff(u(x,t),x)+diff(u(x,t),x$3);
tr1:=G(u)=-6*u(x,t); tr2:=G(u)=6*u(x,t);
tr3:=G(u)=2*a*u(x,t)-3*b*u(x,t)^2;
for i from 1 to 3 do
 Sol||i:=pdsolve(subs(tr||i, KdVF),build);
 Test||i:=pdetest(Sol||i,subs(tr||i, KdVF)); od;
```

Applying the function `TWSolutions` (with various options) of the package `PDEtools`, it is possible to construct different types of traveling wave solutions in *Maple*, e.g., for the Gardner equation we have:

```
infolevel[TWSolutions]:=2;
Eq1:=subs(tr3,KdVF); Fs:=TWSolutions(functions_allowed);
Sol20:=pdsolve(Eq1,HINT='TWS');
Sol21:=TWSolutions(Eq1,singsol=false,functions=Fs):
N:=nops([Sol21]);
for i from 1 to N do op(i,[Sol21]) od;
Sol22:=TWSolutions(Eq1,singsol=false,parameters=[a,b],
       functions=[tan,cot],remove_redundant=true);
Sol23:=TWSolutions(Eq1,function=identity,output=ODE);
```

Adding the function `infolevel` with the values (0–5), we can obtain more and more detailed information on each of the steps of the solving process. Solution `Sol20` is equivalent to `pdsolve(Eq2,build)`. With the aid of the predefined function `TWSolutions` we can find different types of traveling wave solutions (of nonlinear PDEs and their systems) according to a list of functions `Fs` (instead of the `tanh` function, by default). The algorithm is based on the tanh-function method and its various extensions (see Sect. 2.5.2). Including the options `parameters` (splitting into cases with respect to the parameters), `singsol` (avoiding the singular solutions), `functions` (expanding in series of different functions), `remove_redundant=true` (removing all redundant solutions, i.e., all particular cases of more general solutions being constructed), we can obtain numerous traveling wave solutions (`Sol21`). For example, specifying the `cot` and `tan` functions and other options, we can obtain various traveling wave solutions. Sometimes it is useful to obtain a reduction (without computing a solution), that converts the nonlinear PDE into the ODE form, and the corresponding ODE. By default, the predefined function `TWSolutions` generates solutions of the form $f_i(\tau)=\sum_{k=0}^{n_i} A_{ik}\tau^k$, where the n_i are finite, the A_{ik} are constants with respect to the x_j, and $\tau=\tanh(\sum_{i=0}^{j} C_i x_i)$. In `Sol23`, we obtain an ODE using a more simple form, i.e., choosing the identity as the function (the option `function=identity`), where τ takes the form $\tau=\sum_{i=0}^{j} C_i x_i$.
Mathematica:

```
Off[DSolve::"nlpde"];
kdVF=D[u[x,t],t]+g[u[x,t]]*D[u[x,t],x]+D[u[x,t],{x,3}]==0
{tr[1]=g[u[x,t]]->-6*u[x,t], tr[2]=g[u[x,t]]->6*u[x,t],
 tr[3]=g[u[x,t]]->2*a*u[x,t]-3*b*u[x,t]^2}
```

```
Do[sol[i]=DSolve[kdVF/.tr[i],u,{x,t}];
 test[i]=kdVF/.tr[i]/.sol[i]//FullSimplify;
 Print["sol[",i,"]=",sol[i]]; Print["test[",i,"]=",test[i]],
{i,1,3}];
```
□

Problem 1.13 *Ginzburg–Landau equation. Traveling wave solution.* Let us consider the Ginzburg–Landau (GL) equation of the form

$$u_t - a u_{xx} - b u + c|u|^2 u = 0,$$

where $u(x,t)$ is a complex function, $\{x \in \mathbb{R}, t \geq 0\}$, a, c are complex constants, and $b \in \mathbb{R}$. For simplicity, we assume that a, b, c are real positive constants. Applying *Maple* predefined functions, verify that the traveling wave solution (`Sol1`) of the GL equation has the form $u = -\frac{1}{2}\frac{b}{\sqrt{cb}} - \frac{1}{2c}\sqrt{cb}\tanh\Big(-C_1 - \frac{\sqrt{2}}{4}\frac{\sqrt{ab}}{a}x + \frac{3}{4}bt\Big)$. Transforming the solution obtained to the equivalent exponential form (`Sol2`), verify that this solution is exact solution of the given nonlinear equation.

Maple:

```
with(PDEtools); declare(u(x,t)); interface(showassumed=0):
assume(a>0,b>0,c>0); PDE1:=diff(u(x,t),t)-a*diff(u(x,t),x,x)-
 b*u(x,t)+c*abs(u(x,t))^2*u(x,t)=0; Sol1:=pdsolve(PDE1,
 HINT='TWS'); Sol2:=convert(Sol1,exp); Expr1:=expand(map(
 evalc,pdetest(Sol2,PDE1))); Test1:=simplify(Expr1);
```
□

Problem 1.14 *Benjamin–Bona–Mahony equation. Traveling wave solution.* Let us consider the Benjamin–Bona–Mahony (BBM) equation or the regularized long-wave equation

$$u_t + a u_x + b u u_x - c u_{xxt} = 0,$$

where $\{x \in \mathbb{R}, t \geq 0\}$, a, b, and c are real constants. Applying *Maple* and *Mathematica* predefined functions, obtain the traveling wave solution (`Sol1`) of the BBM equation and verify that this solution is exact solution of the BBM equation. Considering a particular set of parameters (`params`), known in scientific literature (see [16], [115]), obtain the traveling wave solution of the BBM equation and visualize it at different times.

Maple:

```
with(PDEtools): with(plots): declare(u(x,t));
PDE1:=diff(u(x,t),t)+a*diff(u(x,t),x)
     +b*u(x,t)*diff(u(x,t),x)-c*diff(u(x,t),x,x,t)=0;
Ops1:=frames=50,numpoints=200,color=blue,thickness=3;
Sol1:=pdsolve(PDE1); pdetest(Sol1,PDE1);
params:={a=1,b=1,c=1,_C1=-1,_C2=-1,_C3=1};
Sol2:=unapply(subs(params,Sol1),x,t);
animate(rhs(Sol2(x,t)),x=-10..10,t=0..5,Ops1);
```

Mathematica:

```
Off[DSolve::"nlpde"]; SetOptions[Plot,ImageSize->300,PlotStyle->
 {Hue[0.7],Thickness[0.01]},PlotPoints->100,PlotRange->All];
pde1=D[u[x,t],t]+a*D[u[x,t],x]+b*u[x,t]*D[u[x,t],x]-c*D[D[u[x,t],
 {x,2}],t]==0; sol1=DSolve[pde1,u,{x,t}]
test1=pde1/.sol1//FullSimplify
params={a->1,b->1,c->1,C[1]->1,C[2]->-1,C[3]->1}
f=(u/.sol1[[1]])/.params; f[x,t]
Animate[Plot[f[x,t],{x,-10,10},PlotRange->{-10,10}],
 {t,0,10,0.01},AnimationRate->0.4]
```

□

Problem 1.15 *Kadomtsev–Petviashvili equation. Traveling wave solution.* Let us consider the Kadomtsev–Petviashvili (KP) equation

$$(u_t+auu_x+u_{xxx})_x+bu_{yy}=0,$$

where $u=u(x,y,t)$ is a real-valued function of two spatial variables, x, y, and the time variable t, $\{x\in\mathbb{R}, y\in\mathbb{R}, t\geq 0\}$, a and b are real constants. If $b=0$, the KP equation reduces to the KdV equation; if $b<0$, the equation is known as the KP-I equation (for strong surface tension); if $b>0$, the equation is known as the KP-II equation (for weak surface tension).

1. Applying *Maple* and *Mathematica* predefined functions, obtain the traveling wave solution (`Sol1`) of the KP equation and verify that this solution is exact solution of the KP equation. Considering a particular set of parameters (`params`), known in scientific literature [157], obtain the traveling wave solution of the KP equation (`Sol2`) and visualize it at a particular time, e.g., $t=1$.

2. Applying *Maple* predefined functions, build the similarity solutions of the KP equation (`Sol3`) and verify that these solutions are exact solutions of the KP equation. Considering a particular set of parameters (params), known in scientific literature [157], obtain the similarity solutions of the KP equation and visualize one of them at a particular time, e.g., t=1.

Maple:

```
with(PDEtools): declare(u(x,y,t)); Ops1:=axes=boxed,grid=[50,50],
 style=patchnogrid,shading=Z,orientation=[-40,50];
PDE1:=diff(diff(u(x,y,t),t)+a*u(x,y,t)*diff(u(x,y,t),x)
     +diff(u(x,y,t),x$3),x)+b*diff(u(x,y,t),y$2)=0;
Sol1:=pdsolve(PDE1); Test1:=pdetest(Sol1,PDE1);
params:={a=6,b=1,_C1=1,_C2=1,_C3=1,_C4=1};
Sol2:=subs(params,Sol1); xR:=-10..10; yR:=-10..10;
plot3d(subs(t=1,rhs(Sol2)),x=xR,y=yR,Ops1);
Sol3:=[SimilaritySolutions(PDE1,removeredundant=true)];
Test2:=map(pdetest,Sol3,PDE1); Sol4:=combine(subs(params,Sol3));
plot3d(subs(t=1,rhs(Sol4[3])),x=xR,y=yR,Ops1);
```

Mathematica:

```
Off[DSolve::nlpde]; pde1=D[D[u[x,y,t],t]+a*u[x,y,t]*
 D[u[x,y,t],x]+D[u[x,y,t],{x,3}],x]+b*D[u[x,y,t],{y,2}]==0;
{sol1=DSolve[pde1,u,{x,y,t}],test1=pde1/.sol1//FullSimplify}
params={a->6,b->1,C[1]->1,C[2]->1,C[3]->1,C[4]->1}
f=(u/.sol1[[1]])/.params; f[x,y,t]
Plot3D[f[x,y,1],{x,-10,10},{y,-10,10},BoxRatios->{1,1,1},
 Mesh->False,PlotPoints->{50,50},PlotRange->All,ViewPoint->
 {-60,90,60},ColorFunction->Function[{u},Hue[0.7+0.15*u]]]
```
□

1.2.2 Nonlinear PDEs with Initial and/or Boundary Conditions

In *Maple* 14, it is possible to construct exact solutions for a (growing) number of linear and nonlinear PDEs subject to initial and/or boundary conditions with the aid of the predefined function `pdsolve` (see `?pdsolve[boundaryconditions]`). Let us show some results.

Problem 1.16 *Burgers equation. Initial value problem (IVP). Exact solution.* Let us consider the Cauchy problem for the Burgers equation,

$$u_t+uu_x-u_{xx}=0, \quad u(x,0)=f(x),$$

where $\{x \in \mathbb{R}, t \geq 0\}$ and $f(x)=8\tan(4x)$. Applying *Maple* predefined functions, solve the Cauchy problem for the Burgers equation and verify that the solution obtained, $u(x,t)=8\tan(4x)$, is exact solution of this initial value problem.

Maple:

```
with(PDEtools): declare(f(x));
PDE1:=diff(u(x,t),t)-diff(u(x,t),x$2)+u(x,t)*diff(u(x,t),x)=0;
IC1:=u(x,0)=f(x); sys1:=[PDE1,IC1]; tr1:=f(x)=8*tan(4*x);
Sol:=[allvalues(pdsolve(sys1))]; map(simplify,Sol);
Sol1:=convert(simplify(algsubs(tr1,op(4,Sol))),tan);
pdetest(Sol1,[PDE1,subs(tr1,IC1)]);
```
□

Problem 1.17 *Nonlinear third-order equation. Boundary value problem (BVP). Exact solution.* Let us consider the following third-order equation with periodic boundary conditions:

$$u_t+uu_{xxx}=0, \quad u(0,t)=0, \quad u(L,t)=0,$$

where $\{x \in [0,L], t \geq 0\}$. Applying *Maple* predefined functions, solve the boundary value problem for the given equation and verify that the solution obtained, $u(x,t)=-\dfrac{x((L+x)\_c_1+3\_C_1)(-x+L)}{6\_c_1 t+6\_C_4}$, is exact solution of the boundary value problem.

Maple:

```
PDE1:=diff(u(x,t),t)+u(x,t)*diff(u(x,t),x$3)=0;
BC1:=u(0,t)=0,u(L,t)=0; Sol1:=pdsolve([PDE1,BC1]);
pdetest(Sol1,[PDE1,BC1]);
```

It should be noted that the same result can be obtained as follows:

```
with(PDEtools): declare(u(x,t)); U:=diff_table(u(x,t)):
PDE1:=U[t]+U[]*U[x,x,x]=0;
BC1:=eval(U[],x=0)=0,eval(U[],x=L)=0;
Sol:=pdsolve([PDE1,BC1]);
```
□

Problem 1.18 *Eikonal equation. Boundary value problem. Exact solution.* Let us consider the boundary value problem for the nonlinear eikonal equation

$$(u_x)^2+(u_y)^2=1, \quad u(x_0,y_0)=0,$$

where $\{x \in \mathbb{R}, y \in \mathbb{R}\}$. Applying *Maple* predefined functions, solve the boundary value problem for the eikonal equation and verify that the solution obtained, $u(x,y)=(x_0-x)\sqrt{1-c_2^2}+c_2(y-y_0)$ (`Sol1`), is exact solution of the boundary value problem. Prove that the complete integral of this equation takes the form $u(x,y)=ay-\sqrt{1-a^2}x+b$ (`Sol2`).

Maple:

```
with(PDEtools): declare(u(x,t));
PDE1:=diff(u(x,y),x)^2+diff(u(x,y),y)^2=1; BC1:=u(x0,y0)=0;
Sol1:=pdsolve([PDE1,BC1]); pdetest(Sol1,[PDE1,BC1]);
Sol2:=subs(_C1=0,_C2=b,_c[2]=a,pdsolve(PDE1,explicit));
pdetest(Sol2,PDE1);
```

These results can be obtained in
Mathematica:

```
Off[DSolve::"nlpde"]; {pde1=D[u[x,y],x]^2+D[u[x,y],y]^2==1,
 bc1=u[x0,y0]==0, sol1=DSolve[pde1,u,{x,y}],
 sol11=u[x,y]/.sol1[[1]], sol12=u[x,y]/.sol1[[2]]}
{tr0={x->x0,y->y0}, tr2={C[1]->b,C[2]->a}}
{eq1=(sol11/.tr0)==0, trC11=Solve[eq1,C[1]]}
sol11F[xN_,yN_]:=(sol11/.trC11//FullSimplify)/.{x->xN,y->yN}//
 First; {eq2=(sol12/.tr0)==0, trC12=Solve[eq2,C[1]]}
sol12F[xN_,yN_]:=(sol12/.trC12//FullSimplify)/.{x->xN,y->yN}//
 First; test11=Map[FullSimplify,{pde1/.{u->sol11F},
 bc1/.{u->sol11F}}]
test12=Map[FullSimplify,{pde1/.{u->sol12F},bc1/.{u->sol12F}}]
sol21F[xN_,yN_]:=sol11/.tr2/.{x->xN,y->yN};
sol22F[xN_,yN_]:=sol12/.tr2/.{x->xN,y->yN};
{test21=pde1/.u->sol21F, test22=pde1/.u->sol22F}
```

□

1.2.3 Nonlinear Systems

Computer algebra system *Maple* has various predefined functions based on symbolic algorithms for constructing analytical solutions of systems of nonlinear PDEs. These functions allow us to solve nonlinear systems and obtain solutions automatically and develop new methods and procedures for constructing new solutions. As before, we will consider the most important functions for finding analytical solutions of a given nonlinear system.

Maple:

```
pdsolve(PDESys,HINT=val,singsol=val,mindim=N,parameters=P);
pdsolve(PDESys,[DepVars]);                    pdetest(Sol,PDESys);
pdsolve([PDESys,ICs],[DepVars],series,order=N); with(PDEtools);
separability(PDESys,{DepVars});                   casesplit(PDESys);
splitsys(PDESys,{DepVars});         ReducedForm(PDESys1,PDESys2);
SimilaritySolutions(PDESys,ops);         TWSolutions(PDESys,ops);
```

NOTE. `ICs` are initial conditions; `DepVars` is a set or list of dependent variables (indicating the solving ordering that is useful for nonlinear PDE systems, where slight changes in the solving ordering can lead to different solutions or make an unsolvable (by `pdsolve`) a solvable equation).

`pdsolve`, finding analytical solutions for a given PDE system; relevant options are: `HINT`, some hints, `singsol`, computing or not singular solutions, `mindim`, a minimum number of the dimension of the solution space, `parameters`, indicating the solving variables with less priority, `series`, computing formal power series solutions,

`PDEtools`, a collection of functions for finding analytical solutions for `PDESys`, e.g.,

`splitsys`, splitting a PDE system into subsets (each one with equations coupled among themselves but not coupled to the equations of the other subsets),

`ReducedForm`, reducing one given PDE system with respect to another given PDE system, etc.

In computer algebra system *Mathematica*, we can obtain only analytical solutions of a single nonlinear partial differential equation (with the aid of the predefined function `DSolve`).

Problem 1.19 *Nonlinear telegraph system. General solution.* Let us consider the nonlinear telegraph system

$$u_t - v_x = 0, \quad v_t - F\big(u(x,t)\big)u_x - G\big(u(x,t)\big) = 0,$$

where $\{x \in \mathbb{R}, t \geq 0\}$. Applying *Maple* predefined functions, solve the nonlinear telegraph system and verify that the general solution (`Sol1`) reads in the *Maple* notation:

$$Sol1 := \left\{u = \_F1\,(x+t)\,, v = \int F\,(\_F1\,(x+t))\,\mathrm{D}\,(\_F1)\,(x+t) + G\,(\_F1\,(x+t))\,dt + \_F2\,(x)\right\}$$

Maple:

```
with(PDEtools): declare((u,v)(x,t),F(u(x,t)),G(u(x,t)));
U,V:=diff_table(u(x,t)),diff_table(v(x,t)):
Sys1:={V[t]-F(u(x,t))*U[x]-G(u(x,t))=0,U[t]-U[x]=0};
Sol1:=pdsolve(Sys1,{u(x,t),v(x,t)}); pdetest(Sol1,Sys1);
```

□

Problem 1.20 *Burgers system. Exact solution.* Let us consider the Burgers system

$$v_x - 2u = 0, \quad v_t - 2u_x + u^2 = 0,$$

where $\{x \in \mathbb{R}, t \geq 0\}$. Applying *Maple* predefined functions, solve the Burgers system and verify that the exact solution (`Sols`) reads in the *Maple* notation:

$$Sols := \left\{ u = \frac{\sqrt{\_c_2}\,(\_C1\ \cos(1/2\sqrt{\_c_2}x) + \_C2\ \sin(1/2\sqrt{\_c_2}x))}{-\_C1\ \sin(1/2\sqrt{\_c_2}x) + \_C2\ \cos(1/2\sqrt{\_c_2}x)}, \right.$$
$$\left. v = 4\ \ln(2) - 2\ \ln\left(\frac{(-\_C1\ \sin(1/2\sqrt{\_c_2}x) + \_C2\ \cos(1/2\sqrt{\_c_2}x))^2}{\_c_2}\right) + \_c_2 t + \_C3 \right\}$$

Maple:

```
with(PDEtools): declare((u,v)(x,t));
U,V:=diff_table(u(x,t)),diff_table(v(x,t)):
sys1:={V[x]-2*U[]=0,V[t]-2*U[x]+U[]^2=0};
Sols:=pdsolve(sys1,{u(x,t),v(x,t)}); pdetest(Sols,sys1);
```

□

1.2.4 Nonlinear Systems with Initial and/or Boundary Conditions

In this section, applying *Maple* predefined functions, we will show how to find exact solutions of the Cauchy problem for the nonlinear first-order system.

Problem 1.21 *Nonlinear first-order system. Cauchy problem. Exact solution.* Let us consider the Cauchy problem for the nonlinear first-order system

$$u_t = vu_x + u + 1, \quad v_t = -uv_x - v + 1; \quad u(x,0) = e^{-x}, \ v(x,0) = e^{x},$$

where $\{x \in \mathbb{R}, t \geq 0\}$. Solve the given nonlinear system and verify that the exact solution (`SolFin`) takes the form $\{u(x,t) = e^{t-x}, v(x,t) = e^{x-t}\}$.

Maple:

```
PDE1:=diff(u(x,t),t)=v(x,t)*diff(u(x,t),x)+u(x,t)+1;
PDE2:=diff(v(x,t),t)=-u(x,t)*diff(v(x,t),x)-v(x,t)+1;
IC1:=u(x,0)=exp(-x); IC2:=v(x,0)=exp(x);
sys1:={PDE1, PDE2}; Sols:=pdsolve(sys1); Sol1:=Sols[2];
sys11:=simplify(subs(IC1,IC2,subs(t=0,Sol1)));
C12:={_C1=-1,_C2=1}; sys12:=eval(subs(C12,sys11));
Sol12:=simplify(solve(sys12,{_C3,_C4}));
SolFin:=combine(subs(C12,Sol12,Sol1));
for i from 1 to 2 do
 simplify(expand(subs(SolFin,sys1[i]))); od;
```

□

Chapter 2
Algebraic Approach

Algebraic approach, based on transformation methods, is the most powerful analytic tool for studying nonlinear partial differential equations. Although the first exact solutions of PDEs have been determined in the 18th century (works by Cauchy, Euler, Hamilton, Jacobi, Lagrange, Monge), the most important results were obtained by S. Lie at the end of 19th century (see [91]). Nowadays, the *continuous transformation groups* (or *symmetries*, or *Lie groups*), proposed by Lie, and other transformations can be computed with the help of computer algebra systems, *Maple* and *Mathematica*.

In general, transformations (with respect to a given nonlinear PDE) can be divided into two parts: transformations of the independent variables, dependent variables; transformations of the independent variables, dependent variables, and their derivatives.

In this chapter, first we will consider various types of transformations, e.g., point and contact transformations, transformations relating differential equations, linearizing and bilinearizing transformations.

Transformation methods allow us to find *transformations* (under which a nonlinear PDE is *invariant*) and *new variables* (independent, dependent), with respect to which the differential equations become more simpler, e.g., linear. Transformations can convert a solution of a nonlinear PDE to the same or another solution of this equation, the invariant solutions can be found by symmetry reductions, rewriting the equation in new variables. Nonlinear PDEs can be written in Cauchy–Kovalevskaya form, or normal form, or canonical form, or linear form after a *point* or *contact transformation*.

Then we will consider the two wide classes of methods for finding exact solutions to nonlinear PDEs and nonlinear systems. The first class of methods, called *reductions*, consists in finding the reductions to a differential equation in a lesser number of independent variables. Among them, we will discuss traveling wave reductions, ansatz methods, self-similar reductions. The second class of methods, called *separation of variables*, allow us to construct an exact solution to a given nonlinear PDE as a combination of yet undetermined functions of fewer variables, and in the solution process to determine these simpler functions. We will consider selective problems in which it is possible to perform ordinary separation of variables, partial separation of variables, generalized separation of variables, and functional separation of variables.

Then, we will consider another class of methods, called the *classical method* of finding symmetries of nonlinear PDEs that allow us to obtain *transformation groups*, under which the PDEs are *invariant*, and *new variables* (independent and dependent), with respect to which differential equations become more simpler. For a class of the nonlinear second-order PDEs of general form in two independent variables, we perform the group analysis (i.e., describe the procedure of finding symmetries of this class of PDEs) and present a procedure for constructing invariant solutions (using symmetries yield by the group analysis).

Finally, we will construct exact solutions of nonlinear systems generalizing the methods considered for scalar nonlinear PDEs.

2.1 Point Transformations

Now we consider the most important transformations of nonlinear PDEs, namely, point transformations (transformations of independent and dependent variables), and then present two particular cases.

Let n be independent variables, $\mathbf{x}=(x_1, \ldots, x_n)$, and m dependent variables, $\mathbf{u}=(u_1, \ldots, u_m)$.

Definition 2.1 A *point transformation* acts on the space of independent and dependent variables of a nonlinear PDE, i.e a point transformation is a one-to-one transformation acting on the $(n+m)$-dimensional space $(\mathbf{x}, \mathbf{u})$.

In particular, if $n=m=1$, a point transformation has the following form: $X=\varphi(x,u)$, $U=\psi(x,u)$, where X is a new independent variable and U is a new dependent variable. The functions φ, ψ are given or need to be found.

2.1.1 Transformations of Independent and/or Dependent Variables

Let us consider transformations of independent and dependent variables; linear point transformations, i.e., translation transformations, scaling transformations, and rotation transformations; nonlinear transformations of the dependent variables. These transformations and their combinations can be applied to simplify nonlinear PDEs, to linearize them, to reduce them to normal, canonical, or invariant forms.

Problem 2.1 *Nonlinear wave-speed equation. Transformations of independent and dependent variables.* Let us consider the nonlinear wave-speed equation [9]:

$$u_{tt}=u^2u_{xx}+u(u_x)^2,$$

where $\{x \in \mathbb{R}, t \geq 0\}$. Applying the transformation of independent and dependent variables $X=ax$, $T=t$, $W(X,T)=au(x,t)$, obtain the invariant wave-speed equation $W_{TT}=W^2W_{XX}+W(W_X)^2$ (Eq3).

Maple:

```
with(PDETools): declare(u(x,t),W(X,T)); U:=diff_table(u(x,t));
tr1:={X=a*x,T=t,W(X,T)=a*u(x,t)}; tr2:=solve(tr1,{x,t,u(x,t)});
Eq1:=U[t,t]=U[]^2*U[x,x]+U[]*U[x]^2;
Eq2:=dchange(tr2,Eq1,[X,T,W(X,T)]); Eq3:=expand(Eq2*a);
```

Mathematica:

```
{tr1={xN==a*x,tN==t}, tr2=w[xN,tN]==a*u[x,t],
 tr11=Solve[tr1,{x,t}], tr12=Solve[tr2,{u[x,t]}]}
eq1[x_,t_]:=D[u[x,t],{t,2}]==u[x,t]^2*D[u[x,t],{x,2}]+u[x,t]*
 D[u[x,t],x]^2; eq1T[v_]:=First[((eq1[x,t]/.u->Function[{x,t},
 u[a*x,t]/a])/.tr11)/.{u->v}]; eq1[x,t]
Expand[Thread[eq1T[w]*a,Equal]]
```

□

It is known that a PDE admits a *Cauchy–Kovalevskaya form*[*] if it can be written in Cauchy–Kovalevskaya form with respect to some independent variable after a point or contact transformation.

Problem 2.2 *Sine–Gordon equation. Transformation of independent variables.* Let us consider the sine–Gordon equation of the form:

$$u_{xt}=\sin u,$$

where $\{x \in \mathbb{R}, t \geq 0\}$. Applying the transformation of independent variables $X=x+t$, $T=t-x$, verify that the sine–Gordon equation admits the Cauchy–Kovalevskaya form $u_{XX}-u_{TT}=\sin\big(u(X,T)\big)$.

[*] A partial differential equation is written in the Cauchy–Kovalevskaya form in terms of a given independent variable if this equation is written in a solved form for a pure derivative of the dependent variable with respect to the given independent variable, and if all other derivatives of the dependent variable are of lower order with respect to that independent variable.

Maple:

```
with(PDETools): declare(u(x,t),w(X,T)); U:=diff_table(u(x,t));
tr1:={T=t-x,X=t+x}; tr2:=solve(tr1,{x,t}); Eq1:=U[x,t]=sin(U[]);
Eq2:=dchange(tr2,Eq1,[X,T]);
```

Mathematica:

```
{tr1={tN==t-x,xN==t+x}, tr2=Solve[tr1,{x,t}]}
eq1[x_,t_]:=D[D[u[x,t],x],t]==Sin[u[x,t]]; eq1[x,t]
eq1T[v_]:=First[((eq1[x,t]/.u->Function[{x,t},u[x+t,-x+t]])/.
 tr2)/.{u->v}]; Simplify[eq1T[u]]
```

□

Let us consider some nonlinear equations and nonlinear systems arising in various problems of physics and *linear point transformations* (of independent or dependent variables) corresponding to the translation of the coordinate system, scaling (expansion or contraction) along the axes, and the rotation of the coordinate system through an angle ξ, i.e., *translation transformations*, *scaling transformations*, and *rotation transformations*. These transformations and their combinations can be applied to simplify nonlinear PDEs, to linearize them, to reduce them to normal, canonical, or invariant forms (see Problems 1.2, 1.3).

Problem 2.3 *KdV equation. Nonlinear heat equation. Nonlinear telegraph system. Translation transformations.* Let us consider the KdV equation (`Eq11`), the nonlinear heat equation (`Eq21`), and the nonlinear telegraph system (`sys1`):

$$u_t+uu_x+u_{xxx}=0, \quad u_t-\big(F(u(x,t))u_x\big)_x=0,$$
$$v_t-F(u(x,t))u_x-G(u(x,t))=0, \quad u_t-v_x=0,$$

where $\{x \in \mathbb{R}, t \geq 0\}$, (x,t) are the independent variables, (u,v) are the dependent variables, and $F(u)$, $G(u)$ are arbitrary functions. Verify that these nonlinear equations and the nonlinear system are invariant under the following linear translation transformations of the independent variables: $X=x+x_0$, $T=t$ (`tr11`), $X=x$, $T=t+t_0$ (`tr21`, `tr3`), i.e., the resulting equations and system take the form (`Eq12`, `Eq22`, `sys2`):

$$u_T+u(X,T)u_X+u_{XXX}, \quad u_T-\big(F(u(X,T))u_X\big)_X=0,$$
$$v_T-F\big(u(X,T)\big)u_X-G(u(X,T))=0, \quad u_T-v_X=0.$$

Maple:

```
with(PDETools): declare((u,v)(x,t)); U,V:=diff_table(u(x,t)),
 diff_table(v(x,t)); tr11:={X=x+x0,T=t};
tr12:=solve(tr11,{x,t}); Eq1:=U[t]+U[]*U[x]+U[x,x,x];
Eq1T:=dchange(tr12,Eq1,[X,T],params=[x0]);
Eq2:=diff(u(x,t),t)-diff(F(u(x,t))*diff(u(x,t),x),x)=0;
tr21:={X=x,T=t+t0}; tr22:=solve(tr21,{x,t});
Eq2T:=dchange(tr22,Eq2,[X,T],params=[t0]);
Sys1:=[V[t]-F(u(x,t))*U[x]-G(u(x,t))=0, U[t]-V[x]=0];
tr3:={X=x,T=t+t0}; tr31:=solve(tr3,{x,t});
Sys1T:=dchange(tr31,Sys1,[X,T],params=[t0]);
```

Mathematica:

```
{tr11={xN==x+x0,tN==t}, tr12=Solve[tr11,{x,t}]}
eq1[x_,t_]:=D[u[x,t],t]+u[x,t]*D[u[x,t],x]+D[u[x,t],
 {x,3}]; eq1T[v_]:=First[((eq1[x,t]/.u->Function[{x,t},
 u[x+x0,t]])/.tr12)/.{u->v}];  {eq1[x,t], Simplify[eq1T[u]]}
{tr21={xN==x,tN==t+t0}, tr22=Solve[tr21,{x,t}]}
eq2[x_,t_]:=D[u[x,t],t]-D[f[u[x,t]]*D[u[x,t],x],x]==0; eq2[x,t]
eq2T[v_]:=First[((eq2[x,t]/.u->Function[{x,t},u[x,t+t0]])/.
 tr22)/.{u->v}]; Simplify[eq2T[u]]
{tr3={xN==x,tN==t+t0}, tr31=Solve[tr3,{x,t}]}
sys1[x_,t_]:={D[v[x,t],t]-f[u[x,t]]*D[u[x,t],x]-g[u[x,t]]==0,
 D[u[x,t],t]-D[v[x,t],x]==0}; sys1[x,t]//Simplify
sys1T[w1_,w2_]:=First[((sys1[x,t]/.u->Function[{x,t},
 u[x,t+t0]]/.v->Function[{x,t},v[x,t+t0]])/.tr31)/.
 {u->w1,v->w2}]; Simplify[sys1T[u,v]]
```

□

Problem 2.4 *Nonlinear wave-speed equation. Scaling transformations.* Let us consider the nonlinear wave-speed equation (as in Problem 2.1):

$$u_{tt}=u^2u_{xx}+u(u_x)^2,$$

where $\{x\in\mathbb{R}, t\geq 0\}$. Applying the linear scaling transformation $X=ax$, $T=at$ of the independent variables, obtain the invariant wave-speed equation $u_{TT}=u(X,T)^2u_{XX}+u(X,T)(u_X)^2$ (Eq3).

Maple:

```
with(PDETools): declare(u(x,t)); U:=diff_table(u(x,t));
tr1:={X=a*x,T=a*t}; tr2:=solve(tr1,{x,t});
Eq1:=U[t,t]=U[]^2*U[x,x]+U[]*U[x]^2;
Eq2:=dchange(tr2,Eq1,[X,T],params=[a]); Eq3:=expand(Eq2/a^2);
```

Mathematica:

```
{tr1={xN==a*x,tN==a*t}, tr2=Solve[tr1,{x,t}]}
eq1[x_,t_]:=D[u[x,t],{t,2}]==u[x,t]^2*D[u[x,t],
 {x,2}]+u[x,t]*D[u[x,t],x]^2; eq1[x,t]
eq1T[v_]:=First[((eq1[x,t]/.u->Function[{x,t},u[a*x,a*t]])/.
 tr2)/.{u->v}]; Expand[Thread[eq1T[u]/a^2,Equal]]
```

□

Problem 2.5 *Nonlinear hyperbolic system. Rotation transformations.* Let us consider the nonlinear system [38]:

$$u_t+\big(uF_1(u,v)\big)_x+\big(uF_2(u,v)\big)_y=0,\quad v_t+\big(vF_1(u,v)\big)_x+\big(vF_2(u,v)\big)_y=0,$$

where $\{x\in\mathbb{R}, y\in\mathbb{R}, t\geq 0\}$, (x,y,t) are the independent variables, (u,v) are the dependent variables, and $F_i(u,v)$ $(i{=}1,2)$ are arbitrary functions. This system can be classified as a nonstrictly hyperbolic system (see Problem 1.5) and it arises in various problems of elastic theory, magnetohydrodynamics. Applying the following rotation transformation $X{=}x\cos\xi{+}y\sin\xi$ and $Y{=}{-}x\sin\xi{+}y\cos\xi$ (`tr1`), obtain the invariant form of the given nonlinear system (`sys2`, `Eqs121`, `Sys2Eq1`, `Eqs122`, `Sys2Eq2`):

$$u_t+\big(uG_1(u,v)\big)_x+\big(uG_2(u,v)\big)_y=0,\quad v_t+\big(vG_1(u,v)\big)_x+\big(vG_2(u,v)\big)_y=0,$$

where $G_1(u,v){=}F_1(u,v)\cos\xi{+}F_2(u,v)\sin\xi$, $G_2(u,v){=}{-}F_1(u,v)\sin\xi+F_2(u,v)\cos\xi$, and $u{=}u(X,Y,t)$, $v{=}v(X,Y,t)$ (`tr3`, `f2`).

Maple:

```
with(PDETools): var:=x,y,t; varN:=X,Y,t; declare((u,v)(var),
 (F1,F2)(u(var),v(var)),(u,v)(varN),(G1,G2)(u(varN),v(varN)));
tr1:={X=x*cos(xi)+y*sin(xi),Y=-x*sin(xi)+y*cos(xi)};
tr2:=simplify(solve(tr1,{x,y})); f1:=u(var),v(var);
f2:=u(varN),v(varN); L1:=[F1(f2),F2(f2)]; L2:=[u(varN),v(varN)];
Sys1:=[diff(u(var),t)+diff(u(var)*F1(f1),x)+diff(u(var)*F2(f1),
 y)=0, diff(v(var),t)+diff(v(var)*F1(f1),x)+diff(v(var)*F2(f1),
 y)=0]; Sys2:=expand(dchange(tr2,Sys1,[X,Y],params=[xi]));
tr3:=[cos(xi)*F1(f2)+sin(xi)*F2(f2)=G1(f2),
     -sin(xi)*F1(f2)+cos(xi)*F2(f2)=G2(f2)];
for i to 2 do tr3||i:=rhs(tr3[i])=lhs(tr3[i]); od;
for i to 2 do Sys2Eq||i:=collect(simplify(Sys2[i],tr3),L1);
              Eq1||i:=diff(L2[i]*subs(tr31,G1(f2)),X);
              Eq2||i:=diff(L2[i]*subs(tr32,G2(f2)),Y);
         Eqs12||i:=simplify(diff(L2[i],t)+Eq1||i+Eq2||i,tr3); od;
Test1:=Eqs121-lhs(Sys2Eq1); Test2:=Eqs122-lhs(Sys2Eq2);
```

Mathematica:

```
{var=Sequence[x,y,t], varN=Sequence[xN,yN,t]}
{tr1={xN==x*Cos[xi]+y*Sin[xi],yN==-x*Sin[xi]+y*Cos[xi]},
 tr2=Simplify[Solve[tr1,{x,y}]], f1=Sequence[u[var],v[var]],
 f2=Sequence[u[varN],v[varN]], l1={fF1[f2],fF2[f2]},
 l2={u[varN],v[varN]}}
sys1[x_,y_,t_]:={D[u[var],t]+D[u[var]*fF1[f1],x]+D[
 u[var]*fF2[f1],y]==0, D[v[var],t]+D[v[var]*fF1[f1],x]+D[
 v[var]*fF2[f1],y]==0}; sys1[x,y,t]//Expand
sys2[w1_,w2_]:=First[((sys1[x,y,t]/.u->Function[{x,y,t},
 u[x*Cos[xi]+y*Sin[xi],-x*Sin[xi]+y*Cos[xi],t]]/.v->Function[
 {x,y,t},v[x*Cos[xi]+y*Sin[xi],-x*Sin[xi]+y*Cos[xi],t]])/.tr2)/.
 {u->w1,v->w2}]; sys2[u,v]//FullSimplify
tr3={Cos[xi]*fF1[f2]+Sin[xi]*fF2[f2]->gG1[f2],
    -Sin[xi]*fF1[f2]+Cos[xi]*fF2[f2]->gG2[f2]}
Do[tTr3[i]=tr3[[i,2]]->tr3[[i,1]]; Print[tTr3[i]],{i,1,2}];
Do[sys2Eq[i]=Collect[Simplify[sys2[u,v][[i]],tr3],l1];
 eq1[i]=D[l2[[i]]*(gG1[f2]/.tTr3[1]),xN];
 eq2[i]=D[l2[[i]]*(gG2[f2]/.tTr3[2]),yN];
 eqs12[i]=Simplify[D[l2[[i]],t]+eq1[i]+eq2[i],tr3];
 Print[sys2Eq[i]]; Print[Eq1[i]]; Print[eq2[i]];
 Print[eqs12[i]],{i,1,2}]; Map[Expand,{test1=
 eqs12[1]-sys2Eq[1][[1]], test2=eqs12[2]-sys2Eq[2][[1]]}]
```
□

Let us consider another type of point transformations, *nonlinear transformations of the dependent variables*, that can be applied to reduce nonlinear equations or systems to linear ones.

Problem 2.6 *Burgers system. Nonlinear transformation of dependent variables.* Let us consider the Burgers system

$$v_x - 2u = 0, \quad v_t - 2u_x + u^2 = 0,$$

where $\{x \in \mathbb{R}, t \geq 0\}$, (x,t) are the independent variables and (u,v) are the dependent variables. The dependent variable u satisfies the Burgers equation $u_t + uu_x - u_{xx} = 0$. Applying the transformation of dependent variables, $w_1 = 2ue^{-v/4}$, $w_2 = 4e^{-v/4}$ (with no change of independent variables), reduce the nonlinear Burgers system to the linear system

$$(w_2)_x + w_1 = 0, \quad (w_2)_t + (w_1)_x = 0.$$

Maple:

```
with(PDEtools): var:=x,t; declare((u,v,w1,w2)(x,t));
U,V:=diff_table(u(x,t)),diff_table(v(x,t)); T:=exp(-V[]/4);
tr1:={w1(var)=2*U[]*T,w2(var)=4*T}; tr2:=solve(tr1,{U[],V[]});
Sys1:={V[x]-2*U[]=0, V[t]-2*U[x]+U[]^2=0};
Eq1:=dchange(tr2,Sys1,[w1(var),w2(var)]);
Sys2:=convert(expand(map(`*`,Eq1,w2(var))),list);
tr3:=isolate(Sys2[1],w1(var));
Sys3:=[Sys2[1],subs(diff(w1(var),x)=W1,Sys2[2])];
Sys4:=simplify([Sys3[1],eval(Sys3[2],tr3)]/(-4));
SysFin:=eval(Sys4,W1=diff(w1(var),x));
```

Mathematica:

```
{var=Sequence[x,t], tN=Exp[-v[x,t]/4], tr1={w1[var]==2*u[x,t]*tN,
 w2[var]==4*tN}, tr2={Reduce[tr1,{u[x,t],v[x,t]}]/.C[1]->0},
 varsN=Sequence[tr2[[1,2,2]],tr2[[1,3,2]]]}
sys1={(D[#2,x]-2*(#1))&[varsN]==0, (D[#2,t]-2*D[#1,x]+
 (#1)^2)&[varsN]==0}; eq1=Expand[sys1]
sys2=Expand[Table[Thread[eq1[[i]]*w2[x,t],Equal],{i,1,2}]]
tr3=Solve[sys2[[1]],w1[var]]
sys3={sys2[[1]],sys2[[2]]/.D[w1[var],x]->wW1}
sys4=Simplify[{sys3[[1]],sys3[[2]]/.tr3}]
sysFin=Flatten[sys4/.wW1->D[w1[var],x]]
```

□

2.1.2 Hodograph Transformation

The *hodograph transformation* is a method for simplifying nonlinear partial differential equations and nonlinear systems or converting them into linear ones, by reversing the roles of the dependent and independent variables.

Definition 2.2 A hodograph transformation for a given nonlinear PDE is a transformation in which one of the new independent variables depends on the old dependent variable.

Problem 2.7 *Nonlinear parabolic equation. Hodograph transformation.* Let us consider the nonlinear PDE [124]

$$u_t u_x^2 = F(t,u) u_{xx},$$

where $\{x \in \mathbb{R}, t \geq 0\}$ and $F(t,u)$ is an arbitrary function. Applying the hodograph transformation $x = x(t,u)$, reduce the given nonlinear PDE to the equation $x_t = F(t,u) x_{uu}$ that is linear PDE for x.

Maple:

```
with(PDEtools): declare(u(x,t),x(t,u)); U,X:=diff_table(u(x,t)),
 diff_table(x(t,u)); Eq1:=U[t]*U[x]^2=F(t,u)*U[x,x];
tr1:=diff(x,x)=X[u]*U[x]; tr2:=diff(x,t)=X[u]*U[t]+X[t];
tr3:=diff(lhs(tr1),x)=X[u,u]*U[x]^2+X[u]*U[x,x];
tr11:=U[x]=solve(tr1,U[x]); tr21:=U[t]=solve(tr2,U[t]);
tr31:=U[x,x]=solve(tr3,U[x,x]); tr32:=lhs(tr31)=subs(tr11,
 rhs(tr31)); HodographTr:={tr11,tr21,tr32};
Eq2:=subs(HodographTr,Eq1); Eq3:=Eq2*(-denom(lhs(Eq2)));
```

Mathematica:

```
eq1=D[u[x,t],t]*D[u[x,t],x]^2==f[t,u]*D[u[x,t],{x,2}]
{tr1=D[x,x]==D[x[t,u],u]*D[u[x,t],x], tr2=D[x,t]==D[x[t,u],u]*
 D[u[x,t],t]+D[x[t,u],t], tr3=D[tr1[[1]],x]==D[x[t,u],{u,2}]*
 D[u[x,t],x]^2+D[x[t,u],u]*D[u[x,t],{x,2}]}
Map[Flatten,{tr11=Solve[tr1,D[u[x,t],x]], tr21=Solve[tr2,
 D[u[x,t],t]],tr31=Solve[tr3,D[u[x,t],{x,2}]]}]
{tr32=tr31/.tr11, hodographTr={tr11,tr21,tr32}//Flatten,
 eq2=eq1/.hodographTr}
eq3=Thread[eq2*Denominator[eq2[[1]]],Equal]//Simplify
```

□

2.2 Contact Transformations

Point transformations that act on the space of the independent and dependent variables of a nonlinear PDE can be extended or prolonged to point transformations that act on the space of the independent variables, dependent variables, and their derivatives to any finite order.

Definition 2.3 A *contact transformation* is equivalent to a point transformation that acts on the space of the independent variables, the dependent variable, and its first derivatives, and can be extended to point transformations that act on the space of the independent variables, the dependent variable, and its derivatives to any finite order.

Sophus Lie considered contact transformations and contact symmetries of PDEs [91]. A *symmetry of a PDE* is a transformation (or mapping) of its solution manifold into itself, i.e., it is a transformation that maps any solution of the PDE into another solution of the same PDE.

For a nonlinear PDE of general form with n independent variables $\mathbf{x}=(x_1,\dots,x_n)$ and one dependent variable $u(\mathbf{x})$ ($m=1$, see Sect. 2.1), a contact transformation is one-to-one transformation on a domain $\mathcal{D}$ in $(\mathbf{x}, u, u_{\mathbf{x}})$-space and can be represented in the form:

$$X_i = \varphi_i(\mathbf{x}, u, u_{\mathbf{x}}), \quad U = \psi(\mathbf{x}, u, u_{\mathbf{x}}), \quad U_{x_i} = \eta_i(\mathbf{x}, u, u_{\mathbf{x}}), \qquad i = 1, \dots, n,$$

where the contact condition $dU = U_{x_i}\, dX^i$ is invariant. It is assumed that φ_i, ψ have an essential dependence on the first derivatives of u (otherwise a contact transformation is a point transformation). In particular case ($i = 1$), a contact transformation has the form: $X = \varphi(x, u, u_x)$, $U = \psi(x, u, u_x)$, $U_x = \eta(x, u, u_x)$.

Let us consider some important contact transformations that can reduce some nonlinear equations of mathematical physics to linear PDEs.

2.2.1 Legendre Transformation

The *Legendre transformation* that belongs to the class of contact transformations, consists in regarding the components of the gradient of a solution as new independent variables.

Definition 2.4 The *Legendre transformation* is defined by the relations $x = w_\xi$, $t = w_\eta$, $u(x, t) = x\xi + t\eta - w(\xi, \eta)$, where w is the new dependent variable of two independent variables ξ and η.

Problem 2.8 *Nonlinear second-order equation. Legendre transformation.* Let us consider the nonlinear second-order PDE [124]

$$F(u_x, u_t)u_{xx} + G(u_x, u_t)u_{xt} + H(u_x, u_t)u_{tt} = 0,$$

where $\{x \in \mathbb{R}, t \geq 0\}$ and $F(u_x, u_t)$, $G(u_x, u_t)$, and $H(u_x, u_t)$ are arbitrary functions. Applying the Legendre transformation, reduce the given nonlinear PDE to the linear PDE: $F(\xi, \eta)w_{\eta\eta} - G(\xi, \eta)w_{\xi\eta} + H(\xi, \eta)w_{\xi\xi} = 0$ (Eq4).

Maple:

```
with(PDEtools): declare(u(x,t),w(xi,eta));
alias(u=u(x,t),w=w(xi,eta));
Eq1:=F(diff(u,x),diff(u,t))*diff(u,x$2)+G(diff(u,x),diff(u,t))
 *diff(u,x,t)+H(diff(u,x),diff(u,t))*diff(u,t$2)=0;
LegendreTr:={x=diff(w,xi),t=diff(w,eta),
             u=-w+diff(w,xi)*xi+diff(w,eta)*eta};
Eq2:=dchange(LegendreTr,Eq1,[xi,eta,w]);
Eq3:=simplify(Eq2); Eq4:=numer(lhs(Eq3))=rhs(Eq1);
```

Mathematica:

```
{var=Sequence[x,t],varN=Sequence[xi,eta]}
j1=D[u[var],{x,2}]*D[u[var],{t,2}]-D[D[u[var],x],t]^2
legTr={x->D[w[varN],xi],t->D[w[varN],eta],u[var]->
 -w[varN]+D[w[varN],xi]*xi+D[w[varN],eta]*eta}
legD1={D[u[var],x]->xi,D[u[var],t]->eta}
legD2={D[u[var],{x,2}]->j1*D[w[varN],{eta,2}],D[D[u[var],x],t]->
 -j1*D[D[w[varN],xi],eta],D[u[var],{t,2}]->j1*D[w[varN],{xi,2}]}
legendreTr={legTr,legD1,legD2}//Flatten
{eq1=f[D[u[var],x],D[u[var],t]]*D[u[var],{x,2}]+g[D[u[var],x],
 D[u[var],t]]*D[u[var],x,t]+h[D[u[var],x],D[u[var],t]]*
 D[u[var],{t,2}]==0, eq2=eq1/.legendreTr, eq3=eq2//FullSimplify}
eq4=eq3[[1,2]]==eq3[[2]]
```

□

2.2.2 Euler Transformation

The *Euler transformation*, closely related to the Legendre transformation, is a special case of contact transformations.

Definition 2.5 The *Euler transformation* is defined by the following relations: $x=w_\xi$, $t=\eta$, $u(x,t)=x\xi-w(\xi,\eta)$, where w is the new dependent variable of two independent variables ξ and η.

Problem 2.9 *Nonlinear second-order equation. Euler transformation.* Let us consider the nonlinear PDE [124]

$$u_t u_{xx}=F(t,u_x),$$

where $\{x\in\mathbb{R}, t\geq 0\}$ and $F(t,u_x)$ is arbitrary function. Applying the Euler transformation, reduce the given nonlinear PDE to the linear partial differential equation $w_\eta=-F(\eta,\xi)w_{\xi\xi}$ (Eq3).

Maple:

```
with(PDEtools): declare(u(x,t),w(xi,eta)); alias(u=u(x,t),
 w=w(xi,eta)); Eq1:=diff(u,t)*diff(u,x$2)=F(t,diff(u,x));
EulerTr:={u=-w+diff(w,xi)*xi,x=diff(w,xi),t=eta};
Eq2:=dchange(EulerTr,Eq1,[xi,eta,w]); Eq3:=simplify(Eq2);
```

Mathematica:

```
{var=Sequence[x,t],varN=Sequence[xi,eta]}
{eTr={u[var]->-w[varN]+D[w[varN],xi]*xi,x->D[w[varN],xi],t->eta},
 eD1={D[u[var],x]->xi, D[u[var],t]->-D[w[varN],eta]},
 eD2={D[u[var],{x,2}]->1/D[w[varN],{xi,2}], D[u[var],x,t]->
 -D[D[w[varN],xi],eta]/D[w[varN],{xi,2}], D[u[var],{t,2}]->
 (D[w[varN],xi,eta]^2-D[w[varN],{xi,2}]*D[w[varN],{eta,2}])/D[
 w[varN],{xi,2}]}, eulerTr={eTr,eD1,eD2}//Flatten}
eq1=D[u[var],t]*D[u[var],{x,2}]==f[t,D[u[var],x]]
eq2=eq1/.eulerTr
```
□

2.3 Transformations Relating Differential Equations

In this section, we will consider the most important transformations (or mappings) relating partial differential equations: the Bäcklund transformations (that consist of two nonlinear PDEs), the Miura transformation (that relates the modified KdV equation and the KdV equation), and the Gardner transformation (that relates the Gardner equation and the KdV equation).

2.3.1 Bäcklund Transformations

Bäcklund transformations, developed in the 1880s for studying differential geometry and differential equations, is one of the powerful methods for constructing exact solutions of nonlinear PDEs. Let us consider scalar nonlinear PDEs of the second-order with respect to one dependent variable and two independent variables.

Definition 2.6 We say that a *Bäcklund transformation* (BT) between two given nonlinear PDEs

$$\mathcal{F}_1(u,x,t,u_x,u_t,u_{xx},u_{xt},u_{tt})=0, \quad \mathcal{F}_2(w,\xi,\eta,w_\xi,w_\eta,w_{\xi\xi},w_{\xi\eta},w_{\eta\eta})=0,$$

is a pair of relations

$$\phi_i(u,x,t,u_x,u_t,w,\xi,\eta,w_\xi,w_\eta)=0 \quad (i=1,2) \tag{2.1}$$

with some transformation between the independent variables (x,t) and (ξ,η), in which ϕ_1, ϕ_2 depend on the derivatives of $u(x,t)$ and $w(\xi,\eta)$, such that the elimination of $u(x,t)$ or $w(\xi,\eta)$ between (ϕ_1,ϕ_2) implies $\mathcal{F}_2(w,\xi,\eta,w_\xi,w_\eta,w_{\xi\xi},w_{\xi\eta},w_{\eta\eta})=0$ or $\mathcal{F}_1(u,x,t,u_x,u_t,u_{xx},u_{xt},u_{tt})=0$.

Definition 2.7 The Bäcklund transformation is called the *auto-Bäcklund transformation* if the two nonlinear PDEs are the same.

Since Bäcklund transformations establish mappings relating differential equations, such transformations can be applied for finding solutions of one equation from solutions of another equation (see Problem 2.14). Auto-Bäcklund transformations can be applied for constructing new exact solutions (see Problem 2.11).

Problem 2.10 *Burgers equation. Bäcklund transformations.* Let us consider the Burgers equation

$$u_t = u_{xx} + u u_x,$$

where $\{x \in \mathbb{R}, t \geq 0\}$. Verify that the Burgers equation (BurEq) is related to the heat equation $w_t = w_{xx}$ (HEq) by the Bäcklund transformations (BT) $w_x - \frac{1}{2} w u = 0$, $w_t - \frac{1}{2}(wu)_x = 0$.

Maple:

```
with(PDEtools): declare((u,w)(x,t)); alias(u=u(x,t),w=w(x,t));
BT:=[diff(w,x)=1/2*w*u,diff(w,t)=1/2*diff(w*u,x)];
Eq1:=isolate(BT[1],u); HEq:=expand(subs(Eq1,BT[2]));
Eq3:=expand(HEq/w); Eq4:=Diff(lhs(Eq3),x)=Diff(diff(w,x)/w,t);
Eq5:=Diff(rhs(Eq3),x); Eq6:=diff(w,x)=solve(BT[1],diff(w,x));
Eq7:=diff(w,x)/w=subs(Eq6,diff(w,x)/w);
Eq8:=subs(lhs(Eq7)^2=rhs(Eq7)^2,diff(Eq7,x));
Eq9:=isolate(Eq8,diff(w,x,x)/w); Eq10:=value(subs(Eq9,Eq5));
Eq11:=value(subs(Eq7,rhs(Eq4))); BurEq:=factor((Eq10=Eq11)*2);
```

Mathematica:

```
var=Sequence[x,t]; trBT={D[w[var],x]==1/2*w[var]*u[var],
                         D[w[var],t]==1/2*D[w[var]*u[var],x]}
{eq1=Solve[trBT[[1]],u[var]], eq11=D[eq1,x]//Flatten,
 heatEq=Simplify[trBT[[2]]/.eq1/.eq11]//First,
 eq3=Thread[heatEq/w[var],Equal]//Expand}
{eq4=HoldForm[D[eq3[[1]],x]==D[D[w[var],x]/w[var],t]],
 eq5=HoldForm[D[eq3[[2]],x]]}
{eq6=D[w[var],x]==Solve[trBT[[1]],D[w[var],x]][[1,1,2]],
 eq7=Thread[eq6/w[var],Equal],
 eq8=Thread[D[eq7,x],Equal]/.eq7[[1]]^2->eq7[[2]]^2,
 eq9=Thread[eq8-eq8[[1,1]],Equal]//ToRules,
 eq10=D[eq3[[2]]/.eq9,x],
 eq11=D[D[w[var],x]/w[var]/.ToRules[eq7],t]}
burgersEq=Factor[Thread[(eq10==eq11)*2,Equal]]
```

□

Problem 2.11 *Sine–Gordon equation (SG). Bäcklund transformations.* Let $u(x,t)$ and $v(x,t)$ $(x \in \mathbb{R}, t \geq 0)$ be nonzero solutions of the sine–Gordon equation

$$u_{xt} = \sin u.$$

Verify that the $\frac{1}{2}(u-v)_x = \beta \sin(\frac{1}{2}(u+v))$, $\frac{1}{2}(u+v)_t = \frac{1}{\beta} \sin(\frac{1}{2}(u-v))$ ($\beta \neq 0$ is an arbitrary constant) is the *auto-Bäcklund transformation* for the sine–Gordon equation. Applying this transformation, solve the sine–Gordon equation and visualize the solutions. Considering the SG equation of the form $u_{XX} - \frac{1}{c^2} u_{TT} = \sin u$, obtain a kink and antikink solutions and visualize them.

1. Auto-Bäcklund transformation. We verify that the sine–Gordon equations $u_{xt} = \sin u$ and $v_{xt} = \sin v$ are related by the Bäcklund transformations, $\frac{1}{2}(u-v)_x = \beta \sin(\frac{1}{2}(u+v))$, $\frac{1}{2}(u+v)_t = \frac{1}{\beta} \sin(\frac{1}{2}(u-v))$. Since both u and v satisfy the same sine–Gordon equation, the pair of Bäcklund transformations is referred to as an *auto-Bäcklund transformation* for the sine–Gordon equation.

Maple:

```
interface(showassumed=0); assume(a<>0); with(PDEtools):
declare((u,v)(x,t)); alias(u=u(x,t),v=v(x,t));
SG:=diff(u,x,t)=sin(u); BT:=[diff(u-v,x)/2=beta*sin(1/2*(u+v)),
    diff(u+v,t)/2=sin(1/2*(u-v))/beta];
Eq1:=subs(BT,diff(BT[1],t)); Eq2:=subs(BT,diff(BT[2],x));
SGEqu:=combine(Eq2+Eq1,trig); SGEqv:=combine(Eq2-Eq1,trig);
```

Mathematica:

```
sGEq[u_]:=D[u,{x,t}]==Sin[u];
trBT={D[u[x,t]-v[x,t],x]/2->beta*Sin[1/2*(u[x,t]+v[x,t])],
      D[u[x,t]+v[x,t],t]/2->Sin[1/2*(u[x,t]-v[x,t])]/beta}
{eq1=D[trBT[[1]],t]/.trBT, eq2=D[trBT[[2]],x]/.trBT}
sGEqu=(eq2[[1]]+eq1[[1]]==eq2[[2]]+eq1[[2]])//FullSimplify
sGEqv=(eq2[[1]]-eq1[[1]]==eq2[[2]]-eq1[[2]])//FullSimplify
```

2. Solution and visualization. Applying the Bäcklund transformations, we solve the sine–Gordon equation and verify that its solution is a *kink* solution and has the form $u(x,t) = 4\arctan(\alpha e^{\beta x + t/\beta})$ (`KinkSol`). We visualize the kink solution in 3D and 2D spaces.

Maple:

```
with(plots): setoptions(plot,scaling=constrained,thickness=3);
Eq3:=subs(v=0,factor(BT*2)); I1:=int(1/sin(U/2),U);
Eq4:=int(2*beta,x)=convert(simplify(convert(I1,tan)),tan)+A(t);
Eq5:=int(2/beta,t)=convert(simplify(convert(I1,tan)),tan)+B(x);
t1:=tan(U/4); Eq6:=expand(isolate(Eq4,t1));
Eq7:=expand(isolate(Eq5,t1)); Eq8:=lhs(Eq6)=rhs(Eq6)*rhs(Eq7);
C:=select(has,rhs(Eq8),[A,B]); Eq9:=lhs(Eq8)=alpha*rhs(Eq8/C);
Eq10:=u=solve(Eq9,U); KinkSol:=unapply(rhs(Eq10),x,t,alpha,beta);
alpha1:=0.1; beta1:=10; xR:=-Pi..Pi; tR:=0..1;
plot3d(evalf(KinkSol(x,t,alpha1,beta1)),x=xR,t=tR,
 orientation=[71,73]); plot(evalf(KinkSol(x,0,alpha1,beta1)),
 x=xR,tickmarks=[spacing(Pi/2),spacing(Pi/2)]);
```

Mathematica:

```
Off[Solve::"ifun"]; SetOptions[Plot,ImageSize->500,PlotStyle->
 {Hue[0.9],Thickness[0.01]},PlotPoints->100,PlotRange->All];
SetOptions[Plot3D,BoxRatios->{1,1,1},PlotRange->All,ViewPoint->
 {-1,2,2}]; eq3=trBT/.{v[x,t]->0,D[v[x,t],x]->0,D[v[x,t],t]->0}
{eq4=Assuming[Cos[uN/4]>0,Simplify[Integrate[2*beta,x]==
 Integrate[1/Sin[uN/2],uN]+a[t]]], eq5=Assuming[Cos[uN/4]>0,
 Simplify[Integrate[2/beta,t]-Integrate[1/Sin[uN/2],uN]==b[x]]]}
{eq6=Solve[eq4,Tan[uN/4]], eq7=Solve[eq5,Tan[uN/4]]}
eq8=eq6[[1,1,1]]==eq6[[1,1,2]]*eq7[[1,1,2]]
const=Exp[Select[eq8[[2,2]],MemberQ[#1,a[t]]||MemberQ[#1,b[x]]&]]
eq9=eq8[[1]]==alpha*(Thread[eq8/const,Equal])[[2]]
eq10=Solve[eq9,uN]
kinkSol[xN_,tN_,alphaN_,betaN_]:=eq10[[1,1,2]]/.{x->xN,t->tN,
 alpha->alphaN,beta->betaN}; {alpha1=0.1, beta1=10}
Plot3D[N[kinkSol[x,t,alpha1,beta1]],{x,-Pi,Pi},{t,0,1}]
Plot[N[kinkSol[x,0,alpha1,beta1]],{x,-Pi,Pi}]
```

3. Kink and antikink solutions. Visualization. For the Cauchy–Kovalevskaya form of the sine–Gordon equation

$$u_{XX}-\frac{1}{c^2}u_{TT}=\sin u$$

with the characteristic coordinate transformations $x=\frac{1}{2}(X+cT)$, $t=\frac{1}{2}(X-cT)$, we obtain the *kink* and *antikink* solutions (`KinkSols`). Finally, we visualize the kink and antikink solutions.

Maple:

```
c1:=1; Eq11:=simplify(subs({x=(X+c*T)/2,t=(X-c*T)/2},rhs(Eq10)));
tr1:=[beta=epsilon*sqrt((1-U)/(1+U)),c=U*m/(2*(beta-1/beta))];
Eq12:=simplify(subs(tr1[1],simplify(subs(tr1,Eq11))));
KinkSols:=unapply(Eq12,m,U,epsilon,T,X,alpha);
m1:=beta->1/2*(beta+1/beta);
U1:=(c,beta)->c*((beta^2-1)/(beta^2+1)); m1(beta1); U1(c1,beta1);
Kink:=plot(evalf(KinkSols(m1(beta1),U1(c1,beta1),1,0,X,alpha1)),
 X=xR,tickmarks=[spacing(Pi/2),spacing(Pi/2)]):
AntiKink:=plot(evalf(KinkSols(m1(beta1),U1(c1,beta1),-1,0,X,
 alpha1)),X=xR,color=blue,tickmarks=[spacing(Pi/2),
 spacing(Pi/2)]): display({Kink,AntiKink});
```

Mathematica:

```
c1=1; {eq11=eq10[[1,1,2]]/.{x->(xN+c*tN)/2,t->(xN-c*tN)/2}
 //FullSimplify, tr1={beta->\[Epsilon]*Sqrt[(1-uN)/(1+uN)],
 c->uN*m/(2*(beta-1/beta))}}
eq12=Assuming[uN>1,Simplify[(eq11/.tr1)/.tr1[[1]]]]
kinkSols[mN1_,uN1_,epsilonN1_,tN1_,xN1_,alphaN1_]:=eq12/.{m->mN1,
 uN->uN1,\[Epsilon]->epsilonN1,tN->tN1,xN->xN1,alpha->alphaN1};
m1[beta_]:=1/2*(beta+1/beta); v1[c_,beta_]:=
 c*((beta^2-1)/(beta^2+1)); {m1[beta1],v1[c1,beta1]}
kink=Plot[N[kinkSols[m1[beta1],v1[c1,beta1],1,0,xN,alpha1]],
 {xN,-Pi,Pi}]; antiKink=Plot[N[kinkSols[m1[beta1],v1[c1,beta1],
 -1,0,xN,alpha1]],{xN,-Pi,Pi},PlotStyle->{Hue[0.7],
 Thickness[0.01]}]; Show[{kink,antiKink}]
```

□

2.3.2 Miura Transformation

In 1968, Miura, Gardner, and Kruskal [103] developed a method for determining an infinite number of conservation laws by introducing the Miura transformation $u=v^2+v_x$, which establishes a connection between the modified KdV equation and the KdV equation, i.e., between two nonlinear PDEs. The Miura transformation is a non-invertible mapping [19].

Problem 2.12 *mKdV and KdV equations. Miura transformation.* Let us consider the mKdV and KdV equations

$$v_t-6v^2v_x+v_{xxx}=0, \quad u_t-6uu_x+u_{xxx}=0,$$

where $\{x \in \mathbb{R}, t \geq 0\}$. Applying the Miura transformation, $u{=}v^2{+}v_x$, show that there is a connection between the mKdV equation and the KdV equation.

Maple:

```
with(PDEtools): declare((u,v)(x,t)); alias(u=u(x,t),v=v(x,t));
mKdVEq:=diff(v,t)-6*v^2*diff(v,x)+diff(v,x$3)=Mv;
KdVEq:=diff(u,t)-6*u*diff(u,x)+diff(u,x$3)=0;
trMiura:=u=v^2+diff(v,x); Eq1:=algsubs(trMiura,KdVEq);
subs(mKdVEq,Eq1); MvL:=lhs(mKdVEq); Eq2:=2*v*Mv +Diff(Mv,x)=0;
Eq3:=expand(2*v*MvL+diff(MvL,x))=0; evalb(Eq1=Eq3);
```

Mathematica:

```
var=Sequence[x,t]
mKdVEq=D[v[var],t]-6*v[var]^2*D[v[var],x]+D[v[var],{x,3}]->mv
kdVEq=D[u[var],t]-6*u[var]*D[u[var],x]+D[u[var],{x,3}]==0
trMiura=u[var]->v[var]^2+D[v[var],x]
trMD={D[trMiura,t],D[trMiura,x],D[trMiura,{x,3}]}
{eq1=kdVEq/.trMiura/.trMD//Expand,eq1/.mKdVEq,mvL=mKdVEq[[1]]}
{eq2=2*v[var]*mv+HoldForm[D[mv,x]]==0,
 eq3=Expand[2*v[var]*mvL+D[mvL,x]]==0, eq1===eq3}
```

□

2.3.3 Gardner Transformation

The *Gardner equation* has been proposed by R. M. Miura, C. S. Gardner, and M. D. Kruskal in 1968 [103] (as a generalization of the KdV equation) by introducing the Gardner transformation $u{=}w{+}\varepsilon^2 w^2{+}\varepsilon w_x$, which establishes a connection between the Gardner equation and the KdV equation.

Problem 2.13 *Gardner and KdV equations. Gardner transformation.* Let us consider the Gardner and KdV equations

$$w_t{-}(6\varepsilon^2 w^2{+}6w)w_x{+}w_{xxx}{=}0, \quad u_t{-}6uu_x{+}u_{xxx}{=}0,$$

where $\{x \in \mathbb{R}, t \geq 0\}$ and ε is an arbitrary real parameter. Applying the Gardner transformation, show that there is a connection between the Gardner equation and the KdV equation.

Maple:

```
with(PDEtools): declare((u,v,w)(x,t)); alias(u=u(x,t),v=v(x,t),
 w=w(x,t)); trMi:=u=v^2+diff(v,x);
KdV:=diff(u,t)-6*u*diff(u,x)+diff(u,x$3)=0;
trv:=v=1/(2*epsilon)+epsilon*w; Eq1:=expand(algsubs(trv,trMi));
trGar:=subs(1/epsilon^2=0,Eq1);
Eq2:=collect(expand(subs(trGar,KdV)),diff);
Eq3:=map(factor,lhs(Eq2)); tr1:=1+2*epsilon^2*w=G;
Eq4:=collect(subs(tr1,Eq3),G);
Eq5:=collect(expand((Eq4-op(1,Eq4))/epsilon),diff);
term1:=op(1,op(2,Eq5))=G1;
EqGar:=subs(G1=lhs(term1),map(int,subs(term1,Eq5),x))=0;
Eq6:=expand(subs(G=lhs(tr1),G*EqGar+epsilon*diff(EqGar,x)));
Test1:=expand(Eq6-Eq2);
Test2:=evalb(KdV=algsubs(w=u,subs(epsilon=0,EqGar)));
```

Mathematica:

```
{var=Sequence[x,t], trMi:=u[var]->v[var]^2+D[v[var],x]}
kdV=D[u[var],t]-6*u[var]*D[u[var],x]+D[u[var],{x,3}]==0
trv=v[var]->1/(2*epsilon)+epsilon*w[var]
{eq1=Expand[trMi/.trv/.D[trv,x]],trGar=eq1/.{1/epsilon^2->0},
 eq2=Expand[kdV/.trGar/.D[trGar,t]/.D[trGar,x]/.D[trGar,
 {x,3}]], tD1=Table[{D[w[x,t],{x,i}],D[w[x,t],{t,i}]},
 {i,1,3}]//Flatten, eq3=Collect[eq2[[1]],tD1]}
{eq31=Map[Factor,eq3], tr1=1+2*epsilon^2*w[var]->g}
{eq4=Collect[eq31/.tr1,g], eq5=Collect[Expand[(eq4-eq4[[3]])
 /epsilon],tD1], term1=eq5[[2,1]]->g1}
eqGar=(Integrate[eq5/.term1,x]/.{g1->term1[[1]]})==0
eq6=Thread[Thread[g*eqGar,Equal]+Thread[epsilon*Thread[
 D[eqGar,x],Equal],Equal],Equal]/.{g->tr1[[1]]}//Expand
test1=Thread[eq6-eq2,Equal]//Expand
test2=kdV===(eqGar/.{epsilon->0}/.w->u)
```

□

2.4 Linearizing and Bilinearizing Transformations

In this section, we will consider the most important linearizing and bilinearizng transformations: the Hopf–Cole and Hopf–Cole-type transformations. These transformations allow us to transform, respectively, a given nonlinear PDE to a linear PDE and bilinear form of the nonlinear PDE (i.e., quadratic form in the dependent variables).

2.4.1 Hopf–Cole Transformation

In 1950–1951, E. F. F. Hopf and J. D. Cole independently showed that there exists a non-invertible mapping $u=-2\nu\phi_x/\phi$, that transforms any solution of the heat equation to the solution of the Burgers equation.

Problem 2.14 *Linear heat equation. Burgers equation. Hopf–Cole transformation.* Let us consider the linear heat equation and the Burgers equation

$$\nu\phi_{xx}=\phi_t, \quad u_t+uu_x=\nu u_{xx},$$

where $\{x \in \mathbb{R}, t \geq 0\}$ and $\phi(x,t)$ is a real function. Applying the Hopf–Cole transformation, $u=\psi_x=-2\nu\phi_x/\phi$ (where $\psi=-2\nu \ln \phi$ is a real function), verify that there exists a connection between the linear heat equation and the Burgers equation, find a particular solution of the Burgers equation.

1. Applying the Hopf–Cole transformation, we verify that there is a connection between the linear heat equation and the Burgers equation.

Maple:

```
with(PDEtools): declare((u,psi,phi)(x,t)); alias(u=u(x,t),
 psi=psi(x,t),phi=phi(x,t));
BEq:=diff(u,t)+u*diff(u,x)=nu*diff(u,x$2);
ConservLaw:=diff(u,t)+Diff((u^2/2-nu*diff(u,x)),x)=0;
Eq1:=u=Diff(psi,x); Eq2:=-op(1,op(2,lhs(ConservLaw)))
 =Diff(psi,t); Eq3:=algsubs(Eq1,Eq2);
tr1:={psi=-2*nu*log(phi)}; tr2:=diff(psi,x)=value(subs(tr1,Eq1));
Eq4:=[lhs(tr2)=rhs(rhs(tr2)),diff(psi,x$2)=diff(rhs(rhs(tr2)),x),
 diff(op(tr1),t)]; Eq5:=expand(subs(Eq4,value(Eq3)));
HEq:=Eq5*(-1/2)*phi/nu;
```

Mathematica:

```
burgersEq=D[u[x,t],t]+u[x,t]*D[u[x,t],x]==nu*D[u[x,t],{x,2}]
{conservLaw=D[u[x,t],t]+Hold[D[(u[x,t])^2/2-nu*D[
 u[x,t],x],x]]==0, eq1=u[x,t]->Hold[D[psi[x,t],x]],
 eq2=conservLaw[[1,1]]==Hold[D[psi[x,t],t]]//Simplify}
{eq3=eq2/.eq1, l3=Level[eq3,{3}], eq31=-l3[[1]]==eq3[[2]]}
{tr1=psi[x,t]->-2*nu*Log[phi[x,t]],
 tr2=D[psi[x,t],x]==ReleaseHold[(eq1/.tr1)]}
{eq4={tr2[[1]]->(tr2[[2]])[[2]],D[psi[x,t],{x,2}]->
 D[(tr2[[2]])[[2]],x],D[tr1,t]}, eq5=(ReleaseHold[eq31]
 /.eq4)//Expand, heatEq=Thread[eq5*(-1/2)*phi[x,t]/nu,Equal]}
```

2. We can find a particular solution of the linear heat equation $F(x,t){=}C_1e^{p(x+pt)}{+}C_2e^{-p(x-pt)}$ (`Sol1`), where $p{=}\sqrt{c_1}$. Therefore, the solution of the Burgers equation is $u{=}-\dfrac{2\nu p(C_1e^{p(x+pt)}{-}C_2e^{-p(x-pt)})}{C_1e^{p(x+pt)}{+}C_2e^{-p(x-pt)}}$ (`SolFin`).

Maple:

```
with(PDEtools): declare((u,phi)(x,t)); alias(u=u(x,t),
 phi=phi(x,t)); BEq:=diff(u,t)+u*diff(u,x)-nu*diff(u,x$2)=0;
tr1:={u=-2*nu*diff(phi,x)/phi}; Eq1:=collect(expand(
 subs(tr1,BEq)),diff); Eq2:=expand(Eq1*phi^2/2);
Eq11:=expand(-diff(1/phi*(diff(phi,t)-nu*diff(phi,x$2)),x)=0);
Eq21:=expand(Eq11*phi^2*nu); Eq2-Eq21;
Eq3:=int(-Diff(1/phi*(diff(phi,t)-nu*diff(phi,x$2)),x),x)=f(t);
Eq4:=Eq3*phi; tr2:=phi=exp(int(-f(t),t))*F(x,t);
Eq5:=expand(algsubs(tr2,Eq4)/op(1,rhs(tr2))); Eq6:=Eq5-rhs(Eq5);
Sol1:=combine(pdsolve(Eq6,F(x,t),explicit));
Sol2:=subs(Sol1,tr2); SolFin:=simplify(subs(Sol2,tr1));
Test1:=pdetest(SolFin,BEq);
```

Mathematica:

```
burgersEq=D[u[x,t],t]+u[x,t]*D[u[x,t],x]-nu*D[u[x,t],{x,2}]==0
{tr1=u[x,t]->-2*nu*D[phi[x,t],x]/phi[x,t], tr11=D[tr1,t],
 tr12=Table[D[tr1,{x,i}],{i,1,2}]//Flatten}
{eq1=(burgersEq/.tr1/.tr11/.tr12)//Expand,
 eq2=Thread[eq1*phi[x,t]^2/2,Equal]//Expand,
 eq11=-D[1/phi[x,t]*(D[phi[x,t],t]-nu*D[phi[x,t],
 {x,2}]),x]==0, eq21=Thread[eq11*phi[x,t]^2*nu,Equal],
 eq22=eq2[[1]]-eq21[[1]]//Expand}
{eq3=Integrate[-D[1/phi[x,t]*(D[phi[x,t],t]-nu*D[phi[x,t],
 {x,2}]),x],x]==f[t], eq4=Thread[eq3*phi[x,t],Equal]//Expand}
tr2=phi[x,t]->Exp[Integrate[-f[t],t]]*fF[x,t]
{eq5=Expand[Thread[(eq4/.tr2/.D[tr2,t]/.D[tr2,{x,2}])
 /tr2[[2,1]],Equal]],eq6=Expand[Thread[eq5-eq5[[2]],Equal]]}
DSolve[eq6,fF,{x,t}]
{sol1=fF[x,t]->c[3]*c[1]*Exp[Sqrt[s[1]]*x+nu*s[1]*t]+
 c[3]*c[2]*Exp[-Sqrt[s[1]]*x+nu*s[1]*t], sol2=ToRules[
 tr2/.Rule->Equal/.sol1], solFin=tr1/.sol2/.D[sol2,x]}
test1=burgersEq/.solFin/.D[solFin,t]/.Table[D[solFin,{x,i}],
 {i,1,2}]//Simplify
```

□

2.4.2 Hopf–Cole-type Transformation

The bilinear formalism, applied for studying nonlinear PDEs, allows us to construct soliton solutions of the complete integrable nonlinear PDEs by using the *Hirota bilinear form* (see Sect. 4.2.3), where soliton solutions appear as polynomials of exponentials in the corresponding new variables. The first step in this process is to find the *bilinearizing transformation*, i.e., to transform the nonlinear PDE to a quadratic form in the dependent variables by using the bilinearizing transformation. This new form is called the *bilinear form* of nonlinear PDE. However, this process is not algorithmic: for some equations we may not find a bilinearizing transformation, some integrable equations (e.g., the KdV and KP equations) can be transformed to a single bilinear form and another equations (e.g., the mKdV and sine–Gordon equations) can be written as a combination of bilinear equations.

Problem 2.15 *KdV equation. Hopf–Cole-type transformation.* Let us consider the KdV equation

$$u_t+6uu_x+u_{xxx}=0,$$

where $\{x\in\mathbb{R},t\geq 0\}$. Applying the bilinearizing transformation $u=w_{xx}$, $w=\alpha\log F$ $(\alpha=2)$, determine the bilinear form.

1. Introducing a new dependent variable $w(x,t)$ defined by $u=w_{xx}$, we rewrite the KdV equation as $w_{xxt}+6w_{xx}w_{xxx}+w_{xxxxx}=0$ (`Eq1`). Integrating this equation with respect to x, we obtain the *potential form* of KdV equation $w_{xt}+3w_{xx}^2+w_{xxxx}=0^*$ (`Eq2`).

2. Introducing a new dependent variable $F(x,t)$ defined by $w=\alpha\log F$ (`tr2`) with an arbitrary parameter α, we transform the potential form of KdV equation to the equation $-F_tF_xF^2+(3\alpha-6)F_x^4+(-6\alpha+12)F_{xx}F_x^2F-4F_{xxx}F_xF^2+F_{xt}F^3+(3\alpha-3)F^2F_{xx}^2+F_{xxxx}F^3=0$ (`Eq3`). Setting $\alpha=2$, we obtain the quadratic (in the dependent variable) equation $FF_{xt}+FF_{xxxx}-F_tF_x-4F_{xxx}F_x+3F_{xx}^2=0$ (`Eq4`).

Maple:

```
with(PDEtools): declare((u,w,F,f,g)(x,t)); alias(u=u(x,t),
 w=w(x,t),F=F(x,t),f=f(x,t),g=g(x,t)); tr1:=u=diff(w,x$2);
PDE1:=u->diff(u,t)+6*u*diff(u,x)+diff(u,x$3)=0;
Eq1:=expand(PDE1(rhs(tr1))); Eq2:=map(int,Eq1,x);
```

*In this case, the form of the new dependent variable $w(x,t)$ allows us to take an arbitrary integration function of t to be equal to 0.

```
tr2:=w=alpha*log(F);
Eq3:=collect(simplify(algsubs(tr2,Eq2))*F^4/alpha,[diff,F]);
Eq4:=expand(subs(alpha=2,Eq3)/F^2);
```

Mathematica:

```
var=Sequence[x,t]; tr1=u[var]->D[w[var],{x,2}]
pde1[u_]:=D[u,t]+6*u*D[u,x]+D[u,{x,3}]==0;
{eq1=pde1[tr1[[2]]]//Expand, eq2=Thread[Integrate[eq1,x],
 Equal], tr2=w[var]->alpha*Log[f[var]]}
eq3=eq2/.tr2/.D[D[tr2,x],t]/.Table[D[tr2,{x,i}],{i,1,4}]
eq31=Expand[Thread[eq3*f[var]^4/alpha,Equal]]
eq4=Thread[(eq3/.alpha->2)*f[x,t]^2/2,Equal]//Expand
```

Thus, we have applied the Hopf–Cole-type transformation $u=2(\log F)_{xx}$ to the Korteweg–de Vries equation and obtained the resulting *bilinear equation* Eq4 (which is not linear but quadratic). This bilinear equation can be written in operator form by using the *Hirota bilinear operator* (see Problem 4.10). □

2.5 Reductions of Nonlinear PDEs

In order to find exact solutions of nonlinear PDEs, there exists a class of methods, called *reductions*, that consists in finding the reductions to a differential equation in a lesser number of independent variables. There are two types of reductions, the *characteristic* and the *noncharacteristic* reductions.

Definition 2.8 A reduction of a nonlinear PDE in n independent variables to an equation in $n-1$ independent variables is called noncharacteristic if it preserves the differential order. In opposite, a reduction is called characteristic.

2.5.1 Traveling Wave Reductions

A traveling wave solution of a given nonlinear PDE is a solution of the reduction $\xi=x-ct$ (see Definition 2.8) if it exists. A traveling wave solution has permanent form moving with a constant velocity. Applying the ansatz $u(x,t)=u(\xi)$, $\xi=x-ct$ (where c is the wave velocity), it is possible to transform the PDE (in x,t) to an ODE (in ξ), which can be solved by appropriate methods.

Traveling wave solutions often occur in various problems of mathematical physics. These solutions are invariant under translation. The traveling wave solution of the linear wave equation was first obtained by d'Alembert in 1747 [39]. Now, this solution can be obtained with *Maple* as follows:

```
LinWaveEq:=diff(u(x,t),x$2)-(1/c^2)*diff(u(x,t),t$2)=0;
IC1:=u(x,0)=f(x); IC2:=D[2](u)(x,0)=g(x);
sysD:=[LinWaveEq,IC1,IC2]; pdsolve(sysD);
```

In practice, the method of finding traveling wave solutions is simple and useful in finding solutions of both linear and nonlinear PDEs. As before, we illustrate the method of finding traveling wave solutions by solving problems.

Problem 2.16 *Burgers equation. Traveling wave solutions.* Let us consider the Burgers equation

$$u_t+uu_x=\nu u_{xx},$$

where $\{x \in \mathbb{R}, t \geq 0\}$. Traveling wave solutions of the Burgers equation are solutions of the form $u(x,t)=U(z)$, $z=x-\lambda t$, where the wave speed λ and the waveform $U(z)$ are to be determined. Investigating the shock-wave structure of the solutions, we assume that there exist the constant values $u_1>0$ and $u_2>0$ $(u_1>u_2)$ such that $\lim\limits_{z\to-\infty} U(z)=u_1$ and $\lim\limits_{z\to+\infty} U(z)=u_2$. We determine traveling wave solution and visualize it varying the diffusion parameter ν.

1. Substituting the ansatz $u(x,t)=U(z)$ (where $z=x-\lambda t$) into the Burgers equation, we have the ODE $-\lambda U_z+UU_z-\nu U_{zz}=0$ (Eq3). Integrating this equation, we have $-\lambda U+\frac{1}{2}U^2-\nu U_z=C$ (Eq4), where C is a constant of integration. The last equation can be written in the equivalent form $U_z=-\frac{1}{2}(2C+2\lambda U-U^2)/\nu$, assuming that u_1, u_2 are the roots of the equation $2C+2\lambda U-U^2=0$ and $\lambda=\frac{1}{2}(u_1+u_2)$ and $C=-\frac{1}{2}u_1u_2$ (tr1).

2. We determine the shape of waveform $U(z)=\dfrac{e^{z(u_1-u_2)/(2\nu)}u_2+u_1}{1+e^{z(u_1-u_2)/(2\nu)}}$ (Sol6) and visualize traveling wave solution $U(z)$ (SolU) varying the diffusion parameter ν.

Maple:

```
with(PDEtools): declare(u(x,t),U(z)); interface(showassumed=0);
assume(nu>0,U(z)>0,u1>0,u2>0,u1>u2); A:=2*nu;
```

```
tr1:={lambda=1/2*(u1+u2),C=-1/2*u1*u2}; tr2:=U(z)-u1=u1-U(z);
tr3:=x-lambda*t=z; Eq1:=u->diff(u,t)+u*diff(u,x)-nu*diff(u,x$2);
Eq2:=Eq1(U(x-lambda*t)); Eq3:=convert(algsubs(tr3,Eq2),diff);
Eq4:=int(Eq3,z)=C; dU:=diff(U(z),z);
Eq41:=simplify(isolate(Eq4,diff(U(z),z)));
Sol1:=solve(Eq4,dU) assuming z<>0;
Sol2:=(dU=factor(subs(tr1,Sol1)))*A; Sol3:=Sol2/rhs(Sol2)/A*dz;
Sol3L:=lhs(Sol3); Sol3R:=rhs(Sol3); Sol4:=combine(int(Sol3L,z),
 symbolic)=int(coeff(Sol3R,dz),z); Sol5:=subs(tr2,Sol4);
Sol6:=solve(1=solve(Sol5,dz),U(z)) assuming z<>0;
SolU:=unapply(Sol6,z,u1,u2,nu); SolU(0,5,1,0.3);
plot(SolU(z,5,1,0.3),z=-5..5,0..6,thickness=3);
plot(SolU(z,5,1,0.001),z=-1..1,0..6,thickness=3);
```

Mathematica:

```
SetOptions[Plot,ImageSize->300,PlotPoints->100,PlotStyle->
 {Hue[0.9],Thickness[0.01]},PlotRange->{All,{0,6}}];
{a=2*nu, tr1={lambda->1/2*(u1+u2), c->-1/2*u1*u2},
 tr2=uN[z]-u1->u1-uN[z], tr3=x-lambda*t->z}
eq1[u_]:=D[u,t]+u*D[u,x]-nu*D[u,{x,2}]; eq2=eq1[uN[x-lambda*t]]
{eq3=eq2/.tr3, eq4=Integrate[eq3,z]==c, duN=D[uN[z],z],
 eq41=Reduce[eq4,D[uN[z],z]]//Simplify, sol1=Assuming[z!=0,
 Solve[eq4,duN]], sol2=Thread[(Factor[sol1/.tr1][[1,1]])*a,Rule]}
sol3=Thread[sol2/sol2[[2]]/a*dz,Rule]
{sol4=Integrate[sol3[[1]],z]==Integrate[Coefficient[
 sol3[[2]],dz],z], sol5=sol4/.tr2, sol6=Assuming[z!=0,
 Solve[1==Solve[sol5,dz][[1,1,2]],uN[z]]]//FullSimplify}
solU[zN_,uN1_,uN2_,nuN_]:=sol6[[1,1,2]]/.{z->zN,u1->uN1,
 u2->uN2,nu->nuN}; solU[0,5,1,0.3]
Show[GraphicsRow[{Plot[solU[zN,5,1,0.999],{zN,-Pi,Pi}],
 Plot[solU[zN,5,1,0.3],{zN,-Pi,Pi}],Plot[solU[zN,5,1,0.09],
 {zN,-Pi,Pi}]}]]
```

□

Problem 2.17 *Nonlinear heat equation. Traveling wave solutions.* Let us consider the nonlinear heat equation of the following form:

$$u_t=(F(u)u_x)_x,$$

where $\{x\in\mathbb{R},t\geq 0\}$ and $F(u)$ is an arbitrary function. Verify that the *implicit form* of the traveling wave solution, $u(x,t)=U(z)$, $z=x+\lambda t$, of the nonlinear heat equation has the form $z+C_2=\int\frac{F(U(z))\,dU(z)}{\lambda U(z)+C_1}$, where C_1 and C_2 are arbitrary constants.

Maple:

```
with(PDEtools): declare(u(x,t),U(z)); tr1:=x+lambda*t=z;
Eq1:=u->-diff(u,t)+diff((F(u)*diff(u,x)),x);
Eq2:=expand(Eq1(U(lhs(tr1)))); Eq3:=algsubs(tr1,Eq2);
Eq31:=map(convert,Eq3,diff); Eq4:=diff(F(U(z))*diff(U(z),z),z);
Eq5:=convert(Eq4,diff); evalb(convert(op(2,Eq31)
 +op(3,Eq31),diff)=Eq5); Eq6:=-op(1,Eq31); Eq7:=int(Eq6,z);
Eq8:=int(diff(F(U(z))*diff(U(z),z),z),z); Eq9:=Eq7+C1=Eq8;
Eq10:=Eq9/(lhs(Eq9));
Eq11:=int(lhs(Eq10),z)+C2=Int(F(U(z))/denom(rhs(Eq10)),dU(z));
```

Mathematica:

```
eq1[u_]:=D[u,t]+D[f[u]*D[u,x],x]; tr1=x+lambda*t->z
{eq2=eq1[uN[tr1[[1]]]]//Expand, eq3=eq2/.tr1}
{eq4=D[f[uN[z]]*D[uN[z],z],z], eq3[[2]]+eq3[[3]]==eq4}
{eq6=-eq3[[1]], eq7=Integrate[eq6,z], eq8=Integrate[
 D[f[uN[z]]*D[uN[z],z],z],z], eq9=eq7+c1==eq8}
{eq10=Thread[eq9/eq9[[1]],Equal], eq11=Integrate[eq10[[1]],
 z]+c2==HoldForm[Integrate[f[uN[z]]/(c1-lambda*uN[z]),uN[z]]]}
```
□

Problem 2.18 *Korteweg–de Vries equation. Traveling wave solutions.* Let us consider the KdV equation of the following form:

$$u_t+auu_x+bu_{xxx}=0,$$

where $\{x\in\mathbb{R}, t\geq 0\}$ and a, b are real constants. It is known that the KdV equation admits a special form of traveling wave solutions, the *soliton solution* (see Sect. 1.1.3). Determine a special type of traveling wave solutions, *one-soliton solution*.

We look for a special type of traveling wave solutions, *one-soliton solution*, $u(x,t)=U(z)$, $z=x-ct$, of the KdV equation, where c is a constant and $U(z)\to 0$ as $|z|\to\infty$.

1. We confirm that substituting this solution form into the KdV equation, we arrive at the nonlinear third-order ODE $-cU_z+aUU_z+bU_{zzz}=0$ (`Eq4`) and integrating this equation with respect to z twice, we obtain the nonlinear first-order ODE $-cU^2+\frac{1}{3}aU^3+bU_z^2-2A_1U=A_2$ (`Eq8`), where A_1 and A_2 are the integration constants.

2. Considering a special case where $U(z)$ and its derivatives tend to zero at infinity and $A_1=A_2=0$, we verify that the exact solution has the form $U(z)=-\frac{3c}{a}(\tanh[\sqrt{cb}\,(-z+C_1)/(2b)]^2-1)$ (Sol1[2]), where C_1 is an arbitrary constant. Setting for example, $a=1$, $b=1$, $c=1$, and $C_1=0$, we confirm that the solution obtained, $U(x-t)=3\,\mathrm{sech}\left[-\frac{1}{2}(x-t)\right]^2$ (Sol2), is indeed exact solution of the KdV equation and it travels at constant velocity without changing the shape.

Maple:

```
with(PDEtools): declare(u(x,t),U(z)); with(plots):
tr1:=x-c*t=z; tr2:={a=1,b=1,c=1}; A12:={A1=0,A2=0};
Eq1:=u->diff(u,t)+a*u*diff(u,x)+b*diff(u,x$3)=0;
Eq2:=expand(Eq1(U(lhs(tr1)))); Eq3:=algsubs(tr1,Eq2);
Eq4:=map(convert,Eq3,diff); Eq5:=map(int,lhs(Eq4),z)-A1=0;
Eq6:=expand(Eq5*2*diff(U(z),z)); Eq7:=map(int,Eq6,z);
Eq8:=lhs(Eq7)=A2; Eq9:=subs(A12,Eq8); Sol1:=[dsolve(Eq9,U(z))];
Sol11:=convert(subs(_C1=0,simplify(Sol1[2])),sech);
Sol2:=eval(subs(z=x-c*t,Sol11),tr2);
animate(plot,[rhs(Sol2),x=-20..20],t=0..20,numpoints=100,
 frames=50,thickness=2,color=blue);
Test1:=pdetest(u(x,t)=rhs(Sol2),subs(tr2,Eq1(u(x,t))));
```

Mathematica:

```
SetOptions[Plot,ImageSize->500,PlotStyle->{Hue[0.9],
 Thickness[0.01]}]; {tr1=x-c*t->z, tr2={a->1,b->1,c->1}}
eq1[u_]:=D[u,t]+a*u*D[u,x]+b*D[u,{x,3}]==0;
{eq2=eq1[uN[tr1[[1]]]]//Expand, eq3=eq2/.tr1,
 eq4=eq3//TraditionalForm, eq5=Integrate[eq3[[1]],z]-c1==0}
{eq6=Thread[eq5*2*D[uN[z],z],Equal]//Expand,
 eq7=Thread[Integrate[eq6,z],Equal], eq8=eq7[[1]]==c2,
 eq9=eq8/.{c1->0,c2->0}, sol1=DSolve[eq9,uN[z],z]}
{sol11=Simplify[sol1[[2]]]/.{C[1]->0}, sol2=(sol11/.{z->x-c*t})/.
 tr2, test1=(eq1[u]/.tr2)/.{u->sol2[[1,2]]}}
sol3[xN_,tN_]:=sol2[[1,2]]/.{x->xN,t->tN}; Animate[Plot[
 sol3[x,t],{x,-20,20},PlotRange->{All,{0,Pi}}],{t,0,20}]
```

□

Problem 2.19 *Monge–Ampère equation. Traveling wave solutions.* Let us consider the homogeneous Monge–Ampère equation

$$u_{xy}^2-u_{xx}\,u_{yy}=0,$$

where $\{x\in\mathbb{R}, y\in\mathbb{R}\}$. Verify that the traveling wave solution of the homogeneous Monge–Ampère equation has the form $u(x,y)=U(z)$, $z=x+\lambda y$, where $U(z)$ and λ are, respectively, arbitrary function and arbitrary constant.

Maple:

```
with(PDEtools): declare(u(x,y),U(z)); tr1:=x+lambda*y=z;
Eq1L:=u->diff(u,x,y)^2;  Eq1R:=u->diff(u,x$2)*diff(u,y$2);
Eq2L:=Eq1L(U(lhs(tr1))); Eq3L:=algsubs(tr1,Eq2L);
Eq2R:=Eq1R(U(lhs(tr1))); Eq3R:=algsubs(tr1,Eq2R);
Eq4:=convert(Eq3L=Eq3R,diff); evalb(Eq4);
```

Mathematica:

```
eq1L[u_]:=D[D[u,x],y]^2; eq1R[u_]:=D[u,{x,2}]*D[u,{y,2}];
{tr1=x+lambda*y->z, eq2L=eq1L[uN[tr1[[1]]]], eq3L=eq2L/.tr1}
{eq2R=eq1R[uN[tr1[[1]]]], eq3R=eq2R/.tr1, eq4=eq3L==eq3R}
```

□

Problem 2.20 *Sine–Gordon equation. Traveling wave solutions.* Let us consider the sine–Gordon equation written in the Cauchy–Kovalevskaya form

$$u_{xx}-\frac{1}{c^2}u_{tt}=\sin u,$$

where $\{x\in\mathbb{R}, t\geq 0\}$. Determine the traveling wave solutions of the sine–Gordon equation, i.e., the kink and antikink solutions, and visualize the results.

We construct traveling wave solutions of the sine–Gordon equation in the form of a *one-soliton solution* and verify that the traveling wave solutions of the sine–Gordon equation take the following form: $u(x,t)=4\arctan\left(\exp\left(\pm\frac{x-Ut}{\sqrt{1-U^2}}\right)\right)$ (`Kink, AntiKink`), where $U=\lambda/c$. These solutions are called, respectively (depending the sign), the soliton or kink and antisoliton or antikink solutions (see Sect. 1.1.3). The solitons propagate, respectively, in the positive or negative x-direction with velocity U. Finally, we visualize the soliton and antisoliton solutions.

Maple:

```
with(PDEtools): with(plots): declare(u(x,t),phi(xi));
alias(u=u(x,t),phi=phi(xi));  interface(showassumed=0);
assume(Phi>Phi0); tr1:=x-lambda*t=xi; tr2:=lambda^2/c^2=U^2;
xR:=-2*Pi..2*Pi; SGEq:=u->diff(u,x$2)-1/c^2*diff(u,t$2)=sin(u);
Eq1:=expand(SGEq(phi(lhs(tr1)))); Eq2:=expand(subs(tr1,Eq1));
```

```
Eq3:=convert(algsubs(tr2,Eq2),diff);
Eq4:=expand(normal(Eq3*diff(phi(xi),xi)/(1-U^2)));
Eq5:=Diff(1/2*(Diff(phi,xi)^2)+cos(phi)/(1-U^2),xi)=0;
Eq41:=lhs(Eq4)-rhs(Eq4)=0; factor(Eq41-value(Eq5));
Eq6:=int(value(op(1,Eq5)),xi)=B; Eq7:=normal(convert(isolate(
 Eq6,diff(phi,xi)),radical)); Eq7R:=rhs(Eq7);
Eq80:=subs(phi=psi,1/sqrt(collect(-numer(op(1,op(2,Eq7R))),B)));
Eq8:=Int(Eq80,psi=Phi0..Phi); Eq90:=op(1,rhs(Eq7))/sqrt(-denom(
 op(1,op(2,rhs(Eq7))))); Eq9:=int(Eq90,eta=xi0..xi);
Eq10:=Eq8=Eq9; Eq11:=subs({B*(1-U^2)=1},Eq10);
Eq12:=simplify(subs(1-cos(psi)=2*sin(psi/2)^2,Eq11),symbolic);
Eq13:=normal(convert(value(Eq12),tan)); Eq13L:=lhs(Eq13);
Eq14:=expand(op(3,op(3,Eq13L))*op(1,Eq13L)*op(2,Eq13L)
     =rhs(Eq13)); Sol:=expand(isolate(Eq14,Phi));
tr3:={lambda=0.1,c=10.}; tr4:={tan(Phi0/4)=1,xi0=0}; tr5:=x-U*t;
Kink:=unapply(subs(xi=tr5,expand(subs(tr4,rhs(Sol)))),x,t,U);
AntiKink:=unapply(subs(xi=-tr5,eval(rhs(Sol),tr4)),x,t,U);
U1:=subs(tr3,lhs(tr2)); K:=plot(Kink(x,0,U1),x=xR):
AK:=plot(AntiKink(x,0,U1),x=xR,color=blue): display({K,AK});
```

Mathematica:

```
trS1[eq_,var_]:=Select[eq,MemberQ[#,var,Infinity]&];
trS3[eq_,var_]:=Select[eq,FreeQ[#,var]&]; Off[Solve::"ifun"];
SetOptions[Plot,ImageSize->500,PlotStyle->{Hue[0.9],
 Thickness[0.01]},PlotRange->{All,{0,2*Pi}}];
sGEq[u_]:=D[u,{x,2}]-1/c^2*D[u,{t,2}]==Sin[u];
{tr1=x-lambda*t->xi, tr2=lambda^2/c^2->uN^2, eq1=sGEq[phi[
 tr1[[1]]]]//Expand, eq2=eq1/.tr1//Expand, eq3=eq2/.tr2,
 eq4=Thread[eq3*D[phi[xi],xi]/(1-uN^2),Equal]//Simplify,
 eq5=Hold[D[1/2*(D[phi[xi],xi]^2)+Cos[phi[xi]]/(1-uN^2),xi]==0],
 eq41=eq4[[1]]-eq4[[2]]==0, eq41==ReleaseHold[Eq5],
 eq51=D[Level[eq5,{3}][[1]],xi]}
{eq6=Integrate[eq51,xi]==b//FullSimplify, eq7=Flatten[Solve[
 eq6,D[phi[xi],xi]]], eq7R=eq7[[2,2]], term7=trS3[eq7R,Sqrt[2]],
 term70=Factor[term7^2], term71=Numerator[term70],
 term72=Denominator[term70], integrd=1/Sqrt[Collect[-term71,b]]/.
 {phi[xi]->psi}, integrd1=integrd/.{b*(1-uN^2)->1}}
{integrd2=Assuming[Sin[psi/2]>0,PowerExpand[integrd1/.
 {1-Cos[psi]->2*Sin[psi/2]^2}]], eq8=Assuming[phi0<phi &&
 (phi<=0||phi0>0) && (phi<=-2*Pi||phi0>=-2*Pi) &&
 (phi<=2*Pi||phi0>=2*Pi),Integrate[integrd2,{psi,phi0,phi}]]}
eq90=trS1[eq7R,Sqrt[2]]/Sqrt[-term72]
```

```
{eq9=Integrate[eq90,{eta,xi0,xi}], eq10=eq8==eq9,
 sol=First[Solve[eq10,phi]], tr3={lambda->0.1,c->10.},
 tr4={Cot[phi0/4]->1,xi0->0,C[1]->0}, tr5=x-uN*t}
kink[xN_,tN_,uNU_]:=Expand[(sol[[1,2]]/.tr4)/.{xi->tr5}]/.{x->xN,
 t->tN,uN->uNU}; antiKink[xN_,tN_,uNU_]:=Expand[(sol[[1,2]]/.
 tr4)/.{xi->-tr5}]/.{x->xN,t->tN,uN->uNU}; uN1=tr2[[1]]/.tr3
kS=Plot[kink[x,0,uN1],{x,-2*Pi,2*Pi}]; aKS=Plot[
 antiKink[x,0,uN1],{x,-2*Pi,2*Pi},PlotStyle->{Hue[0.7],
 Thickness[0.01]}]; Show[{kS,aKS}]
```

□

2.5.2 Ansatz Methods

Now we consider the most important *ansatz methods* (the tanh-function method, sine-cosine method, and Exp-function method) for constructing traveling wave solutions of nonlinear PDEs. These methods allow us to find exact solutions of nonlinear nonintegrable PDEs. Recently developed, these methods became the most powerful and effective algebraic methods for finding exact solutions of nonlinear PDEs partly due to the modern computer algebra systems *Maple* and *Mathematica*, which allows us to perform a lot of cumbersome analytical calculations.

In recent years there have been numerous published scientific papers, describing the construction of *new exact solutions* of various nonlinear PDEs, various applications and generalizations of these methods, their implementations in computer algebra systems. However, some exact solutions are often *not new* or are *redundant* (see Sect. 1.1.3). As shown N. A. Kudryashov [86], many ansatz methods can be considered as a consequence of the truncated expansion method. Exact solutions obtained by applying these methods are often not new, i.e., there exist coincidence between different forms of solutions (for details, see [144]).

The tanh-function expansion method has been developed by W. Malfliet [94]. The main idea of the tanh-method is based on the assumption that the traveling wave solutions can be expressed in terms of the tanh function, i.e., a new variable, e.g., $Y=\tanh(z)$, $z=\mu(x+ct)$, can be introduced and all derivatives of Y are expressed in terms of tanh function. Then, the tanh-method has been modified, extended, and generalized for finding more and more exact solutions. In recent years, various forms of the tanh-method have been developed, e.g., the tanh-coth method has been proposed by A. M. Wazwaz [158]. Many researchers developed symbolic programs and packages to deal with the tedious algebraic computations that arise in the solution process (e.g., see [118]).

Problem 2.21 *Klein–Gordon equation. The tanh-function method.* Let us consider the Klein–Gordon equation of the form

$$u_{tt}-u_{xx}+u-u^3=0,$$

where $\{x \in \mathbb{R}, t \geq 0\}$. Applying the tanh-function method, construct traveling wave solutions of the Klein–Gordon equation.

1. We look for traveling wave solutions of the Klein–Gordon equation, i.e., solutions of the form $u(x,t)=U(z)$, $z=\mu(x+ct)$. First, we convert the PDE into the following ODE: $\mu^2c^2U_{zz}-\mu^2U_{zz}+U-U^3=0$ (`Eq2`).*

2. Introducing a new variable $Y=\tanh(z)$ or $Y=\coth(z)$, we propose the following ansatz: $U(z)=S(Y)=\sum_{i=0}^{M} a_iY^i$ (`Ansatz1`), where M $(M \in \mathbb{N})$ have to be determined. Substituting this series expansion into the ODE, we obtain an equation in powers of Y:

$$(-\mu^2S_{YY}+\mu^2c^2S_{YY})Y^4+(2\mu^2c^2S_Y-2\mu^2S_Y)Y^3+(-2\mu^2c^2S_{YY}+2\mu^2S_{YY})Y^2$$
$$+(2\mu^2S_Y-2\mu^2c^2S_Y)Y+S(Y)+\mu^2c^2S_{YY}-\mu^2S_{YY}-S(Y)^3=0.$$

3. To determine the parameter M, we balance the linear terms of highest order in the resulting equation with the highest-order nonlinear terms according to the following formulas for the highest exponents of the function $U(z)$ and its derivatives: $U(z) \to M$, $U^n(z) \to nM$, $U'(z) \to M+1$, $U''(z) \to M+2$, $U^{(k)}(z) \to M+k$. In our case, we have $3M=M+2$, $M=1$ and $S(Y)=a_0+a_1Y$.

4. Collecting all coefficients of powers of Y in the equation `Eq41`

$$(2\mu^2c^2a_1-a_1^3-2\mu^2a_1)Y^3-3a_0a_1^2Y^2+(-3a_0^2a_1+2\mu^2a_1+a_1-2\mu^2c^2a_1)Y-a_0^3+a_0=0,$$

where these coefficients have to vanish, we obtain the system of algebraic equations (`sys1`) with respect to the unknowns a_i $(i=0,\ldots,M)$, μ, and c:

$$-3a_0a_1^2=0,\ -a_0^3+a_0=0,\ 2\mu^2c^2a_1-a_1^3-2\mu^2a_1=0,\ -3a_0^2a_1+2\mu^2a_1+a_1-2\mu^2c^2a_1=0.$$

Finally, determining these unknowns, μ, a_0, a_1 (`Sols`), and using the series expansion (`Ansatz1`), we obtain the exact solutions (`SolsTF`)

$$u(x,t)=\pm\tanh\left[\sqrt{1/(-2+2c^2)}(x+ct)\right]$$

and verify that these solutions are exact solutions.

*If all terms of this ODE contain derivatives in z, the ODE should be simplified (by integrating).

Maple:

```
with(PDEtools): declare(u(x,t),U(z));
alias(u=u(x,t),U=U(z)); interface(showassumed=0);
assume(n>1); f:=tanh; tr1:=mu*(x+c*t)=z; tr2:=f(z)=Y;
PDE1:=u->diff(u,t$2)-diff(u,x$2)+u-u^3=0;
Eq1:=expand(PDE1(U(lhs(tr1)))); Eq2:=convert(expand(
 subs(tr1,Eq1)),diff); tr3:=U(z)=S(lhs(tr2));
Ansatz1:=S(Y)=Sum(a[i]*Y^i,i=0..M);
Eq3:=convert(algsubs(tr2,algsubs(tr3,Eq2)),diff);
Eq31:=collect(Eq3,Y); tr5:=isolate(3*M=M+2,M);
tr6:=value(subs(tr5,Ansatz1)); Eq4:=algsubs(tr6,Eq3);
Eq41:=collect(Eq4,Y); sys1:={}; SolsT:={}:
for i from 0 to 3 do sys1:=sys1 union {coeff(lhs(Eq4),Y,i)
 =0}; od: sys1; vars:=indets(sys1) minus {c};
Sols:=[allvalues([solve(sys1,vars)])]; NSols:=nops(Sols);
for i from 1 to NSols do Op||i:=op(i,Sols): Nops:=nops(Op||i):
 for j from 1 to Nops do SolsT:=SolsT union {u=subs(
  op(j,Op||i),subs(z=lhs(tr1),subs(Y=lhs(tr2),rhs(tr6))))};
od: od: SolsT; SolsTF:=select(has,SolsT,f);
for i from 1 to nops(SolsTF) do
 simplify(PDE1(rhs(SolsTF[1]))); od;
```

Mathematica:

```
Off[Solve::"svars"]; pde1[u_]:=D[u,{t,2}]-D[u,{x,2}]+u-u^3==0;
{tr1=mu*(x+c*t)->z,tr2=Tanh[z]->yN, sys1={}, solsT={}}
{eq1=pde1[uN[tr1[[1]]]]//Expand, eq2=eq1/.tr1//Expand,"eq2="eq2}
{tr3=uN[z]->s[tr2[[1]]], tr31=D[uN[z],{z,2}]->D[s[tr2[[1]]],
 {z,2}], ansatz1=s[yN]->Sum[a[i]*yN^i,{i,0,mN}],
 "ansatz1="ansatz1, eq3=(eq2/.tr3)/.tr31, tr32=Table[
 (Sech[z]^2)^i->(1-Tanh[z]^2)^i,{i,1,2}]}
{eq31=(eq3/.tr32)/.tr2, eq32=Collect[eq31,yN], tr5=Solve[
 3*mN==mN+2,mN]//First, tr6=ansatz1/.tr5, tr61=Table[D[ansatz1,
 {yN,i}]/.tr5,{i,1,2}], eq4=(eq32/.tr6)/.tr61,"eq4="eq4}
Do[sys1=Union[sys1,{Coefficient[eq4[[1]],yN,i]==0}],{i,0,3}];
{sys1, vars=Complement[Variables[Table[sys1[[i]]//First,
 {i,1,Length[sys1]}]],{c}], sols=Solve[sys1,vars],
 nSols=Length[sols]}
Do[solsT=Union[solsT,{u[x,t]->((tr6[[2]]/.yN->tr2[[1]])/.z->
 tr1[[1]])/.sols[[i]]}],{i,nSols-3,nSols}]; solsT
Table[pde1[solsT[[i]][[2]]]//FullSimplify,{i,1,Length[solsT]}]
```
□

The sine-cosine method. The main idea of the sine-cosine method is based on the assumption that the traveling wave solutions can be expressed in terms of the sine or cosine functions, i.e., a new variable $Y = \cos(z)$ or $Y = \sin(z)$ can be introduced. As in the tanh method considered above, the main advantage of the sine-cosine method consists in simplifying the solution process and reducing the size of computational work (compared to modern analytical methods), i.e., we have to solve a system of algebraic equations instead of solving nonlinear differential equation. Moreover, we can apply computer algebra systems, *Maple* and *Mathematica*, for this purpose.

Problem 2.22 *Klein–Gordon equation. The sine-cosine method.* Let us consider the Klein–Gordon equation (see Problem 2.21)

$$u_{tt} - u_{xx} + u - u^3 = 0,$$

where $\{x \in \mathbb{R}, t \geq 0\}$. Applying the sine-cosine method, find traveling wave solutions of the Klein–Gordon equation, i.e., solutions of the form $u(x,t) = U(z)$, $z = x + ct$.

1. First, we convert the nonlinear PDE into the following ODE: $c^2 U_{zz} - U_{zz} + U - U^3 = 0$ (`Eq2`).*

2. According to the the sine-cosine method, the traveling wave solutions can be expressed in the form $u(x,t) = \lambda \cos^{\beta}(\mu z)$ for $|z| \leq \frac{1}{2}\pi/\mu$ or in the form $u(x,t) = \lambda \sin^{\beta}(\mu z)$ for $|z| \leq \pi/\mu$, where the parameters λ, μ (wave number), and β have to be determined. Introducing a new variable $Y = \cos(z)$ or $Y = \sin(z)$, we propose the following ansatz: $U(z) = \lambda \cos^{\beta}(\mu z)$ (`AnsatzC`) or $U(z) = \lambda \sin^{\beta}(\mu z)$ (`AnsatzS`).

3. Substituting this ansatz (`AnsatzC`) into the ODE, we obtain the trigonometric equation (`Eq3`). Then simplifying this equation and introducing a new variable $Y = \cos(\mu z)$, we obtain (`Eq33`)

$$-\lambda^3 Y^{3\beta} + \lambda\beta^2\mu^2 c^2 Y^{\beta-2} - \lambda\beta^2\mu^2 c^2 Y^{\beta} - \lambda\beta\mu^2 c^2 Y^{\beta-2} - \lambda\beta^2\mu^2 Y^{\beta-2}$$
$$+ \lambda\beta^2\mu^2 Y^{\beta} + \lambda\beta\mu^2 Y^{\beta-2} + \lambda Y^{\beta} = 0.$$

4. To determine the parameter β, we balance the exponents of each pair of Y. Then, to determine λ and μ, we collect all coefficients of the same power of Y, where these coefficients have to vanish. This gives the

*If all terms of this ODE contain derivatives in z, the ODE should be simplified (by integrating).

system of algebraic equations with respect to the unknown parameters β, λ, and μ: $Y^{3\beta}{=}Y^{\beta-2}$ (Eqbeta), $-\lambda\beta^2\mu^2c^2 + \lambda\beta^2\mu^2 + \lambda = 0$ (Eqmu), $-\lambda^3 + \lambda\beta^2\mu^2c^2 - \lambda\beta\mu^2c^2 - \lambda\beta^2\mu^2 + \lambda\beta\mu^2 = 0$ (Eqlambda).

5. Finally, determining the parameters $\beta{=}-1$ (trbeta), $\mu{=}1/\sqrt{c^2{-}1}$ (trmu), and $\lambda{=}\sqrt{2}$ (trlambda), and using the proposed ansatz (AnsatzC, AnsatzS), we obtain the exact solutions and verify that these solutions are exact solutions. If $c^2{-}1{>}0$ and $d{=}\sqrt{c^2{-}1}$, we have (SolC, SolS):

$$u_1(x,t) = \sqrt{2}\,\sec\big(d^{-1}(x+ct)\big) \quad \text{for} \quad |d^{-1}z| \le \tfrac{1}{2}\pi,$$
$$u_2(x,t) = \sqrt{2}\,\csc\big(d^{-1}(x+ct)\big) \quad \text{for} \quad 0 < |d^{-1}z| \le \pi.$$

Maple:

```
with(PDEtools): with(plots): declare(u(x,t),U(z));
alias(u=u(x,t),U=U(z)); interface(showassumed=0); f:=cos:
PDE1:=u->diff(u,t$2)-diff(u,x$2)+u-u^3=0; tr1:=x+c*t=z;
tr2:=f(z)=Y; Eq1:=expand(PDE1(U(lhs(tr1))));
Eq2:=convert(expand(subs(tr1,Eq1)),diff);
AnsatzC:=U(z)=lambda*cos(mu*z)^beta; AnsatzS:=U(z)=
 lambda*sin(mu*z)^beta; Eq3:=algsubs(AnsatzC,Eq2);
Eq31:=expand(simplify(Eq3,trig),cos); Eq32:=simplify(Eq31,
 power); Eq33:=combine(subs(cos(mu*z)=Y,lhs(Eq32)));
term11:=select(has,Eq33,Y^(3*beta)); term12:=select(has,
 term11,Y); term21:=select(has,Eq33,Y^(beta-2));
term22:=select(has,op(1,term21),Y); trbeta:=isolate(
 term12=term22,beta);
trmu:=sort(subs(trbeta,[solve(coeff(Eq33,Y^beta),mu)]))[1];
Eqlambda:=coeff(Eq33,Y^(3*beta))+coeff(Eq33,Y^(beta-2))=0;
Eqlambda1:=subs(mu=trmu,trbeta,Eqlambda);
Solslambda:=expand([solve(Eqlambda1,lambda)]);
trlambda:=sort(convert(Solslambda,set) minus {0})[1];
SolC:=simplify(subs(z=lhs(tr1),trbeta,lambda=trlambda,mu=trmu,
 u=rhs(AnsatzC))); SolS:=simplify(subs(z=lhs(tr1),trbeta,
 lambda=trlambda,mu=trmu,u=rhs(AnsatzS))); SolC:=convert(SolC,
 sec);  SolS:=convert(SolS,csc); simplify([PDE1(rhs(SolC)),
 PDE1(rhs(SolS))],trig); Sol1G:=simplify(subs(c=2,n=3,SolC));
Sol2G:=simplify(subs(c=1/2,n=3,SolC));
animate(rhs(Sol1G),x=0..Pi,t=0..5,view=[default,-200..200]);
animate(rhs(Sol2G),x=-3..3,t=0..1,frames=50);
```

Mathematica:

```
pde1[u_]:=D[u,{t,2}]-D[u,{x,2}]+u-u^3==0; {tr1=x+c*t->z,
 tr2=Cos[z]->yN, eq1=pde1[uN[tr1[[1]]]]//Expand, eq2=eq1/.tr1,
 "eq2="eq2, ansatzC=uN[z]->lambda*Cos[mu*z]^beta, ansatzS=
 uN[z]->lambda*Sin[mu*z]^beta, ansatzC1=Table[D[uN[z],{z,i}]->
 D[lambda*Cos[mu*z]^beta,{z,i}],{i,1,2}]}
{eq3=(eq2/.ansatzC)/.ansatzC1, eq31=eq3/.{Sin[x_]^2:>1-
 Cos[x]^2}//Factor, eq32=eq31[[1]]/.{Cos[mu*z]->yN}//Expand}
{y3b=yN^(3*beta), yb2=yN^(beta-2), term11=Select[eq32,MemberQ[
 #1,y3b]&], term12=Select[term11,MemberQ[#1,yN]&], term21=
 Select[eq32,MemberQ[#1,yb2]&], term22=Select[Factor[term21],
 MemberQ[#1,yN]&], trbeta=Solve[term12[[2]]==term22[[2]],beta]}
{eqmu=Reduce[Select[eq32,MemberQ[#1,yN^(beta)]&]==0,mu],
 eqmu1=eqmu[[3]], trmu=(eqmu1[[2,2]]//ToRules)/.trbeta}
{eqlambda=Select[eq32,MemberQ[#1,y3b]&]+Select[eq32,MemberQ[
 #1,yb2]&]==0, eqlambda1=(eqlambda/.trbeta)/.trmu//First}
{solslambda=Solve[eqlambda1,lambda]//Simplify,trlambda=
 solslambda[[3]], solC=(u[x,t]->ansatzC[[2]]/.trmu/.trlambda/.
 trbeta/.z->tr1[[1]]), solS=(u[x,t]->ansatzS[[2]]/.trmu/.
 trlambda/.trbeta/.z->tr1[[1]])}
Map[Simplify[pde1[#1[[1,1,2]]]]&,{solC,solS}]
{sol1G=solC/.{c->2,n->3}, sol2G=solC/.{c->1/2,n->3}}
f1[xN_,tN_]:=sol1G[[1,1,2]]/.x->xN/.t->tN; f2[xN_,tN_]:=
 sol2G[[1,1,2]]/.x->xN/.t->tN; Animate[Plot[f1[x,t],{x,0,Pi},
 PlotRange->{{0,Pi},{-200,200}},PlotStyle->Hue[0.7]],{t,0,5}]
Animate[Plot[f2[x,t],{x,-3,3},PlotRange->{{-3,3},{0,2}},
 PlotStyle->Hue[0.7]],{t,0,5}]
```
□

The Exp-function method has been proposed by J. H. He [46] to obtain exact solutions of nonlinear evolution equations. Then, the method has been applied to various nonlinear PDEs of mathematical physics, e.g., Burgers and KdV equations, Kuramoto–Sivashinsky (KS) and Boussinesq equations.

Problem 2.23 *Klein–Gordon equation. Exp-function method.* Let us consider the Klein–Gordon equation (see Problems 2.21, 2.22)

$$u_{tt}-u_{xx}+u-u^3=0,$$

where $\{x \in \mathbb{R}, t \geq 0\}$. Find exact solutions and present the solution procedure of the Exp-function method with the aid of computer algebra systems *Maple* and *Mathematica*.

1. We look for traveling wave solutions, i.e., solutions of the form $u(x,t)=U(z)$, $z=\mu(x+ct)$, where μ and c are constants to be determined. As before, we convert the nonlinear PDE into the following ODE: $\mu^2c^2U_{zz}-\mu^2U_{zz}+U-U^3=0$ (`Eq2`).

2. According to the Exp-function method [47], we propose the ansatz, i.e we assume that the traveling wave solutions can be expressed in the form $U(z)=\frac{\sum_{k=-r}^{s} a_k e^{kz}}{\sum_{j=-p}^{q} b_j e^{jz}}=\frac{a_r e^{rz}+\cdots+a_{-s}e^{-sz}}{a_p e^{pz}+\cdots+a_{-q}e^{-qz}}$ (`tr3`), where r, s, p and q are unknown positive integers that have to be determined, and a_k, b_j are unknown constants.

3. Considering the resulting ODE, we can determine values of r, p and s, q. First, balancing the highest-order linear term with the highest-order nonlinear term, we obtain $-3r-3p=-r-5p$, i.e., $r=p$. Similarly, balancing the lowest-order linear term with the lowest-order nonlinear term, we have $3s+3q=s+5q$, i.e $s=q$. These computations can be performed with *Maple* and *Mathematica* (see the symbolic solution described below).

4. Let us consider a particular case, $r=p=1$ and $s=q=1$. Then the ansatz takes the form (`tr4`): $U(z)=\frac{a_{-1}e^{-z}+a_0+a_1e^z}{b_{-1}e^{-z}+b_0+b_1e^z}$. Substituting this expression into the ODE and equating the coefficients of all powers of e^{kz} to zero, we generate the system of algebraic equations (`sys1`) with respect to the unknowns c, μ, a_{-1}, a_0, a_1, b_{-1}, b_0, b_1.

5. To determine all unknown constants, we solve this system of algebraic equations and obtain 128 solutions (`Sols`). Without full analysis of all solutions, let us compare some solutions with the exact solutions obtained by the tanh-function method. So we choose the following solutions (`Sol1`, `Sol2`):

$$c=c_0,\ \mu=-\sqrt{\frac{1}{2(c_0^2-1)}},\ a_{-1}=b_{-1},\ a_0=0,\ a_1=-b_1,\ b_{-1}=b_{-1},\ b_0=0,\ b_1=b_1;$$

$$c=c_0,\ \mu=-\sqrt{\frac{1}{2(c_0^2-1)}},\ a_{-1}=-b_{-1},\ a_0=0,\ a_1=b_1,\ b_{-1}=b_{-1},\ b_0=0,\ b_1=b_1.$$

Let $d=\sqrt{1/(c^2-1)}$. The corresponding traveling wave solutions take the form:

$$u(x,t)=\pm\frac{b_{-1}+b_{-1}\tanh[\frac{\sqrt{2}}{2}\,d(x+ct)]-b_1+b_1\tanh[\frac{\sqrt{2}}{2}\,d(x+ct)]}{b_{-1}+b_{-1}\tanh[\frac{\sqrt{2}}{2}\,d(x+ct)]+b_1-b_1\tanh[\frac{\sqrt{2}}{2}\,d(x+ct)]}.$$

Setting $b_1=-1$ and $b_{-1}=-1$, we rewrite the above solutions in the form:

$$u(x,t) = \pm\tanh[\tfrac{\sqrt{2}}{2}\,d(x+ct)],$$

that coincide to the exact solutions obtained above by applying the tanh-function method. It is possible to find another types of exact solutions analyzing all sets of parameters or considering another particular cases, e.g., $r=p=2$, $s=q=2$.

Maple:

```
with(PDEtools): declare(u(x,t),U(z)); alias(U=U(z),u=u(x,t));
PDE1:=u->diff(u,t$2)-diff(u,x$2)+u-u^3=0; tr1:=mu*(x+c*t)=z;
Eq1:=expand(PDE1(U(lhs(tr1)))); Eq2:=convert(expand(subs(tr1,
 Eq1)),diff); tr3:=U(z)=Sum(a[k]*exp(k*z),k=-r..s)/Sum(
 b[j]*exp(j*z),j=-p..q); trpc:=isolate(-3*r-3*p=-r-5*p,p);
trqd:=isolate(3*s+3*q=s+5*q,q); params:={p=1,r=1,q=1,s=1};
tr4:=value(subs(params,tr3)); Eq3:=factor(value(algsubs(
 subs(params,tr3),Eq2))); Eq31:=simplify(op(2,lhs(Eq3)));
for i from 1 to 3 do E||i:=coeff(Eq31,exp(i*z));
 E||(i+3):=coeff(Eq31,exp(-i*z)); od;
E7:=remove(has,Eq31,exp); E8:=c=c0; sys1:={seq(E||i,i=1..8)};
vars:={mu,c,seq(a[i],i=-1..1),seq(b[j],j=-1..1)};
Sols:=[allvalues([solve(sys1,vars)])]; NSols:=nops(Sols);
Sol1:=op(11,op(7,Sols)); Sol2:=op(15,op(7,Sols));
SolF1:=U=subs(c0=c,subs(Sol1,subs(z=lhs(tr1),rhs(tr4))));
SolF2:=U=subs(c0=c,subs(Sol2,subs(z=lhs(tr1),rhs(tr4))));
SolF12:=simplify(convert(convert(SolF1,trig),tanh),tanh);
SolF22:=simplify(convert(convert(SolF2,trig),tanh),tanh);
SolF13:=collect(subs(b[1]=-1,b[-1]=-1,SolF12),tanh);
SolF23:=collect(subs(b[1]=-1,b[-1]=-1,SolF22),tanh);
factor(PDE1(rhs(SolF13))); factor(PDE1(rhs(SolF23)));
```

Mathematica:

```
pde1[u_]:=D[u,{t,2}]-D[u,{x,2}]+u-u^3==0; tr1=mu*(x+c*t)->z
{eq1=pde1[uN[tr1[[1]]]]//Expand, eq2=eq1/.tr1}
{tr3=uN[z]->Sum[a[k]*Exp[k*z],{k,-r,s}]/Sum[b[j]*Exp[j*z],
 {j,-p,q}], trpc=Solve[-3*r-3*p==-r-5*p,p], trqd=Solve[
 3*s+3*q==s+5*q,q], params={p->1,r->1,q->1,s->1},
 tr4=tr3/.params, tr41=Table[D[tr4[[1]],{z,i}]->D[tr4[[2]],
 {z,i}],{i,1,2}]//Expand, eq3=((eq2/.tr4)/.tr41)//Factor}
eq31=eq3[[1,3]]
```

```
Do[eq[i]=Coefficient[eq31,Exp[i*z]]==0;Print[eq[i]],{i,1,6}];
s1=0; nL=Length[eq31];
Do[If[D[eq31[[i]],z]==0,s1=s1+eq31[[i]],s1=s1+0],{i,1,nL}];
{eq[7]=s1==0, eq[8]=c==c0}; sys1=Table[eq[i],{i,1,8}]
vars={mu,c,Table[a[i],{i,-1,1}],Table[b[j],{j,-1,1}]}//Flatten
sols=Reduce[sys1//FullSimplify,vars]
sol12[s_]:=Map[ToRules,{sols[[s,2,1]],sols[[s,3]],sols[[s,5]],
 sols[[s,6]],sols[[s,7]],sols[[s,8]]}]//Flatten;
{nSols=Length[sols], sol1=sol12[45], sol2=sol12[46]}
solF1=u[x,t]->((tr4[[2]]/.{z->tr1[[1]]})/.sol1)/.{c0->c}
solF2=u[x,t]->((tr4[[2]]/.{z->tr1[[1]]})/.sol2)/.{c0->c}
ruleSCH={Sinh[x_]:>Tanh[x]/(Sqrt[1-Tanh[x]^2]),Cosh[x_]:>1/
 (Sqrt[1-Tanh[x]^2])}; solF12=(solF1//ExpToTrig)/.ruleSCH//Factor
{solF22=(solF2//ExpToTrig)/.ruleSCH//Factor, tr5={a[1]->-1,
 a[-1]->-1}, tr6={a[1]->b[1],a[-1]->-b[-1]}, tr7={a[1]->-b[1],
 a[-1]->b[-1]}, tr8={b[1]->-1,b[-1]->-1}, solF13=solF12/.tr5,
 solF23=solF22/.tr5, solF14=solF12/.tr6, solF24=solF22/.tr7,
 solF15=solF14/.tr8, solF25=solF24/.tr8}
{pde1[solF15[[2]]],pde1[solF25[[2]]]}//FullSimplify
```

6. Finally, let us present the procedure of balancing the highest-order linear term with the highest-order nonlinear term and the lowest-order linear term with the lowest-order nonlinear term in the Exp-function method. These computations are performed with *Maple* and *Mathematica* as follows:

Maple:

```
tr31:=U=(a[-c]*exp(-c*z)+a[d]*exp(d*z))/(a[-p]*exp(-p*z)
 +a[q]*exp(q*z)); Ex1:=subs(tr31,U^3); NEx1:=nops(Ex1);
N1:=simplify(expand(numer(Ex1))); D1:=simplify(expand(
 denom(Ex1))); N1/D1; Ex11:=combine(factor(expand(N1*D1)))
 /combine(factor(expand(D1*D1)));
Ex2:=algsubs(tr31,diff(U,z$2)); NEx2:=nops(Ex2);
for i from 1 to NEx2 do
  N2||i:=simplify(expand(numer(op(i,Ex2))));
  D2||i:=simplify(expand(denom(op(i,Ex2)))); N2||i/D2||i;
  Ex2||i:=combine(factor(expand(N2||i*D2||i)))/combine(factor(
   expand(D2||i*D2||i))); od;
E1L:=expand(op(1,select(has,numer(Ex11),p))/z);
E1R:=expand(op(1,select(has,numer(Ex23),-c*z))/z);
isolate(E1L=E1R,c); Ex12:=combine(expand(N1*D1)/expand(D1*D1));
NumEx12:=numer(Ex12); NNumEx12:=nops(NumEx12);
```

```
E2L:=expand(op(1,select(has,op(NNumEx12,NumEx12),3*q*z))/z);
for i from 1 to NEx2 do
  N2||i:=simplify(expand(numer(op(i,Ex2))));
  D2||i:=simplify(expand(denom(op(i,Ex2)))); N2||i/D2||i;
  Ex22||i:=combine(expand(N2||i*D2||i)/expand(D2||i*D2||i)); od;
NumEx223:=numer(Ex223); NNumEx223:=nops(NumEx223);
E2R:=expand(op(1,select(has,op(NNumEx223,NumEx223),5*q*z))/z);
isolate(E2L=E2R,d);
```

Mathematica:

```
{tr31=uN[z]->(a[-c]*Exp[-c*z]+a[d]*Exp[d*z])/(a[-p]*Exp[-p*z]+
 a[q]*Exp[q*z]), ex1=uN[z]^3/.tr31, nEx1=Length[ex1]}
{n1=Expand[Numerator[ex1]], d1=Expand[Denominator[ex1]], n1/d1}
ex11=Expand[n1*d1]/Expand[d1*d1]
{ex20=D[uN[z],{z,2}]/.tr31/.D[tr31,{z,1}]/.D[tr31,{z,2}],
 ex2=ex20[[1]]+ex20[[2]]+ex20[[3,1]]*ex20[[3,2,1]]+ex20[[3,1]]*
 ex20[[3,2,2]], nEx2=Length[ex2]}
Do[n2[i]=Simplify[Expand[Numerator[ex2[[i]]]]]; d2[i]=Simplify[
 Expand[Denominator[ex2[[i]]]]]; n2[i]/d2[i]; ex2N[i]=Expand[
 n2[i]*d2[i]]/Expand[d2[i]*d2[i]]; Print["ex2N",i," ",ex2N[i]],
 {i,1,nEx2}];
{e1L=Expand[Factor[Numerator[ex11]][[1,2]]/z],
 e1R=Expand[Factor[Numerator[ex2N[3]]][[2,2]]/z],
 Solve[e1L==e1R,c]}
{ex12=Expand[n1*d1]/Expand[d1*d1], numEx12=Numerator[ex12],
 nnumEx12=Length[numEx12], e2L=Expand[numEx12[[nnumEx12,1,2]]/z]}
Do[n2[i]=Expand[Numerator[ex2[[i]]]]; d2[i]=Expand[Denominator[
 ex2[[i]]]]; n2[i]/d2[i]; ex22N[i]=Expand[n2[i]*d2[i]]/Expand[
 d2[i]*d2[i]]; Print["ex22N",i," ",ex22N[i]],{i,1,nEx2}];
{numEx22N3=Numerator[ex22N[3]], nnumEx223=Length[numEx22N3],
 e2R=Expand[numEx22N3[[nnumEx223,2,2]]/z],
 Solve[e2L==e2R,d]//Expand}
```

□

2.5.3 Self-Similar Reductions

Self-similar (or automodel) solutions, obtained by reduction through invariance under scalings of the variables, are often occur in various problems of mathematical physics. The method of similarity is based on some symmetrical properties of a physical system and the algebraic symmetry of a nonlinear PDE. Self-similar solutions can be obtained by solving an associated ODE. These solutions are invariant under a *scaling* or *similarity transformation* (see Sect. 1.1.3). For example, the self-similar solution of the problem of expanding blast wave was first obtained (based on dimensional analysis) by G. I. Taylor [151] and L. I. Sedov [132] in the middle of the 20th century. The method of similarity is important and useful in finding solutions of both linear and nonlinear PDEs. As before, we illustrate the method of finding self-similar solutions by solving problems.

Problem 2.24 *Nonlinear diffusion equation. Self-similar reduction.* Let us consider the nonlinear diffusion equation of the following form:

$$u_t = a u_{xx} + b u^n,$$

where $\{x \in \mathbb{R}, t \geq 0\}$ and a, b are real constants. Let $t = CT$, $x = C^k X$, and $W = C^m W$, be a *scaling transformation*, where C $(C \neq 0)$ is an arbitrary constant (or parameter) and m, k are some unknown constants. Assuming that this equation is invariant under the scaling transformation for suitable values of m and k, find the self-similar reduction.

1. Applying the scaling transformation, we reduce the nonlinear diffusion equation to the equation (`Eq1`)

$$C^{m-1} W_T = a C^{m-2k} W_{XX} + b C^{mn} W^n.$$

2. We show that $k = \frac{1}{2}$ and $m = \frac{1}{1-n}$ and the *self-similar variables* take the form $u(x,t) = t^{\alpha} U(\xi)$, $\xi = x t^{\beta}$, where $\alpha = m$ and $\beta = -k$.

3. We show that the nonlinear diffusion equation reduces to the ordinary differential equation $a U_{\xi\xi} + \dfrac{1}{2} \xi U_{\xi} + \dfrac{1}{n-1} U + b U^n = 0$.

Maple:

```
with(PDEtools): declare(u(x,t),W(X,T),U(xi));
alias(u=u(x,t),W=W(X,T),U=U(xi)); interface(showassumed=0);
assume(k>0,m>0,n>0,C>0,t>0);
DiffusEq:=(t,x,u)->diff(u,t)=a*diff(u,x$2)+b*u^n;
tr1:={t=T*C,x=X*C^k,u=C^m*W};
Eq1:=combine(dchange(tr1,DiffusEq(t,x,u),[T,X,W]));
Ex21:=select(has,lhs(Eq1),C); Ex22:=select(has,rhs(Eq1),k);
Ex23:=select(has,expand(rhs(Eq1)),n); Ex31:=op(2,Ex21);
Ex32:=op(2,select(has,Ex22,k));
Ex33:=op(2,select(has,Ex23,C)); Eqs:={Ex31=Ex32,Ex32=Ex33};
tr2:=convert((solve(Eqs,{k,m}) assuming n<>1),list);
```

```
alpha:=rhs(tr2[2]); beta:=-rhs(tr2[1]);
tr3:={xi=x*(t^beta),u=U(xi)*t^(alpha)}; tr31:=x*(t^beta)=xi;
Eq21:=DiffusEq(t,x,U(lhs(tr31))*t^alpha);
ODE1:=convert(algsubs(tr31,Eq21),diff);
c1:=select(has,op(2,lhs(ODE1)),[t,n]);
ODE11:=expand(simplify(ODE1/c1*t));
ODEFin:=map(factor,lhs(collect(factor(ODE11-rhs(ODE11)),xi)))=0;
```

Mathematica:

```
diffusEq[x_,t_]:=D[u[x,t],t]==a*D[u[x,t],{x,2}]+b*u[x,t]^n;
{tr11={x->xN*c^k,t->tN*c}, tr12=u->w*c^m, tr13={(c^m*w)[xN,tN]->
 c^m*w[xN,tN], D[(c^m*w)[xN,tN],{xN,2}]->c^m*D[w[xN,tN],{xN,2}],
 D[(c^m*w)[xN,tN],tN]->c^m*D[w[xN,tN],tN]}}
eq1T[v_]:=((Simplify[diffusEq[x,t]/.u->Function[{x,t},
 u[x/c^k,t/c]]])/.tr11//ExpandAll)/.{u->v};
{eq1=eq1T[w*c^m]/.tr13//PowerExpand, ex21=Select[eq1[[2]],
 MemberQ[#1,c]&], ex22=Select[eq1[[1]],MemberQ[#1,a]&],
 ex23=Select[eq1[[1]],MemberQ[#1,b]&]}
{ex31=ex21[[2]], ex32=ex22[[2,2]], ex33=ex23[[2,2]]}
tr2=Assuming[n!=1,Solve[{ex31==ex32,ex32==ex33},{m,k}]]
{alpha=tr2[[1,1,2]], beta=-tr2[[1,2,2]]}
{tr3={xi->x*t^(beta),u->uN[xi]*t^(alpha)},tr31=x*t^(beta)->xi}
dEq[x_,t_,u_]:=D[u,t]==a*D[u,{x,2}]+b*u^n;
eq21=dEq[x,t,uN[tr31[[1]]]*t^(alpha)]//PowerExpand
{ode1=eq21/.tr31,c1=Select[ode1[[2,1]],MemberQ[#1,t]&]}
ode11=Thread[ode1/c1,Equal]/.tr31//FullSimplify
odeFin=ode11/.tr31
```

□

Problem 2.25 *Nonlinear wave equation. Self-similar reduction.* Let us consider the nonlinear wave equation of the form

$$u_{tt}=a(u^n u_x)_x,$$

where $\{x\in\mathbb{R}, t\geq 0\}$, a is a real constant, and $t=CT$, $x=C^kX$, $W=C^mW$ ($C\neq 0$ is an arbitrary constant) is the scaling transformation. Assuming that this equation is invariant under the scaling transformation for suitable values of m and k, determine the self-similar reduction.

1. Applying the scaling transformation, we reduce the given nonlinear wave equation to the equation

$$C^{m-2}W_{TT}=aC^{mn+m-2k}(nW^{n-1}W_X^2+W_{XX}W^n).$$

2. We verify that $k=\frac{1}{2}mn+1$ and m is arbitrary constant and show that self-similar variables take the form $u(x,t)=t^{\alpha}U(\xi)$, $\xi=xt^{\beta}$, where $\alpha=m$ and $\beta=-k$.

3. We verify that the given nonlinear equation reduces to the following ODE: $(-4aU^n+q^2\xi^2)U_{\xi\xi}+U_{\xi}(q+2-4m)q\xi-4anU_{\xi}^2U^{n-1}+4Um(m-1)=0$, and $q=2+mn$.

Maple:

```
with(PDEtools): declare(u(x,t),W(X,T),U(xi));
alias(u=u(x,t),W=W(X,T),U=U(xi)); interface(showassumed=0);
assume(k>0,m>0,n>0,C>0,t>0);
NPDE:=(t,x,u)->diff(u,t$2)=a*diff(u^n*diff(u,x),x);
tr1:={t=T*C, x=X*C^k, u=C^m*W};
Eq1:=factor(combine(expand(dchange(tr1,NPDE(t,x,u),[T,X,W]))));
Ex21:=select(has,lhs(Eq1),[C]); Ex22:=select(has,rhs(Eq1),[k]);
Ex31:=op(2,Ex21); Ex32:=op(2,Ex22);
tr2:=solve(Ex31=Ex32,k) assuming k<>0; alpha:=m; beta:=-tr2;
tr3:={xi=x*(t^beta),u=U(xi)*t^alpha};
NPDE1:=u->diff(u,t$2)=a*diff(u^n*diff(u,x),x);
tr31:=x*(t^beta)=xi; Eq21:=NPDE1(U(lhs(tr31))*t^alpha);
ODE1:=convert(algsubs(tr31,Eq21),diff);
ODE12:=combine(expand((lhs(ODE1)-rhs(ODE1))*4*U*t^2/(t^m*U)));
ODE13:=map(factor,collect(factor(ODE12),[xi^2,U]));
ODEFin:=collect(ODE13,diff);
```

Mathematica:

```
npde[x_,t_]:=D[u[x,t],{t,2}]==a*D[u[x,t]^n*D[u[x,t],x],x];
{tr11={t->tN*c,x->xN*c^k},tr12={u->c^m*w},
 tr13={(c^m*w)[xN,tN]->c^m*w[xN,tN], Table[D[(c^m*w)[xN,tN],
 {xN,i}]->c^m*D[w[xN,tN],{xN,i}],{i,1,2}], D[(c^m*w)[xN,tN],
 {tN,2}]->c^m*D[w[xN,tN],{tN,2}]}//Flatten}
eq1T[v_]:=((Simplify[npde[x,t]/.u->Function[{x,t},
 u[x/c^k,t/c]]])/.tr11//ExpandAll)/.{u->v};
eq1=eq1T[w*c^m]/.tr13//PowerExpand//Simplify
{ex21=Select[eq1[[1]],MemberQ[#1,c]&], ex22=Select[eq1[[2]],
 MemberQ[#1,c]&], ex31=ex21[[2]], ex32=ex22[[2]]}
{tr2=Assuming[k!=0,Solve[ex31==ex32,k]]//Expand, alpha=m,
 beta=-tr2[[1,1,2]], tr3={xi->x*(t^beta), u->uN[xi]*t^alpha}}
npde1[u_]:=D[u,{t,2}]==a*D[u^n*D[u,x],x];
{tr31=x*(t^beta)->xi, tr32=x->xi/t^beta}
eq21=npde1[uN[tr31[[1]]]*t^alpha]
```

```
{ode1=eq21/.tr31/.tr32, ode12=Assuming[{m>0,n>0,t>0},
 (ode1[[1]]-ode1[[2]])*4*uN[xi]*t^2/(t^m*uN[xi])//
 PowerExpand]//ExpandAll}
ode13=Map[Factor,Collect[Factor[ode12],{xi^2,uN[xi]}]]
odeFin=Collect[ode13/.{m*n->q-2},{uN'[xi],uN''[xi]}]
```

□

Problem 2.26 *Sine–Gordon equation. Self-similar reduction.* Let us consider the sine–Gordon equation of the form

$$u_{xt}=\sin u,$$

where $\{x\in\mathbb{R}, t\geq 0\}$ and $t=C^nT$, $x=C^mX$, $W=C^kW$ ($C\neq 0$ is an arbitrary constant) is the scaling transformation. Assuming that this equation is invariant under the scaling transformation for suitable values of n, m, and k, find the self-similar reduction.

1. Applying the scaling transformation, we reduce the sine–Gordon equation to the equation $C^{-k-n+m}W_{XT}=\sin(C^mW)$.

2. We verify that $m=-n$ and $k=0$ and show that self-similar variables take the form $u(x,t)=t^{\alpha}U(\xi)$, $\xi=x/t^{\beta}$, where $\alpha=-k/(2n)$ and $\beta=n/m$.

3. We show that the sine–Gordon equation reduces to the ordinary differential equation $\xi U_{\xi\xi}+U_{\xi}=\sin U$.

Maple:

```
with(PDEtools): declare(u(x,t),W(X,T),U(xi));
alias(u=u(x,t),W=W(X,T),U=U(xi)); interface(showassumed=0);
assume(C>0,t>0); SGEq:=(t,x,u)->diff(u,x,t)=sin(u);
tr1:={t=T*C^n,x=X*C^m,u=C^k*W};
Eq1:=combine(dchange(tr1,SGEq(t,x,u),[T,X,W]));
Ex21:=select(has,lhs(Eq1),C);
Ex22:=select(has,op(rhs(Eq1)),C); Ex31:=op(2,Ex21);
Ex32:=op(2,Ex22); tr2:=solve(Ex31=Ex32,{n,k});
alpha:=subs(k=0,-k/(2*n)); beta:=subs(tr2,n/m);
tr3:={xi=x/(t^beta),u=U(xi)*t^(alpha)}; tr31:=x/(t^beta)=xi;
Eq21:=SGEq(t,x,U(lhs(tr31))*t^alpha);
ODE1:=convert(algsubs(tr31,Eq21),diff);
```

Mathematica:

```
sGEq[x_,t_]:=D[u[x,t],x,t]==Sin[u[x,t]];
{tr11={t->tN*c^n,x->xN*c^m},tr12=u->c^k*w,
tr13={(c^k*w)[xN,tN]->c^k*w[xN,tN], D[(c^k*w)[xN,tN],xN,tN]->
 c^k*D[w[xN,tN],xN,tN]}}
eq1T[v_]:=((Simplify[sGEq[x,t]/.u->Function[{x,t},u[x/c^m,
 t/c^n]]])/.tr11//ExpandAll)/.{u->v};
{eq1=eq1T[w*c^k]/.tr13//PowerExpand, ex21=Select[eq1[[2]],
 MemberQ[#1,c]&], ex22=Select[eq1[[1,1]],MemberQ[#1,c]&],
 ex31=ex21[[2]]==0, ex32=ex22[[2]]==0}
tr2=Solve[{ex31,ex32},{n,k}]
{alpha=(-k/(2*n)/.tr2)[[1]], beta=(n/m/.tr2)[[1]]}
{tr3={xi->x/t^(beta),u->uN[xi]*t^(alpha)},tr31=x/t^(beta)->xi}
sGEq1[x_,t_,u_]:=D[u,x,t]==Sin[u];
ode1=sGEq1[x,t,uN[tr31[[1]]]*t^alpha]//.tr31
```

□

2.6 Separation of Variables

Separation of variables is one of the oldest and most widely used methods for constructing explicit analytical solutions to partial differential equations.

Definition 2.9 A *separable PDE* is an equation that can be reduced to a set of separate equations of lower dimensionality (fewer independent variables) by a method of separation of variables. The separable PDE can be solved by finding solutions of a set of simpler PDEs or ODEs.

For linear PDEs, separation of variables is one of the most important methods, in which the structure of a PDE allows us to look for multiplicative or additive separable exact solutions, e.g., $u(x,t)=\phi(x)\circ\psi(t)$ (where the multiplication or addition is denoted by $\circ$). Although for some specific nonlinear PDEs it is possible to apply this idea of the separation of variables as for linear PDEs, the wide classes of nonlinear PDEs cannot allow us to follow this idea. Additionally, in some cases it is possible to perform the separation of variables, but in more complicated form. In the last decades this theme has been studied extensively with respect to nonlinear PDEs and new methods have been developed, which form a family of methods of separation of variables such as *ordinary* separation of variables, *partial* separation of variables, *generalized* separation of variables, *functional* separation of variables. In this section, we will solve nonlinear PDEs in which it is possible to make these types of separation of variables.

2.6.1 Ordinary Separation of Variables

Now we consider the first method (most simple) of the family of methods of separation of variables and nonlinear equations, where it is possible to make the ordinary separation of variables (as in the linear case).

Problem 2.27 *Nonlinear heat equation. Ordinary separation of variables.* Let us consider the nonlinear heat equation of the form

$$u_t = a(u^k u_x)_x,$$

where $\{x \in \mathbb{R}, t \geq 0\}$ and a is a real constant. Applying the method of ordinary separation of variables and searching for exact solutions in the form $u(x,t) = \phi(x)\psi(t)$, prove that the exact solution of this equation describes by the solutions of the two ODEs (`Eq6`):

$$\psi^{-k-1}\psi_t = C, \qquad \frac{a(k\phi^{k-1}\phi_x^2 + \phi^k \phi_{xx})}{\phi} = C.$$

Maple:

```
with(PDEtools): declare((u,W)(x,t),phi(x),psi(t));
interface(showassumed=0): assume(k>0,phi(x)>0,psi(t)>0):
tr1:=phi(x)*psi(t);
PDE1:=u->diff(u(x,t),t)=a*diff(u(x,t)^k*diff(u(x,t),x),x);
Eq2:=expand(PDE1(W)); Eq3:=subs(W(x,t)=tr1,Eq2);
Eq4:=simplify(collect(Eq3,[phi(x),psi(t)]));
Eq5:=Eq4/(phi(x)*psi(t)^(k+1)); Eq6:=map(simplify,Eq5);
Sol:=[dsolve(lhs(Eq6)=C,psi(t)),dsolve(rhs(Eq6)=C,phi(x))];
```

Mathematica:

```
tr1=w[x,t]->phi[x]*psi[t]
pde1[u_]=D[u[x,t],t]==a*D[u[x,t]^k*D[u[x,t],x],x];
{eq2=Expand[pde1[w]], eq3=eq2/.tr1/.D[tr1,t]/.Table[
 D[tr1,{x,i}],{i,1,2}], eq4=PowerExpand[eq3]}
eq5=Thread[eq4/(phi[x]*psi[t]^(k+1)),Equal]//Expand
eq6=Map[Simplify,eq5]
sol={DSolve[eq6[[1]]==c,psi,t],DSolve[eq6[[2]]==c,phi,x]}
```

□

Problem 2.28 *Nonlinear wave equation. Ordinary separation of variables.* Let us consider the nonlinear wave equation of the form

$$u_{tt}=a(e^{\lambda u}u_x)_x,$$

where $\{x \in \mathbb{R}, t \geq 0\}$ and a is a real constant. Applying the method of separation of variables and searching for exact solutions in the form $u(x,t)=\phi(x)+\psi(t)$, prove that the exact solution of this equation has the form

$$u(x,t)=\frac{1}{\lambda}\ln\left(\frac{\lambda C_1 C x^2 - 2\lambda C_1^2 x a + 2\lambda C_1 C_2 a}{2Ca\cos(t\sqrt{\lambda C_1} + C_2\sqrt{\lambda C_1}) + 2Ca}\right),$$

where we assume that $t>0$, $C>0$, $a>0$, $C_1>0$, and $C_2>0$.

Maple:

```
with(PDEtools): declare((u,W)(x,t),phi(x),psi(t));
interface(showassumed=0): assume(lambda>0):
tr1:=phi(x)+psi(t); PDE1:=u->diff(u(x,t),t$2)=
         a*diff(exp(lambda*u(x,t))*diff(u(x,t),x),x);
Eq2:=expand(PDE1(W)); Eq3:=expand(subs(W(x,t)=tr1,Eq2));
Eq4:=factor(Eq3/(exp(lambda*psi(t)))); Eq5:=map(simplify,Eq4);
Sol:=[dsolve(lhs(Eq5)=C,psi(t)),dsolve(rhs(Eq5)=C,phi(x))];
SolFin:=u(x,t)=factor(rhs(Sol[1])+rhs(Sol[2]));
SolFin1:=map(combine,simplify(combine(SolFin)
         assuming t>0,C>0,a>0,_C1>0,_C2>0));
Test1:=pdetest(SolFin1,PDE1(u));
```

Mathematica:

```
Off[Solve::"ifun"]; tr1=w[x,t]->phi[x]+psi[t]
tr1D[v_]:=Table[D[tr1,{v,i}],{i,1,2}];
pde1[u_]:=D[u,{t,2}]==a*D[Exp[lambda*u]*D[u,x],x];
{eq2=Expand[pde1[w[x,t]]], eq3=PowerExpand[eq2/.tr1/.tr1D[t]
 /.tr1D[x]], eq4=Factor[Thread[eq3/(Exp[lambda*psi[t]]),
 Equal]], eq5=Map[Simplify,eq4]}
{sol1=DSolve[eq5[[1]]==c,psi,t], sol2=DSolve[eq5[[2]]==c,phi,x]}
solFin=(psi[t]/.sol1)+(phi[x]/.sol2)//Simplify//First
{solFin1=Assuming[{t>0,c>0,x>0,a>0,C[1]>0,C[2]>0},FullSimplify[
 Map[ExpandAll,solFin]]], pde1[solFin1]//FullSimplify}
```

□

2.6.2 Partial Separation of Variables

Now we consider the second method of the family of methods of separation of variables and nonlinear equations, where it is possible to make the partial separation of variables (different from the linear case). For a given nonlinear equation, we perform an analysis of the equation and look for all the special cases where the equation admits an ordinary separation of variables.

Problem 2.29 *Nonlinear parabolic equation. Partial separation of variables.* We consider the nonlinear second-order parabolic equation [124]

$$u_t = F(t)u_{xx} + uu_x^2 - au^3,$$

where $\{x \in \mathbb{R}, t \geq 0\}$, $F(t)$ is an arbitrary function and a is a real constant ($a > 0$). Applying the method of separation of variables and searching for exact solutions in the form $u(x,t) = \phi(x)\psi(t)$, we obtain the functional-differential equation $\frac{\psi_t}{F(t)\psi} = (\phi_x^2 - a\phi^2)\frac{\psi^2}{F(t)} + \frac{\phi_{xx}}{\phi}$ (Eq7). Prove that the exact solutions of this equation have the following forms:

Case 1. If $(\phi_x^2 - a\phi^2) = 0$, the separable solutions are (SolFin1, SolFin2) $u(x,t) = Ce^{\pm x\sqrt{a}} \exp\Big(\int aF(t)\,dt\Big)$.

Case 2. If $\frac{\psi_t}{\psi} = \text{const}$ and $(\phi_x^2 - a\phi^2) = \text{const}$ simultaneously, the separable solutions are $u(x,t) = \pm(C_1 e^{x\sqrt{a}} + C_2 e^{-x\sqrt{a}})e^G\Big(\int 8aC_1C_2e^{2G}\,dt + C_3\Big)^{-1/2}$ (see SolFin31, SolFin41). Here $G = a\int F(t)\,dt$, C is an arbitrary constant.

Maple:

```
with(PDEtools): declare((u,W)(x,t),phi(x),psi(t));
interface(showassumed=0): assume(a>0): tr1:=phi(x)*psi(t);
PDE1:=u->diff(u(x,t),t)=F(t)*diff(u(x,t),x$2)
     +u(x,t)*diff(u(x,t),x)^2-a*u(x,t)^3;
Eq2:=expand(PDE1(W)); Eq3:=expand(subs(W(x,t)=tr1,Eq2));
Eq4:=collect(Eq3,[phi,psi]); Eq5:=expand(Eq4/F(t)/psi(t)/phi(x));
Eq6:=map(expand,Eq5); Eq7:=collect(Eq6,psi(t)^2);
Sol1:=[dsolve(op(1,op(1,rhs(Eq7)))=0,phi(x))];
Sol11:=subs(exp(_C1*sqrt(a))=C,Sol1);
Eq8:=expand(subs(Sol11[1],Eq7)); Sol2:=dsolve(Eq8,psi(t));
SolFin1:=algsubs(_C1/C=C,u(x,t)=factor(rhs(Sol11[1])*rhs(Sol2)));
```

```
SolFin2:=algsubs(_C1*C=C,u(x,t)=factor(rhs(Sol11[2])*rhs(Sol2)));
T1:=pdetest(SolFin1, PDE1(u)); T2:=pdetest(SolFin2, PDE1(u));
Sol3:=dsolve(op(2,rhs(Eq7))=a,phi(x));
K:=expand(algsubs(Sol3,F(t)*op(1,op(1,rhs(Eq7)))));
Eq9:=subs({op(1,op(1,rhs(Eq7)))=K/F(t),op(2,rhs(Eq7))=a},Eq7);
Eq10:=expand(Eq9*F(t)*psi(t)); Sol4:=[dsolve(Eq10,psi(t))];
Sol41:=combine(algsubs(a*(Int(F(t),t))=G,Sol4[1]) assuming G>0);
Sol42:=combine(algsubs(a*(Int(F(t),t))=G,Sol4[2]) assuming G>0);
SolFin3:=u(x,t)=rhs(Sol3)*rhs(Sol41);
SolFin4:=u(x,t)=rhs(Sol3)*rhs(Sol42);
SolFin31:=subs(G=a*(Int(F(t),t)),SolFin3);
SolFin41:=subs(G=a*(Int(F(t),t)),SolFin3);
T3:=pdetest(SolFin31,PDE1(u)); T4:=pdetest(SolFin41,PDE1(u));
```

Mathematica:

```
tr1=w[x,t]->phi[x]*psi[t]; trD[u_,var_]:=Table[D[u,{var,i}],
 {i,1,2}]//Flatten; pde1[u_]:=D[u[x,t],t]==f[t]*D[u[x,t],{x,2}]+
 u[x,t]*D[u[x,t],x]^2-a*u[x,t]^3; {eq2=Expand[pde1[w]],
 eq3=Expand[eq2/.tr1/.trD[tr1,t]/.trD[tr1,x]], eq4=Collect[eq3,
 {phi,psi}], eq5=Expand[Thread[eq4/f[t]/psi[t]/phi[x],Equal]]}
{eq6=Map[Expand,eq5], eq7=Collect[eq6,psi[t]^2]}
sol1=DSolve[eq7[[2,1,2]]==0,phi[x],x]
eq8=Expand[eq7/.sol1[[1]]]/.trD[sol1[[1]],x]
{sol2=DSolve[eq8,psi[t],t], solFin12=Table[u[x,t]->
 sol1[[i,1,2]]*sol2[[1,1,2]]/.C[1]^2->C[1],{i,1,2}]}
test12=Table[pde1[u]/.solFin12[[i]]/.trD[solFin12[[i]],x]/.
 trD[solFin12[[i]],t],{i,1,2}]
sol3=DSolve[eq7[[2,2]]==a,phi[x],x]
k=Expand[eq7[[2,1,2]]*f[t]/.sol3/.trD[sol3,x]]//First
eq9=eq7/.{eq7[[2,1,2]]->k/f[t],eq7[[2,2]]->a}
eq10=Expand[Thread[eq9*f[t]*psi[t],Equal]]
{sols4=DSolve[eq10,psi[t],t], solsFin34=Table[u[x,t]->
 sol3[[1,1,2]]*sols4[[i,1,2]],{i,1,2}],
test34=Table[pde1[u]/.solsFin34[[i]]/.trD[solsFin34[[i]],x]/.
 trD[solsFin34[[i]],t]//FullSimplify,{i,1,2}]}
```

□

Problem 2.30 *Nonlinear third-order equation. Partial separation of variables.* Let us consider the nonlinear third-order equation [124]

$$u_t u_{xx} + a u_x u_{tt} = b u_{xxx} + c u_{ttt},$$

where $\{x \in \mathbb{R}, t \geq 0\}$ and a, b, and c are real constants. Applying the method of separation of variables and searching for exact solutions in the form $u(x,t)=\phi(x)+\psi(t)$, verify that the exact solutions of this equation take the following forms:

Case 1. If $\psi_t=C_1$, then $\psi(t)=C_1t+C_2$, $\phi(x)=C_2+C_3x+C_4e^{C_1x/b}$, and $u(x,t)=C_1t+C_2+C_3x+C_4e^{C_1x/b}$;

Case 2. If $\phi_x=C_1$, then $\phi(x)=C_1x+C_2$, $\psi(t) = C_2+C_3t+C_4e^{aC_1t/c}$, and $u(x,t)=C_1t+C_2+C_3x+C_4e^{C_1x/b}$;

Case 3. If we look for the solution in the form $\phi(x)=C_1e^{-A_1\lambda x}+A_2\lambda x$ and $\psi(t)=C_2e^{A_3\lambda t}+A_4\lambda t-C_3$ with the unknown coefficients A_1, A_2, A_3, and A_4, then we can find these coefficients ($A_3=1$, $A_2=c/a$, $A_1=a$, and $A_4=-ab$), and the separable solution is $\phi(x) = C_1e^{-a\lambda x} + c\lambda x/a$, $\psi(t)=e^{\lambda t}C_2-ab\lambda t+C_3$, and $u(x,t)=C_1e^{-a\lambda x}+c\lambda x/a+e^{\lambda t}C_2-ab\lambda t+C_3$.

Maple:

```
with(PDEtools): declare((u,W)(x,t),phi(x),psi(t));
interface(showassumed=0): assume(a>0,b>0,c>0):
tr1:=phi(x)+psi(t); PDE1:=u->diff(u(x,t),t)*diff(u(x,t),x$2)
  +a*diff(u(x,t),x)*diff(u(x,t),t$2)=b*diff(u(x,t),x$3)
  +c*diff(u(x,t),t$3); Eq2:=expand(PDE1(W));
Eq3:=expand(subs(W(x,t)=tr1,Eq2));
Sol1:=dsolve(op(1,op(1,lhs(Eq3)))=_C1,psi(t));
Eq4:=algsubs(Sol1,Eq3); Sol2:=dsolve(Eq4,phi(x));
SolFin1:=u(x,t)=subs(2*_C2=_C2,rhs(Sol1)+rhs(Sol2));
Sol3:=dsolve(op(2,op(2,lhs(Eq3)))=_C1,phi(x));
Eq5:=algsubs(Sol3,Eq3); Sol4:=dsolve(Eq5,psi(t));
SolFin2:=u(x,t)=subs(2*_C2=_C2,rhs(Sol3)+rhs(Sol4));
T1:=pdetest(SolFin1,PDE1(u)); T2:=pdetest(SolFin2,PDE1(u));
tr2:=[phi(x)=_C1*exp(-A1*lambda*x)+A2*lambda*x,
      psi(t)=_C2*exp( A3*lambda*t)+A4*lambda*t+_C3];
Eq6:=expand(algsubs(tr2[2],algsubs(tr2[1],Eq3)));
Eq61:=expand(Eq6/lambda^3); Eq62:=collect(Eq61,
 [exp(A1*lambda*x),exp(A3*lambda*t)]); C3:=A3=1;
Eq63:=subs(C3,Eq62); C2:=isolate(op(2,rhs(Eq63))/
 op(1,lhs(Eq63))=1,A2); Eq64:=expand(subs(C2,Eq63));
C1:=A1=a; C4:=subs(C1,expand(isolate(Eq64,A4)));
tr21:=subs({C1,C2,C3,C4},tr2); SolFin3:=u(x,t)=
 rhs(tr21[1])+rhs(tr21[2]); T3:=pdetest(SolFin3,PDE1(u));
```

Mathematica:

```
tr1=w[x,t]->phi[x]+psi[t]; trD[u_,var_]:=Table[D[u,{var,i}],
 {i,1,3}]//Flatten; pde1[u_]:=D[u[x,t],t]*D[u[x,t],{x,2}]+
  a*D[u[x,t],x]*D[u[x,t],{t,2}]==b*D[u[x,t],{x,3}]+
  c*D[u[x,t],{t,3}]; {eq2=Expand[pde1[w]], eq3=Expand[
  eq2/.tr1/.trD[tr1,x]/.trD[tr1,t]]}
sol1=DSolve[eq3[[1,1,1]]==C[1],psi[t],t]//First
{eq4=eq3/.sol1/.trD[sol1,t], sol2=DSolve[eq4,phi[x],x]//First}
{solFin1=u[x,t]->tr1[[2]]/.sol1/.sol2, sol3=DSolve[eq3[[
 1,2,2]]==C[1],phi[x],x]//First, eq5=eq3/.sol3/.trD[sol3,x]}
{sol4=DSolve[eq5,psi[t],t]//First, solFin2=u[x,t]->tr1[[2]]/.
 sol3/.sol4, solFin12={solFin1,solFin2}}
test12=Table[pde1[u]/.solFin12[[i]]/.trD[solFin12[[i]],x]/.
 trD[solFin12[[i]],t],{i,1,2}]
tr2={phi[x]->C[1]*Exp[-a1*lambda*x]+a2*lambda*x,
     psi[t]->C[2]*Exp[a3*lambda*t]+a4*lambda*t+C[3]}
eq6=Expand[eq3/.tr2/.trD[tr2,x]/.trD[tr2,t]]
eq61=Expand[Thread[eq6/lambda^3,Equal]]
eq62=Collect[eq61,{Exp[a1*lambda*x],Exp[a3*lambda*t]}]
{c3=a3->1, eq63=eq62/.c3, c2=Solve[eq63[[2,2]]/eq63[[1,1]]==1,
 a2], eq64=Expand[eq63/.c2], c1=a1->a, c4=Expand[Solve[
 eq64,a4]]/.c1, tr21=tr2/.c1/.c2/.c3/.c4//Flatten}
{solFin3=u[x,t]->tr1[[2]]/.tr21, test3=pde1[u]/.solFin3/.
 trD[solFin3,x]/.trD[solFin3,t]//FullSimplify}
```

□

2.6.3 Generalized Separation of Variables

For nonlinear PDEs in two independent variables, e.g., x, t, and a dependent variable u, a generalization to the ordinary separable solutions can be written in the form

$$u(x,t)=\phi_1(x)\psi_1(t)+\cdots+\phi_n(x)\psi_n(t),$$

where $n \in \mathbb{N}$ and the functions $\phi_i(x)\psi_i(t)$ are particular solutions of the given nonlinear PDE. The solution $u(x,t)$ is called the *generalized separable solution*. Opposite to linear PDEs, in nonlinear PDEs the functions $\phi_i(x)$ (with different subscripts i) are usually related to one another and to functions $\psi_j(t)$. In general, the functions $\phi_i(x)$ and $\psi_j(t)$ are unknown functions that have to be determined.

The concept of the so-called *nonlinear separation of variables* was first introduced by V. A. Galaktionov (see [61], [62]) in the study of blow-up of nonlinear parabolic equations, where the solution is given in the form $u(x,t)=\phi(t)\psi(x)+\eta(t)$. A great number of nonlinear PDEs of various types that admit the generalized separable solutions were presented by A. D. Polyanin and V. F. Zaitsev [124], [122]. Now will find generalized separable solutions to various problems by differentiating and applying simplified approach.

Problem 2.31 *Nonlinear parabolic equation. Generalized separation of variables by differentiating.* Let us consider the nonlinear second-order parabolic PDE [124]

$$u_t=auu_{xx}+b(u_x)^2+c,$$

where $\{x\in\mathbb{R}, t\geq 0\}$ and a, b, and c are real constants. Applying the method of generalized separation of variables and searching for exact solutions in the form $u(x,t)=\psi_1(t)+\psi_2(t)\phi_2(x)$, find the exact solution of this equation.

1. Applying the method of generalized separation of variables, we arrive at the following ODEs:

$$\phi_{2xxx}=C\phi_{2x}, \quad -2(\psi_{2t})^2+\psi_2\psi_{2tt}=C(\psi_2 a(-\psi_1\psi_{2t}+\psi_{1t}\psi_2)).$$

2. If we consider the case $C=0$, we obtain the following exact solution $\phi_2(x)=A_1x^2+A2_x+A_3$, $\psi_2(t)=B/(t+C_1)$, where $\psi_1(t)$ is an arbitrary function and A_1, A_2, A_3, B, C_1 are arbitrary constants. Then, substituting the solution obtained into the equation with respect to $\psi_1(t)$, $\psi_2(t)$, and $\phi_2(x)$, and determining the arbitrary function $\psi_1(t)$ (adding the arbitrary constant C_2) and finding some constants ($A_1=1$, $A_2=2C_3$, $A_3=C_3^2$, $B=-1/(2q)$, where $q=a+2b$), we arrive at the final form of the exact solution $u(x,t)=\dfrac{cq}{2(a+b)}(t+C_1)+C_2(t+C_1)^{-a/q}-\dfrac{1}{2q}\dfrac{(x+C_3)^2}{t+C_1}$.

Maple:

```
with(PDEtools): declare((u,W)(x,t),(psi1,psi2)(t),phi2(x));
tr1:=psi1(t)+phi2(x)*psi2(t);
PDE1:=u->diff(u(x,t),t)=a*u(x,t)*diff(u(x,t),x$2)
        +b*(diff(u(x,t),x))^2+c;
Eq2:=expand(PDE1(W)); Eq3:=expand(subs(W(x,t)=tr1,Eq2));
Eq4:=expand(Eq3/psi2(t)^2); Eq5:=diff(diff(Eq4,t),x);
Eq51:=collect(lhs(Eq5),diff(phi2(x),x))=
      collect(rhs(Eq5),diff(phi2(x),x$3));
TermX:=select(has,rhs(Eq51),a);
Eq6:=evala(Eq51/TermX/diff(phi2(x),x));
Eq71:=rhs(Eq6)=C; Eq72:=factor(lhs(Eq6)=C);
```

```
SolPhi2:=dsolve(subs(C=0,Eq71),phi2(x));
SolPhi21:=sort(subs(_C2=A2,_C3=A3,algsubs(coeff(
 rhs(SolPhi2),x^2)=A1,SolPhi2)));
SolPsi2:=dsolve(expand(subs(C=0,Eq72)),psi2(t));
SolPsi21:=lhs(SolPsi2)=subs(_C1=1,_C2=C1,rhs(SolPsi2)*(-B));
Eq8:=algsubs(SolPhi21,algsubs(SolPsi21,Eq3));
SolPsi1:=subs(_C1=C2,dsolve(Eq8,psi1(t)));
SolFin1:=rhs(SolPsi1); SolFin2:=rhs(SolPhi21)*rhs(SolPsi21);
tr2:=A1*x^2+A2*x+A3=(x+C3)^2; tr3:={A1=1,A2=2*C3,A3=C3^2};
tr4:=B=-1/(2*(a+2*b)); SolFin21:=factor(subs(tr2,tr4,SolFin2));
SolFin12:=sort(collect(expand(subs(t+C1=T,simplify(subs(
  tr3,tr4,SolFin1)))),t)); SolFin13:=map(factor,collect(
  SolFin12,[C1,C2])); SolFin14:=factor(select(has,SolFin13,c))
  +select(has,SolFin13,T);
SolFin15:=u(x,t)=subs(T=t+C1,SolFin14+SolFin21);
Test1:=pdetest(SolFin15,PDE1(u));
```

Mathematica:

```
tr1=w[x,t]->psi1[t]+phi2[x]*psi2[t]; trD[u_,var_]:=Table[D[u,
 {var,i}],{i,1,2}]//Flatten; pde1[u_]:=D[u[x,t],t]==a*u[x,t]*D[
 u[x,t],{x,2}]+b*(D[u[x,t],x])^2+c; {eq2=Expand[pde1[w]],
 eq3=Expand[eq2/.tr1/.trD[tr1,x]/.trD[tr1,t]], eq4=Expand[Thread[
 eq3/psi2[t]^2,Equal]], eq5=Thread[D[Thread[D[eq4,t],Equal],x],
 Equal], eq51=Collect[eq5[[1]],D[phi2[x],x]]==Collect[eq5[[2]],
 D[phi2[x],{x,3}]], termX=Cases[eq51[[2]],_Plus]//First}
eq6=Expand[Thread[eq51/termX/D[phi2[x],x],Equal]]
{eq71=eq6[[2]]==c, eq72=Factor[eq6[[1]]]==c}
{solPhi2=DSolve[eq71/.c->0,phi2[x],x]//First, solPhi21=solPhi2/.
 Coefficient[solPhi2[[1,2]],x^2]->a1/.{C[2]->a2,C[1]->a3},
 solPsi2=DSolve[Expand[eq72/.c->0],psi2[t],t]//First,
 solPsi21=solPsi2[[1,1]]->solPsi2[[1,2]]*bN/.C[2]->1/.C[1]->c1}
eq8=eq3/.solPsi21/.solPhi21/.trD[solPsi21,t]/.trD[solPhi21,x]
solPsi1=DSolve[eq8,psi1[t],t]/.C[1]->c2//First
{tr2=a1*x^2+a2*x+a3->(x+c3)^2, tr3={a1->1,a2->2*c3,a3->c3^2},
 tr4=bN->-1/(2*(a+2*b)), tr5=c1+t->tN}
{solFin11=tr1[[2]]/.solPsi1/.solPhi21/.solPsi21/.tr2/.tr3/.
 tr4/.tr5, solFin12=u[x,t]->Map[FullSimplify,Collect[solFin11,
 {tN,c}]]/.tN->tr5[[1]], test1=pde1[u]/.solFin12/.
 trD[solFin12,x]/.trD[solFin12,t]//FullSimplify}
```

□

Problem 2.32 *Sine–Gordon equation. Generalized separation of variables by differentiating.* Let us consider the sine–Gordon equation

$$u_{xx} - u_{tt} = \sin u,$$

where $\{x \in \mathbb{R}, t \geq 0\}$. Applying the method of generalized separation of variables, determine various types of one-soliton solutions and two-soliton solutions of the sine–Gordon equation and visualize them.

1. Let us apply the transformation $u(x,t){=}4\arctan\big(U(x,t)\big)$ and the method of generalized separation of variables. Searching for exact solutions in the form $u(x,t){=}\phi(x)/\psi(t)$, we arrive at the following equation:

$$2(\phi_x^2{-}\psi_t^2){+}(\psi^2{+}\phi^2)\left(\frac{\phi_{xx}}{\phi}{+}\frac{\psi_{tt}}{\psi}\right){=}\psi^2{-}\phi^2.$$

Then differentiating this equation with respect to x and t, we separate the variables and determine the following ODEs:

$$\frac{\phi_{xxx}}{\phi^2\phi_x}{-}\frac{\phi_{xx}}{\phi^3}{=}C, \quad \frac{\psi_{ttt}}{\psi^2\psi_t}{-}\frac{\psi_{tt}}{\psi^3}{=}C,$$

where C is a separation constant. These ODEs can be integrated twice:

$$\phi_x^2{=}\tfrac{1}{4}C\phi^4{+}A_1\phi^2{+}A_2, \quad \psi_t^2{=}\tfrac{1}{4}C\psi^4{+}B_1\psi^2{+}B_2,$$

where A_1, A_2, B_1, B_2 are arbitrary constants. Substituting these equations into the above ordinary differential equation with respect to $\phi(x)$ and $\psi(t)$, we can remove some constants, i.e., we obtain the relations $B_1{=}1{-}A_1$, $B_2{=}{-}A_2$ and the equations become

$$\phi_x^2{=}{-}\tfrac{1}{4}C\phi^4{+}A_1\phi^2{+}A_2, \quad \psi_t^2{=}\tfrac{1}{4}C\psi^4{+}(A_1{-}1)\psi^2{-}A_2.$$

Maple:

```
with(PDEtools): declare((u,U,W)(x,t),phi(x),psi(t));
tr1:=4*arctan(U(x,t)); tr2:=phi(x)/psi(t);
PDE1:=u->diff(u(x,t),x$2)-diff(u(x,t),t$2)=sin(u(x,t));
Eq2:=expand(PDE1(W)); Eq3:=expand(algsubs(W(x,t)=tr1,Eq2));
Eq31:=collect(normal(subs(U(x,t)=tr2,Eq3)),diff);
Eq32:=normal(map(expand,Eq31)); Eq33:=map(expand,Eq32/4*
 denom(rhs(Eq32))/phi(x)/psi(t)); Eq34:=collect(Eq33,diff);
Eq35:=map(normal,lhs(Eq34))=rhs(Eq34); Eq4:=diff(Eq35,x,t);
Eq41:=expand(Eq4/2/psi(t)/diff(psi(t),t)/phi(x)/diff(phi(x),x));
```

```
ODEs:=[selectremove(has,lhs(Eq41),phi(x))]; ODE1:=-ODEs[1]=C;
ODE2:=ODEs[2]=C; Eq51:=expand(ODE1*phi(x)*diff(phi(x),x));
Eq52:=int(lhs(Eq51),x)=int(rhs(Eq51),x)+A1;
Eq53:=expand(Eq52*phi(x)*diff(phi(x),x));
Eq54:=int(lhs(Eq53),x)=int(rhs(Eq53),x)+A2;
Eq55:=algsubs(-A1=A1,algsubs(-2*A2=A2,Eq54*(-2)));
Eq61:=expand(ODE2*psi(t)*diff(psi(t),t));
Eq62:=int(lhs(Eq61),t)=int(rhs(Eq61),t)+B1;
Eq63:=expand(Eq62*psi(t)*diff(psi(t),t));
Eq64:=int(lhs(Eq63),t)=int(rhs(Eq63),t)+B2;
Eq65:=subs(2*B2=B2,Eq64*2); Eq7:=collect(subs(Eq55,Eq65,Eq35),
 [A1,B1,A2,B2,C]); Eq71:=select(has,lhs(Eq7),[B1,A1])=select(
 has,rhs(Eq7),[phi,psi]); Eq72:=collect(lhs(Eq71)-rhs(Eq71),
 [phi,psi]); Eq73:=(coeff(Eq72,phi(x),2)-coeff(Eq72,
 psi(t),2))/2; Eq74:=select(has,lhs(Eq7)/2,[B2,A2])=0;
Consts:=[isolate(Eq73,B1),isolate(Eq74,B2)];
Eqs:=collect(subs(Consts,[Eq55,Eq65]),psi(t)^2);
```

Mathematica:

```
tr1=w[x,t]->4*ArcTan[uN[x,t]]; tr2=uN[x,t]->phi[x]/psi[t];
trD[u_,var_]:=Table[D[u,{var,i}],{i,1,2}]//Flatten;
trS[eq_,var_]:=Select[eq,MemberQ[#,var,Infinity]&];
pde1[u_]:=D[u[x,t],{x,2}]-D[u[x,t],{t,2}]==Sin[u[x,t]];
{eq2=Expand[pde1[w]], eq3=FunctionExpand[eq2/.tr1/.trD[tr1,x]/.
 trD[tr1,t]]//Expand, eq31=Map[Simplify,eq3/.tr2 /.trD[tr2,x]/.
 trD[tr2,t]], eq33=Map[Expand,Thread[eq31/4*Denominator[
 eq31[[2]]]/phi[x]/psi[t],Equal]], eq34=Collect[eq33,
 {psi''[t],phi''[x]}], eq35=Map[Factor,eq34[[1]]]==eq34[[2]]}
{eq4=Thread[D[eq35,x,t],Equal], eq41=Expand[Thread[eq4/2/
 psi[t]/psi'[t]/phi[x]/phi'[x],Equal]]}
{odes=Level[eq41,{2}], ode1=-(Plus@@trS[odes,phi[x]])==c,
 ode2=Plus@@trS[odes,psi[t]]==c}
eq51=Expand[Thread[ode1*phi[x]*phi'[x],Equal]]
eq52=Integrate[eq51[[1]],x]==Integrate[eq51[[2]],x]+a1
eq53=Expand[Thread[eq52*phi[x]*phi'[x],Equal]]
eq54=Integrate[eq53[[1]],x]==Integrate[eq53[[2]],x]+a2
eq55=(Thread[eq54*(-2),Equal]//Expand)/.-a1->a1/.-2*a2->a2
eq61=Expand[Thread[ode2*psi[t]*psi'[t],Equal]]
eq62=Integrate[eq61[[1]],t]==Integrate[eq61[[2]],t]+b1
eq63=Expand[Thread[eq62*psi[t]*psi'[t],Equal]]
eq64=Integrate[eq63[[1]],t]==Integrate[eq63[[2]],t]+b2
eq65=(Thread[eq64*2,Equal]//Expand)/.-2*b2->b2
```

```
{tr55=ToRules[eq55]//First, tr65=ToRules[eq65]//First,
 eq701=eq35/.tr55, eq702=eq701/.tr65, eq7=Collect[eq702,
 {a1,b1,a2,b2,c}], eq71=Plus@@{trS[eq7[[1]],b1],
 trS[eq7[[1]],a1]}==Plus@@{trS[eq7[[2]],phi[x]],
 trS[eq7[[2]],psi[t]]}, eq72=Collect[eq71[[1]]-eq71[[2]],
 {phi[x],psi[t]}], eq73=(Coefficient[eq72,phi[x],2]-
 Coefficient[eq72,psi[t],2])/2//Expand}
eq74=Plus @@ {trS[eq7[[1]],b2],trS[eq7[[1]],a2]}/4//Expand
consts={Solve[eq73==0,b1],Solve[eq74==0,b2]}//Flatten
eqs=Collect[{eq55,eq65}/.consts,psi[t]^2]
```

2. Soliton solutions. Now let us consider some important particular cases, and obtain and visualize soliton solutions. Among them we will find the *one-soliton solutions* that can be classified into two different cases, kink and antikink.

A kink is a traveling wave solution whose boundary values at the minus infinity is 0 and at the plus infinity is 2π, whereas an antikink is a solution whose the boundary values are 0 and -2π, respectively (see Sect. 1.1.3).

Also, the one-soliton solution produces solitary wave solutions, i.e., hump-shaped solitons or antisolitons moving without changing their forms in opposite directions.

The *two-soliton solutions* can be classified into several different cases: interaction of two kinks, interaction of two antikinks, interaction of a kink and antikink, interaction of an antikink and kink, and a special solution called breather solution.

Also, the two-soliton solutions produce another solutions that represent the interaction of two equal hump-shaped solitons or antisolitons and the interaction of the soliton and antisoliton or the interaction the antisoliton and soliton.

Case 1. Setting $C{=}0$, $A_2{=}0$, $A_1{>}1$, we obtain the following exact solutions:

(a) $\phi(x){=}D_1e^{xk_1}$, $\psi(t){=}E_1e^{tk_2}$, and $u_1(x,t){=}4\arctan(Ae^{xk_1-tk_2})$, where $k_1{=}\sqrt{A_1}$, $k_2{=}\sqrt{A_1-1}$, $D_1{=}e^{-C_1k_1}$, $E_1{=}e^{-C_1k_2}$, $A{=}D_1/E_1$, and $A_1{>}1$ is an arbitrary constant.

(b) $\phi(x){=}D_1e^{xk_1}$, $\psi(t){=}E_2e^{-tk_2}$, and $u_2(x,t){=}4\arctan(Ae^{xk_1+tk_2})$. Here $D_1{=}e^{-C_1k_1}$, $E_2{=}e^{C_1k_2}$, $A{=}D_1/E_2$, and $A_1{>}1$ is an arbitrary constant. These solutions represent the one-soliton solution, the case kink solution of the sine–Gordon equation.

(c) $\phi(x){=}D_2e^{-xk_1}$, $\psi(t){=}E_1e^{tk_2}$, $u_3(x,t){=}4\arctan(Ae^{-xk_1-tk_2})$, where $D_2{=}e^{C_1k_1}$, $E_1{=}e^{-C_1k_2}$, $A{=}D_2/E_1$, and $A_1{>}1$ is an arbitrary constant.

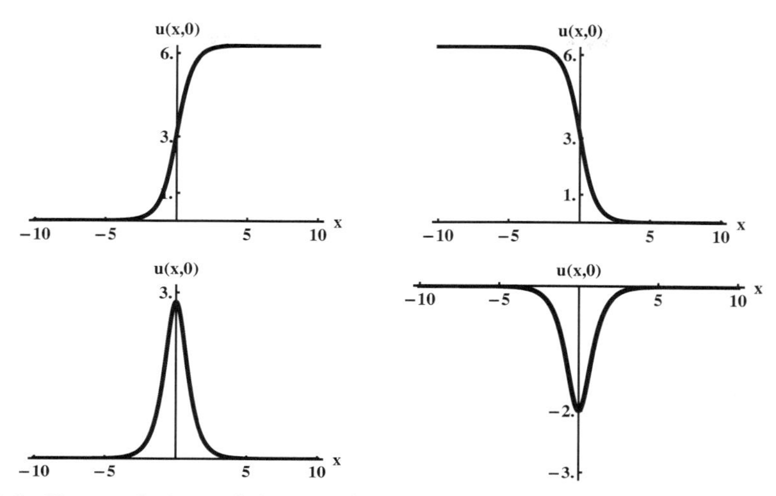

Fig. 2.1. Exact solutions of the sine–Gordon equation: one-soliton solutions (kink, antikink), solitary wave solutions (hump-shaped soliton and antisoliton)

(d) $\phi(x)=D_2e^{-xk_1}$, $\psi(t)=E_2e^{-tk_2}$, $u_4(x,t)=4\arctan(Ae^{-xk_1+tk_2})$, where $D_2=e^{C_1k_1}$, $E_2=e^{C_1k_2}$, $A=D_2/E_2$, and $A_1>1$ is an arbitrary constant.

These solutions represent the one-soliton solution, the case antisoliton or antikink solution of the sine–Gordon equation. We verify that these solutions are, indeed, exact solutions and plot the corresponding graphs and animations of these solutions.

The above soliton and antisoliton solutions produce u_{ix} and u_{it} (for $i=1,\ldots,4$), which represent *solitary wave solutions*, i.e., hump-shaped solitons or antisolitons moving without changing their forms in opposite directions. We plot the corresponding animations of these solitary wave solutions. These exact solutions of the sine–Gordon equation are shown in Fig. 2.1.

Maple:

```
with(plots): AG:=array(1..2): BG:=array(1..2): CG:=array(1..4):
setoptions(animate,thickness=3,scaling=constrained):
Eqs1:=subs(C=0,A2=0,Eqs); Sol11:=expand([dsolve(Eqs1[1],
 phi(x))]); Sol12:=expand([dsolve(Eqs1[2],psi(t))]);
tr3:=[select(has,rhs(Sol11[1]),_C1)=D1,select(has,
 rhs(Sol11[2]),_C1)=D2,select(has,rhs(Sol12[1]),_C1)=E1,
 select(has,rhs(Sol12[2]),_C1)=E2];
for i from 1 to 2 do Solphi1||i:=subs(tr3,Sol11[i]);
                     Solpsi1||i:=subs(tr3,Sol12[i]); od;
```

```
SolFin10:=subs(U(x,t)=tr2,tr1);
SolFin11:=u(x,t)=subs(Solphi11,Solpsi11,SolFin10);
SolFin12:=algsubs(D1/E1=A,combine(SolFin11,exp));
SolFin21:=u(x,t)=subs(Solphi11,Solpsi12,SolFin10);
SolFin22:=algsubs(D1/E2=A,combine(SolFin21,exp));
SolFin31:=u(x,t)=subs(Solphi12,Solpsi11,SolFin10);
SolFin32:=algsubs(D2/E1=A,combine(SolFin31,exp));
SolFin41:=u(x,t)=subs(Solphi12,Solpsi12,SolFin10);
SolFin42:=algsubs(D2/E2=A,combine(SolFin41,exp));
Params:=[A1=2,A=1]; xR:=-10..10; tR:=-10..10; N:=200; P:=200:
Op1:=frames=N,numpoints=P;
for i from 1 to 4 do
 plot(subs(Params,t=0,rhs(SolFin||i||2)),x=xR);
 pdetest(SolFin||i||1,PDE1(u)); od;
for i from 1 to 4 do  CG[i]:=animate(subs(Params,
                      rhs(SolFin||i||2)),x=xR,t=tR,Op1): od:
for i from 1 to 2 do
 SolFin||i||2||Dx:=diff(rhs(SolFin||i||2),x);
 SolFin||i||2||Dt:=diff(rhs(SolFin||i||2),t); od;
for i from 1 to 2 do
 AG[i]:=animate(subs(Params,SolFin||i||2||Dx),x=xR,t=tR,Op1):
 BG[i]:=animate(subs(Params,SolFin||i||2||Dt),x=xR,t=tR,Op1):
od: display(AG); display(BG); display(CG);
```

Mathematica:

```
p=10; SetOptions[Plot,ImageSize->300,PlotStyle->{Hue[0.9],
 Thickness[0.01]},PlotRange->All]; SetOptions[Animate,
AnimationRate->0.9]; {eqs1=eqs/.c->0/.a2->0,
sol11=Expand[DSolve[eqs1[[1]],phi[x],x]], sol12=Expand[
 DSolve[eqs1[[2]],psi[t],t]],solphi12=phi[x]->sol11[[1,1,2]]/.
 C[1]->d2, solphi11=phi[x]->sol11[[2,1,2]]/.C[1]->d1,
 solpsi12=psi[t]->sol12[[1,1,2]]/.C[1]->e2,
 solpsi11=psi[t]->sol12[[2,1,2]]/.C[1]->e1, solFin10=tr1/.tr2}
{solFin11=u[x,t]->solFin10[[2]]/.solphi11/.solpsi11,
 solFin[1]=solFin11/.d1->e1*a, solFin21=u[x,t]->solFin10[[2]]/.
 solphi11/.solpsi12,solFin[2]=solFin21/.d1->e2*a}
{solFin31=u[x,t]->solFin10[[2]]/.solphi12/.solpsi11,
 solFin[3]=solFin31/.d2->e1*a, solFin41=u[x,t]->solFin10[[2]]/.
 solphi12/.solpsi12, solFin[4]=solFin41/.d2->e2*a,
 pars={a1->2,a->1}}
GraphicsRow[Table[Plot[Evaluate[solFin[i][[2]]/.pars/.t->0],
 {x,-p,p}],{i,1,4}]]
```

```
{Table[pde1[u]/.solFin[i]/.trD[solFin[i],x]/.trD[solFin[i],t]//
 FullSimplify,{i,1,4}], Table[solFinDx[i]=D[solFin[i][[2]],x],
 {i,1,2}],Table[solFinDt[i]=D[solFin[i][[2]],t],{i,1,2}]}
Do[g11[i_,xN_,tN_]:=solFin[i][[2]]/.pars/.{x->xN,t->tN},{i,1,4}];
Do[g12[i_,xN_,tN_]:=solFinDx[i]/.pars/.{x->xN,t->tN},{i,1,2}];
Do[g13[i_,xN_,tN_]:=solFinDt[i]/.pars/.{x->xN,t->tN},{i,1,2}];
Manipulate[Plot[Evaluate[g11[1,x,t]],{x,-p,p},PlotRange->All],
 {t,-p,p}]
Animate[Plot[g11[3,x,t],{x,-p,p},PlotRange->{0,2*Pi}],{t,-p,p}]
Animate[Plot[g11[4,x,t],{x,-p,p},PlotRange->{0,2*Pi}],{t,-p,p}]
Animate[Plot[g12[1,x,t],{x,-p,p},PlotRange->{0,Pi}],{t,-p,p}]
Animate[Plot[g13[1,x,t],{x,-p,p},PlotRange->{-Pi,0}],{t,-p,p}]
```

Case 2. Setting $C{=}0$, $A_2{\neq}0$, $A_1{>}1$, we obtain, for example, the following more complicated exact solutions:

$$u_1(x,t){=}-4\arctan\left(\frac{k_2(A_2e^{2k_1C_1}-e^{2k_1x})e^{k_2(t+C_1)-k_1(x+C_1)}}{k_1(A_2e^{2k_2C_1}+e^{2k_2t})}\right),$$
$$u_2(x,t){=}-4\arctan\left(\frac{k_2(A_2e^{2k_1x}-e^{2k_1C_1})e^{k_2(t+C_1)-k_1(x+C_1)}}{k_1(A_2e^{2k_2t}+e^{2k_2C_1})}\right), \quad (2.2)$$

where $k_1{=}\sqrt{A_1}$, $k_2{=}\sqrt{A_1-1}$, C_1, A_1, and A_2 are arbitrary constants. These solutions belong to the class of two-soliton solutions and were first investigated numerically by Perring and Skyrme [120]. These solutions describe the interaction of two equal kinks or antikinks. We verify that these solutions are, indeed, exact solutions and plot the corresponding animations of these solutions. The above two-soliton solutions, the case of the interaction of two equal kinks or antikinks, produce u_{1x} and u_{1t}, which represent, respectively the interaction of two equal hump-shaped solitons or antisolitons and the interaction of the soliton and antisoliton or the interaction of the antisoliton and soliton. We plot the corresponding animations of these interactions. The class of two-soliton solutions of the sine–Gordon equation is shown in Fig. 2.2.

Maple:

```
DG:=array(1..2): EG:=array(1..2): FG:=array(1..2):
Eqs2:=subs(C=0,Eqs); Sol21:=expand([dsolve(Eqs2[1],phi(x))]);
Params:=[A1=2,A2=1,_C1=0]; xR:=-10..10; tR:=-10..10;
Sol22:=expand([dsolve(Eqs2[2],psi(t))]);
SolFin20:=subs(U(x,t)=tr2,tr1);
SolFin21:=u(x,t)=simplify(subs(Sol21[3],Sol22[3],SolFin20));
```

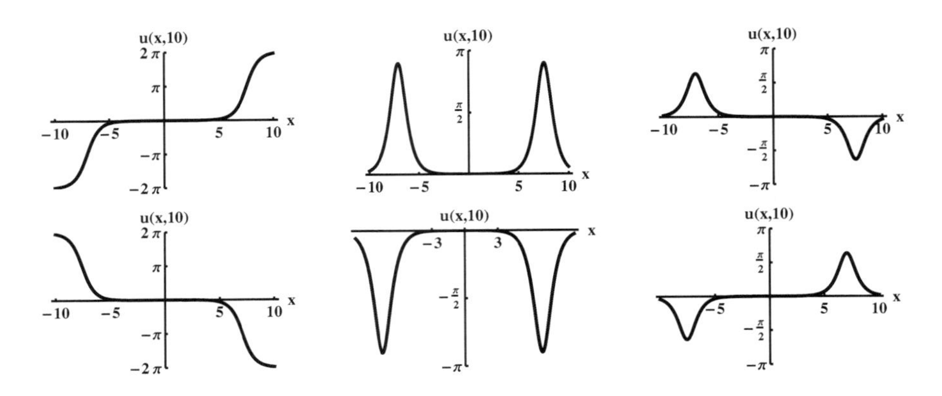

Fig. 2.2. Two-soliton solutions of the sine–Gordon equation: the interaction of two kinks (antikinks), the interaction of two equal hump-shaped solitons (antisolitons), the interaction of the soliton and antisoliton (or vice versa)

```
SolFin22:=u(x,t)=simplify(subs(Sol21[4],Sol22[4],SolFin20));
pdetest(SolFin21,PDE1(u));  pdetest(SolFin22,PDE1(u));
SolFin21Dx:=simplify(diff(SolFin21,x)); SolFin22Dx:=simplify(
 diff(SolFin22,x)); SolFin21Dt:=simplify(diff(SolFin21,t));
SolFin22Dt:=simplify(diff(SolFin22,t));
for i from 1 to 2 do
 DG[i]:=animate(expand(subs(Params,rhs(SolFin2||i))),x=xR,t=tR,
  color=blue,Op1): EG[i]:=animate(expand(subs(Params,
  rhs(SolFin2||i||Dx))),x=xR,t=tR,color=green,Op1):
 FG[i]:=animate(expand(subs(Params,rhs(SolFin2||i||Dt))),x=xR,
  t=tR,color=red,Op1): od: display(DG); display(EG); display(FG);
```

Mathematica:

```
{eqs2=eqs/.c->0/.a1->2, sol21=Flatten[Expand[DSolve[eqs2[[1]],
 phi[x],x]]], sol22=Expand[DSolve[eqs2[[2]],psi[t],t]]//Flatten}
{solFin20=tr1/.tr2, solFin2[1]=u[x,t]->Simplify[solFin20[[2]]/.
 sol21[[1]]/.sol22[[1]]], solFin2[2]=u[x,t]->Simplify[
 solFin20[[2]]/.sol21[[2]]/.sol22[[2]]]}
test12={pde1[u]/.solFin2[1]/.trD[solFin2[1],x]/.trD[solFin2[1],
 t]//FullSimplify, pde1[u]/.solFin2[2]/.trD[solFin2[2],x]/.
 trD[solFin2[2],t]//FullSimplify}
{solFinDx2[1]=Simplify[D[solFin2[1],x]], solFinDx2[2]=Simplify[
 D[solFin2[2],x]], solFinDt2[1]=Simplify[D[solFin2[1],t]],
 solFinDt2[2]=Simplify[D[solFin2[2],t]]}
```

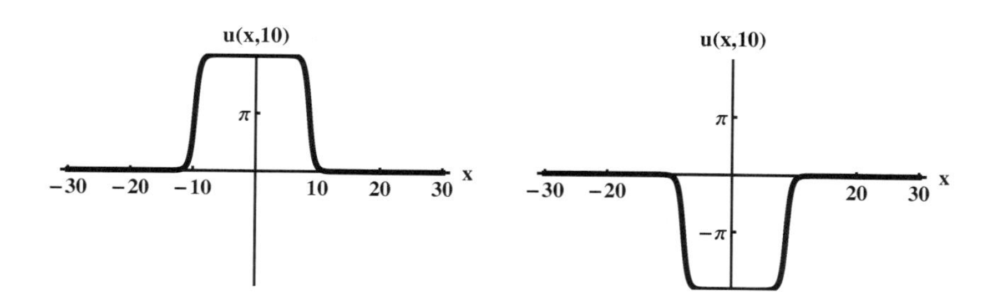

Fig. 2.3. Two-soliton solutions of the sine–Gordon equation: the kink-antikink and antikink-kink interactions

```
params={a1->2,a2->1,C[1]->0}
Do[g21[i_,xN_,tN_]:=solFin2[i][[2]]/.params/.{x->xN,t->tN},
 {i,1,2}]; Do[g22[i_,xN_,tN_]:=solFinDx2[i][[2]]/.params/.{x->xN,
 t->tN},{i,1,2}]; Do[g23[i_,xN_,tN_]:=solFinDt2[i][[2]]/.params/.
 {x->xN,t->tN},{i,1,2}]; p=2*Pi;
Animate[Plot[g21[1,x,t],{x,-10,10},PlotRange->{-p,p}],{t,-10,10}]
Animate[Plot[g22[1,x,t],{x,-10,10},PlotRange->{-p,p}],{t,-10,10}]
Animate[Plot[g23[1,x,t],{x,-10,10},PlotRange->{-p,p}],{t,-10,10}]
```

Case 3. Setting $C\neq0$, $A_2=0$, $A_1>1$ ($A1=4$, $C=4$, $C_1=0$, $C_2=0$) we obtain the exact solutions that describe an antikink (or kink) and kink (or antikink) moving toward to the boundary, $u_x(0,t)=0$, and then reflecting from the boundary as a kink (or antikink) and antikink (or kink), respectively. In other words, we can say that these two-soliton solutions represent the interaction of an antikink with a kink or the interaction of a kink with an antikink, respectively. We verify that these solutions are, indeed, exact solutions and plot the corresponding animations of these solutions. We note that u_{1x} and u_{1t} do not have real values in this case. The two-soliton solutions that represent the kink-antikink and antikink-kink interactions are shown in Fig. 2.3.

Mathematica:

```
consts={C[1]->0,C[2]->0}; params1={a2->0,c->2}; params2=a1->4;
{eqs3=eqs/.params1, sols1=(DSolve[eqs3[[1]],phi[x],x]/.consts//
 Flatten)/.params2, sols2=(DSolve[eqs3[[2]],psi[t],t]/.consts//
 Flatten)/.params2, sol3[1]=sols1[[1]], sol3[2]=sols1[[2]],
 sol3[3]=sols2[[1]], sol3[4]=sols2[[2]], solFin30=tr1[[2]]/.tr2}
solFin3[1]=u[x,t]->solFin30/.sol3[1]/.sol3[3]//FullSimplify
```

```
solFin3[2]=u[x,t]->solFin30/.sol3[2]/.sol3[4]//FullSimplify
p3=2*Pi; xR=10*Pi; tR=10;
Map[FullSimplify,{pde1[u]/.solFin3[1]/.trD[solFin3[1],x]/.
 trD[solFin3[1],t], pde1[u]/.solFin3[2]/.
 trD[solFin3[2],x]/.trD[solFin3[2],t]}]
g31[xN_,tN_]:=solFin3[1][[2]]/.{x->xN,t->tN};
g32[xN_,tN_]:=solFin3[2][[2]]/.{x->xN,t->tN};
Animate[Plot[g31[x,t],{x,-xR,xR},PlotRange->{-p3,p3}],{t,-tR,tR}]
Animate[Plot[g32[x,t],{x,-xR,xR},PlotRange->{-p3,p3}],{t,-tR,tR}]
```

Case 4. Setting $C\neq 0$, $A_2=0$, $0<A_1<1$, $A_1=1-\omega^2$, $\omega=\sqrt{1-0.9}$, $C=4$, $C_1=0$, $C_2=0$, we obtain the following special solutions:

$$u_{1,2}(x,t)=\pm 4\arctan\left(\frac{(2k^2(-1+e^{2xk}\sqrt{e^{-4xk}}))e^{xk}|\cos(\omega t)|\sin(\omega t)}{(-4e^{2xk}+4\omega^2e^{2xk}-1)\omega\cos(\omega t)}\right),$$

where we denote $k=\sqrt{1-\omega^2}$. These solutions are called the breather solutions of the sine–Gordon equation and describe a pulse-type structure of a hump-shaped antisoliton or soliton, respectively. Therefore, in this case the solutions are periodic functions of time t with frequency ω and for fixed x. We plot the corresponding animations of these solutions for the frequency $\omega=\sqrt{1-0.9}$. If we consider the case $\omega\approx 1$, e.g., $\omega=\sqrt{1-0.1}$, then the solutions obtained describe the small-amplitude breather solutions.

Maple:

```
interface(showassumed=0):assume(omega,'real',omega>0,t,'real');
omega1:=sqrt(1-0.9); HG:=array(1..2): xR:=-20..20; tR:=-150..150;
Params:=[C=4,A2=0]; Consts:=[_C1=0,_C2=0];
Eqs4:=subs(A1=1-omega^2,Params,Eqs); Sols:=[dsolve(Eqs4)];
N1:=nops(Sols);for i from 1 to N1 do Sols[i]; od;
Sol41:=subs(Consts,[op(op(1,Sols[N1]))]); Sol42:=subs(Consts,
 [op(op(2,Sols[N1]))]); SolFin40:=subs(U(x,t)=tr2,tr1);
for i from 1 to 2 do for j from 1 to 2 do
 SolFin4||i:=u(x,t)=subs(Sol41[2],Sol42[j],SolFin40); od; od;
for i from 1 to 2 do SolFin4||i; od;
for i from 1 to 2 do Z||i:=subs(omega=omega1,omega^2=omega1^2,
 subs(omega*t=OmegaT,rhs(SolFin4||i))); od;
for i  to 2 do HG[i]:=animate(evalf(Z||i),x=xR,OmegaT=tR,
 scaling=unconstrained,numpoints=70): od:
display(HG); for i to 2 do convert(SolFin4||i,abs); od;
```

□

Problem 2.33 *Nonlinear n-th order equation. Generalized separation of variables. Simplified approach.* Let us consider a generalization of the boundary layer equation [124], i.e., the nonlinear n-th order equation

$$u_t u_{xt} - u_x u_{tt} = F(x) u_t^{(n)},$$

where $\{x \in \mathbb{R}, t \geq 0\}$ and $F(x)$ is an arbitrary function. Applying the method of generalized separation of variables and a simplified approach, i.e., specifying one of the systems of coordinate functions $\{\psi_i(t)\}$ and searching for exact solutions in a more simplified form, for example $u(x,t) = \phi_1(x) e^{\lambda t} + \phi_2(x)$, determine the exact solution of the nonlinear equation.

1. According to the method of generalized separation of variables, we arrive at the following ODE $\phi_{2x} \lambda^2 + F(x) \lambda^n = 0$, whose solution is $\phi_2 = -\lambda^{n-2} \int F(x)\, dx + C_1$.

2. Finally, the exact solution of the given nonlinear PDE of the desired simplified form is $u(x,t) = \phi_1 e^{\lambda t} - \lambda^{n-2} \int F(x)\, dx + C_1$, where C_1, λ are arbitrary constants and $\phi_1(x)$ is an arbitrary function.

Maple:

```
with(PDEtools): declare((u,W)(x,t),(phi1,phi2)(x));
interface(showassumed=0): assume(n,'integer',n>0);
assume(lambda,'integer',lambda>0);
tr1:=phi1(x)*exp(lambda*t)+phi2(x);
PDE1:=u->diff(u(x,t),t)*diff(u(x,t),x,t)-diff(u(x,t),x)
 *diff(u(x,t),t$2)-F(x)*diff(u(x,t),t$n)=0;
Eq2:=expand(PDE1(W)); Eq3:=expand(subs(W(x,t)=tr1,Eq2));
Eq4:=factor(Eq3); ODE1:=select(has,lhs(Eq4),F(x))=0;
Solphi2:=simplify(dsolve(ODE1,phi2(x)));
SolFin1:=u(x,t)=subs(Solphi2,tr1); pdetest(SolFin1,PDE1(u));
```

Mathematica:

```
tr1=w[x,t]->phi1[x]*Exp[lambda*t]+phi2[x];
trD[u_,var_]:=Table[D[u,{var,i}],{i,1,10}]//Flatten;
trDM[u_,var1_,var2_]:=D[u,var1,var2]//Flatten;
trS[eq_,var_]:=Select[eq,MemberQ[#,var,Infinity]&];
pde1[u_]:=D[u[x,t],t]*D[u[x,t],x,t]-D[u[x,t],x]*D[u[x,t],
 {t,2}]-f[x]*D[u[x,t],{t,10}]==0;
eq2=Expand[pde1[w]]
```

```
{eq3=Expand[eq2/.tr1/.trD[tr1,t]/.trD[tr1,x]/.
 trDM[tr1,x,t]/.{10->n}], eq4=Factor[eq3]}
{ode1=trS[eq4[[1]],f[x]]==0, solphi2=Simplify[DSolve[ode1,
 phi2[x],x]]//First, solFin1=u[x,t]->tr1[[2]]/.solphi2}
pde1[u]/.solFin1/.trD[solFin1,t]/.trD[solFin1,x]/.
 trDM[solFin1,x,t]/.{10->n}//FullSimplify
```

□

Problem 2.34 *Nonlinear n-th order equation. Generalized separation of variables. Simplified approach.* Let us consider a generalization of the nonlinear PDE arising in hydrodynamics [124], i.e., the nonlinear n-th order equation

$$u_{xt}+(u_x)^2-uu_{xx}=F(t)u_x^{(n)},$$

where $\{x\in\mathbb{R}, t\geq 0\}$ and $F(t)$ is an arbitrary function. Applying the method of generalized separation of variables and a simplified approach, i.e., specifying one of the systems of coordinate functions $\{\phi_i(x)\}$ and searching for exact solutions in a more simplified form, for example $u(x,t)=\psi_1(t)e^{\lambda x}+\psi_2(t)$, determine the exact solution of the nonlinear PDE.

1. According to the method of generalized separation of variables, we arrive at the equation $\psi_{1t}\lambda-\psi_1\lambda^2\psi_2-\lambda^n\psi_1 F(t)=0$ with respect to $\psi_2(t)$, whose solution is $\psi_2(t)=\psi_{1t}/(\psi_1\lambda)-\lambda^n F(t)/\lambda^2$.

2. Finally, the exact solution of the given nonlinear PDE of the desired simplified form is $u(x,t)=\psi_1 e^{\lambda x}+\psi_{1t}/(\psi_1\lambda)-\lambda^n F(t)/\lambda^2$, where λ is an arbitrary constant and $\psi_1(t)$ is an arbitrary function.

Maple:

```
with(PDEtools): declare((u,W)(x,t),(psi1,psi2)(t));
interface(showassumed=0): assume(n,'integer',n>0);
assume(lambda,'integer',lambda>0);
tr1:=psi1(t)*exp(lambda*x)+psi2(t);
PDE1:=u->diff(u(x,t),x,t)+(diff(u(x,t),x))^2-u(x,t)
        *diff(u(x,t),x$2)-F(t)*diff(u(x,t),x$n)=0;
Eq2:=expand(PDE1(W)); Eq3:=expand(subs(W(x,t)=tr1,Eq2));
Eq4:=factor(Eq3); Solpsi2:=expand(isolate(Eq4,psi2(t)));
SolFin1:=u(x,t)=subs(Solpsi2,tr1); pdetest(SolFin1,PDE1(u));
```

Mathematica:

```
tr1=w[x,t]->psi1[t]*Exp[lambda*x]+psi2[t];
trD[u_,var_]:=Table[D[u,{var,i}],{i,1,10}]//Flatten;
trDM[u_,var1_,var2_]:=D[u,var1,var2]//Flatten;
pde1[u_]:=D[u[x,t],x,t]+(D[u[x,t],x])^2-u[x,t]*D[u[x,t],
 {x,2}]-f[t]*D[u[x,t],{x,10}]==0; {eq2=Expand[pde1[w]],
 eq3=Expand[eq2/.tr1/.trD[tr1,t]/.trD[tr1,x]/.
 trDM[tr1,x,t]/.{10->n}], eq4=Factor[eq3]}
{solpsi2=Expand[Solve[eq4,psi2[t]]]//First, solFin1=u[x,t]->
 tr1[[2]]/.solpsi2, pde1[u]/.solFin1/.trD[solFin1,t]/.
 trD[solFin1,x]/.trDM[solFin1,x,t]/.{10->n}//FullSimplify}
```
□

2.6.4 Functional Separation of Variables

For nonlinear PDEs, it is of great interest to find functional separable solutions, i.e., exact solutions of the form

$$u(x,t)=F(z) \text{ or } F(u(x,t))=z, \quad \text{where } z=\sum_{i=1}^{n}\phi_i(x)\psi_i(t),$$

this solutions can be found if there exist the functions $F(z)$ (or $F(u)$), $\phi_i(x)$, and $\psi_i(t)$ $(i=1,\dots,n)$. In recent years, many approaches have been proposed for studying the functional separation of variables, e.g., by specifying some functions $\phi_i(x)$ or $\psi_i(t)$ and finding special functional separable solutions, by differentiating and splitting, by extending the Lie theory, and other approaches. We note that the classical additive separable solutions and multiplicative separable solutions and the generalized separable solutions are particular cases of the above functional separable solution if $u(x,t)=F(z)=z$ (or $F(u(x,t))=u(x,t)=z$). It has been shown by A. D. Polyanin and V. F. Zaitsev [124] that a great number of nonlinear PDEs of various types admit the functional separable solutions. We will find functional separable solutions to various problems by differentiating and splitting and applying the simplified approach.

Problem 2.35 *Nonlinear parabolic equation. Functional separation of variables. Simplified approach.* Let us now consider a generalization of the nonlinear second-order parabolic equation, considered in Problem 2.31, i.e., the equation

$$u_t=a(t)u_{xx}+b(t)u_x+c(t)F(u),$$

where $\{x\in\mathbb{R}, t\geq 0\}$, $F(u)$, $a(t)$, $b(t)$, and $c(t)$ are arbitrary functions. Applying the method of functional separation of variables, $u(x,t)=W(z)$, and a simplified approach, i.e., searching for exact solutions in a more simplified form, e.g., $z=x\psi_1(t)+\psi_2(t)$, determine the exact solution of the given PDE.

1. ODE. According to the method of functional separation of variables, we arrive at the following ODE:

$$-\psi_{1t}\frac{z}{\psi_1}+\psi_{1t}\frac{\psi_2}{\psi_1}-\psi_{2t}+\frac{a(t)W_{zz}}{W_z}\psi_1^2+b(t)\psi_1+\frac{c(t)F(W(z))}{W_z}=0.$$

Maple:

```
with(PDEtools): declare(psi1(t),psi2(t),W(z));
interface(showassumed=0); assume(z>0);
for i from 1 to 4 do assume(A||i,constant); od:
 tr1:=psi1(t)*x+psi2(t)=z; tr2:=expand(isolate(tr1,x));
PDE1:=u->diff(u,t)=a(t)*diff(u,x$2)+b(t)*diff(u,x)+c(t)*F(u);
Eq2:=expand(PDE1(W(lhs(tr1)))); Eq3:=convert(
     algsubs(tr1,Eq2),diff); Eq4:=expand(Eq3/diff(W(z),z));
Eq5:=expand(subs(tr2,Eq4)); Eq51:=-Eq5+rhs(Eq5);
```

Mathematica:

```
{tr1=psi1[t]*x+psi2[t]==z, tr2=Expand[Solve[tr1,x]]//First}
pde1[u_]:=D[u,t]==a[t]*D[u,{x,2}]+b[t]*D[u,x]+c[t]*f[u];
{eq2=pde1[w[tr1[[1]]]]//Expand, eq3=eq2/.ToRules[tr1]}
{eq4=Expand[Thread[eq3/w'[z],Equal]], eq5=Expand[eq4/.tr2]}
eq51=Thread[Thread[(-1)*eq5,Equal]+eq5[[2]],Equal]
```

2. Functional equation. Then, applying the splitting procedure, i.e rewriting this equation in the form of the functional equation, we obtain $\sum_{i=1}^{4}\Phi_i(x)\Psi_i(t)=0$, where $\Phi_1=1$, $\Psi_1=\frac{\psi_{1t}\psi_2}{\psi_1}-\psi_{2t}+b(t)\psi_1$, $\Phi_2=z$, $\Psi_2=-\frac{\psi_{1t}}{\psi_1}$, $\Phi_3=\frac{W_{zz}}{W_z}$, $\Psi_3=a(t)\psi_1^2$, $\Phi_4=\frac{F(W(z))}{W_z}$, $\Psi_4=c(t)$. Knowing the solutions of this functional equation, e.g., we arrive at the following system of ODEs:

$$\frac{\psi_{1t}\psi_2}{\psi_1}-\psi_{2t}+b(t)\psi_1=A_1a(t)\psi_1^2+A_2c(t), \quad -\frac{\psi_{1t}}{\psi_1}=A_3a(t)\psi_1^2+A_4c(t),$$
$$\frac{W_{zz}}{W_z}=-A_1-A_3z, \quad \frac{F(W(z))}{W_z}=-A_2-A_4z,$$

where A_i $(i=1,\dots,4)$ are arbitrary constants. Integrating these equations, we can find the unknown functions $\psi_1(t)$, $\psi_2(t)$, $W(z)$, $F(W(z))$ with four integrating constants, B_i $(i=1,\dots,4)$, and three arbitrary functions, $a(t)$, $b(t)$, $c(t)$, and the exact solution takes the following form: $u(x,t)=W(x\psi_1(t)+\psi_2(t))$.

Maple:

```
FunDiffEq1:=add(Phi[i]*Psi[i],i=1..4)=0; SolFunDiffEq1:=
 [Psi[1]=A1*Psi[3]+A2*Psi[4], Psi[2]=A3*Psi[3]+A4*Psi[4],
 Phi[3]=-A1*Phi[1]-A3*Phi[2], Phi[4]=-A2*Phi[1]-A4*Phi[2]];
L11:=[selectremove(has,op(2,lhs(Eq51)),[psi1,psi2])];
L12:=[selectremove(has,op(3,lhs(Eq51)),[-1,psi2])];
L13:=[selectremove(has,op(5,lhs(Eq51)),[1])];
L2:=[selectremove(has,op(1,lhs(Eq51)),[-1,psi1])];
L3:=[selectremove(has,op(4,lhs(Eq51)),[a(t),psi1])];
L4:=[selectremove(has,op(6,lhs(Eq51)),[c(t)])];
tr3:=[Phi[1]=L11[2]*L12[2]*L13[1],Psi[1]=L11[1]+L12[1]+L13[2],
 Phi[2]=L2[2],Psi[2]=L2[1],Phi[3]=L3[2],Psi[3]=L3[1],
 Phi[4]=L4[2],Psi[4]=L4[1]]; sys1:=subs(tr3,SolFunDiffEq1);
S2:=subs(_C1=B1,[dsolve(sys1[2],psi1(t))]); S2[1];
S1[1]:=simplify(subs(S2[1],subs(_C1=B2,map(expand,dsolve(
       sys1[1],psi2(t)))))); S1[2]:=simplify(subs(S2[2],
       subs(_C1=B2,map(expand,dsolve(sys1[1],psi2(t))))));
Eq3sys1:=int(lhs(sys1[3]),z)=int(rhs(sys1[3]),z);
Eq3sys11:=isolate(Eq3sys1,diff(W(z),z));
S3:=int(lhs(Eq3sys11),z)=B3*Int(rhs(Eq3sys11),z)+B4;
Eq4sys1:=expand(subs(S3,(subs(F(W(z))=F1,sys1[4]))));
S4:=F(W(z))=rhs(simplify(isolate(Eq4sys1,F1)));
Sol1:=[S1[1],S2[1],S3,S4]; Sol2:=[S1[2],S2[2],S3,S4];
```

Mathematica:

```
funDiffEq1=Sum[phi[i]*psi[i],{i,1,4}]==0;
trS1[eq_,var_]:=Select[eq,MemberQ[#,var,Infinity]&];
trS2[eq_,var1_,var2_]:=Select[eq,MemberQ[#,var1,Infinity] &&
 MemberQ[#,var2,Infinity]&]; trS3[eq_,var_]:=Select[eq,
 FreeQ[#,var]&]; solFunDiffEq1={psi[1]==a1*psi[3]+a2*psi[4],
 psi[2]==a3*psi[3]+a4*psi[4],phi[3]==-a1*phi[1]-a3*phi[2],
 phi[4]==-a2*phi[1]-a4*phi[2]}
{l110=trS2[eq51[[1]],psi1[t],psi2[t]],l11={l110,trS3[l110,t]}}
{l120=trS1[eq51[[1]],psi2'[t]], l12={l120,-trS3[l120,t]}}
{l130=trS1[eq51[[1]],b[t]], l13={trS3[l130,t],l130}}
```

```
{l20=trS2[eq51[[1]],psi1'[t],z],l2={trS3[l20,z],-trS3[l20,t]}}
{l30=trS2[eq51[[1]],a[t],z], l3={trS3[l30,z],trS3[l30,t]}}
{l40=trS1[eq51[[1]],c[t]], l4={trS3[l40,z],trS3[l40,t]}}
tr3={phi[1]->l11[[2]]*l12[[2]]*l13[[1]], psi[1]->l11[[1]]+
 l12[[1]]+l13[[2]],phi[2]->l2[[2]], psi[2]->l2[[1]], phi[3]->
 l3[[2]],psi[3]->l3[[1]], phi[4]->l4[[2]],psi[4]->l4[[1]]}
{sys1=solFunDiffEq1/.tr3, s2=DSolve[sys1[[2]],psi1[t],t]/.
 C[1]->b1, s1[1]=DSolve[sys1[[1]],psi2[t],t]/.C[1]->b2/.
 s2[[1]]//Simplify//First, s1[2]=DSolve[sys1[[1]],psi2[t],t]/.
 C[1]->b2/.s2[[2]]//Simplify//First}
eq3sys1=Integrate[sys1[[3,1]],z]==Integrate[sys1[[3,2]],z]
eq3sys11=Solve[eq3sys1,w'[z]]//First
{s3=Integrate[eq3sys11[[1,1]],z]->b3*Hold[Integrate[
 eq3sys11[[1,2]],z]]+b4, eq4sys1=Expand[sys1[[4]]/.f[w[z]]->
 f1/.s3/.D[s3,z]]//ReleaseHold}
s4=f[w[z]]->(Solve[eq4sys1,f1]//First)[[1,2]]//Simplify
Print["sol1="]; sol1={s1[1],s3,s4}//Flatten
Print["sol2="]; sol2={s1[2],s3,s4}//Flatten
```

3. Special cases. Let us consider some special cases.

(a) Setting $A_1{=}-1$, $A_2{\neq}0$, $A_3{=}0$, $A_4{\neq}0$, $B_1{\neq}0$, $B_2{\neq}0$, $B_3{=}1$, $B_4{=}0$, $a(t){=}1$, $b(t){=}0$, $c(t){=}1$, we obtain that the exact solutions will coincide with the solutions obtained for the nonlinear diffusion equation $u_t{=}u_{xx}{+}F(u)$.

(b) Setting $A_1{=}-1$, $A_2{\neq}0$, $A_3{=}0$, $A_4{\neq}0$, $B_1{\neq}0$, $B_2{\neq}0$, $B_3{=}1$, $B_4{=}0$, $a(t){=}1$, $b(t){=}1$, $c(t){=}1$, we obtain that the exact solutions take the form

$$u(x,t) = W(z) = e^z = e^{x\psi_1(t)+\psi_2(t)}, \; F(W) = -W\Big(A_2 + A_4\ln(W)\Big),$$
$$\psi_1(t) = \pm\frac{q}{B_1}, \; \psi_2(t) = -\frac{1}{2}\frac{1 \pm \ln kq + 2k + 2A_2B_1 \mp 2B_2A_4q}{B_1A_4},$$

where $k{=}e^{-2A_4t}$, $q{=}\sqrt{B_1k}$.

Maple:

```
Consts1:=[A1=-1,A3=0,B3=1,B4=0,a(t)=1,b(t)=0,c(t)=1];
Sol11:=factor(value(subs(Consts1,Sol1)));
Sol12:=factor(value(subs(Consts1,Sol2)));
tr41:=expand(ln(rhs(Sol11[3])))=combine(ln(lhs(Sol11[3])));
F1:=F(W(z))=factor(expand(subs(tr41,rhs(Sol11[4]))));
Sol11:=subsop(4=F1,Sol11); Sol12:=subsop(4=F1,Sol12);
```

```
Consts2:=[A1=-1,A3=0,B3=1,B4=0,a(t)=1,b(t)=1,c(t)=1];
Sol21:=simplify(value(subs(Consts2,Sol1)));
Sol22:=simplify(value(subs(Consts2,Sol2)));
tr42:=isolate(subs(W(z)=W,Sol21[3]),z);
F2:=F(W)=factor(expand(subs(tr42,rhs(Sol21[4]))));
Sol21:=simplify(subsop(4=F2,Sol21),symbolic);
Sol22:=simplify(subsop(4=F2,Sol22),symbolic);
```

Mathematica:

```
{consts1={a1->-1,a3->0,b3->1,b4->0,a[K[1]]->1,b[K[1]]->0,
 c[K[1]]->1}, solpsi11=s2[[1]]/.consts1, solpsi12=s2[[2]]/.
 consts1, sol11=Factor[sol1/.consts1/.psi1[K[1]]->
 solpsi11[[1,2]]]//FullSimplify, sol12=Factor[sol2/.consts1/.
 psi1[K[1]]->solpsi11[[1,2]]]//FullSimplify}
{tr41=Log[sol11[[2,2]]]==Simplify[Log[sol11[[2,1]]]]//
 PowerExpand, f1=sol11[[3,2]]/.ToRules[tr41]}
sol111={sol11/.{sol11[[3,2]]->f1},solpsi11}//Flatten
sol112={sol12/.{sol12[[3,2]]->f1},solpsi12}//Flatten
{consts2={a1->-1,a3->0,b3->1,b4->0,a[K[1]]->1,b[K[1]]->1,
 c[K[1]]->1}, sol21=Factor[sol1/.consts2/.psi1[K[1]]->
 solpsi11[[1,2]]]//FullSimplify, sol22=Factor[sol2/.consts2/.
 psi1[K[1]]->solpsi11[[1,2]]]//FullSimplify}
{tr42=Log[sol21[[2,2]]]==Simplify[Log[sol21[[2,1]]]]//
 PowerExpand, f2=sol21[[3,2]]/.ToRules[tr42]}
sol211={sol21/.{sol21[[3,2]]->f2},solpsi11}//Flatten
sol212={sol22/.{sol22[[3,2]]->f2},solpsi12}//Flatten
sol21[[2]]=w[z]->Integrate[eq3sys11[[1,2]]/.consts1,z]; sol21
sol22[[2]]=w[z]->Integrate[eq3sys11[[1,2]]/.consts2,z]; sol22
```
□

Problem 2.36 *Nonlinear heat equation. Functional separation of variables. Simplified approach.* Let us consider a generalization of the nonlinear heat equation considered in Problem 2.27, i.e., the equation of the form

$$u_t{=}(G(u)u_x)_x{+}F(u),$$

where $\{x \in \mathbb{R}, t \geq 0\}$, and $F(u)$, $G(u)$ are arbitrary functions. Applying the method of functional separation of variables, $u(x,t){=}W(z)$, and searching for exact solutions in the form, $z{=}x\psi_1(t){+}\psi_2(t)$, determine the exact solution of the given equation.

1. System of ODEs. According to the method of functional separation of variables, we can obtain the ODE, which can be solved by splitting (see the previous problem). Knowing the solutions of the functional equation obtained, we can arrive at the system of ODEs.

Maple:

```
with(PDEtools): declare(psi1(t),psi2(t),W(z));
interface(showassumed=0); assume(z>0);
for i from 1 to 4 do assume(A||i,constant); od:
tr1:=psi1(t)*x+psi2(t)=z; tr2:=expand(isolate(tr1,x));
PDE1:=u->diff(u,t)=diff(G(u)*diff(u,x),x)+F(u);
Eq2:=expand(PDE1(W(lhs(tr1)))); Eq3:=algsubs(tr1,Eq2);
Eq4:=expand(Eq3/diff(W(z),z)); Eq5:=expand(subs(tr2,Eq4));
Eq51:=map(convert,Eq5,diff); Eq52:=-Eq51+rhs(Eq51);
```

Mathematica:

```
{tr1=psi1[t]*x+psi2[t]==z, tr2=Expand[Solve[tr1,x]]//First}
pde1[u_]:=D[u,t]==D[g[u]*D[u,x],x]+f[u];
{eq2=pde1[w[tr1[[1]]]]//Expand, eq3=eq2/.ToRules[tr1]}
{eq4=Expand[Thread[eq3/w'[z],Equal]], eq5=Expand[eq4/.tr2]}
eq51=Thread[Thread[(-1)*eq5,Equal]+eq5[[2]],Equal]
```

2. Solution of the system. Integrating these equations, we can find the unknown functions $\psi_1(t)$ and $\psi_2(t)$. One of the functions $F(W(z))$ and $G(W(z))$ can be defined as an arbitrary function and the other is determined in the solution process. Finding the function $W(z)$, we arrive at the exact solution $u(x,t)=W(x\psi_1(t)+\psi_2(t))$.

Maple:

```
FunDiffEq1:=add(Phi[i]*Psi[i],i=1..4)=0; SolFunDiffEq1:=
 [Psi[1]=A1*Psi[3]+A2*Psi[4], Psi[2]=A3*Psi[3]+A4*Psi[4],
  Phi[3]=-A1*Phi[1]-A3*Phi[2], Phi[4]=-A2*Phi[1]-A4*Phi[2]];
L11:=[selectremove(has,op(2,lhs(Eq52)),[psi1,psi2])];
L12:=[selectremove(has,op(3,lhs(Eq52)),[-1,psi2])];
L2:=[selectremove(has,op(1,lhs(Eq52)),[-1,psi1])];
L31:=[selectremove(has,op(4,lhs(Eq52)),[psi1])];
L32:=[selectremove(has,op(5,lhs(Eq52)),[psi1])];
L4:=[selectremove(has,op(6,lhs(Eq52)),[1])];
tr3:=[Phi[1]=L11[2]*L12[2],Psi[1]=L11[1]+L12[1],
 Phi[2]=L2[2],Psi[2]=L2[1],Phi[3]=L31[2]+L32[2],
 Psi[3]=L31[1],Phi[4]=L4[2],Psi[4]=L4[1]];
tr31:=subs(rhs(tr3[5])=expand(diff(G(W(z))*diff(W(z),z),z)
      /diff(W(z),z)),tr3); sys0:=subs(tr31,SolFunDiffEq1);
```

Mathematica:

```
funDiffEq1=Sum[phi[i]*psi[i],{i,1,4}]==0; trS1[eq_,var_]:=
 Select[eq,MemberQ[#,var,Infinity]&]; trS2[eq_,var1_,var2_]:=
 Select[eq,MemberQ[#,var1,Infinity]&&MemberQ[#,var2,Infinity]&];
trS3[eq_,var_]:=Select[eq,FreeQ[#,var]&]; solFunDiffEq1=
{psi[1]==a1*psi[3]+a2*psi[4],psi[2]==a3*psi[3]+a4*psi[4],
 phi[3]==-a1*phi[1]-a3*phi[2],phi[4]==-a2*phi[1]-a4*phi[2]}
{l110=trS2[eq51[[1]],psi1[t],psi2[t]], l11={l110,trS3[l110,t]},
 l120=trS1[eq51[[1]],psi2'[t]], l12={l120,-trS3[l120,t]},
 l20=trS2[eq51[[1]],psi1'[t],z], l2={trS3[l20,z],-trS3[l20,t]},
 l310=trS2[eq51[[1]],psi1[t],g'[w[z]]], l31={trS3[l310,z],
 trS3[l310,t]}, l320=trS2[eq51[[1]],psi1[t],w''[z]]}
{l32={trS3[l320,z],trS3[l320,t]}, l40=trS1[eq51[[1]],f[w[z]]],
 l4={trS3[l40,z],trS3[l40,t]}}
tr3={phi[1]->l11[[2]]*l12[[2]], psi[1]->l11[[1]]+l12[[1]],
 phi[2]->l2[[2]], psi[2]->l2[[1]], phi[3]->l31[[2]]+l32[[2]],
 psi[3]->l31[[1]], phi[4]->l4[[2]], psi[4]->l4[[1]]}
sys0=solFunDiffEq1/.tr3
```

3. Special case. Setting $A_1{=}-1$, $A_2{\neq}0$, $A_3{=}0$, $A_4{\neq}0$, $B_1{\neq}0$, $B_2{\neq}0$, $B_3{=}1$, $B_4{=}0$, $G(W(z)){=}1$, we obtain the following explicit form of the exact solution

$$u(x,t) = W(z) = B_1 + B_2 e^z = B_1 + B_2 e^{x\psi_1(t)+\psi_2(t)},$$

$$F(W(z)){=}\left(-A_2{-}A_4 \ln\left(\frac{W{-}B_1}{B_2}\right)\right)(W{-}B_1),\ \psi_1(t){=}B_1 e^{-A_4 t},$$

$$\psi_2(t) = -B_1 e^{-A_4 t}\left(\frac{B_1}{A_4}e^{-A_4 t} + \frac{A_2}{B_1 A_4}e^{A_4 t}\right) + B_1 e^{-A_4 t} B_3.$$

Maple:

```
Consts1:=[A1=-1,A3=0,B3=1,B4=0]; sys1:=subs(Consts1,sys0);
tr4:=G(W(z))=1; Eq3sys1:=subs(tr4,(D(G))(W(z))=0,sys1[3]);
S3:=subs(_C1=B1,_C2=B2,dsolve(Eq3sys1,W(z)));
Eq4sys1:=algsubs(S3,subs(F(W(z))=F(W),sys1[4]));
S4:=isolate(Eq4sys1,F(W)); tr5:=expand(isolate(
    subs(W(z)=W,S3),z)); F1:=combine(subs(tr5,S4));
S2:=subs(_C1=B1,dsolve(sys1[2],psi1(t))); S2;
S1:=value(subs(S2,subs(_C1=B3,map(expand,
  dsolve(sys1[1],psi2(t)))))); Sol1:=[S1,S2,S3,F1,tr4];
```

Mathematica:

```
Off[Solve::ifun]; consts1={a1->-1,a3->0,b3->1,b4->0}
{sys1=sys0/.consts1, tr4=g[w[z]]->1,
 eq3sys1=sys1[[3]]/.tr4/.g'[w[z]]->0,
 s3=DSolve[eq3sys1,w[z],z]/.C[1]->b2/.C[2]->b1//First}
eq4sys1=sys1[[4]]/.f[w[z]]->f[w]/.s3/.D[s3,z]
{s4=Solve[eq4sys1,f[w]], tr5=Expand[Solve[s3/.w[z]->w1/.
 Rule->Equal,z]]//First, f1=s4/.tr5//First}
s2=DSolve[sys1[[2]],psi1[t],t]/.C[1]->b1//First
{s1=DSolve[sys1[[1]],psi2[t],t]/.C[1]->b3/.psi1[K[1]]->
 s2[[1,2]]/.s2//FullSimplify, sol1={s1,s2,s3,f1,tr4}//Flatten}
```

□

Problem 2.37 *Nonlinear m-th order equation. Functional separation of variables. Simplified approach.* Let us consider the nonlinear m-th order partial differential equation [124]

$$u_y u_{xy} - u_x u_{yy} = F(x)(u_{yy})^{n-1} u_y^{(m)},$$

where $\{x \in \mathbb{R}, y \in \mathbb{R}\}$ and $F(x)$ is an arbitrary function. If $F(x)$=const and m=3, the nonlinear equation describes a boundary layer of a power-law fluid on a flat plate. Applying the method of functional separation of variables, $u(x,t)=W(z)$, and searching for exact solutions in the form $z=\psi_1(x)y+\psi_2(x)$, obtain the exact solution of the nonlinear PDE.

1. According to the method of functional separation of variables, we arrive at the ODE:

$$W_z^2 \psi_{1x} = W_{zz}^{n-1} \psi_1^{-3+2n+m} f(x) W_z^{(m)},$$

which is independent of $\psi_2(x)$.

2. Separating the variables and integrating, we find the exact solution:

$$u(x,y) = W(z) = W(\psi_1(x)y + \psi_2(x)),$$
$$\psi_1(x) = \left(\int f(x)\, dx + B_1\right)^{1/(4-2n-m)}, \quad \psi_2(x) \text{ is arbitrary},$$

where the function $W(z)$ is described by the ordinary differential equation $(4-2n-m)W_{zz}^{n-1}W_z^{(m)}=W_z^2$.

Maple:

```
with(PDEtools):declare((psi1,psi2)(x));interface(showassumed=0);
assume(n,'integer',n>0,m,'integer',m>0,C,'integer',C>0):
tr1:=psi1(x)*y+psi2(x)=z; tr2:=expand(isolate(tr1,y));
PDE1:=u->diff(u,y)*diff(u,x,y)-diff(u,x)*diff(u,y$2)=f(x)*
 diff(u,y$2)^(n-1)*diff(u,y$m); Eq2:=expand(PDE1(W(lhs(tr1))));
Eq21:=lhs(Eq2)=rhs(Eq2)*psi1(x)^m; Eq3:=subs(y=z,subs(tr1,Eq21));
Eq4:=expand(Eq3/psi1(x)); Eq51:=combine(map(convert,Eq4,diff));
Eq61:=[selectremove(has,lhs(Eq51),[psi1])];
Eq62:=[selectremove(has,rhs(Eq51),[psi1,f])];
Eq7:=expand(Eq51/Eq61[2]/Eq62[1]);
Eq8:=subs(_C1=B1,factor(combine(dsolve(lhs(Eq7)=C,psi1(x)))));
Eq9:=combine(rhs(Eq7)=C); Eq10:=simplify(Eq9*denom(lhs(Eq9))/C);
EqC:=[selectremove(has,op(1,op(1,op(1,rhs(Eq8)))),[C,n,m])];
trC:=isolate(EqC[1]=EqC[2]/f(x),C); Eqpsi1:=subs(trC,Eq8);
EqW:=subs(trC,Eq10);
```

Mathematica:

```
trS1[eq_,var_]:=Select[eq,MemberQ[#,var,Infinity]&];
{tr1=psi1[x]*y+psi2[x]==z, tr11=psi1[x]*z+psi2[x]->z}
tr2=Expand[Solve[tr1,y]]//First
pde1[u_]:=D[u,y]*D[u,x,y]-D[u,x]*D[u,{y,2}]==f[x]*D[u,
 {y,2}]^(n-1)*D[u,{y,m}]; {eq2=pde1[w[tr1[[1]]]]//Expand,
 eq21=eq2[[1]]==eq2[[2]]*psi1[x]^m, eq3=eq21/.y->z/.tr11}
{eq4=Expand[Thread[eq3/psi1[x],Equal]], eq51=eq4//PowerExpand//
 Together, eq61={trS1[eq51[[1]],x],trS1[eq51[[1]],z]},
 eq62={trS1[eq51[[2]],x],trS1[eq51[[2]],z]}}
eq7=Expand[Thread[eq51/eq61[[2]]/eq62[[1]],Equal]]
eq8=DSolve[eq7[[1]]==c,psi1[x],x]/.C[1]->b1//First
{eq9=eq7[[2]]==c//Together, eq10=Thread[eq9*Denominator[
 eq9[[1]]]/c,Equal], eq80=Level[eq8,{3}], eqC={c*eq80[[2,1]],
 -f[x]}, trC=Solve[eqC[[1]]==eqC[[2]]/f[x],c], eqpsi1=eq8/.
 trC//First, eqw=Thread[(eq10/.trC//First)/w''[z],Equal]}
```

□

Sometimes it is possible to find exact solutions applying the method of functional separation of variables and a simplified approach, i.e., looking for exact solutions in the form $u(x,t)=W(z)$, where the function $W(z)$ is chosen (this function can be given to reduce the original equation to an equation with a power nonlinearity), usually the function $W(z)$ can be defined as $W(z)=e^{\lambda z}$, $W(z)=z^{\lambda}$, or $W(z)=\lambda \ln z$. The equation obtained (with a power nonlinearity) can be solved by the method of generalized separation of variables and a simplified approach, i.e., looking for exact solutions in the form $z=\sum_{i=1}^{n}\phi_i(x)\psi_i(t)$ and specifying some functions $\phi_i(x)$ or $\psi_i(t)$ (for details, see [60]).

Problem 2.38 *Nonlinear heat equation. Functional separation of variables. Simplified approach.* Let us consider the nonlinear heat equation with a logarithmic nonlinearity [124]

$$u_t=au_{xx}+F(t)u\ln u+G(t)u,$$

where $\{x\in\mathbb{R}, t\geq 0\}$ and $F(t)$, $G(t)$ are arbitrary functions. Applying the approach described above, i.e., specifying $u(x,t)=W(z)=e^z$, determine the exact solution of this nonlinear equation.

1. According to this approach, we arrive at the following ODE: $z_t-az_{xx}-az_x^2-F(t)z-G(t)=0$.

2. This equation has a quadratic nonlinearity and can be solved by the method of generalized separation of variables. Specifying the form $z=x^2\psi_1(t)+x\psi_2(t)+\psi_3(t)$, we can determine the unknown functions $\psi_i(t)$ $(i=1,2,3)$.

Maple:

```
with(PDEtools): declare(psi1(t),psi2(t),psi3(t));
interface(showassumed=0);  assume(z(x,t)>0):
PDE1:=u->diff(u,t)-a*diff(u,x$2)-F(t)*u*ln(u)-G(t)*u=0;
Eq2:=combine(PDE1(exp(z(x,t)))); Eq3:=expand(Eq2/exp(z(x,t)));
tr1:=x^2*psi1(t)+x*psi2(t)+psi3(t); PDE2:=unapply(Eq3,z);
Eq4:=PDE2(z); Eq5:=expand(subs(z(x,t)=tr1,Eq4));
Eq6:=collect(Eq5,[x,a]);
EDOs:=[select(has,op(1,lhs(Eq6)),t)=0, select(has,op(2,
      lhs(Eq6)),t)=0,convert([op(3..6,lhs(Eq6))],`+`)=0];
dsolve(EDOs,{psi1(t),psi2(t),psi3(t)});
```

Mathematica:

```
pde1[u_]:=D[u,t]-a*D[u,{x,2}]-f[t]*u*Log[u]-g[t]*u==0;
trD[u_,var_]:=Table[D[u,{var,i}],{i,1,2}]//Flatten;
trS1[eq_,var_]:=Select[eq,MemberQ[#,var,Infinity]&];
{eq2=pde1[Exp[z[x,t]]]//Together, eq3=Expand[Thread[eq2/Exp[
 z[x,t]],Equal]]//PowerExpand}
```

```
tr1=z[x,t]->x^2*psi1[t]+x*psi2[t]+psi3[t]
{eq4=eq3/.tr1/.trD[tr1,t]/.trD[tr1,x], eq5=Collect[eq4,{x,a}]}
odes={Coefficient[eq5[[1]],x^2]==0,Coefficient[eq5[[1]],x]==0,
 Coefficient[eq5[[1]],x,0]==0}
{DSolve[odes[[1]],psi1[t],t], DSolve[odes[[2]],psi2[t],t],
 DSolve[odes[[3]],psi3[t],t]}//Flatten
```

□

Sometimes applying the method of functional separation of variables, we can arrive at a more complicated functional-differential equation with three variables. In some cases, it is possible to reduce by differentiation this complicated equation to a standard functional-differential equation with two variables (where one of variables is eliminated).

Problem 2.39 *Nonlinear heat equation. Functional separation of variables. Differentiation approach.* Let us consider the nonlinear heat equation

$$u_t=(F(u)u_x)_x,$$

where $\{x \in \mathbb{R}, t \geq 0\}$ and $F(u)$ is an arbitrary function. Applying the method of functional separation of variables, and looking for solutions in the form $u(x,t)=W(z)$, $z=\phi(x)+\psi(t)$, find the exact solution of the nonlinear heat equation.

1. Systems of ODEs. According to the method of functional separation of variables, we arrive at the functional-differential equation with three variables

$$\psi_t=F(W)\phi_{xx}+H(z)\phi_x^2, \quad \text{where} \quad H(z)=F(W)_z+F(W)\frac{W_{zz}}{W_z}.$$

Differentiating this equation with respect to x, we obtain the functional-differential equation with two variables

$$F(W)\phi_{xxx}+H_z\phi_x^3+\big(F(W)_z+2H(z)\big)\phi_x\phi_{xx}=0,$$

which can be solved by splitting procedure. This equation has two solutions that can be determined by solving the following systems of ODEs:

$$\phi_x\phi_{xx}=A_2\phi_x^3, \ \phi_{xxx}=A_1\phi_x^3, \ H_z=-A_1F-A_2\big(F_z+2H\big),$$
$$F_z+2H=A_1F, \ H_z=A_2F, \ \phi_{xxx}=-A_1\phi_x\phi_{xx}-A_2\phi_x^3.$$

Maple:

```
with(PDEtools): declare(phi(x),psi(t)); tr1:=phi(x)+psi(t)=z;
PDE1:=u->diff(u,t)=diff(F(u)*diff(u,x),x);
phi1:=diff(phi(x),x)^2; Eq2:=expand(PDE1(W(lhs(tr1))));
Eq3:=subs(tr1,Eq2); Eq4:=expand(Eq3/diff(W(z),z));
trH:=H(z)=expand(convert([op(1..2,rhs(Eq4))],`+`)/phi1);
trH1:=H(z)=map(convert,op(1,rhs(trH)),diff)
         +map(convert,op(2,rhs(trH)),diff);
trH2:=H(z)=subs(F(W(z))=F1,subsop(1=diff(F1(z),z),
      rhs(trH1))); trH3:=algsubs(F1=F1(z),trH2);
Eq5:=lhs(Eq4)=op(3,rhs(Eq4))+lhs(trH1)*phi1; Eq6:=convert(
     simplify(Eq5),diff); Eq7:=subs(z=z(x),Eq6);
Eq8:=diff(Eq7,x); Eq9:=subs(diff(z(x),x)=diff(lhs(tr1),x),Eq8);
Eq10:=convert(subs(W(z(x))=W,z(x)=z,Eq9),diff);
termPhi:=select(has,op(1,rhs(Eq10)),phi);
Eq11:=subsop(1=diff(F1(z),z)*termPhi,rhs(Eq10))=0;
Eq12:=subs(F(W)=F1(W),Eq11);
FunDiffEq1:=add(Phi[i]*Psi[i],i=1..3)=0;
SolFun1:=[Psi[1]=A2*Psi[2],Psi[3]=A1*Psi[2],Phi[2]=-A1*Phi[3]
 -A2*Phi[1]]; SolFun2:=[Phi[1]=A1*Phi[3],Phi[2]=A2*Phi[3],
  Psi[3]=-A1*Psi[1]-A2*Psi[2]];
L11:=[selectremove(has,op(1,lhs(Eq12)),[phi])];
L12:=[selectremove(has,op(4,lhs(Eq12)),[phi])];
L2:=[selectremove(has,op(3,lhs(Eq12)),[phi])];
L3:=[selectremove(has,op(2,lhs(Eq12)),[phi])];
tr3:=[Phi[1]=L11[2]+L12[2],Phi[2]=L2[2],Phi[3]=L3[2],
 Psi[1]=L11[1], Psi[2]=L2[1],Psi[3]=L3[1]];
sys01:=subs(tr3,SolFun1); sys02:=subs(tr3,SolFun2);
```

Mathematica:

```
trS1[eq_,var_]:=Select[eq,MemberQ[#,var,Infinity]&];
trS3[eq_,var_]:=Select[eq,FreeQ[#,var]&];
pde1[u_]:=D[u,t]==D[(f[u]*D[u,x]),x]; {tr1=phi[x]+psi[t]==z,
 phi1=(phi'[x])^2, eq2=pde1[w[tr1[[1]]]]//Expand,
 eq3=eq2/.ToRules[tr1], eq4=Expand[Thread[eq3/w'[z],Equal]]}
trH=h[z]->Expand[(eq4[[2,1]]+eq4[[2,3]])/phi1]
trH2=trH/.f[w[z]]->f1[z]/.trH[[2,1]]->f1'[z]
{eq5=eq4[[1]]==eq4[[2,2]]+trH2[[1]]*phi1, eq6=eq5/.z->z[x],
 eq7=Thread[D[eq6,x],Equal], eq8=eq7/.z'[x]->D[tr1[[1]],x]}
{eq9=eq8/.w[z[x]]->w/.z[x]->z, termphi=trS1[trS1[eq9[[2]],
 h[z]],x], termf=trS1[eq9[[2]],f'[w]]}
```

```
{eq10=eq9/.termf->f1'[z]*termphi, eq12=eq10/.f[w]->f1[w]}
funDiffEq1=Sum[phi[i]*psi[i],{i,1,3}]==0
{solFun1={psi[1]==a2*psi[2], psi[3]==a1*psi[2], phi[2]==
 -a1*phi[3]-a2*phi[1]}, solFun2={phi[1]==a1*phi[3], phi[2]==
 a2*phi[3], psi[3]==-a1*psi[1]-a2*psi[2]}}
{l110=trS1[eq12[[2]],f1'[z]], l11={trS3[l110,z],trS3[l110,x]},
 l120=trS1[eq12[[2]],h[z]], l12={trS3[l120,z]/2,trS3[l120,x]},
 l20=trS1[eq12[[2]],h'[z]], l2={trS3[l20,z],trS3[l20,x]},
 l30=trS1[eq12[[2]],f1[w]], l3={trS3[l30,w],trS3[l30,x]}}
tr3={phi[1]->l11[[2]]+l12[[2]],phi[2]->l2[[2]],phi[3]->l3[[2]],
 psi[1]->l11[[1]],psi[2]->l2[[1]],psi[3]->l3[[1]]}
{sys01=solFun1/.tr3, sys02=solFun2/.tr3}
```

2. Solution of one system. Special case. Without full analysis, we solve the first system of ODEs for one special case, $A_1{=}A_2{=}0$. In this case we find $\phi{=}B_1x{+}B_2$. We can observe that the function $F(W)$ is arbitrary, so we choose the function $F(z){=}e^z$ and can determine the functions $H(z){=}B_1$ and $W(z){=}B_2{+}e^{-e^{-z}B_1}B_3$. Then, eliminating z, we obtain $F(W){=}-\dfrac{B_1}{\ln(-(-W{+}B_2)/B_3)}$.

Substituting all our results into the original functional-differential equation with three variables, we arrive at the equation for $\psi(t)$, $\psi_t{=}B_1^3$. Its general solution is given by $\psi(t){=}B_1^3t{+}B_4$, where B_1 and B_4 are arbitrary constants.

Finally, we can obtain (by specifying $B_2{=}0$, $B_4{=}0$, $B_1^3{=}B_2$) the traveling wave solution, $W{=}W(z){=}W(B_1x{+}B_2t)$, of the nonlinear heat equation.

Maple:

```
Consts1:={A1=0,A2=0}; sys11:=subs(Consts1,{sys01[1],sys01[2]});
Solphi1:=subs(_C1=B1,_C2=B2,op(op(dsolve(sys11,phi(x))[2])));
trF:=F1(z)=exp(z); Eq3sys01:=algsubs(trF,subs(F1(W)=F1(z),
     sys01[3])); SolH:=subs(_C1=B1,dsolve(Eq3sys01,H(z)));
SolH1:=eval(SolH,Consts1); EqW:=subs(SolH1,algsubs(trF,trH3));
SolW:=simplify(value(subs(Consts1,_C1=B2,_C2=B3,dsolve(EqW,
      W(z))))); sysFW:=subs(W(z)=W1,F1(z)=F2,{trF,SolW});
SolF1:=eliminate(sysFW,z); SolF2:=F(W)=subs(W1=W,solve(
       op(SolF1[2]),F2));
Eqpsi:=eval(subs(Solphi1,algsubs(z=lhs(tr1),subs(SolH,algsubs(
       Solphi1,Eq6)))),Consts1); Solpsi1:=subs(_C1=B4,
       dsolve(Eqpsi,psi(t)));
z=subs(B2=0,B4=0,B1^3=B2,rhs(Solphi1+Solpsi1));
```

Mathematica:

```
Off[Solve::ifun]; trD[u_,var_]:=Table[D[u,{var,i}],{i,1,2}]//
 Flatten; {consts1={a1->0,a2->0}, sys11={sys01[[1]],sys01[[2]]}/.
 consts1, solphi1=(DSolve[sys11[[1]],phi[x],x]/.C[2]->b1/.
 C[1]->b2)[[2]], trF=f1[z]->Exp[z], eq3sys01=sys01[[3]]/.trF/.
 f1[w]->f1[z], solH=DSolve[eq3sys01,h[z],z]/.C[1]->b1//First,
 solH1=solH/.consts1, eqW=trH2/.solH1/.trF/.D[trF,z],
 solW=DSolve[eqW/.Rule->Equal,w[z],z]/.C[2]->b2/.C[1]->b3/.
 consts1//First, sysFW={trF,solW}/.w[z]->w1/.f1[z]->f2//Flatten}
{solF1=Reduce[sysFW/.Rule->Equal,z], termw1=trS1[solF1,b3]}
solF2=f[w]->Solve[termw1,f2][[1,1,2]]/.w1->w
eqpsi=eq5/.solH/.z->tr1[[1]]/.solphi1/.trD[solphi1,x]/.consts1
solpsi1=DSolve[eqpsi,psi[t],t]/.C[1]->b4//First
solFin=tr1/.solphi1/.solpsi1/.b2->0/.b4->0/.b1^3->b2
```

□

2.7 Transformation Groups

Continuous groups (or Lie groups) have been introduced by Sophus Lie in 1881, for unifying and extending various methods for solving ordinary differential equations. The approach proposed by Lie is based on finding symmetries of differential equations (see Sect. 1.1.3), i.e., transformation groups that depend on continuous parameters and consist of point or contact transformations (see Sect. 2.1–2.2), for example translations, rotations, and scalings are Lie groups (see Sect. 2.1.1). Lie also established that for linear PDEs, invariance under a Lie group leads to superpositions of solutions in terms of transforms.

In this section, we will consider the *classical method* of finding symmetries of nonlinear PDEs (also known as the *Lie group analysis of differential equations* or the *classical method of symmetry reductions*) that allow us to obtain *transformation groups*, under which PDEs are *invariant*, and *new variables* (independent and dependent), with respect to which differential equations become more simpler. The transformations can convert a solution of a nonlinear PDE to the same or another solution of this equation, the invariant solutions can be found by symmetry reductions, rewriting the equation in new variables. For a class of the nonlinear second-order PDEs of general form in two independent variables, we will perform the group analysis (i.e., describe the procedure of finding symmetries of this class of PDEs) and describe the procedure of constructing invariant solutions (using symmetries obtained with the group analysis).

It should be noted that the classical method of symmetry analysis of nonlinear PDEs is related to the similarity method (described in

Sect. 2.5.3). Birkhoff [18] was the first to discover that the Boltzmann method for solving the diffusion equation is based on the algebraic symmetry of the equation, and special solutions, called *similarity solutions* of this equation, can be obtained by solving an associated ODE. Moreover, on the basis of the similarity method it is possible to find a group of transformations under which a given PDE is invariant.

2.7.1 One-Parameter Groups of Transformations

Let us introduce a *one-parameter set of transformations* in the (x, t, u)-space defined by

$$T_a : \mathbb{R}^3 \to \mathbb{R}^3 \ \forall a \in I, \quad T_a = \left\{ x = \frac{X}{a^\alpha}, t = \frac{T}{a^\beta}, u(x,t) = \frac{U(X,T)}{a^\zeta} \right\}, \quad (2.3)$$

under which the partial differential equation is invariant. Here $a \in \mathbb{R}$ is a parameter which belongs to an open interval I containing the value $a=1$, and α, β, ζ are constants. The set of all transformations $\{T_a\}$ form a *Lie group* on $\mathbb{R}^3$ with the identity element T_1, i.e., $\{T_a\}$ obeys the following laws:

$$T_a T_b = T_{ab}, \ T_{ab} = T_{ba}, \ T_a(T_b T_c) = (T_a T_b)T_c, \ \forall a, b \in I,$$
$$T_1 T_a = T_a, \ T_a T_{a^{-1}} = T_{a^{-1}} T_a, \ \forall a \neq 0.$$

Problem 2.40 *Linear and nonlinear heat equations. One-parameter group of transformations.* Let us consider the classical linear heat equation and the nonlinear heat equation

$$u_t = k u_{xx}, \quad u_t = (u u_x)_x,$$

where $\{x \in \mathbb{R}, t \geq 0\}$. Verify that the set of all transformations $\{T_a\}$ form a Lie group on $\mathbb{R}^3$ with the identity element T_1 and prove that the linear heat equation is invariant under the set of transformations $T_a = \{X = a^\alpha x, T = a^{2\alpha} t, U = a^\zeta u\}$ and the nonlinear heat equation is invariant under the set of transformations $T_a = \{X = a^\alpha x, T = a^\beta t, U = a^{2\alpha-\beta} u\}$, where α and β are arbitrary constants.

1. Let us verify that the set of all transformations $\{T_a\}$ form a *Lie group* on $\mathbb{R}^3$ with the identity element T_1.

Maple:

```
with(PDEtools): declare((u,U)(x,t)); interface(showassumed=0):
assume(alpha>0,beta>0,zeta>0,a>0,b>0,c>0,k>0);
T:=(a)->[a^alpha,a^beta,a^zeta]; T(a); T(b);
```

```
for i from 1 to 3 do
 expand(T(a)[i]*T(b)[i]*x=T(a*b)[i]*x);
 expand(T(a)[i]*(T(b)[i]*T(c)[i])*x=T(a*b*c)[i]*x);
 expand(T(1)[i]*T(a)[i]*x=T(1*a)[i]*x);
 simplify(T(a)[i]*T(a^(-1))[i]*x=T(a*a^(-1))[i]*x)=T(1)[i]*x; od;
```

Mathematica:

```
tT[a_]:={a^alpha,a^beta,a^zeta}; tT[a]; tT[b];
Map[PowerExpand, {tT[a]*tT[b]*x==tT[a*b]*x,
  tT[a]*(tT[b]*tT[c])*x==tT[a*b*c]*x, tT[1]*tT[a]*x==tT[1*a]*x,
  tT[a]*tT[a^(-1)]*x==tT[a*a^(-1)]*x==tT[1]*x}]
```

2. We show that the classical linear heat equation is invariant under the set of transformations $T_a = \{X = a^\alpha x, T = a^{2\alpha} t, U = a^\zeta u\}$ and the nonlinear heat equation is invariant under the set of transformations $T_a = \{X = a^\alpha x, T = a^\beta t, U = a^{2\alpha-\beta} u\}$, where α and β are arbitrary constants.

Maple:

```
PDEL1:=diff(u(x,t),t)-k*diff(u(x,t),x$2)=0; tr1:={x=X/a^alpha,
 t=T/a^beta,u(x,t)=U(X,T)/a^zeta}; PDEL2:=combine(dchange(tr1,
 PDEL1,[X,T,U])); Eq1:=map(lhs,PDEL1=PDEL2);
termL1:=select(has,select(has,select(has,rhs(Eq1),a),beta),a);
termL2:=expand(rhs(Eq1)/termL1); Eq2:=select(has,select(has,
 termL2,a),a)=a^0; tr2:=isolate(Eq2,beta); tr3:=subs(tr2,tr1);
PDEL3:=dchange(tr3,PDEL1,[X,T,U]); termL3:=select(has,op(1,
 lhs(PDEL3)),a); PDEL4:=expand(PDEL3/termL3); PDEL1=PDEL4;
PDENL1:=diff(u(x,t),t)-k*diff(u(x,t)*diff(u(x,t),x),x)=0;
PDENL2:=combine(dchange(tr1,PDENL1,[X,T,U]));
Eq1:=map(lhs,PDENL1=PDENL2);
termNL1:=select(has,select(has,select(has,rhs(Eq1),a),beta),a);
termNL2:=expand(rhs(Eq1)/termNL1);
Eq2:=select(has,select(has,select(has,op(1,termNL2),a),a),a)=a^0;
tr4:=simplify(isolate(Eq2,zeta)); tr5:=subs(tr4,tr1);
PDENL3:=expand(dchange(tr5,PDENL1,[X,T,U]));
termNL3:=select(has,op(1,lhs(PDENL3)),a);
PDENL4:=collect(expand(PDENL3/termNL3),k); PDENL1=PDENL4;
```

Mathematica:

```
Off[Solve::ifun]; pdeL1[x_,t_]:=D[u[x,t],t]-k*D[u[x,t],{x,2}]==0;
trS1[eq_,var_]:=Select[eq,MemberQ[#,var,Infinity]&];
trS3[eq_,var_]:=Select[eq,FreeQ[#,var]&]; pdeL1[x,t]
{tr11={x->xN/a^alpha,t->tN/a^beta},tr12=u[x,t]->uN[xN,tN]/a^zeta}
pdeL2[v_]:=First[((pdeL1[x,t]/.u->Function[{x,t},
 u[a^alpha*x,a^beta*t]/a^zeta])/.tr11)/.{u->v}]==0; pdeL2[uN]
{eq1=Map[First,pdeL1[x,t]==pdeL2[uN]], termL1=trS3[trS1[eq1[[2]],
 beta],xN], termL2=Thread[eq1[[2]]/termL1,Equal]//Simplify}
{eq2=trS1[trS1[termL2,a],a]==a^0, tr2=Solve[eq2,beta]//First,
 tr31=tr11/.tr2, Solve[tr31/.Rule->Equal,{xN,tN}]}
pdeL3[v_]:=First[((pdeL1[x,t]/.u->Function[{x,t},u[a^alpha*x,
 a^(2*alpha)*t]/a^zeta])/.tr31)/.{u->v}]==0; pdeL3[uN]
{termL3=trS1[trS3[pdeL3[uN][[1]],k],a], pdeL4=Thread[
 pdeL3[uN]/termL3,Equal]//Expand, pdeL1[x,t]==pdeL4}
pdeNL1[x_,t_]:=D[u[x,t],t]-k*D[u[x,t]*D[u[x,t],x],x]==0;
pdeNL2[v_]:=First[((pdeNL1[x,t]/.u->Function[{x,t},
 u[a^alpha*x,a^beta*t]/a^zeta])/.tr11)/.{u->v}]==0; pdeNL2[uN]
{eq1NL=Map[First,pdeNL1[x,t]==pdeNL2[uN]], termNL1=trS3[trS1[
 eq1NL[[2]],beta],xN], termNL2=Thread[eq1NL[[2]]/termNL1,Equal]//
 Expand, eq2=-trS3[trS3[termNL2[[2]],k],xN]==a^0}
{tr4=Solve[eq2,zeta]//First, tr5=tr11/.tr4, tr51=tr12/.tr4}
{Solve[tr5/.Rule->Equal,{xN,tN}], Solve[tr51/.Rule->Equal,
 uN[xN,tN]]}
pdeNL3[v_]:=First[((pdeNL1[x,t]/.u->Function[{x,t},u[a^alpha*x,
 a^beta*t]/a^(2*alpha-beta)])/.tr11)/.{u->v}]==0; pdeNL3[uN]
{termNL3=trS3[trS3[pdeNL3[uN][[1]],k],xN], pdeNL4=Collect[
 Expand[Thread[pdeNL3[uN]/termNL3,Equal]],k]}
pdeNL1[x,t]==pdeNL4
```
□

Problem 2.41 *Family of KdV equations. One-parameter group of transformations.* Let us consider one of the principal model equations for free surface water waves, the *KdV equation*

$$u_t+6uu_x+u_{xxx}=0,$$

and the *modified KdV equation* (mKdV), the *KdV-type equation*, and the *modified KdV-type equation*

$$u_t+6u^2u_x+u_{xxx}=0, \quad u_t+uu_x+u_{xxx}=0, \quad u_t+u^pu_x+u_{xxx}=0,$$

where $\{x \in \mathbb{R}, t \geq 0\}$. Introducing a one-parameter set of transformations (2.3), find the values of α and β for which the considered above equations are invariant under the transformations T_a. As we noted above, the set of all transformations $\{T_a\}$ form an infinite *Lie group* with parameter a on $\mathbb{R}^3$ with the identity element T_1. Prove that the KdV, KdV-type, mKdV, and mKdV-type equations, are invariant, respectively, under the group of transformations:

$$\begin{aligned}
T_a &= \{x = X\sqrt{a}, \quad t = Ta^{3/2}, \quad u(x,t) = U(X,T)/a\}, \\
T_a &= \{x = X\sqrt{a}, \quad t = Ta^{3/2}, \quad u(x,t) = U(X,T)/a\}, \\
T_a &= \{x = Xa, \quad t = Ta^3, \quad u(x,t) = U(X,T)/a\}, \\
T_a &= \{x = X/a^{-(1/2)p}, \quad t = T/a^{-(3/2)p}, \quad u(x,t) = U(X,T)/a\}.
\end{aligned}$$

Maple:

```
with(PDEtools): declare(u(x,t),U(X,T));interface(showassumed=0):
assume(a>0,p>0); W:=diff_table(u(x,t));
KdV:=W[t]+6*W[]*W[x]+W[x,x,x]=0; KdVT:=W[t]+W[]*W[x]+W[x,x,x]=0;
mKdV:=W[t]+6*W[]^2*W[x]+W[x,x,x]=0;
mKdVT:=W[t]+W[]^p*W[x]+W[x,x,x]=0;
InvKdV:=proc(PDE)
 local Eq1,Eq2,Eq3,Eq4,tr1,tr2,tr3,tr4,term1,term2,term3,
 term4,sys1; tr1:={x=X/a^alpha,t=T/a^beta,u(x,t)=U(X,T)/a^zeta};
 tr2:=eval(tr1,zeta=1); Eq1:=combine(dchange(tr2,PDE,[X,T,U]));
 Eq2:=map(lhs,PDE=Eq1);
 term1:=select(has,select(has,select(has,rhs(Eq2),a),beta),a);
 term2:=expand(rhs(Eq2)/term1);
 term3:=select(has,select(has,term2,a),a);
 sys1:={select(has,op(1,term3),a)=a^0,
        select(has,op(2,term3),a)=a^0};
 tr3:=solve(sys1,{alpha,beta}); tr4:=subs(tr3,tr2);
 print(tr3,tr4); Eq3:=dchange(tr4,PDE,[X,T,U]);
 term4:=select(has,op(1,lhs(Eq3)),a);
 Eq4:=expand(Eq3/term4); PDE=simplify(Eq4); end proc:
InvKdV(KdV);  InvKdV(KdVT); InvKdV(mKdV); InvKdV(mKdVT);
```

Mathematica:

```
kdV=D[u[x,t],t]+6*u[x,t]*D[u[x,t],x]+D[u[x,t],{x,3}]==0
kdVT=D[u[x,t],t]+u[x,t]*D[u[x,t],x]+D[u[x,t],{x,3}]==0
mKdV=D[u[x,t],t]+6*u[x,t]^2*D[u[x,t],x]+D[u[x,t],{x,3}]==0
mKdVT=D[u[x,t],t]+u[x,t]^p*D[u[x,t],x]+D[u[x,t],{x,3}]==0
Off[Solve::ifun];
```

```
invKdV[pde_]:=Module[{eq1,eq2,eq3,eq4,tr1,tr2,tr3,tr4,term1,
 term2,term3,term4,sys1},
 trS1[eq_,var_]:=Select[eq,MemberQ[#,var,Infinity]&];
 trS3[eq_,var_]:=Select[eq,FreeQ[#,var]&]; tr11={x->xN/a^alpha,
  t->tN/a^beta}; tr12=u[x,t]->uN[xN,tN]/a^zeta; tr2=tr12/.
  zeta->1; eq1[v_]:=First[((pde/.u->Function[{x,t},u[a^alpha*x,
  a^beta*t]/a])/.tr11)/.{u->v}]==0; eq2=Map[First,
  pde==eq1[uN]]; Print[eq2]; term1=trS1[trS1[eq2[[2]],beta],a];
 term2=Thread[eq2[[2]]/term1,Equal]//Expand; term3=trS1[term2,
  a]//PowerExpand; sys1={trS1[term3[[1]],a]==a^0,trS1[
  term3[[2]],a]==a^0}; tr3=Solve[sys1,{alpha,beta}]//First;
 tr41=tr11/.tr3; tr42=tr2/.tr3; Print[{tr3,tr41,tr42}//
  Flatten]; alpha1=tr3[[1,2]]; beta1=tr3[[2,2]];
 eq3[v_]:=First[((pde/.u->Function[{x,t},u[a^alpha1*x,
  a^beta1*t]/a])/.tr41)/.{u->v}]==0; term4=trS1[eq3[uN][[1,1]],
  a]; eq4=Thread[eq3[uN]/term4,Equal]//Expand;
 pde==PowerExpand[eq4]];
{invKdV[kdV], invKdV[kdVT], invKdV[mKdV], invKdV[mKdVT]}
```
□

Problem 2.42 *Nonlinear Schrödinger equation. One-parameter group of transformations.* Let us consider the nonlinear Schrödinger (NLS) equation

$$iu_t + u_{xx} + \gamma u|u|^2 = 0,$$

where u is a complex function of real variables x and t: $\{x \in \mathbb{R}, t > 0\}$, and $\gamma \in \mathbb{R}$ is a constant. Introducing the two *one-parameter sets of transformations* in the (x, t, u)-space defined by

$$T_a^1 : \mathbb{R}^3 \to \mathbb{R}^3 \ \forall a \in I, a \in \mathbb{R}, \ T_a^1 = \left\{ x = \frac{X}{a^\alpha}, t = \frac{T}{a^\beta}, u(x,t) = \frac{U(X,T)}{a^\zeta} \right\},$$

$$T_a^2 : \mathbb{R}^3 \to \mathbb{R}^3 \ \forall a \in I, a \in \mathbb{R}, \ T_a^2 = \left\{ x = X - a^\alpha, t = \frac{T}{a^\beta}, u(x,t) = \frac{U(X,T)}{a^\zeta} \right\},$$

(where $\alpha = 1$, and β, ζ are still arbitrary constants), find the values of β and ζ for which the NLS equation is invariant under the transformations T_a^1 and T_a^2. The set of all transformations $\{T_a\}$ form an infinite *Lie group* with parameter a on $\mathbb{R}^3$ with the identity element T_1. Prove that the NLS equation is invariant under the transformations:

$$T_a^1 = \{x = X/a, \quad t = T/a^2, \quad u(x,t) = U(X,T)a\},$$
$$T_a^2 = \{x = X - a, \quad t = T, \quad u(x,t) = U(X,T)\}.$$

Maple:

```
with(PDEtools): declare(u(x,t),U(X,T));interface(showassumed=0):
assume(a>0,zeta<0); W:=diff_table(u(x,t));
NLS:=I*W[t]+W[x,x]+gamma*W[]*abs(W[])^2=0;
tr1:={x=X/a^alpha,t=T/a^beta,u(x,t)=U(X,T)/a^zeta};
tr3:={x=X-a^alpha,t=T/a^beta,u(x,t)=U(X,T)/a^zeta};
tr2:=eval(tr1,alpha=1); tr4:=eval(tr3,alpha=1);
InvNLS:=proc(PDE,tr)
 local Eq1,Eq2,Eq3,Eq4,tr1,tr2,tr3,term1,term2,term3,term4,sys1;
 Eq1:=combine(dchange(tr,PDE,[X,T,U])); Eq2:=map(lhs,PDE=Eq1);
 term1:=select(has,select(has,select(has,rhs(Eq2),a),beta),a);
 term2:=expand(rhs(Eq2)/term1); term3:=select(has,select(has,
  term2,a),a); sys1:={select(has,op(1,term3),a)=a^0, select(has,
 op(2,term3),a)=a^0}; tr1:=solve(sys1,{beta,zeta}) assuming z<0;
 print(tr1); tr2:=subs(tr1,tr); print(tr2); Eq3:=dchange(tr2,
  PDE,[X,T,U]); term4:=select(has,op(1,lhs(Eq3)),a);
 Eq4:=expand(Eq3/term4); PDE=Eq4; end proc:
InvNLS(NLS,tr2); InvNLS(NLS,tr4);
```

Mathematica:

```
nLS=I*D[u[x,t],t]+D[u[x,t],{x,2}]+gamma*u[x,t]*Abs[u[x,t]]^2==0
Off[Solve::ifun]; invNLS[pde_]:=Module[{eq1,eq11,eq2,eq3,eq31,
 eq4,tr1,tr3,term1,term2,term3,term4,sys1,tr11,tr2,tr31,t1,t2},
 trS1[eq_,var_]:=Select[eq,MemberQ[#,var,Infinity]&];
 trS3[eq_,var_]:=Select[eq,FreeQ[#,var]&]; tr11={x->xN/a^alpha,
  t->tN/a^beta}/.alpha->1; tr2=u[x,t]->uN[xN,tN]/a^zeta;
 eq1[v_]:=First[((pde/.u->Function[{x,t},
 u[a*x,a^beta*t]/a^zeta])/.tr11)/.{u->v}]==0;
 eq11=Assuming[{a\[Element]Reals,a<0,zeta\[Element]Reals},
  ExpandAll[PowerExpand[FullSimplify[PowerExpand[eq1[uN]]]]]];
 eq2=Map[First,pde==eq11]; term1=trS1[trS1[eq2[[2]],beta],a];
 term2=Thread[eq2[[2]]/term1,Equal]//Expand; term3=trS1[term2,
  a]//PowerExpand; sys1={trS1[term3[[1]],a]==a^0,trS1[term3[[2]],
  a]==a^0}; tr3=Solve[sys1,{beta,zeta}]//First; tr41=tr11/.tr3;
 tr42=tr2/.tr3; Print[{tr3,tr41,tr42}//Flatten];
 beta1=tr3[[1,2]]; zeta1=tr3[[2,2]]; eq3[v_]:=First[((pde/.u->
 Function[{x,t},u[a*x,a^beta1*t]/a^zeta1])/.tr41)/.{u->v}]==0;
 eq31=Assuming[{a\[Element]Reals,a< 0,zeta\[Element]Reals},
  ExpandAll[PowerExpand[FullSimplify[PowerExpand[eq3[uN]]]]]];
 pde==PowerExpand[eq31]]; invNLS[nLS]
```

It should be noted that in order to prove that the NLS equation is invariant under the second transformation, it is possible to replace the three statements in this *Mathematica* solution by the following statements:

```
tr11={x->xN-a^alpha,t->tN/a^beta}/.alpha->1;
eq1[v_]:=First[((pde/.u->Function[{x,t},u[a+x,a^beta*t]/
 a^zeta])/.tr11)/.{u->v}]==0; eq3[v_]:=First[((pde/.u->
 Function[{x,t},u[a+x,a^beta1*t]/a^zeta1])/.tr41)/.{u->v}]==0;
```
□

2.7.2 Group Analysis

Sophus Lie showed that for a given differential equation (linear or nonlinear), the admitted continuous group of point transformations can be determined by an explicit computational algorithm (or the Lie group analysis, or the Lie algorithm). The procedure of finding symmetries of differential equations, i.e., finding transformations that preserve the form of the equation, consists of four steps. Let us describe this algorithm for a class of the nonlinear second-order PDEs in two independent variables of the general form (1.3)

$$\mathcal{F}(x, y, u, u_x, u_y, u_{xx}, u_{xy}, u_{yy})=0.$$

1. We introduce the *one-parameter set of transformations*:

$$T_\varepsilon : \mathbb{R}^3 \to \mathbb{R}^3 \quad \forall \varepsilon \in I, \quad \varepsilon \in \mathbb{R},$$
$$T_\varepsilon = \{X=\varphi_1(x, y, u, \varepsilon), Y=\varphi_2(x, y, u, \varepsilon), U=\psi(x, y, u, \varepsilon)\}, \tag{2.4}$$

where $X=\varphi_1(x, y, u, 0)=x$, $Y=\varphi_2(x, y, u, 0)=y$, and $U=\psi(x, y, u, 0)=u$. The original nonlinear PDE must be invariant under these transformations, i.e.,

$$\mathcal{F}(X, Y, U, U_X, U_Y, Y_{XX}, U_{XY}, U_{YY})=0.$$

We look for linear first-order differential operators, called the *infinitesimal operators* X, admitted by this equation and which correspond to the infinitesimal transformation, in the form

$$X=\xi_1(x, y, u)\partial_x+\xi_2(x, y, u)\partial_y+\eta(x, y, u)\partial_u,$$

where the coordinates $\xi_1=\xi_1(x, y, u)$, $\xi_2=\xi_2(x, y, u)$, $\eta=\eta(x, y, u)$ are still unknown and are to be determined in the presented analysis. Expanding this equation into a series in ε about the point $\varepsilon=0$ and retaining the terms to the first-order of ε, we obtain the *invariance condition*

$$X_2\mathcal{F}(x, y, u, u_x, u_y, u_{xx}, u_{xy}, u_{yy})\Big|_{\mathcal{F}=0}=0.$$

Here $X_2\mathcal{F}=\xi_1\mathcal{F}_x+\xi_2\mathcal{F}_y+\eta\mathcal{F}_u+\zeta_1\mathcal{F}_x+\zeta_2\mathcal{F}_y+\zeta_{11}\mathcal{F}_{xx}+\zeta_{12}\mathcal{F}_{xy}+\zeta_{22}\mathcal{F}_{yy}$ is the *second prolongation operator.* The coordinates ζ_1, ζ_2 of the first prolongation and the coordinates ζ_{11}, ζ_{12}, ζ_{22} of the second prolongation are defined by the formulas:

$$\zeta_1=D_x(\eta)-u_xD_x(\xi_1)-u_yD_x(\xi_2),\quad \zeta_2=D_y(\eta)-u_xD_y(\xi_1)-u_yD_y(\xi_2),$$
$$\zeta_{11}=D_x(\zeta_1)-u_{xx}D_x(\xi_1)-u_{xy}D_x(\xi_2),\quad \zeta_{12}=D_y(\zeta_1)-u_{xx}D_y(\xi_1)-u_{xy}D_y(\xi_2),$$
$$\zeta_{22}=D_y(\zeta_2)-u_{xy}D_y(\xi_1)-u_{yy}D_y(\xi_2),$$

where D_x and D_y are the total differential operators with respect to x and y. Then the derivative containing in the original equation (e.g., u_{xx} or u_{yy} for the equation $u_{xx}+u_{yy}-G(u)=0$) is eliminated from the invariance condition using the original nonlinear PDE.

2. The invariance condition can be represented as a polynomial of the form $\sum C_{ijkl}\, u_x^i u_y^j u_{xx}^k u_{xy}^l=0$ with respect to the "independent variables" u_x, u_y, u_{xx}, and u_{xy}, where C_{ijkl} are the functional coefficients depended on x, y, u, ξ_1, ξ_2, and η and the derivatives of ξ_1, ξ_2, and η. This polynomial equation is valid if all the coefficients C_{ijkl} are zero. So, the invariance condition is split to an overdetermined *determining system.*

3. The overdetermined *determining system* obtained can be solved and found the admissible coordinates ξ_1, ξ_2, and η of the infinitesimal operator.

4. It is possible to consider some special cases to verify if this *determining system* can admit others solutions that will result in other infinitesimal operators.

Problem 2.43 *Nonlinear Poisson equation. Group analysis.* Applying *Maple* functions, carry out the group analysis, i.e., perform step-by-step the described above procedure for the nonlinear Poisson equation

$$u_{xx}+u_{yy}-G(u)=0,$$

where $\{x\in\mathbb{R}, y\in\mathbb{R}\}$ and $G(u)$ is an arbitrary function (nonlinear source).

1. In this case we have $\mathcal{F}=u_{xx}+u_{yy}-G(u)$ and the invariance condition takes the form: $-\eta G_u+\zeta_{11}+\zeta_{22}=0$. Substituting the expressions for the coordinates of the second prolongation, ζ_{11} and ζ_{22}, and replacing the derivative u_{yy} by $G(u)-u_{xx}$, as a result we obtain the final form of the invariance condition:

$$
\begin{aligned}
&-2u_yu_x\xi_{1uy} - 2u_x\xi_{1u}u_{xx} - 2u_xu_y\xi_{2ux} - 2u_{xy}u_x\xi_{2u} + 2u_yu_{xx}\xi_{2u}\\
&-u_x^2u_y\xi_{2uu} - u_y^2u_x\xi_{1uu} - 3u_y\xi_{2u}G(u) - u_x\xi_{1u}G(u) + 2u_x\eta_{ux} + u_x^2\eta_{uu}\\
&-2u_{xx}\xi_{1x} - u_x\xi_{1xx} - 2u_x^2\xi_{1ux} - u_x^3\xi_{1uu} - 2u_{xy}\xi_{2x} - u_y\xi_{2xx} + 2u_y\eta_{uy}\\
&+u_y^2\eta_{uu} - 2u_{xy}\xi_{1y} - u_x\xi_{1yy} - u_y\xi_{2yy} - 2u_y^2\xi_{2uy} - u_y^3\xi_{2uu} - \eta G_u\\
&+\eta_uG(u) - 2\xi_{2y}G(u) + 2\xi_{2y}u_{xx} + \eta_{xx} + \eta_{yy} - 2u_{xy}u_y\xi_{1u}.
\end{aligned}
$$

Maple:

```
with(PDEtools): declare(u(x,y),(xi1,xi2,eta,zeta1,zeta2,
 H,Q)(x,y,u)); U:=diff_table(u(x,y)); DepVars:=u(x,y);
PDE1:=U[x,x]+U[y,y]-G(u)=0; F:=U[x,x]+U[y,y]-G(u);
UF,Uxi1,Uxi2,Ueta:=diff_table(F),diff_table(xi1(x,y,u)),
 diff_table(xi2(x,y,u)),diff_table(eta(x,y,u));
DT:=(Q,Z)->Q[Z]+U[Z]*Q[u];
DTM:=(Q,Z)->diff(Q,Z)+diff(u(x,y),Z)*diff(Q,u);
zeta1:=expand(DT(Ueta,x)-U[x]*DT(Uxi1,x)-U[y]*DT(Uxi2,x));
zeta2:=expand(DT(Ueta,y)-U[x]*DT(Uxi1,y)-U[y]*DT(Uxi2,y));
zeta11:=expand(map(DTM,zeta1,x)-U[x,x]*DT(Uxi1,x)-U[x,y]*
 DT(Uxi2,x)); zeta12:=expand(map(DTM,zeta1,y)-U[x,x]*DT(Uxi1,y)
 -U[x,y]*DT(Uxi2,y)); zeta22:=expand(map(DTM,zeta2,y)-U[x,y]*
 DT(Uxi1,y)-U[y,y]*DT(Uxi2,y));
tr1:={F_x=0,F_y=0,F_ux=0,F_uy=0,F_uxx=1,F_uxy=0,F_uyy=1};
tr2:={Zeta1=zeta1,Zeta2=zeta2,Zeta11=zeta11,Zeta12=zeta12,
 Zeta22=zeta22}; InvCond:=xi1*F_x+xi2*F_y+eta*UF[u]+Zeta1*F_ux+
 Zeta2*F_uy+Zeta11*F_uxx+Zeta12*F_uxy+Zeta22*F_uyy;
Eq1:=eval(InvCond,tr1); Eq2:=eval(Eq1,tr2);
Eq3:=expand(subs(U[y,y]=G(u)-U[x,x],Eq2));
```

2. Let us obtain the determining system. For the nonlinear Poisson equation, we obtain, respectively, the following combinations of the derivatives and the equations of the determining system:

$u_{xx}u_x, E_1: -2\xi_{1u}=0,\quad u_{xx}u_y, E_2: 2\xi_{2u}=0,\quad u_xu_y, E_3: -2\xi_{2ux}-2\xi_{1uy}=0,$
$u_{xx}, E_4: -2\xi_{1x}+2\xi_{2y}=0,\quad u_{xy}, E_5: -2\xi_{2x}-2\xi_{1y}=0,\quad u_x^2, E_6: \eta_{uu}-2\xi_{1ux}=0,$
$u_y^2, E_7: \eta_{uu}-2\xi_{2uy}=0,\quad u_x, E_8: -\xi_{1u}G(u)+2\eta_{ux}-\xi_{1xx}-\xi_{1yy}=0,$
$u_y, E_9: -3\xi_{2u}G(u)-\xi_{2xx}+2\eta_{uy}-\xi_{2yy}=0,\quad u_{xy}u_x, E_{10}: -2\xi_{2u}=0,$
$u_{xy}u_x, E_{11}: -2\xi_{2u}=0,\quad u_x^2u_y, E_{12}: -\xi_{2uu}=0,\quad u_y^2u_x, E_{13}: -\xi_{1uu}=0,$
$u_x^3, E_{14}: -\xi_{1uu}=0,\quad u_y^3, E_{15}: -\xi_{2uu}=0.$

Maple:

```
LD:=[U[x],U[y],U[x,y],U[x,x]]; LE1:=[[U[x,x],U[x]],[U[x,x],
 U[y]],[U[x],U[y]]]; LE2:=[U[x,x],U[x,y],U[x]^2,U[y]^2,U[x],
 U[y]]; E00:=sort(remove(has,Eq3,LD)=0); E0:=subsop(
 1=-eta(x,y,u)*diff(G(u),u),lhs(E00))=0; k1:=nops(LE1);
k2:=nops(LE2); S1:=NULL: for i from 1 to k1 do
 T||i:=expand(select(has,Eq3,LE1[i])/LE1[i][1]/LE1[i][2]):
 E||i:=remove(has,T||i,LD)=0; print(cat(E,i),LE1[i],E||i); od:
for i from 1 to k2 do
 T||(i+k1):=expand(select(has,Eq3,LE2[i])/LE2[i]):
 E||(i+k1):=remove(has,T||(i+k1),LD)=0;
 print(cat(E,i+k1),LE2[i],E||(i+k1));  od:
LE1D:=[[U[x,y],U[x]],[U[x,y],U[x]],[U[x]^2,U[y]],[U[y]^2,
 U[x]]]; LE2D:=[U[x]^3,U[y]^3]; k3:=nops(LE1D); k4:=nops(LE2D);
for i from 1 to k3 do
 T||(i+k1+k2):=expand(select(has,Eq3,LE1D[i])/LE1D[i][1]
   /LE1D[i][2]): E||(i+k1+k2):=remove(has,T||(i+k1+k2),LD)=0;
 print(cat(E,i+k1+k2),LE1D[i], E||(i+k1+k2)); od:
for i from 1 to k4 do
 T||(i+k1+k2+k3):=expand(select(has,Eq3,LE2D[i])/LE2D[i]):
 E||(i+k1+k2+k3):=remove(has,T||(i+k1+k2+k3),LD)=0;
 print(cat(E,i+k1+k2+k3),LE2D[i], E||(i+k1+k2+k3)); od:
for i from 1 to 1+k1+k2+k3+k4 do E||(i-1); S1:=S1,E||(i-1) od:
Sys1:=[S1];
```

3. Let us solve this overdetermined *determining system* and find the admissible coordinates ξ_1, ξ_2, and η of the infinitesimal operator. First, applying the *Maple* predefined function `pdsolve`, we can find the following expressions for the admissible coordinates: $\eta=0$, $\xi_1=-C_1y+C_3$, $\xi_2=C_1x+C_2$. Then, by setting one of the constants to unity and the others to zero, we can find that the original equation admits three different operators: $X_1=\partial_x, X_2=\partial_y, X_3=y\partial_x-x\partial_y$.

Then let us perform the procedure of finding these three operators step-by-step, analyzing each equation of the determining system. First, we solve the equations E_2, E_3, and E_7 with respect to ξ_1, ξ_2, η, and obtain that $\xi_1=\xi_1(x,y)$, $\xi_2=\xi_2(x,y)$, and $\eta=a(x,y)u+b(x,y)$. Then, from the equations E_5 and E_6 we obtain $\xi_{1xx}+\xi_{1yy}=0$ and $\xi_{2yy}+\xi_{2xx}=0$, and from the equations E_8 and E_9 and the previous relations we have that

$a(x, y)=A=\text{const}$. Thus, the determining system becomes

$$\begin{aligned}&-2\xi_{1x} + 2\xi_{2y} = 0, \quad -2\xi_{2x} - 2\xi_{1y} = 0,\\ &-2\xi_{2y}G(u) + b_{xx} - G_u uA + G(u)A - G_u b(x,y) + b_{yy} = 0. \end{aligned} \tag{2.5}$$

We can observe from this system that $A=b(x,y)=\eta_y=0$ for arbitrary function $G(u)$, and as a result, we obtain the same three operators.

Maple:

```
Sol0:=pdsolve(Sys1,{eta(x,y,u),xi1(x,y,u),xi2(x,y,u)});
L10:=subs(_C1=0,_C2=0,_C3=1,Sol0); L20:=subs(_C3=0,_C2=1,_C1=0,
 Sol0); L30:=subs(_C1=-1,_C2=0,_C3=0,Sol0);
X0:=f->xi1(x,y,u)*diff(f(x,y,u),x)+xi2(x,y,u)*diff(f(x,y,u),y)
 +eta(x,y,u)*diff(f(x,y,u),u);
X10:=eval(X0(f),L10); X20:=eval(X0(f),L20); X30:=eval(X0(f),L30);
A1:=pdsolve({Sys1[2],Sys1[3],Sys1[7]},{eta(x,y,u),xi1(x,y,u),
 xi2(x,y,u)}); A2:=convert(subs(_F1(x,y)=xi1(x,y),
 _F2(x,y)=xi2(x,y),_F3(x,y)=a(x,y),_F4(x,y)=b(x,y),A1),list);
A560:=[expand((diff(Sys1[5],x)+diff(Sys1[6],y))/(2)),
       expand((diff(Sys1[5],y)-diff(Sys1[6],x))/2)];
A56:=subs(xi1(x,y,u)=xi1(x,y),xi2(x,y,u)=xi2(x,y),A560);
A890:={Sys1[8],Sys1[9]};A891:=algsubs(A2[3],algsubs(A2[2],A890));
A892:=simplify(algsubs(A2[1],A891),{A56[1]});
A89:=subs(_C1=A,pdsolve(A892 union {diff((a(x,y),y))=0},a(x,y)));
Sys20:=convert(simplify(algsubs(A89[1],algsubs(A2[3],
 algsubs(A2[2],algsubs(A2[1],Sys1)))),convert(A56,set)),set);
Sys2:=convert(Sys20 minus {0=0},list); Sys21:=algsubs(diff(
 eta(x,y),y)=0,algsubs(b(x,y)=0,subs(A=0,Sys2)));
Sol1:=pdsolve(Sys21,{xi1(x,y),xi2(x,y)}) union {eta(x,y)=0};
L1:=subs(_C1=0,_C2=0,_C3=1,Sol1); L2:=subs(_C3=0,_C2=1,_C1=0,
 Sol1); L3:=subs(_C1=-1,_C2=0,_C3=0,Sol1);
X:=f->xi1(x,y)*diff(f(x,y,u),x)+xi2(x,y)*diff(f(x,y,u),y)+
 eta(x,y)*diff(f(x,y,u),u); X1:=eval(X(f),L1); X2:=eval(X(f),L2);
X3:=eval(X(f),L3);
```

4. We consider two *special cases* to verify if this determining system can admit others solutions that will result in other infinitesimal operators: $A\neq 0$ and $A=0$.

(a) If $A\neq 0$, $b(x, y)=B=\text{const}$, and $\xi_{2y}=M=\text{const}$, then the solution of the third equation of the system (2.5) gives $G(u)=C_1(uA+B)^{-2M/A+1}$. Thus, for a special case of the function $G(u)$, e.g., $G(u)=u^k$ (where

$C_1=1$, $B=0$, $A=1$, $-2M+1=k$), we can find another solution of the determining system $\xi_1(x,y)=x$, $\xi_2(x,y)=y$, $\eta(x,y)=-2u/(k-1)$, and the corresponding infinitesimal operator $X_{41}=x\partial_x+y\partial_y-\frac{2}{k-1}u\partial_u$.

(b) If $A=0$, $b(x,y)=B$=const, and $\xi_{2_y}=M$=const, then the solution of the third equation of the system (2.5) gives $G(u)=C_1e^{-2uM/B}$. Thus, for a special case of the function $G(u)$, e.g., $G(u)=e^u$ (where $C_1=1$, $B=-2$, $A=0$, $M=1$), we can find another solution of the determining system $\xi_1(x,y)=x$, $\xi_2(x,y)=y$, $\eta(x,y)=-2$, and the corresponding infinitesimal operator $X_{42}=x\partial_x+y\partial_y-2\partial_u$.

Maple:

```
Equat1:=algsubs(b(x,y)=3,Sys2[3]); Equat2:=simplify(dsolve(
 Equat1,G(u))); Equat3:=algsubs(diff(xi2(x,y),y)=M,Equat2);
Equat31:=subs({_C1=1,B=0,A=1},Equat3); trG:=subs(-2*M+1=k,
 Equat31); trA0:=A=solve(algsubs(trG,subs(B=0,algsubs(
 diff(xi2(x,y),y)=M,Equat1))),A); trM:=M=1;
trA:=simplify(subs(trM,trA0)); Solxi1:=subs(_F1(y)=0,
 pdsolve(subs(trM,algsubs(diff(xi2(x,y),y)=M,Sys2[1])),
 xi1(x,y))); Sys2[2]; Solxi2:=subs(_F1(y)=y,
 pdsolve(algsubs(Solxi1,Sys2[2]),xi2(x,y)));
test1:=algsubs(xi2(x,y)=y,algsubs(xi1(x,y)=x,{Sys2[1]/2,
 Sys2[2]/(-2)})); L40:=subs(a(x,y)=rhs(trA),b(x,y)=0,xi1(x,y)=
 rhs(Solxi1),xi2(x,y)=rhs(Solxi2),A2);
L41:=subs(eta(x,y,u)=eta(x,y),xi1(x,y,u)=xi1(x,y),xi2(x,y,u)=
 xi2(x,y),L40); X41:=eval(X(f),L41);
Equat11:=subs(A=0,algsubs(b(x,y)=B,Sys2[3]));
Equat21:=simplify(dsolve(Equat11,G(u))); Equat311:=algsubs(
 diff(xi2(x,y),y)=M,subs(_C1=1,Equat21));
trG:=subs(-2*u*M/B=u,Equat31); trB0:=isolate(-2*u*M/B=u,B);
trB:=subs(M=1,trB0); L402:=subs(a(x,y)=0,b(x,y)=rhs(trB),
 xi1(x,y)=rhs(Solxi1),xi2(x,y)=rhs(Solxi2),A2);
L42:=subs(eta(x,y,u)=eta(x,y),xi1(x,y,u)=xi1(x,y),
 xi2(x,y,u)=xi2(x,y),L402); X42:=eval(X(f),L42);
```
□

Problem 2.44 *Nonlinear Poisson equation. Group analysis.* Let us consider the nonlinear Poisson equation (as in the previous problem)

$$u_{xx}+u_{yy}-G(u)=0,$$

where $\{x\in\mathbb{R}, y\in\mathbb{R}\}$ and $G(u)$ is an arbitrary function. Applying various *Maple* predefined functions (contained in `PDEtools` package and related to symmetries of PDEs), find the admissible coordinates ξ_1, ξ_2, and η and the infinitesimal operators X_i $(i=1,\dots,3)$.

In order to determine the additional infinitesimal operators X_{41} and X_{42}, we have to know (from the previous analysis) the special value of the function $G(u)$. Substituting it into the original equation, we can obtain the infinitesimal operators X_{41} and X_{42}.

Maple:

```
with(PDEtools): declare(u(x,y)); U:=diff_table(u(x,y));
interface(showassumed=0): assume(k>0,u>0);
PDE1:=U[x,x]+U[y,y]-G(u)=0; show; DepVars:=u(x,y);
Infs:=Infinitesimals(PDE1,DepVars,split=false,
      displayfunction=false);
Infs1:=eval(Infs,{_C1=1,_C2=0,_C3=0}); Infs2:=eval(Infs,
 {_C1=0,_C2=1,_C3=0}); Infs3:=eval(Infs,{_C1=0,_C2=0,_C3=1});
G1:=InfinitesimalGenerator(Infs1,DepVars,expanded);
G2:=InfinitesimalGenerator(Infs2,DepVars,expanded);
G3:=InfinitesimalGenerator(Infs3,DepVars,expanded);
PDE2:=subs(G(u)=u^k,PDE1);
Infs2:=simplify(Infinitesimals(PDE2,DepVars,split=false,
       displayfunction=false));
Infs41:=eval(Infs2,{_C1=0,_C2=0,_C3=1,_C4=0});
G41:=InfinitesimalGenerator(Infs41,DepVars,expanded);
PDE3:=subs(G(u)=exp(u),PDE1);
Infs3:=simplify(Infinitesimals(PDE3,DepVars,split=false,
       displayfunction=false));
Infs42:=eval(Infs3,{_C1=0,_F2=0}); G42:=simplify(
        InfinitesimalGenerator(Infs42,DepVars,expanded));
```

We note that the operator X_{41} coincides to the operator X_{41} obtained in the previous problem, but the operator X_{42} is represented in the equivalent complex form, because the first two equations of the determining system (2.5) coincide with the Cauchy–Riemann conditions for analytic functions. □

Problem 2.45 *Nonlinear heat equation. Group analysis.* Let us consider the nonlinear heat equation of the form

$$u_t-(G(u)u_x)_x=0,$$

where $\{x\in\mathbb{R}, t\geq 0\}$ and $G(u)$ is an arbitrary function. Applying *Maple* predefined functions (related to symmetries of PDEs), verify that the admissible coordinates ξ_1, ξ_2, and η and the corresponding infinitesimal operators X_i $(i=1,\ldots,3)$ that admits this equation for an arbitrary function $G(u)$, respectively, have the following values:

$$\xi_1=\tfrac{1}{2}x,\ \ \xi_2=t,\ \ \eta=0,\quad X_1=\frac{1}{2}x\partial_x+t\partial_t;$$
$$\xi_1=0,\ \xi_2=1,\ \eta=0,\quad X_2=\partial_t;\quad \xi_1=1,\ \xi_2=0,\ \eta=0,\quad X_3=\partial_x.$$

Verify that the admissible coordinates ξ_1, ξ_2, and η and the corresponding infinitesimal operators X_{41} and X_{42} that additionally admits this equation for the two special cases of the function $G(u)$, $G(u)=u^2$ and $G(u)=e^u$, respectively, have the following values:

$$\xi_1=x,\ \ \xi_2=0,\ \ \eta=u,\quad X_{41}=x\partial_x+u\partial_u;$$
$$\xi_1=\tfrac{1}{2}x,\ \ \xi_2=0,\ \ \eta=1,\quad X_{42}=\tfrac{1}{2}x\partial_x+\partial_u.$$

Maple:

```
with(PDEtools): declare((u,G)(x,t)); U,GU:=diff_table(u(x,t)),
 diff_table(G(u(x,t))); PDE1:=U[t]-G(u)*U[x,x]-GU[x]*U[x]=0;
show; DepVars:=u(x,t); Op1:=split=false,displayfunction=false;
Infs:=Infinitesimals(expand(PDE1),DepVars,Op1); Infs1:=eval(Infs,
 {_C1=1,_C2=0,_C3=0}); Infs2:=eval(Infs,{_C1=0,_C2=1,_C3=0});
Infs3:=eval(Infs,{_C1=0,_C2=0,_C3=1});
G1:=InfinitesimalGenerator(Infs1,DepVars,expanded);
G2:=InfinitesimalGenerator(Infs2,DepVars,expanded);
G3:=InfinitesimalGenerator(Infs3,DepVars,expanded);
PDE2:=diff(u(x,t),t)-u(x,t)^2*diff(u(x,t),x$2)-diff(u(x,t)^2,x)*
 diff(u(x,t),x)=0; Infs2:=Infinitesimals(PDE2,DepVars,Op1);
Infs41:=eval(Infs2,{_C1=0,_C2=0,_C3=1,_C4=0});
G41:=InfinitesimalGenerator(Infs41,DepVars,expanded);
PDE3:=diff(u(x,t),t)-exp(u(x,t))*diff(u(x,t),x$2)-
 diff(exp(u(x,t)),x)*diff(u(x,t),x)=0;
Infs3:=simplify(Infinitesimals(PDE3,DepVars,Op1));
Infs42:=eval(Infs3,{_C1=0,_C2=0,_C3=1,_C4=0});
G42:=InfinitesimalGenerator(Infs42,DepVars,expanded);
```
□

Problem 2.46 *Nonlinear wave equation. Group analysis.* Let us consider the nonlinear wave equation of the form

$$u_{tt}-(G(u)u_x)_x=0,$$

where $\{x \in \mathbb{R}, t \geq 0\}$ and $G(u)$ is an arbitrary function. Applying *Maple* predefined functions (related to symmetries of PDEs), verify that the admissible coordinates ξ_1, ξ_2, and η and the corresponding infinitesimal operators X_i (i=1, ..., 3) that admits this equation for an arbitrary function $G(u)$, respectively, have the following values:

$$\xi_1 = x, \quad \xi_2 = t, \quad \eta = 0, \quad X_1 = x\partial_x + t\partial_t;$$
$$\xi_1 = 1, \ \xi_2 = 0, \ \eta = 0, \quad X_2 = \partial_x; \quad \xi_1 = 0, \ \xi_2 = 1, \ \eta = 0, \quad X_3 = \partial_t.$$

Verify that the admissible coordinates ξ_1, ξ_2, and η and the corresponding infinitesimal operators X_{41} and X_{42} that additionally admits this equation for the two special cases of the function $G(u)$, $G(u)=u^2$ and $G(u)=e^u$, respectively, have the following values:

$$\xi_1 = x, \quad \xi_2 = 0, \quad \eta = u, \quad X_{41} = x\partial_x + u\partial_u;$$
$$\xi_1 = x, \quad \xi_2 = 0, \quad \eta = 2, \quad X_{42} = x\partial_x + 2\partial_u.$$

Maple:

```
with(PDEtools): declare(u(x,t),G(u(x,t))); DepVars:=u(x,t);
U,GU:=diff_table(u(x,t)),diff_table(G(u(x,t)));
PDE1:=U[t,t]-G(u)*U[x,x]-GU[x]*U[x]=0; show;
Op1:=split=false,displayfunction=false;
Infs:=Infinitesimals(expand(PDE1),DepVars,Op1);
Infs1:=eval(Infs,{_C1=1,_C2=0,_C3=0});
Infs2:=eval(Infs,{_C1=0,_C2=1,_C3=0});
Infs3:=eval(Infs,{_C1=0,_C2=0,_C3=1});
G1:=InfinitesimalGenerator(Infs1,DepVars,expanded);
G2:=InfinitesimalGenerator(Infs2,DepVars,expanded);
G3:=InfinitesimalGenerator(Infs3,DepVars,expanded);
PDE2:=diff(u(x,t),t$2)-u(x,t)^2*diff(u(x,t),x$2)
 -diff(u(x,t)^2,x)*diff(u(x,t),x)=0;
Infs2:=Infinitesimals(PDE2,DepVars,Op1);
Infs41:=eval(Infs2,{_C1=1,_C2=0,_C3=0,_C4=0});
G41:=InfinitesimalGenerator(Infs41,DepVars,expanded);
PDE3:=diff(u(x,t),t$2)-exp(u(x,t))*diff(u(x,t),x$2)
 -diff(exp(u(x,t)),x)*diff(u(x,t),x)=0;
Infs3:=simplify(Infinitesimals(PDE3,DepVars,Op1));
Infs42:=eval(Infs3,{_C1=1,_C2=0,_C3=0,_C4=0});
G42:=InfinitesimalGenerator(Infs42,DepVars,expanded);
```

□

2.7.3 Invariant Solutions

An *invariant* of the infinitesimal operator X_i that admits the equation $\mathcal{F}(x, y, u, u_x, u_y, u_{xx}, u_{xy}, u_{yy})=0$ under one-parameter set of transformations (2.4) is a function $\mathcal{I}(x, y, u)$ that satisfies the condition

$$\mathcal{I}(X, Y, U)=\mathcal{I}(x, y, u).$$

We can say that an invariant solution is converted to itself under this set of transformations (see Sect. 1.1.3). Let us now consider the procedure of constructing invariant solutions applying symmetries obtained with the group analysis (described above). This procedure consists of the three steps:

1. We note that the admissible coordinates ξ_1, ξ_2, and η and the infinitesimal operators are known as a result of the group analysis of the equation. *Invariant solutions* (or invariants) of the known infinitesimal operator X_i are determined in the implicit form $X_i\mathcal{I}=0$, i.e., we have to solve the linear first-order PDE. Thus, we write out the characteristic system of ODEs: $dx/\xi_1=dy/\xi_2=du/\eta$ (see Chap. 3).

2. We find two functionally independent *first integrals* of the characteristic system of ODEs: $\mathcal{I}_i(x, y, u)=C_i$ $(i=1, 2)$, where C_i $(i=1, 2)$ are arbitrary constants. The general solution of the equation $X_i\mathcal{I}=0$ is determined by the formula $\mathcal{I}=\Phi(\mathcal{I}_1, \mathcal{I}_2)$, where the functions $\mathcal{I}_i(x, y, u)$ $(i=1, 2)$ are known and $\Phi(\mathcal{I}_1, \mathcal{I}_2)$ is a function to be determined.

3. We solve the equation $\mathcal{I}=\Phi(\mathcal{I}_1, \mathcal{I}_2)=0$ with respect to the invariant $\mathcal{I}_2$ and obtain $\mathcal{I}_2=\Phi(\mathcal{I}_1)$. Then solving this equation with respect to u and substituting the resulting expression into the original PDE, we arrive at an ODE for Φ.

Problem 2.47 *Nonlinear Poisson equation. Invariant solutions.* Let us now construct invariant solutions for the nonlinear Poisson equation

$$u_{xx}+u_{yy}-G(u)=0,$$

where $\{x \in \mathbb{R}, y \in \mathbb{R}\}$ and $G(u)$ is an arbitrary function. If we take $G(u)=u^k$, the equation admits the additional operator X_{41} (see Problem 2.43)

$$X_{41} = x\partial_x + y\partial_y - \frac{2}{-1+k}u\partial_u.$$

Let us perform step-by-step the procedure of finding invariant solutions.

1. Following the procedure described above for constructing invariant solutions, we verify that the invariants of the operator X_{41} are $\mathcal{I}_1 = y/x$ and $\mathcal{I}_2 = x^{2/(k-1)}u$.

2. Solving the equation $\mathcal{I}_2 = \Phi(\mathcal{I}_1)$ with respect to u, we verify that the form of the invariant solution is $u = \Phi(y/x)x^{-2/(k-1)}$.

3. Substituting this form of the invariant solution into the original equation, we obtain the ODE for $\Phi(z)$ $(z = y/x)$:

$$2(k+1)\Phi + 2z(k^2-1)\Phi_z + (z^2+1)(k-1)^2\Phi_{zz} - \Phi^k(k^2-1)^2.$$

4. Integrating this equation for Φ, we verify that the solution of the nonlinear Poisson equation with $G(u) = u^k$, which is invariant with respect to operator X_{41}, has the form $u = x^{-2/K}\left(\frac{4x^2}{K^2(y^2+x^2)}\right)^{1/K}$, where $K = k-1$. We verify that the solution obtained is a solution of the original nonlinear Poisson equation with $G(u) = u^k$.

Maple:

```
with(PDEtools): interface(showassumed=0): assume(k>0,x>0);
declare(u(x,y),In(x,y,u),Phi(z)); U:=diff_table(u(x,y));
PDE1:=U[x,x]+U[y,y]-F(u)=0; show; DepVars:=u(x,y);
term0:=2/(1-k); X41:=f->x*diff(f,x)+y*diff(f,y)+term0*u*
 diff(f,u); X41(In(x,y,u))=0; ODEs:=dx/x=(dy/y=1/term0*du/u);
Eq1:=lhs(ODEs)=lhs(rhs(ODEs)); Eq2:=(lhs(ODEs)=rhs(rhs(ODEs)))*
 term0; FI10:=int(lhs(Eq1)/dx,x)=int(rhs(Eq1)/dy,y)+C1;
FI11:=subs(1/exp(C1)=C1,simplify(map(exp,FI10-lhs(FI10)))/
 exp(C1)); FI20:=int(lhs(Eq2)/dx,x)=int(rhs(Eq2)/du,u)+C2;
FI21:=subs(exp(C2)=C2,simplify(map(exp,FI20)));
FI22:=simplify(subs(1/C2=C2,(FI21/lhs(FI21))/C2));
I1:=rhs(FI11); I2:=rhs(FI22);
InvSol:=u(x,y)=simplify(solve(I2=Phi(I1),u));
PDE2:=u->diff(u,x$2)+diff(u,y$2)-u^k=0; OdePhi:=PDE2(rhs(
 InvSol)); OdePhi1:=expand(algsubs(y/x=z,OdePhi));
term1:=select(has,op(2,lhs(OdePhi1)),x);
OdePhi2:=map(simplify,expand(lhs(OdePhi1)/term1));
term2:=select(has,op(2,OdePhi2),k); OdePhi3:=map(simplify,
 OdePhi2/term2); OdePhi4:=convert(collect(OdePhi3,[Phi,k,z]),
 diff); SolPhi:=[dsolve(OdePhi4,Phi(z))];
SolPhi1:=unapply(SolPhi[2],z);
InvSolFin:=simplify(subs(SolPhi1(y/x),InvSol),exp);
expand(pdetest(subs(k=3,InvSolFin),subs(k=3,PDE2(u(x,y)))));
```

Mathematica:

```
trS1[eq_,var_]:=Select[eq,MemberQ[#,var,Infinity]&];
trS3[eq_,var_]:=Select[eq,FreeQ[#,var]&]; {term0=2/(1-k),
 pde1=D[u[x,y],{x,2}]+D[u[x,y],{y,2}]-f[u]==0}
x41[f_]:=x*D[f,x]+y*D[f,y]+term0*u*D[f,u]; x41[iN[x,y,u]]==0
{odes=dx/x==(dy/y==1/term0*du/u), eq1=odes[[1]]==odes[[2, 1]]}
eq2=Thread[(odes[[1]]==odes[[2,2]])*term0,Equal]//Expand
fI10=Integrate[eq1[[1]]/dx,x]==Integrate[eq1[[2]]/dy,y]+c1
{fI11=Solve[Map[Exp,Thread[fI10-fI10[[2]],Equal]]/.Exp[c1_]->
 -c1,c1]//First, termeq21=trS1[Thread[eq2[[1]]/dx,Equal],x],
 termeq22=trS3[Thread[eq2[[1]]/dx,Equal],x]}
{fI20=termeq22*Integrate[Thread[termeq21,Equal],x]==Integrate[
 Thread[eq2[[2]]/du,Equal],u]+c2, fI21=Map[Exp,fI20]/.
 Exp[c2]->c2, fI22=Thread[fI21/fI21[[1]]/c2,Equal]/.
 {1/c2->c2}//Simplify, i1=fI11[[1,2]], i2=fI22[[2]]}
invSol=Solve[i2==phi[i1],u]//First
pde2[u_]:=D[u,{x,2}]+D[u,{y,2}]-u^k==0;
{odePhi=pde2[invSol[[1,2]]], odePhi1=odePhi/.{y->x*z}//Expand}
{term1=trS1[odePhi1[[1,1]],x], odePhi2=Thread[odePhi1[[1]]/
 term1,Equal]//Expand, term2=trS1[odePhi2[[1]],k]}
{odePhi3=Thread[odePhi2/term2,Equal]//FullSimplify//Together//
 PowerExpand, odePhi4=Collect[odePhi3,{phi[z],k,z}]//
 FullSimplify, solPhi=DSolve[odePhi4==0,phi[z],z]}
```

We note that the *Mathematica* predefined function `DSolve` does not allow us to integrate the equation for Φ, to obtain the final form of the solution, and verify the result.

Now, applying various *Maple* predefined functions (`Infinitesimals`, `InfinitesimalGenerator`, `DeterminingPDE`, `Invariants`, etc.), let us show:

1. how to obtain the admissible coordinates ξ_1, ξ_2, η, the infinitesimal operator X_{41}, and the determining system;

2. how to verify if these infinitesimals represent a symmetry of the nonlinear PDE;

3. how to compute the symmetry transformations under which the equation is invariant, the similarity transformations, the invariant transformations;

4. how to determine the differential invariants of the *one-parameter Lie group* (for our case, it is $S{=}[x, y, 2u/(1{-}k)]$);

5. how to compute *canonical coordinates* for the symmetry.

Maple:

```
Infs:=Infinitesimals(PDE2(u(x,y)),DepVars,split=false,
      displayfunction=false);
Infs1:=simplify(Infs,{_C1=0,_C2=0,_C3=1,_C4=0}) assuming u>0;
X41M:=InfinitesimalGenerator(Infs1,DepVars,expanded);
DetSys:=DeterminingPDE(PDE2(u(x,y)));
SymmetryTest(Infs1,PDE2(u(x,y)));
SymmetryTransformation(Infs1,DepVars,V(X,T));
SimilarityTransformation(Infs1,DepVars,V(X,T));
InvariantTransformation(Infs1,u(x,t),V(X,T));
S:=[x,y,2/(1-k)*u];
CC:=CanonicalCoordinates(S,DepVars,V(X,T));
Invs:=[Invariants(S,u(x,t),jetnotation=false)];
```

□

Problem 2.48 *Nonlinear wave equation. Invariant solutions.* Let us now construct invariant solutions for the nonlinear wave equation

$$u_{tt}-(G(u)u_x)_x=0,$$

where $\{x \in \mathbb{R}, t \geq 0\}$ and $G(u)$ is an arbitrary function.

1. Applying *Maple* predefined functions, let us verify that if the function $G(u)$ is arbitrary, the equation admits the infinitesimal operator $X_3=x\partial_x+t\partial_t$ (see Problem 2.46). We find the invariants of this operator (using the function `Invariants`).

2. Following the procedure (described in Sect. 2.7.3) for constructing invariant solutions, verify that the invariants of the operator X_3 are $\mathcal{J}_1=t/x$ and $\mathcal{J}_2=u$.

3. Solving the equation $\mathcal{J}_2=\Phi(\mathcal{J}_1)$ with respect to u, verify that the form of the invariant solution is $u=\Phi(t/x)$.

4. Substituting this form of the invariant solution into the original equation, obtain the ODE for $\Phi(z)$ ($z=t/x$):

$$(-G\Phi_{zz}-G_z\Phi_z)z^2-2G\Phi_z z+\Phi_{zz}=0.$$

Maple:

```
with(PDEtools): declare(u(x,t),G(u(x,t)),In(x,t,u),Phi(z));
DepVars:=u(x,t); U,GU:=diff_table(u(x,t)),diff_table(G(u(x,t)));
PDE1:=U[t,t]-G(u)*U[x,x]-GU[x]*U[x]=0; show;
Infs:=Infinitesimals(expand(PDE1),DepVars,split=false,
      displayfunction=false);
```

```
Infs1:=eval(Infs,{_C1=1,_C2=0,_C3=0});
X3:=InfinitesimalGenerator(Infs1,DepVars,expanded); S:=[x,t,0];
Invs:=[Invariants(S,u(x,t),jetnotation=false)];
expand(map(X3,Invs)); X3(f(x,t)); X3(In(x,t,u))=0;
ODEs:=dx/x=(dt/t=du/zero); Eq1:=lhs(ODEs)=lhs(rhs(ODEs));
Eq2:=numer(rhs(rhs(ODEs)))=0;
FI10:=int(lhs(Eq1)/dx,x)=int(rhs(Eq1)/dt,t)+C1;
FI11:=subs(1/exp(C1)=C1,simplify(map(exp,
      FI10-lhs(FI10)))/exp(C1));
FI20:=int(lhs(Eq2)/du,u)=int(rhs(Eq2)/du,u)+C2; I1:=rhs(FI11);
I2:=lhs(FI20); InvSol:=u(x,t)=simplify(solve(I2=Phi(I1),u));
PDE2:=u->diff(u,t$2)-diff(G(u)*diff(u,x),x)=0;
OdePhi:=PDE2(rhs(InvSol));
OdePhi1:=expand(algsubs(t/x=z,OdePhi));
term1:=select(has,op(1,lhs(OdePhi1)),x);
OdePhi2:=map(simplify,expand(lhs(OdePhi1)/term1));
```

Mathematica:

```
trS1[eq_,var_]:=Select[eq,MemberQ[#,var,Infinity] &];
pde2[u_]:=D[u,{t,2}]-D[g[u]*D[u,x],x]==0;
{odes=dx/x==(dt/t==du/zero), eq1=odes[[1]]==odes[[2,1]]}
eq2=Numerator[odes[[2,2]]]==0
fI10=Integrate[eq1[[1]]/dx,x]==Integrate[eq1[[2]]/dt,t]+c1
fI11=Solve[Map[Exp,Thread[fI10-fI10[[2]],Equal]]/.
 Exp[c1_]->-c1,c1]//First
fI20=Integrate[eq2[[1]]/du,u]==Integrate[eq2[[2]]/du,u]+c2
{i1=fI11[[1,2]], i2=fI20[[1]], invSol=Solve[i2==phi[i1],u]//
 First, odePhi=pde2[invSol[[1,2]]], odePhi1=odePhi/.t->x*z//
 Expand, term1=trS1[odePhi1[[1,1]],x]}
odePhi2=Collect[Thread[odePhi1[[1]]/term1,Equal]//Expand,z]
```

□

Problem 2.49 *Nonlinear heat equation. Invariant solutions.* Let us now construct invariant solutions for the nonlinear heat equation

$$u_t-(G(u)u_x)_x=0,$$

where $\{x\in\mathbb{R}, t\geq 0\}$ and $G(u)$ is an arbitrary function.

1. Applying *Maple* predefined functions, let us verify that if the function $G(u)$ is arbitrary, the equation admits the infinitesimal operator $X_3=\frac{1}{2}x\partial_x+t\partial_t$ (see Problem 2.45). We find the invariants of this operator (using the function `Invariants`).

2. Following the procedure (described in Sect. 2.7.3) for constructing invariant solutions, verify that the invariants of the operator X_3 are $\mathcal{I}_1=t/x^2$ and $\mathcal{I}_2=u$.

3. Solving the equation $\mathcal{I}_2=\Phi(\mathcal{I}_1)$ with respect to u, verify that the form of the invariant solution is $u=\Phi(t/x^2)$.

4. Substituting this form of the invariant solution into the original equation, obtain the ODE for $\Phi(z)$ $(z=t/x^2)$:

$$-4(G\Phi_{zz}+G_z\Phi_z)z^2-6G\Phi_z z+\Phi_z=0.$$

Maple:

```
with(PDEtools): declare(u(x,t),G(u(x,t)),In(x,t,u),Phi(z));
DepVars:=u(x,t); U,GU:=diff_table(u(x,t)),diff_table(G(u(x,t)));
PDE1:=U[t]-G(u)*U[x,x]-GU[x]*U[x]=0; show;
Infs:=Infinitesimals(expand(PDE1),DepVars,split=false,
 displayfunction=false); Infs1:=eval(Infs,{_C1=2,_C2=0,_C3=0});
S:=[x,2*t,0]; X3:=InfinitesimalGenerator(Infs1,DepVars,expanded);
Invs:=[Invariants(S,u(x,t),jetnotation=false)];
X3(f(x,t)); X3(In(x,t,u))=0; ODEs:=dx/x=(dt/(2*t)=du/zero);
Eq1:=lhs(ODEs)=lhs(rhs(ODEs)); Eq2:=numer(rhs(rhs(ODEs)))=0;
FI10:=(int(lhs(Eq1)/dx,x)=int(rhs(Eq1)/dt,t)+C1)*2;
FI11:=simplify(map(exp,FI10-lhs(FI10)));
termC1:=select(has,rhs(FI11),C1); FI12:=subs(1/termC1=C1,
 FI11/termC1); FI20:=int(lhs(Eq2)/du,u)=int(rhs(Eq2)/du,u)+C2;
I1:=rhs(FI12); I2:=lhs(FI20); InvSol:=u(x,y)=simplify(solve(
 I2=Phi(I1),u)); PDE2:=u->diff(u,t)-diff(G(u)*diff(u,x),x)=0;
OdePhi:=PDE2(rhs(InvSol)); OdePhi1:=expand(algsubs(t/x^2=z,
 OdePhi)); term1:=select(has,op(1,lhs(OdePhi1)),x);
OdePhi2:=map(simplify,expand(lhs(OdePhi1)/term1));
```

Mathematica:

```
trS1[eq_,var_]:=Select[eq,MemberQ[#,var,Infinity]&];
pde2[u_]:=D[u,t]-D[g[u]*D[u,x],x]==0;
{odes=dx/x==(dt/(2*t)==du/zero),eq1=odes[[1]]==odes[[2,1]]}
eq2=Numerator[odes[[2,2]]]==0
fI10=Integrate[eq1[[1]]/dx,x]==Integrate[eq1[[2]]/dt,t]+c1
fI11=Solve[Map[Exp,Thread[fI10-fI10[[2]],Equal]]/.
 Exp[c1_]->-c1,c1]//First
fI20=Integrate[eq2[[1]]/du,u]==Integrate[eq2[[2]]/du,u]+c2
{i1=fI11[[1,2]]^2, i2=fI20[[1]]}
```

```
invSol=Solve[i2==phi[i1],u]//First
{odePhi=pde2[invSol[[1,2]]], odePhi1=odePhi/.t->x^2*z//Expand,
 term1=trS1[odePhi1[[1,1]],x]}
odePhi2=Collect[Thread[odePhi1[[1]]/term1,Equal]//Expand,z]
```

□

Problem 2.50 *Nonlinear heat equation. Invariant solutions. Special case of the function* $G(u)=u^k$. Let us now construct invariant solutions for the nonlinear heat equation

$$u_t-(G(u)u_x)_x=0,$$

where $\{x \in \mathbb{R}, t \geq 0\}$ and $G(u)=u^k$.

1. Applying *Maple* predefined functions, let us verify that if $G(u)=u^k$, the equation admits the infinitesimal operator $X_4=\frac{1}{2}kx\partial_x+u\partial_u$. We find the invariants of this operator (using the function `Invariants`).

2. Following the procedure (described in Sect. 2.7.3) for constructing invariant solutions, verify that the invariants of the operator X_4 are $\mathcal{J}_1=t$ and $\mathcal{J}_2=x^{-2/k}u$.

3. Solving the equation $\mathcal{J}_2=\Phi(\mathcal{J}_1)$ with respect to u, verify that the form of the invariant solution is $u=\Phi(t)x^{2/k}$.

4. Substituting this form of the invariant solution into the original equation, obtain the ODE for $\Phi(t)$: $k^2\Phi_t-2(k+2)\Phi^{k+1}=0$.

5. Integrating this equation for Φ, verify that the solution of the nonlinear heat equation with $G(u)=u^k$, which is invariant with respect to operator X_4, has the form $u=k^{1/k}(-2kt-4t+Ak)^{-1/k}x^{2/k}$. Verify that the solution obtained is exact solution of the original nonlinear heat equation with $G(u)=u^k$.

Maple:

```
with(PDEtools): interface(showassumed=0): assume(k>1,x>0,u>0);
declare(u(x,t),In(x,t,u),Phi(z)); DepVars:=u(x,t);
PDE1:=u->diff(u,t)-diff(u^k*diff(u,x),x)=0;
Infs:=Infinitesimals(PDE1(u(x,t)),DepVars,split=false,
 displayfunction=false); Infs1:=eval(Infs,{_C1=0,_C2=0,_C3=1,
 _C4=0}); S:=[k*x/2,0,u];
X4:=InfinitesimalGenerator(Infs1,DepVars,expanded);
Invs:=[Invariants(S,u(x,t),jetnotation=false)];
X4(f(x,t)); X4(In(x,t,u))=0; ODEs:=dx/(k*x/2)=(dt/zero=du/u);
```

```
Eq1:=lhs(ODEs)=rhs(rhs(ODEs)); Eq2:=numer(lhs(rhs(ODEs)))=0;
FI10:=(int(lhs(Eq1)/dx,x)=int(rhs(Eq1)/du,u)+C1);
FI11:=expand(map(exp,FI10-lhs(FI10))); termC1:=select(has,
 rhs(FI11),C1); FI12:=subs(1/termC1=C1,FI11/termC1);
FI20:=int(lhs(Eq2)/dt,t)=int(rhs(Eq2)/du,u)+C2;
I2:=rhs(FI12); I1:=lhs(FI20);
InvSol:=u(x,t)=simplify(solve(I2=Phi(I1),u)) assuming u>0;
OdePhi:=expand(PDE1(rhs(InvSol))); term1:=select(has,op(2,
 lhs(OdePhi)),x); OdePhi1:=map(simplify,expand(
 lhs(OdePhi)/term1)); SolPhi:=[dsolve(OdePhi1,Phi(t))];
SolPhi1:=lhs(SolPhi[1])=subs(_C1=A,combine(rhs(SolPhi[1])));
SolPhi2:=unapply(SolPhi1,t); SolPhi2(t);
InvSolFin:=factor(subs(SolPhi2(t),InvSol));
simplify(pdetest(subs(k=3,InvSolFin),subs(k=3,PDE1(u(x,t)))));
```

Mathematica:

```
Off[Solve::ifun]; trS1[eq_,var_]:=Select[eq,MemberQ[#,var,
 Infinity]&]; pde1[u_]:=D[u,t]-D[u^k*D[u,x],x]==0;
{odes=dx/(k*x/2)==(dt/zero==du/u),
 eq1=odes[[1]]==odes[[2,2]],eq2=Numerator[odes[[2,1]]]==0}
{fI10=Integrate[eq1[[1]]/dx,x]==Integrate[eq1[[2]]/du,u]+
 c1//Simplify, fI11=Solve[(Map[Exp,Thread[fI10-fI10[[1]],
 Equal]]//ExpandAll)/.Exp[-c1]->c1//Expand,c1]//First}
fI20=Integrate[eq2[[1]]/dt,t]==Integrate[eq2[[2]]/du,u]+c2
{i2=fI11[[1,2]], i1=fI20[[1]]}
invSol=Solve[i2==phi[i1],u]//First
odePhi=pde1[invSol[[1,2]]]//PowerExpand//ExpandAll
term1=trS1[odePhi[[1,1]],x]
odePhi1=Collect[Thread[odePhi[[1]]/term1,Equal]//Expand,z]
solPhi=DSolve[odePhi1==0,phi[t],t]//ExpandAll//First
{solPhi1=solPhi/.C[1]->a, invSolFin=invSol/.solPhi1}
test1=pde1[invSolFin[[1,2]]]//PowerExpand//FullSimplify
```

□

If an equation admits the infinitesimal operators X_i $(i = 1, \ldots, n)$, then we have n different invariant solutions. Now we are interested in finding solutions that are invariant under a *linear superposition* of the infinitesimal operators, such solutions may have a significantly different form. Therefore, for determining all types of invariant solutions, it is necessary investigate all possible combinations of the infinitesimal operators.

Problem 2.51 *Nonlinear heat equation. Invariant solutions. Linear combinations of admissible operators.* Let us now construct invariant solutions induced by linear combinations of admissible operators for the nonlinear heat equation

$$u_t - (G(u)u_x)_x = 0,$$

where $\{x \in \mathbb{R}, t \geq 0\}$ and $G(u)$ is an arbitrary function.

1. Applying *Maple* predefined functions, let us verify that if the function $G(u)$ is arbitrary, the equation admits the infinitesimal operators X_i $(i{=}1, 2, 3)$

$$X_1{=}\partial_t, \quad X_2{=}\partial_x, \quad X_3{=}\tfrac{1}{2}x\partial_x{+}t\partial_t.$$

2. Let us consider the following linear combination: $X_{1,2}{=}X_1{+}aX_2$ $(a{\neq}0)$. Following the procedure (described in Sect. 2.7.3) for constructing invariant solutions, verify that the invariants of the operator $X_{1,2}$ are $\mathcal{I}_1{=}x{-}at$ and $\mathcal{I}_2{=}u$.

3. Solving the equation $\mathcal{I}_2{=}\Phi(\mathcal{I}_1)$ with respect to u, verify that the form of the invariant solution is $u{=}\Phi(x{-}at)$. It should be noted that these solutions (called *traveling wave solutions*) are not contained in the family of invariant solutions induced by the operators X_i $(i{=}1, 2, 3)$.

Maple:

```
with(PDEtools): declare(u(x,t),G(u(x,t)),In(x,t,u),Phi(z));
DepVars:=u(x,t); U,GU:=diff_table(u(x,t)),diff_table(G(u(x,t)));
PDE1:=U[t]-G(u)*U[x,x]-GU[x]*U[x]=0; Infs:=Infinitesimals(PDE1,
 DepVars,split=false,displayfunction=false); Infs1:=eval(Infs,
 {_C1=0,_C2=1,_C3=0}); Infs2:=eval(Infs,{_C1=0,_C2=0,_C3=1});
Infs3:=eval(Infs,{_C1=1,_C2=0,_C3=0});
for i from 1 to 3 do
 X||i:=InfinitesimalGenerator(Infs||i,DepVars,expanded); od;
collect(X1(In(x,t,u))+a*X2(In(x,t,u)),diff)=0;
ODEs:=dx/a=(dt/1=du/zero); Eq1:=lhs(ODEs)=lhs(rhs(ODEs));
Eq2:=numer(rhs(rhs(ODEs)))=0;
FI10:=expand((int(lhs(Eq1)/dx,x)=int(rhs(Eq1)/dt,t)+C1)*a);
termC1:=select(has,rhs(FI10),C1); FI11:=isolate(subs(termC1=C1,
 FI10),C1); FI20:=int(rhs(Eq2)/du,u)+C2=int(lhs(Eq2)/du,u);
I1:=rhs(FI11); I2:=rhs(FI20);
InvSol:=u(x,t)=simplify(solve(I2=Phi(I1),u));
```

Mathematica:

```
Off[Solve::ifun]; pde1[u_]:=D[u,t]-g[u]*D[u,{x,2}]-
 D[g[u],x]*D[u,x]==0; {odes=dx/a==(dt/1==du/zero),
 eq1=odes[[1]]==odes[[2,1]], eq2=Numerator[odes[[2,2]]]==0}
{fI10=Integrate[eq1[[1]]/dx,x]==Integrate[eq1[[2]]/dt,t]+c1//
 Simplify, fI11=Solve[fI10,c1]//Expand//First}
fI20=Integrate[eq2[[2]]/du,u]+c2==Integrate[eq2[[1]]/du,u]
{i1=fI11[[1,2]]*a//Expand, i2=fI20[[2]]}
invSol=Solve[i2==phi[i1],u]//First
```

□

Problem 2.52 *Nonlinear heat equation. Invariant solutions. Special case of the function $G(u)=e^u$.* We construct invariant solutions induced by linear combinations of admissible operators for the nonlinear heat equation

$$u_t-(G(u)u_x)_x=0,$$

where $\{x\in\mathbb{R}, t\geq 0\}$ and $G(u)=e^u$.

1. Let us verify that if $G(u)=e^u$, apart from the infinitesimal operators X_i $(i=1,2,3)$ obtained in the previous problem, the equation admits additional infinitesimal operator $X_4=\frac{1}{2}x\partial_x+\partial_u$.

2. Consider the following linear combination: $X_{3,4}=X_3+aX_4$ $(a\neq 0)$. Following the procedure (described in Sect. 2.7.3) for constructing invariant solutions, verify that the invariants of the operator $X_{3,4}$ are $\mathcal{I}_1=x^2t^{-a-1}$ and $\mathcal{I}_2=-u+a\ln t$.

3. Solving the equation $\mathcal{I}_2=\Phi(\mathcal{I}_1)$ with respect to u, verify that the form of the invariant solution is $u=-\Phi(x^2t^{-a-1})+a\ln t$.

4. Substituting this form of the invariant solution into the original equation, obtain the ODE for $\Phi(t)$:

$$((\Phi_z z+1)a+\Phi_z z)e^{\Phi}+(-4\Phi_z^2+4\Phi_{zz})z+2\Phi_z=0.$$

These solutions are not contained in the family of invariant solutions induced by the operators X_i $(i=1,\dots,4)$.

Maple:

```
with(PDEtools): declare(u(x,t),In(x,t,u),Phi(z));
DepVars:=u(x,t); PDE1:=u->diff(u,t)-diff(exp(u)*diff(u,x),x)=0;
Infs:=Infinitesimals(PDE1(u(x,t)),DepVars,split=false,
 displayfunction=false);
```

```
Infs3:=eval(Infs,{_C1=1,_C2=0,_C3=0,_C4=0});
Infs4:=eval(Infs,{_C1=0,_C2=0,_C3=1,_C4=0});
for i from 3 to 4 do
 X||i:=InfinitesimalGenerator(Infs||i,DepVars,expanded); od;
collect(X3(In(x,t,u))+a*X4(In(x,t,u)),diff)=0;
ODEs:=dx/(1/2*x*(a+1))=(dt/t=du/a);
Eq1:=lhs(ODEs)=lhs(rhs(ODEs)); Eq2:=rhs(ODEs);
FI10:=expand((int(lhs(Eq1)/dx,x)=int(rhs(Eq1)/dt,t)+C1)*(a+1));
FI11:=expand(map(exp,FI10-lhs(FI10))); termC1:=select(has,
 rhs(FI11),C1); FI12:=combine(subs(1/termC1=1/C1,FI11/termC1));
FI13:=isolate(FI12,C1);
FI20:=simplify((int(lhs(Eq2)/dt,t)=int(rhs(Eq2)/du,u)+C2)*a);
termC2:=select(has,rhs(FI20),C2);
FI21:=subs(termC2=C2,FI20); FI22:=isolate(FI21,C2);
I1:=combine(rhs(FI13)); I2:=rhs(FI22);
InvSol:=u(x,t)=simplify(solve(I2=Phi(I1),u));
OdePhi:=expand(PDE1(rhs(InvSol)));
OdePhi1:=simplify(algsubs(I1=z,OdePhi));
term1:=denom(lhs(OdePhi1)); OdePhi2:=expand(OdePhi1*term1);
OdePhi3:=numer(lhs(factor(OdePhi2)))=0;
OdePhi4:=convert(collect(OdePhi3,[exp,a,z]),diff);
```

Mathematica:

```
Off[Solve::ifun]; trS1[eq_,var_]:=Select[eq,MemberQ[#,var,
 Infinity]&]; pde1[u_]:=D[u,t]-D[Exp[u]*D[u,x],x]==0;
trS3[eq_,var_]:=Select[eq,FreeQ[#,var]&];
{odes=dx/(1/2*x*(a+1))==(dt/t==du/a), eq1=odes[[1]]==
 odes[[2,1]],eq2=odes[[2]], fI10=Thread[(Integrate[
 eq1[[1]]/dx,x]==Integrate[eq1[[2]]/dt,t]+c1)*(a+1),
 Equal]//FullSimplify, fI11=Map[Exp,fI10]//ExpandAll}
{termC1=trS1[fI11[[1]],a]/.t->1, fI12=Thread[fI11/termC1,
 Equal]/.Exp[-c1-a*c1]->1/c1, fI13=Solve[fI12,c1]//First}
{fI20=Thread[(Integrate[eq2[[1]]/dt,t]==Integrate[
 eq2[[2]]/du,u]+c2)*a,Equal]//Expand, termC2=trS1[
 fI20[[2]], c2], fI21=fI20/.termC2->c2, fI22=Solve[
 fI21,c2]//First, i1=fI13[[1,2]]/(a+1)^2, i2=fI22[[1,2]]}
invSol=Solve[i2==phi[i1],u]//First
odePhi=pde1[invSol[[1,2]]]//PowerExpand//ExpandAll
{odePhi1=odePhi/.x^2->z/t^(-1-a), term1=trS1[odePhi1[[1,1]],t],
 odePhi2=Thread[odePhi1/term1,Equal]//Expand,
 odePhi3=Numerator[Factor[odePhi2[[1]]]]==0}
Collect[Thread[odePhi3,Equal],{Exp[phi[z]],a,z}]
```
□

2.8 Nonlinear Systems

In this section, we will obtain traveling wave solutions of nonlinear systems of first-order and second-order equations and exact solutions for a class of nonlinear systems that can be reduced to an ordinary differential equation, and finally we will find generalized separable solutions of nonlinear second-order systems.

2.8.1 Traveling Wave Reductions

Let us consider nonlinear systems of the fist-order PDEs of the form

$$u_x = F(u, v), \quad v_t = G(u, v),$$

where $\{x \in \mathbb{R}, t \geq 0\}$. It is known that such systems can admit traveling wave solutions, i.e., we look for a solution in the form $u = W_1(\xi)$, $v = W_2(\xi)$, where $\xi = kx - \lambda t$ and k, λ are arbitrary constants.

Problem 2.53 *Nonlinear system of first-order equations. Traveling wave solutions.* For the nonlinear system,

$$u_x = a_1 u \ln v, \quad v_t = a_2 v \ln u,$$

where $\{x \in \mathbb{R}, t \geq 0\}$ and a_1, a_2 are real constants, determine the traveling wave solution.

1. System of ODEs. We look for a solution in the form $u = W_1(\xi)$ and $v = W_2(\xi)$ ($\xi = kx - \lambda t$) and establish that the functions $W_1(\xi)$ and $W_2(\xi)$ are described by the autonomous system of ordinary differential equations $kW_{1\xi} - a_1 W_1 \ln(W_2) = 0$, $-\lambda W_{2\xi} - a_2 W_2 \ln(W_1) = 0$ (`Sys4`).

2. Solution of system. We find the traveling wave solution of the given nonlinear system (`trW1`, `trW2`): $W_1(\xi) = \exp\left[a_2^{-1} \lambda \sin[\sqrt{a_1 a_2}\, \xi / \sqrt{k\lambda}]\right]$ and $W_2(\xi) = \exp\left[(a_1 a_2)^{-1/2} \sqrt{k\lambda} \cos[\sqrt{a_1 a_2}\, \xi / \sqrt{k\lambda}]\right]$, and verify that this solution is exact solution of the given nonlinear system.

Maple:

```
with(PDEtools): interface(showassumed=0):
assume(k>0,lambda>0,x>0,t>0,xi>0); tr1:=k*x-lambda*t=xi;
declare((W1,W2)(xi),(w1,w2,u,v)(x,t),(F,G)(u,v));
U,V:=diff_table(u(x,t)),diff_table(v(x,t));
F:=(u,v)->a1*u*ln(v); G:=(u,v)->a2*v*ln(u);
Sys1:=(w1,w2)->[diff(w1,x)-F(w1,w2)=0,diff(w2,t)-G(w1,w2)=0];
tr1:=k*x-lambda*t=xi;
```

```
Sys2:=expand(Sys1(W1(lhs(tr1)),W2(lhs(tr1))));
Sys3:=algsubs(tr1,Sys2); Sys4:=map(convert,Sys3,diff);
Sol1:=combine(dsolve(Sys4,{W1(xi),W2(xi)}));
Sol11:=simplify(subs(_C1=0,_C2=1,Sol1));
trW1:=expand(subs(Sol11[1],Sol11[2])); trW2:=Sol11[1];
test1:=simplify(expand(subs(trW1,trW2,Sys4)),symbolic);
```

Mathematica:

```
Off[InverseFunction::"ifun"]; f[u_, v_]:=a1*u*Log[v];
g[u_,v_]:=a2*v*Log[u]; sys1[w1_,w2_]:={D[w1,x]-f[w1, w2]==0,
 D[w2,t]-g[w1,w2]==0}; {tr1=k*x-lambda*t->xi,
 sys2=sys1[w1[tr1[[1]]],w2[tr1[[1]]]]//Expand, sys3=sys2/.tr1}
sol1=DSolve[sys3,{w1[xi],w2[xi]},xi];
{sol2=sol1/.{C[1]->1,C[2]->0}//PowerExpand, n=Length[sol2]}
sols=Table[sol2[[i]]//FullSimplify,{i,1,n}]
{trW1=sols[[1,2]], trW2=sols[[1,1]], test1=sys3/.trW1/.
 D[trW1,xi]/.trW2/.D[trW2,xi]//PowerExpand//FullSimplify}
```

It should be noted that this *Mathematica* solution depends on the internal algorithms for solving ordinary differential equations (the predefined function `DSolve`) and has been prepared for Ver. 8. If we change the three lines, `sol1=DSolve[sys3,w2[xi],w1[xi],xi]`, `trW1=sols[[1,2]]]`, `trW2=sols[[1,1]]`, we can obtain the correct solution for Ver. 7. □

Let us consider nonlinear systems of the second-order PDEs of the form

$$u_t = a_1 u_{xx} + F(u,v), \quad v_t = a_2 v_{xx} + G(u,v),$$

where $\{x \in \mathbb{R}, t \geq 0\}$ and a_1, a_2 are real constants. These systems can admit traveling wave solutions, i.e., we look for a solution in the form $u = W_1(\xi)$, $v = W_2(\xi)$, $\xi = kx - \lambda t$, where k and λ are arbitrary constants.

Problem 2.54 *Nonlinear system of second-order equations. Traveling wave reduction.* For the nonlinear system,

$$u_t = a_1 u_{xx} + u \ln(v), \quad v_t = a_2 v_{xx} + u v^n,$$

where $\{x \in \mathbb{R}, t \geq 0\}$ and a_1, a_2 are real constants, verify that this system admits exact traveling wave solution and the functions $W_1(\xi)$ and $W_2(\xi)$ are described by the autonomous system of ordinary differential equations

$$-\lambda W_{1\xi} - a_1 k^2 W_{1\xi\xi} - W_1 \ln(W_2) = 0, \quad -\lambda W_{2\xi} - a_2 k^2 W_{2\xi\xi} - W_1 W_2^n = 0.$$

Maple:

```
with(PDEtools): interface(showassumed=0):
assume(k>0,lambda>0,x>0,t>0,xi>0,a1>0,a2>0);
declare((W1,W2)(xi),(w1,w2,u,v)(x,t),(F,G)(u,v));
U,V:=diff_table(u(x,t)),diff_table(v(x,t));
F:=(u,v)->u*ln(v); G:=(u,v)->u*v^n; tr1:=k*x-lambda*t=xi;
Sys1:=(w1,w2)->[diff(w1,t)-a1*diff(w1,x$2)-F(w1,w2)=0,
                diff(w2,t)-a2*diff(w2,x$2)-G(w1,w2)=0];
Sys2:=expand(Sys1(W1(lhs(tr1)),W2(lhs(tr1))));
Sys3:=algsubs(tr1,Sys2); Sys4:=map(convert,Sys3,diff);
```

Mathematica:

```
f[u_,v_]:=u*Log[v]; g[u_,v_]:=u*v^n; tr1=k*x-lambda*t->xi
sys1[w1_,w2_]:={D[w1,t]-a1*D[w1,{x,2}]-f[w1,w2]==0,
                D[w2,t]-a2*D[w2,{x,2}]-g[w1,w2]==0};
sys2=sys1[w1[tr1[[1]]],w2[tr1[[1]]]]//Expand
sys3=sys2/.tr1
```

□

2.8.2 Special Reductions

Let us consider a special class of nonlinear systems of the fist-order PDEs of the form [124]

$$u_x = uF(v), \quad v_t = u^k G(v),$$

where $\{x \in \mathbb{R}, t \geq 0\}$, $k \neq 0$, $F(v) \neq \text{const}$, $u(x,t)$ and $v(x,t)$ are the unknown functions to be determined.

First, introducing the transformation $u{=}U^{1/k}$, $G(v){=}v_t/W_t$, where $U(x,t)$ and $W(x,t)$ are the new functions, we can rewrite the nonlinear original system in a more simple form $U_x{=}kF(v)U$, $W_t{=}U$. Then, substituting the second equation of the system $U{=}W_t$ into the first equation, we can obtain the second-order PDE for the new function $W(x,t)$: $W_{tx}{=}W_t\Phi(W)$, where $\Phi(W)$ is defined parametrically by the two equations $\Phi(W){=}kF(v)$ and $W_t{=}v_t/G(v)$.

Integrating this equation, we can obtain the first-order equation for W: $W_x{=}\int \Phi(W)\,dW{+}\theta(x)$, where $\theta(x)$ is an arbitrary function. Finally, applying the transformation $W{=}\int dv/G(v)$, we can arrive at the more simple equation $v_x{=}\int kF(v)\,dv{+}G(v)\theta(x)$ that can be treated as a first-order ODE with respect to x.

Problem 2.55 *System of first-order equations. Special reduction.* For the nonlinear system,

$$u_x = 2a_1 uv, \quad v_t = a_2 u^k v,$$

where $\{x \in \mathbb{R}, t \geq 0\}$, $k \neq 0$, determine the exact solution applying a special reduction described above.

1. Reducing to ODE. According to the above described procedure, we reduce this system to the ODE $v_x = a_1 v^2 + a_2 v\theta(x)$ with respect to $v(x)$, whose general solution has the form $v(x) = -\frac{\phi_x}{a_1\phi(x) + a_2\psi(t)}$, where $\phi(x) = -\int \xi(x)\,dx$, $\xi(x) = \exp\left(\int a_2\theta(x)\,dx\right)$, and $\psi(t)$ and $\theta(x)$ are arbitrary functions.

Maple:

```
with(PDEtools): interface(showassumed=0): assume(k>0,x>0,t>0,
 U(x,t)>0,W(x,t)>0); declare((u,v)(x,t),(xi,phi)(x),psi(t));
Sys1:=[diff(u(x,t),x)=u(x,t)*F(v(x,t)),
       diff(v(x,t),t)/G(v(x,t))=u(x,t)^k];
tr1:={u(x,t)=U(x,t)^(1/k),
      G(v(x,t))=diff(v(x,t),t)/diff(W(x,t),t)};
Sys2:=dchange(tr1,Sys1,[U(x,t),W(x,t)]);
Sys3:=expand(simplify(Sys2));
Sys31:=[Sys3[1]*k/U(x,t)^(1/k)*U(x,t),Sys3[2]];
tr2:={rhs(Sys31[2])=lhs(Sys31[2])}; Eq1:=sort(subs(tr2,
 Sys31[1])); tr3:=k*F(v(x,t))=Phi(W);
Eq2:=algsubs(k*F(v(x,t))=Phi(W),Eq1);
Eq3:=int(lhs(Eq2),t)=int(Phi(W),W)+theta(x);
tr4:={W(x,t)=Int(1/G(v),v)}; Eq4:=expand(subs(rhs(tr3)=lhs(tr3),
   Eq3*G(v(x,t)))); Eq5:=subs(tr4,subs(W=W(x,t),subs(tr4,Eq4)));
Eq6:=subs(lhs(Eq5)=diff(v(x,t),x),Eq5); Eq7:=combine(Eq6);
F:=v->2*a1*v; G:=v->a2*v; Eq8:=expand(value(subs(k=1,Eq7)));
term1:=select(has,rhs(Eq8),int); tr5:=term1=a1*v^2;
Eq9:=subs(tr5,Eq8); Eq10:=lhs(Eq9)=algsubs(v=v(x,t),rhs(Eq9));
Solv:=pdsolve(Eq10); term2:=select(has,numer(rhs(Solv)),a2);
tr6:={term2=xi(x)}; Solv1:=expand(subs(_F1(t)=a2*psi(t),
   subs(tr6,Solv))); term3:=select(has,denom(rhs(Solv1)),xi(x));
tr7:={-term3=phi(x)*a1}; Solv2:=expand(subs(tr7,Solv1));
tr8:={diff(term3/a1,x)=-diff(phi(x),x)};
Solv3:=expand(subs(tr8,Solv2));
```

Mathematica:

```
trS1[eq_,var_]:=Select[eq,MemberQ[#,var,Infinity]&];
trS3[eq_,var_]:=Select[eq,FreeQ[#,var]&]; fF[v_]:=2*a1*v;
gG[v_]:=a2*v; sys1={D[u[x,t],x]==u[x,t]*f[v[x,t]],
 D[v[x,t],t]/g[v[x,t]]==u[x,t]^k}
tr1={u[x,t]->uN[x,t]^(1/k),g[v[x,t]]->D[v[x,t],t]/D[w[x,t],t]}
{sys2=sys1/.tr1/.D[tr1[[1]],x], sys3=sys2//PowerExpand}
sys31={Thread[sys3[[1]]*k/uN[x,t]^(1/k)*uN[x,t],Equal],sys3[[2]]}
{tr2=sys31[[2,2]]->sys31[[2,1]], eq1=sys31[[1]]/.tr2/.D[tr2,x]}
{tr3=k*f[v[x,t]]->phi[w], eq2=eq1/.k*f[v[x,t]]->phi[w]}
eq3=Integrate[eq2[[1]],t]==Integrate[phi[w],w]+theta[x]
{tr4=w[x,t]->Hold[Integrate[1/g[v],v]], eq4=Thread[eq3*g[v[x,t]],
 Equal]/.tr3[[2]]->tr3[[1]]//Expand, eq5=(eq4/.w->w[x,t])/.tr4}
{eq6=eq5/.eq5[[1]]->D[v[x,t],x], eq70=eq6//PowerExpand}
{termk=trS1[eq70[[2]],k], eq71=eq70/.termk->trS3[termk,g]}
{trk=k*f[v]->Integrate[k*f[v],v], eq7=eq71[[1]]==eq71[[2]]/.
 v[x,t]->v/.trk, eq8=eq7/.k->1/.f[v]->fF[v]/.g[v]->gG[v]}
{eq9=eq8[[1]]==(eq8[[2]]/.v->v[x,t]), solv=DSolve[eq9,v[x,t],
 {x,t}]//First, term2=-Numerator[solv[[1,2]]], tr6=term2->xi[x],
 tr61=tr6/.x->K[2], solv1=solv/.tr6/.tr61/.C[1][t]->-a2*psi[t]}
term3=term3=trS1[Denominator[solv1[[1,2]]],K[2]]
{tr7=term3->phi[x]*a1, solv2=solv1/.tr7,
 tr8=D[term3/a1,x]->phi'[x], solv3=solv2/.tr8}
```

2. Solution of ODE. We obtain that the general solution of the first equation of the original nonlinear system has the form $u=\frac{F_1(t)}{(a\phi+b\psi)^2}$, where the arbitrary function $F_1(t)$ can be determined (substituting it into the original nonlinear system) and has the form $F_1(t)=-\frac{\phi_x v_t(a\phi+b\psi)}{bv^2}$. We find the solution $u=-\frac{\psi_t}{a\phi+b\psi}$, $v=-\frac{\phi_x}{a\phi+b\psi}$, which is exact solution of the original nonlinear system.

Maple:

```
Eq11:=subs(Solv3,Sys1[1]); Solu:=pdsolve(Eq11,u(x,t));
tr9:=diff(isolate(Solv3,psi(t)),t); simplify(expand(subs(Solv3,
 tr9))); EqF1:=simplify(expand(subs(Solu,Solv3,k=1,Sys1)));
SolF1:=factor(solve(EqF1,_F1(t))) assuming a>0,b>0;
SolF11:=subs(tr9,SolF1); Solu1:=simplify(expand(subs(Solv3,subs(
 SolF11,Solu)))); simplify(expand(subs(Solu1,Solv3,k=1,Sys1)));
```

Mathematica:

```
trF=f[v[x,t]]->fF[v[x,t]];  trG=g[v[x,t]]->gG[v[x,t]];
{eq11=(sys1[[1]]/.trF)/.solv3, solu=DSolve[eq11,u[x,t],
 {x,t}]//First}
tr9=D[Solve[solv3/.Rule->Equal,psi[t]],t]//First//Simplify
eqf1=sys1/.k->1/.solu/.D[solu,x]/.trF/.trG/.solv3/.D[solv3,t]
{solf1=Solve[eqf1,C[1][t]]//First, solf11=solf1/.tr9}
solu1=solu/.solf11/.solv3/.D[solv3,t]
sys1/.k->1/.solu1/.D[solu1,x]/.trF/.trG/.solv3/.D[solv3,t]
```
□

2.8.3 Separation of Variables

Let us consider the systems of nonlinear second-order PDEs of the following form [124]

$$u_t=u_{xx}+a_1uF(u-v)+a_2G_1(u-v),\ \ v_t=v_{xx}+a_1vF(u-v)+a_2G_2(u-v),$$

where $\{x \in \mathbb{R}, t \geq 0\}$ and F, G_1 and G_2 are arbitrary functions of the argument $(u-v)$ and a_1, a_2 are real constants. Applying the method of generalized separation of variables (see, e.g., Problem 2.31) and searching for exact solutions in the form

$$u(x,t)=\phi_1(t)\theta(x,t)+\psi_1(t),\ \ v(x,t)=\phi_2(t)\theta(x,t)+\psi_2(t), \qquad (2.6)$$

we can find the exact solution of this nonlinear system. We assume that the functions $\phi_i(t)$ and $\psi_i(t)$ $(i=1,2)$ are chosen so that the two equations of the original nonlinear system can be reduced to a single equation for the function $\theta(x,t)$. Also we assume that the argument $(u-v)$ of the functions F, G_1, and G_2 in the original system depends only on t, i.e., $(u-v)_x=(\phi_1-\phi_2)\theta_x=0$. Introducing the new function $\phi(t)$, we can obtain the following conditions on $\phi_i(t)$ $(i=1,2)$: $\phi_1=\phi_2=\phi$.

Problem 2.56 *Nonlinear system of second-order equations. Generalized separation of variables.* For the system of nonlinear second-order PDEs,

$$u_t=u_{xx}+(a_1u+a_2)\ln(u-v),\ \ v_t=v_{xx}+(a_1v+a_2)\ln(u-v),$$

where $F=G_1=G_2=\ln(u-v)$, apply the method of generalized separation of variables, determine the exact solution, and verify that the solution obtained is exact solution.

1. System of ODEs. Substituting the relations (2.6) describing the solution form and the above derived conditions $\phi_i(t)$ $(i=1,2)$ into the original system, we reduce the original system to the form (Sys6):

$$(\phi^{-1}\phi_t - a_1 \ln z)\theta + \theta_t - \theta_{xx} + \phi^{-1}(\psi_{1t} - a_1\psi_1 \ln z - a_2 \ln z) = 0,$$
$$(\phi^{-1}\phi_t - a_1 \ln z\Big)\theta + \theta_t - \theta_{xx} + \phi^{-1}(\psi_{2t} - a_1\psi_2 \ln z - a_2 \ln z) = 0, \qquad (2.7)$$

where $z=\psi_1-\psi_2$. Solving the resulting system of equations, we obtain the following system of ordinary differential equations for the functions $\phi(t)$ and $\psi_i(t)$ $(i=1,2)$ (Eq1, Eq21, Eq31):

$$\phi_t/\phi = a_1 \ln z, \quad \psi_{1t} = \ln(z)(a_1\psi_1 + a_2), \quad \psi_{2t} = \ln z(a_1\psi_2 + a_2),$$

and the linear heat equation $\theta_t=\theta_{xx}$ for the function $\theta(x,t)$.

2. Solution of system. We determine the exact solution of the original nonlinear system: $u=\phi\theta+\psi_1$ and $v=\phi\theta+\psi_2$, where the function $\phi(t)$ has the form $\phi = \exp\Big(\int a_1 \ln(\psi_1 - \psi_2\, dt\Big)$ (Solphi), the functions $\psi_i(t)$ $(i=1,2)$ are described by the ODEs obtained above, and the function $\theta(x,t)$ is a solution of the linear heat equation. Applying *Maple* predefined functions, we solve the system of ODEs for $\psi_1(t)$, $\psi_2(t)$ and the linear heat equation for $\theta(x,t)$, obtain the solution (Solu, Solv), and verify that this solution is exact solution of the given nonlinear system.

Maple:

```
with(PDEtools): declare((u,v,W1,W2)(x,t),(phi1,phi2,psi1,psi2,
 phi)(t),theta(x,t)); tr1:=phi1(t)*theta(x,t)+psi1(t);
tr2:=phi2(t)*theta(x,t)+psi2(t); Sys1:=(u,v)->[diff(u(x,t),t)-
 diff(u(x,t),x$2)-u(x,t)*a1*ln(u(x,t)-v(x,t))-a2*ln(u(x,t)-
 v(x,t)),diff(v(x,t),t)-diff(v(x,t),x$2)-a1*v(x,t)*ln(u(x,t)-
 v(x,t))-a2*ln(u(x,t)-v(x,t))]; Sys2:=expand(Sys1(W1,W2));
Sys3:=expand(subs(W1(x,t)=tr1, W2(x,t)=tr2,Sys2));
Cond1:=diff(u(x,t)-v(x,t),x); Cond11:=collect(expand(subs(
 u(x,t)=tr1,v(x,t)=tr2,Cond1)),diff)=0; Sol1:=isolate(lhs(
 Cond11)=0,phi2(t)); tr31:={phi1(t)=phi(t)}; tr32:={subs(tr31,
 Sol1)}; Sys4:=collect(expand(subs(tr31,tr32,Sys3)),diff);
Sys5:=expand([Sys4[1]/phi(t),Sys4[2]/phi(t)]);
Sys6:=collect(expand(factor(Sys5)),theta);
Eq1:=remove(has,expand(select(has,Sys6[1],ln)/theta(x,t)),
 theta); Eq2:=remove(has,Sys6[1],[theta]); Eq3:=remove(has,
 Sys6[2],theta); Eq21:=simplify(isolate(Eq2,diff(psi1(t),t)));
Eq31:=simplify(isolate(Eq3,diff(psi2(t),t)));
Eq40:=[selectremove(has,Sys6[1],[a1,a2,phi])];
Eq4:=op(Eq40[2]); Solphi:=subs(_C1=1,dsolve(Eq1,phi(t)));
```

```
Solu:=u(x,t)=subs(Solphi,subs(tr31,tr1)); Solv:=v(x,t)=subs(
 Solphi,subs(tr32,tr2)); Consts:={_C1=0,_C2=1,_C3=1,_c[1]=1};
solpsi:=dsolve({Eq21,Eq31},{psi1(t),psi2(t)}); soltheta:=pdsolve(
 diff(theta(x,t),t)-diff(theta(x,t),x$2)=0,build);
solpsi1:=[solpsi[1,1],subs(solpsi[1,1],op(solpsi[2]))];
SoluF:=rhs(simplify(subs(Consts,subs(solpsi1,soltheta,Solu))));
SolvF:=rhs(simplify(subs(Consts,subs(solpsi1,soltheta,Solv))));
uv:=SoluF-SolvF; Test1:=simplify(diff(SolvF,t)-diff(SolvF,x$2)-
 a1*SolvF*ln(uv)-a2*ln(uv))=0; Test2:=simplify(diff(SoluF,t)-
 diff(SoluF,x$2)-a1*SoluF*ln(uv)-a2*ln(uv))=0;
```

Mathematica:

```
trS1[eq_,var_]:=Select[eq,MemberQ[#,var,Infinity]&];
trS3[eq_,var_]:=Select[eq,FreeQ[#,var]&]; trD[u_,var_]:=Table[
 D[u,{var,i}],{i,1,2}]//Flatten; sys1[u_,v_]:={D[u,t]-D[u,{x,2}]-
 a1*u*Log[u-v]-a2*Log[u-v], D[v,t]-D[v,{x,2}]-a1*v*Log[u-v]-
 a2*Log[u-v]}; {tr1=phi1[t]*theta[x,t]+psi1[t], tr2=phi2[t]*
 theta[x,t]+psi2[t], sys2=sys1[w1,w2]//Expand, sys3=sys1[tr1,
 tr2]//Expand, cond1=D[u[x,t]-v[x,t],x], cond11=(cond1/.
 {D[u[x,t],x]->D[tr1,x],D[v[x,t],x]->D[tr2,x]}//Expand)==0}
{sol1=Solve[cond11,phi2[t]], tr31=phi1[t]->phi[t],
 tr32=sol1/.tr31, sys4=(sys3/.tr31/.tr32/.D[tr31,t]/.D[tr32,t]//
 Simplify)[[1,1]], sys5={sys4[[1]]/phi[t],sys4[[2]]/phi[t]}//
 Expand, sys6=Collect[Expand[Factor[sys5]],theta[x,t]]}
eq1=Coefficient[trS1[sys6[[1]],a1],theta[x,t]]==0
{eq2=trS3[sys6[[1]],theta], eq3=trS3[sys6[[2]],theta]}
{Solve[eq2==0,psi1'[t]], Solve[eq3==0,psi2'[t]]}
{eq4=trS3[trS3[trS3[sys6[[1]],a1],a2],phi], solphi=DSolve[eq1,
 phi[t],t]/.C[1]->1//First, solu=u[x,t]->tr1/.tr31/.solphi,
 solv=v[x,t]->tr2/.tr32/.solphi}
consts={C[1]->0,C[2]->0,C[3]->1,c1->1}; solpsi=DSolve[{eq2==0,
 eq3==0},{psi1[t],psi2[t]},t]//FullSimplify//First
tr4={psi1[K[1]]->solpsi[[2,2]],psi2[K[1]]->solpsi[[1,2]]};
DSolve[eq4==0,theta,{x,t}]
soltheta=theta[x,t]->Exp[Sqrt[c1]*x]*C[3]*Exp[c1*t]*C[1]+C[3]*
 Exp[c1*t]*C[2]/Exp[Sqrt[c1]*x];
{soluF=solu/.solpsi/.soltheta/.tr4/.consts//Factor, solvF=solv/.
 solpsi/.soltheta/.tr4/.consts//Factor//First}
Assuming[{t>0,a1>0},sys1[u[x,t],v[x,t]]/.soluF/.solvF/.trD[soluF,
 x]/.trD[soluF,t]/.trD[solvF,x]/.trD[solvF,t]//FullSimplify]
```

□

Chapter 3
Geometric-Qualitative Approach

In this chapter, following a geometric-qualitative approach to partial differential equations, we will consider important methods and concepts concerning quasilinear and nonlinear PDEs (in two independent variables) and solutions of classical and generalized Cauchy problems (with continuous and discontinuous initial data), namely, the Lagrange method of characteristics and its generalizations, the concepts of solution surfaces (or integral surfaces), general solutions, discontinuous or weak solutions, solution profiles at infinity, complete integrals, the Monge cone, characteristic directions. The method of characteristics can be applied to first-order PDEs in n ($n>2$) independent variables, higher-order PDEs, and general nonlinear PDEs [13]. The method of characteristics is applicable to all hyperbolic systems. Finally, we perform the standard qualitative analysis for studying geometric properties of the solutions of nonlinear PDEs and nonlinear systems.

3.1 Method of Characteristics

Now we consider the geometrical interpretation of the quasi-linear first-order equations and the method of finding their solutions, called the *method of characteristics*. This method allows us to reduce a PDE to a system of ODEs along which the given PDE with some initial data is integrable. Once the system of ODEs is found, it can be solved along the characteristics and transformed into a general solution for the original PDE. This method was firstly proposed by Lagrange in 1772 and 1779 for solving first-order linear and nonlinear PDEs.

3.1.1 Characteristic Directions. General Solution

In this section, we will construct integral surfaces and characteristic curves of first-order equations, which arise in a variety of physical theories, e.g., fluid and analytical dynamics, continuum mechanics, geometrical optics.

Definition 3.1 The solution $u(x, y)$ of the first-order equation of the general form $\mathcal{F}(x, y, u, u_x, u_y)=0$ in two independent variables (x, y) can be visualized geometrically as a surface, called an *integral surface* in the (x, y, u)-space.

For example, for the equation

$$A(x, y, u)u_x+B(x, y, u)u_y-C(x, y, u)=0, \tag{3.1}$$

a solution has an explicit form $u=u(x, y)$ or an implicit form $f(x, y, u)= u(x, y)-u=0$ and represents a solution surface (or integral surface) in (x, y, u)-space. At a point (x, y, u) on the integral surface, the gradient vector $\nabla f=(f_x, f_y, f_u)=(u_x, u_y, -1)$ is normal to the integral surface. Therefore, Eq. (3.1) can be written in the form

$$Au_x+Bu_y-C=(A, B, C)\cdot(u_x, u_y, -1)=0.$$

Definition 3.2 A *direction vector field* or the *characteristic direction*, (A, B, C), is a *tangent vector* of the integral surface $f(x, y, u)=0$ at the point (x, y, u).

Definition 3.3 A *characteristic curve* is a curve in (x, y, u)-space such that the tangent at each point coincides with the characteristic direction field (A, B, C).

The parametric equations of the characteristic curve can be written in the form $x=x(t)$, $y=y(t)$, $u=u(t)$, and the tangent vector to this curve $(dx/dt, dy/dt, du/dt)$ is equal to (A, B, C).

Definition 3.4 The *characteristic equations* of Eq. (3.1) in parametric form is the system of ODEs: $dx/dt = A(x, y, u)$, $dy/dt = B(x, y, u)$, $du/dt = C(x, y, u)$. The *characteristic equations* of Eq. (3.1) in nonparametric form are: $dx/A(x, y, u)=dy/B(x, y, u)=du/C(x, y, u)$.

Definition 3.5 A *characteristic* is the projection on $u=0$ of a characteristic curve.

The slopes of the characteristics are determined by the equation $dy/dx=B(x, y, u)/A(x, y, u)$ (about the slopes of the characteristics for nonlinear equations see Problem 3.15).

The geometrical interpretation and these results can be generalized for the case of $n > 2$ independent variables and for higher-order PDEs. For example, generalizing Eq. (3.1) to n independent variables, we have the equation $\sum_{i=1}^{n} A_i(x_1, \ldots, x_n, u)u_{x_i}{=}C(x_1, \ldots, x_n, u)$, that can be reduced to the system of ODEs $dx_1/A_1{=}\ldots{=}dx_n/A_n{=}du/C$ for an n parameter family of characteristics. In this case an $n{-}1$ parameter family of solutions of this equation generates an n-dimensional surface in the $(n{+}1)$-dimensional space (with Cartesian coordinates $x_1, \ldots, x_n$, and u), and in nonparametric form this solution takes the form $u{=}u(x_1, \ldots, x_n)$. In this case, the general solution of this equation depends on one arbitrary function of $n{-}1$ variables.

For nonlinear PDEs, the geometrical interpretation is more complicated: the normals to integral surfaces at a point do not lie in a plane, the tangent planes do not intersect along one straight line, they envelope along a curved surface, the Monge cone (see Sect. 3.2).

Problem 3.1 *First-order equations. Direction vector fields.* Let us consider the first-order PDEs of the form

$$uu_x{+}u_t{+}u{=}0, \quad xu_x{+}tu_t{+}(x^2 + t^2){=}0,$$

where $\{x \in \mathbb{R}, t \geq 0\}$. Construct the direction vector fields for the given first-order partial differential equations.

Maple:

```
with(plots): R1:=-Pi..Pi: N1:=10: V1:=[-24,72]; V2:=[-59,-29];
A1:=`THICK`;setoptions3d(fieldplot3d,grid=[N1,N1,N1],axes=boxed);
r1:=(x,t,u)->[u,1,-u]; r2:=(x,t,u)->[x,t,-(x^2+t^2)];
fieldplot3d(r1(x,t,u),x=R1,t=R1,u=R1,arrows=SLIM,orientation=V1);
fieldplot3d(r2(x,t,u),x=R1,t=R1,u=R1,arrows=A1,orientation=V2);
```

Mathematica:

```
r1[x_,t_,u_]:={u,1,-u}; r2[x_,t_,u_]:={x,t,-(x^2+t^2)};
{n1=10, p=Pi, v1={1,-3,1}, v2={1,2,3}}
SetOptions[VectorPlot3D,VectorColorFunction->Hue,VectorPoints->
 {n1,n1,n1},VectorStyle->Arrowheads[0.02],PlotRange->All];
VectorPlot3D[r1[x,t,u],{x,-p,p},{t,-p,p},{u,-p,p},ViewPoint->v1]
VectorPlot3D[r2[x,t,u],{x,-p,p},{t,-p,p},{u,-p,p},
 ViewPoint->v2]
```

□

General solution (or general integral) of the given first-order PDE is an equation of the form $f(\phi, \psi)=0$, where f is an arbitrary function of the known functions $\phi=\phi(x, y, u)$, $\psi=\psi(x, y, u)$ and provides a solution of this partial differential equation. The functions $\phi=C_1$, $\psi=C_2$ are solution curves of the characteristic equations or the families of characteristic curves of Eq. (3.1). Here C_1 and C_2 are constants.

Problem 3.2 *First-order equation. General solution.* Let us consider the first-order PDE of the form

$$u(x+y)u_x+u(x-y)u_y=x^2+y^2,$$

where $\{x \in \mathbb{R}, y \in \mathbb{R}\}$. Prove that the general solution of the given first-order PDE takes the form $f(2xy-u^2, x^2-y^2-u^2)$.

Maple:

```
with(PDEtools): declare(u(x,y)); U:=diff_table(u(x,y));
F:=U[]*(x+y); G:=U[]*(x-y); H:=x^2+y^2;
PDE:=F*U[x]+G*U[y]=H; CharEqs:=[dx/F,dy/G,du/H];
Eq1:=expand([U[]*CharEqs[1],U[]*CharEqs[2],-U[]*CharEqs[3]]);
Eq21:=numer(Eq1[1])*y+numer(Eq1[2])*x+numer(Eq1[3]);
Eq22:=expand(denom(Eq1[1])*y+denom(Eq1[2])*x-denom(Eq1[3]));
Eq31:=numer(Eq1[1])*x-numer(Eq1[2])*y+numer(Eq1[3]);
Eq3:=expand(denom(Eq1[1])*x-denom(Eq1[2])*y-denom(Eq1[3]));
I11:=(x*y-int(u,u))=C1; I1:=subs(2*C1=C1,I11*2);
I21:=int(op(1,Eq31)/dx,x)+int(op(2,Eq31)/dy,y)-int(u,u)=C2;
I2:=subs(2*C2=C2,I21*2);  f(lhs(I1),lhs(I2));
```

Mathematica:

```
{fU=u[x,y], fF=u[x,y]*(x+y), fG=u[x,y]*(x-y), fH=x^2+y^2}
{pde=fF*D[fU,x]+fG*D[fU,y]==fH,charEqs={dx/fF,dy/fG,du/fH}}
{eq1={fU*charEqs[[1]],fU*charEqs[[2]],-fU*charEqs[[3]]}//
 Expand, eq21=Numerator[eq1[[1]]]*y+Numerator[eq1[[2]]]*x+
 Numerator[eq1[[3]]], eq22=Denominator[eq1[[1]]]*y+
 Denominator[eq1[[2]]]*x-Denominator[eq1[[3]]]//Expand}
{eq31=Numerator[eq1[[1]]]*x-Numerator[eq1[[2]]]*y+
 Numerator[eq1[[3]]], eq3=Denominator[eq1[[1]]]*x-
 Denominator[eq1[[2]]]*y-Denominator[eq1[[3]]]//Expand}
{i11=(x*y-Integrate[u,u])==c1, i1=(Thread[i11*2, Equal]//
 Expand)/.{2*c1->c1}, i21=Integrate[eq31[[1]]/dx,x]+
 Integrate[eq31[[2]]/dy,y]-Integrate[u,u]==c2}
i2=Thread[i21*2,Equal]//Expand/.{2*c2->c2}
f[i1[[1]],i2[[1]]]
```

□

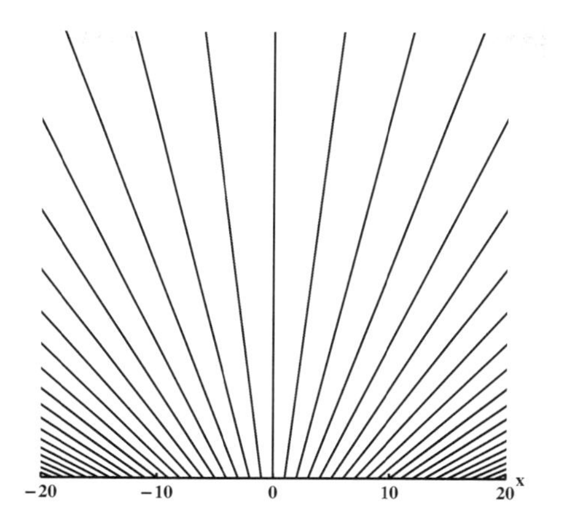

Fig. 3.1. Characteristics for a rarefaction wave

3.1.2 Integral Surfaces. Cauchy Problem

In many problems in applied sciences and engineering we have to obtain solutions of nonlinear PDEs that satisfy some conditions. In this section, we will determine exact solutions of initial value problems or Cauchy problems.

Problem 3.3 *Inviscid Burgers equation. Integral surfaces. Cauchy problem.* Let us consider the following classical Cauchy problem for the inviscid Burgers equation:

$$u_t + uu_x = 0, \quad u(x,0) = x+1,$$

where $\{x \in \mathbb{R}, t \geq 0\}$. Applying the method of characteristics, prove that the solution of this Cauchy problem is $u(x,t) = (x+1)/(t+1)$ and plot characteristics. It can be observed that the solution is well-defined at all future times. Physically, such solutions represent *rarefaction waves*, which gradually spread out as time increases (see Fig. 3.1).

Maple:

```
with(plots); tR:=0..5; xR:=-20..20; ODE:=diff(U(t),t)=0;
Sol_Ch:=dsolve({ODE,U(0)=X[0]}); Eq_Ch:=diff(x(t),t)=U(t);
Eq_Ch:=subs(Sol_Ch,Eq_Ch); Cur_Ch:=dsolve({Eq_Ch, x(0)=X[0]});
display([seq(plot([subs(X[0]=x,eval(x(t),Cur_Ch)),t,t=tR],
         color=blue,thickness=2),x=xR)],view=[xR,tR]);
```

```
u:=unapply(subs(X[0]=solve(subs(x(t)=x+1,Cur_Ch),X[0]),
   eval(U(t),Sol_Ch)),x,t);
```

Mathematica:

```
SetOptions[Plot,ImageSize->500,PlotStyle->{Hue[0.7],
 Thickness[0.001]}]; {ode=D[uN[t],t]==0, solCh=DSolve[{ode,
 uN[0]==xN[0]},uN[t],t], eqCh=D[x[t],t]==uN[t]/.solCh[[1]]}
curCh=DSolve[{eqCh,x[0]==xN[0]},x[t],t]//Simplify
g=Table[ParametricPlot[{(curCh[[1,1,2]]/.xN[0]->x),t},
 {t,0,5}],{x,-20,20}]; Show[g,PlotRange->{{-20,20},{0,5}},
 AspectRatio->1]
u1=solCh[[1]]/.Solve[curCh[[1,1,2]]==x+1,xN[0]]
u[xN_,tN_]:=u1[[1,1,2]]/.{x->xN,t->tN}; u[x,t]
```
□

Problem 3.4 *Inhomogeneous inviscid Burgers equation. Integral surfaces. Cauchy problem.* Let us consider the initial value problem for the inhomogeneous inviscid Burgers equation

$$u_t+uu_x=g(x,t), \quad u(x,0)=f(x),$$

where $\{x \in \mathbb{R}, t \geq 0\}$, $g(x,t)=x$, $f(x)=1$ or $f(x)=x$. Applying the method of characteristics, prove that the solution of this Cauchy problem is equal to $u(x,t)=x\tanh t+\operatorname{sech} t$ (`SolFin1`) for $f(x)=1$ and $u(x,t)=\pm x$ (`SolFin2`) for $f(x)=x$. Verify that the results obtained are solutions of the given Cauchy problem.

Maple:

```
PDE:=u->diff(u(x,t),t)+u(x,t)*diff(u(x,t),x)=x;
ODEs:=dt/1=(dx/u=du/x); f1:=x->1; f2:=x->x;
Eq2:=lhs(rhs(ODEs))+rhs(rhs(ODEs))=d(x+u)/(x+u);
Eq3:=rhs(Eq2)=lhs(ODEs); Eq4:=log(x+u)-log(C1)=t;
tr1:=isolate(Eq4,C1); Eq5:=lhs(rhs(ODEs))=rhs(rhs(ODEs));
Eq6:=Eq5*x*u; Eq61:=op(1,lhs(Eq6))=op(1,rhs(Eq6));
Eq7:=int(lhs(Eq61),x)+C2/2=int(rhs(Eq61),u);
Eq71:=simplify(Eq7*2); tr2:=isolate(Eq7,C2);
sys1:=simplify(subs(t=0,u=f1(x),[tr1,tr2]));
tr3:=isolate(sys1[1],x); tr21:=simplify(subs(tr3,sys1[2]));
Sol1:=subs(tr1,subs(tr21,tr2)); Sol11:=lhs(Sol1)=factor(
 rhs(Sol1)); Sol12:=normal(Sol11/(x+u));
Sol13:=collect(lhs(Sol12),[exp(t),u])=rhs(Sol12);
Sol14:=combine(convert(combine(Sol13),trigh));
```

```
Sol15:=expand(Sol14); Sol16:=u=solve(Sol15,u);
Sol17:=collect(Sol16,x); SolFin1:=expand(simplify(Sol17,trig));
sys2:=simplify(subs(t=0,u=f2(x),[tr1,tr2]));
tr3:=isolate(sys2[1],x); tr21:=sys2[2];
Sol1:=subs(tr1,subs(tr21,tr2)); SolFin2:=u=[solve(Sol1,u)];
u1:=unapply(rhs(SolFin1),x,t);expand(PDE(u1));
u21:=unapply(rhs(SolFin2)[1],x,t); expand(PDE(u21));
u22:=unapply(rhs(SolFin2)[2],x,t); expand(PDE(u22));
```

Mathematica:

```
pde[u_]:=D[u,t]+u*D[u,x]==x; f1[x_]:=1; f2[x_]:=x;
{odes=dt/1==(dx/u==du/x), eq2=odes[[2,1]]+odes[[2,2]]==
 HoldForm[d(x+u)/(x+u)], eq3=eq2[[2]]==odes[[1]]}
{eq4=Log[x+u]-Log[c1]==t, tr1=Solve[eq4,c1]//First}
{eq5=odes[[2,1]]==odes[[2,2]], eq6=Thread[eq5*x*u,Equal]}
eq7=Integrate[eq6[[1]]/dx,x]+c2/2==Integrate[eq6[[2]]/du,u]
{eq71=Thread[eq7*2,Equal]//Expand,tr2=Solve[eq7,c2]//First}
{sys1={tr1,tr2}/.t->0/.u->f1[x]//Flatten, tr3=Solve[
 sys1[[1]]/.Rule->Equal,x], tr21=sys1[[2]]/.tr3}
sol1=tr2/.tr21/.tr1/.Rule->Equal//Expand//First
{sol11=sol1[[1]]==Factor[sol1[[2]]], sol12=Thread[
 sol11/(x+u),Equal]//Factor, sol13=Collect[sol12[[1]],
 {Exp[t],u}]==sol12[[2]], sol14=sol13//ExpToTrig}
{sol15=sol14//FullSimplify//ExpToTrig, sol16=Solve[
 sol15,u]//First, solFin1=Collect[sol16,x]}
sys2={tr1,tr2}/.t->0/.u->f2[x]//Flatten
tr3=Solve[sys2[[1]]/.Rule->Equal,x]//First
{tr21=sys2[[2]], sol1=tr2/.tr21/.tr1/.Rule->Equal}
solFin2=Solve[sol1,u]//Flatten
test1=Map[FullSimplify,{pde[solFin1[[1,2]]],
 pde[solFin2[[1,2]]],pde[solFin2[[2,2]]]}]
```

□

Problem 3.5 *Inhomogeneous inviscid Burgers equation. Integral surfaces. Generalized Cauchy problem.* Let us consider the equation

$$u_t+uu_x=1,$$

where $\{x\in\mathbb{R}, t\geq 0\}$. We formulate the Cauchy problem for this equation to determine the family of curves in the parametric form

$$x=x(s,r), \quad t=t(s,r), \quad u=u(s,r).$$

Let us find a solution surface $u{=}u(x,t)$ with the given initial data at $s{=}0$: $x(0,r)=r$, $t(0,r)=2r$, $u(0,r)=r$ (r is a parameter) or $u(r,2r)=r$.

1. Applying the method of characteristics, prove that the integral surface of the given Cauchy problem has the form $u(x,t){=}\dfrac{2x{-}2t{+}t^2}{2(t{-}1)}$, where $t{\neq}1$.

2. Visualize the characteristics defined by the equation $-(t{-}r)^2{+}2x{-}2r+r^2{=}0$ in the (x,t)-plane for a given parameter r.

3. Visualize the solution obtained $u(x,t)$ at various moments of time t ($t{\neq}1$).

Maple:

```
with(PDEtools): with(Student[Precalculus]): with(plots):
setoptions(implicitplot,numpoints=100): declare(u(x,t));
alias(u=u(x,t)); Chars:=NULL: GrU1:=NULL: GrU2:=NULL:
PDE1:=u->diff(u(x,t),t)+u(x,t)*diff(u(x,t),x)=1; PDE1(u);
IniCurve1:=u(r,2*r)=r; IniCurve2:=[T(0,r)=2*r,X(0,r)=r,U(0,r)=r];
CharEqs:=[diff(T(s),s)=1,diff(X(s),s)=U(s),diff(U(s),s)=1];
sys1:={subs(_C1=rhs(IniCurve2[3]),dsolve(CharEqs[3],U(s))),
       subs(_C1=rhs(IniCurve2[1]),dsolve(CharEqs[1],T(s)))};
sys11:=subs(_C1=rhs(IniCurve2[2]),dsolve(subs(U(s)=rhs(sys1[2]),
       CharEqs[2]),X(s))); tr2:={X(s)=X,T(s)=T,U(s)=U};
CharsEq:=eliminate(subs(tr2,{sys1[1],sys11}),s);
CharsEq1:=subs(r=R,CompleteSquare(op(CharsEq[2])));
for R from -10 to 10 by 0.5 do Chars:=Chars,CharsEq1; od:
Chars; implicitplot([Chars],X=-10..10,T=0..2,color=blue);
sys3:=subs(tr2,sys1 union {sys11}); Ch1:=op(op(2,CharsEq));
r1:=isolate(Ch1,r); s1:=simplify(isolate(subs(r1,sys3[1]),s));
SolFin:=simplify(subs(r1,s1,sys3[2])); U1:=rhs(SolFin);
for i from 0 to 0.999 by 0.1 do
 GU1||round(i*10):=plot(subs(T=i,U1),X=-10..10,-10..10,
   color=[blue,blue]): GrU1:=GrU1,GU1||round(i*10): od:
for j from 1.001 to 2 by 0.1 do
 GU2||round(j*10):=plot(subs(T=j,U1),X=-10..10,-10..10,
   color=[blue,blue]): GrU2:=GrU2,GU2||round(j*10) od:
 display({GrU1,GrU2});
```

Mathematica:

```
cS[a_.x_^2+b_.x_+c_.]:=a*((x+b/(2*a))^2-(b^2-4*a*c)/(4*a^2));
completeSquare[x_]:=If[TrueQ[x==Expand[x]],x,cS[Expand[x]]];
```

```
p=10; SetOptions[ContourPlot,Frame->True,ImageSize->500,
ContourShading->False,ContourStyle->Blue]; guN1={}; guN2={};
SetOptions[Plot,Frame->True,ImageSize->500,PlotRange->
 {{-p,p},{-p,p}},PlotStyle->Blue]; chars={}; gC={};
pde1[u_]:=D[u[x,t],t]+u[x,t]*D[u[x,t],x]==1; pde1[u]
{iniCurve1=u[r,2*r]->r, iniCurve2={tN[0,r]->2*r,xN[0,r]->r,
 uN[0,r]->r}, charEqs={tN'[s]==1,xN'[s]==uN[s],uN'[s]==1}}
sys1={DSolve[charEqs[[3]],uN[s],s]/.C[1]->iniCurve2[[3,2]],
 DSolve[charEqs[[1]],tN[s],s]/.C[1]->iniCurve2[[1,2]]}//Flatten
sys11=DSolve[(charEqs[[2]]/.uN[s]->sys1[[1,2]]),xN[s],s]/.
 C[1]->iniCurve2[[2,2]]//Flatten
tr2={xN[s]->xN,tN[s]->tN,uN[s]->uN}
{charsEq=Eliminate[Flatten[{sys1[[2]],sys11}/.tr2/.Rule->Equal],
 s]//Together, charsEq1=Map[Factor,completeSquare[charsEq[[1]]-
 charsEq[[2]]]]/.r->rN}
Do[chars=Append[chars,charsEq1],{rN,-p,p,0.5}]; chars
Do[gC=Append[gC,ContourPlot[chars[[i]]==0,{xN,-p,p},{tN,0,2}]],
 {i,1,Length[chars]}]; Show[gC]
{sys3={sys1,sys11}/.tr2//Flatten, ch1=charsEq[[1]]-
 charsEq[[2]]==0, r1=Solve[ch1,r]//First, s1=Solve[sys3[[2]]/.
 Rule->Equal/.r1,s]//First, solFin=sys3[[1]]/.Rule->Equal/.
 r1/.s1//Simplify, uN1=solFin[[2]]}
Do[guN1=Append[guN1,Plot[uN1/.tN->i,{xN,-p,p}]],{i,0,0.999,0.1}];
Do[guN2=Append[guN2,Plot[uN1/.tN->i,{xN,-p,p}]],{i,1.001,2,0.1}];
Show[{guN1,guN2}]
```
□

Now, applying the method of characteristics, let us investigate the first-order quasilinear PDEs, which can be written as *conservation laws* (see Sect. 4.2.1), and solve Cauchy problems (with continuous and discontinuous initial data).

Problem 3.6 *Kinematic wave equation. Characteristics. Cauchy problem.* Let us consider the kinematic wave equation with the initial curve

$$u_t+c(u)u_x=0, \quad u(x,0)=f(x),$$

where $\{x\in\mathbb{R}, t\geq 0\}$. This equation represents a homogeneous case of the *conservation law* $u_t+c(u)u_x=g(x,t,u)$, where $c(u)$ and $f(x)\in C^1(\mathbb{R})$.

1. Applying the method of characteristics, prove that the solution of the given Cauchy problem can be written in the *parametric form* $u(x,t)=f(\xi)$, $\xi=x-c(f(\xi))t$.

2. Solve the given Cauchy problem for $c(u)=u$, $g(x,t,u)=0$, and $u(x,0)=f(x)=x^2$.

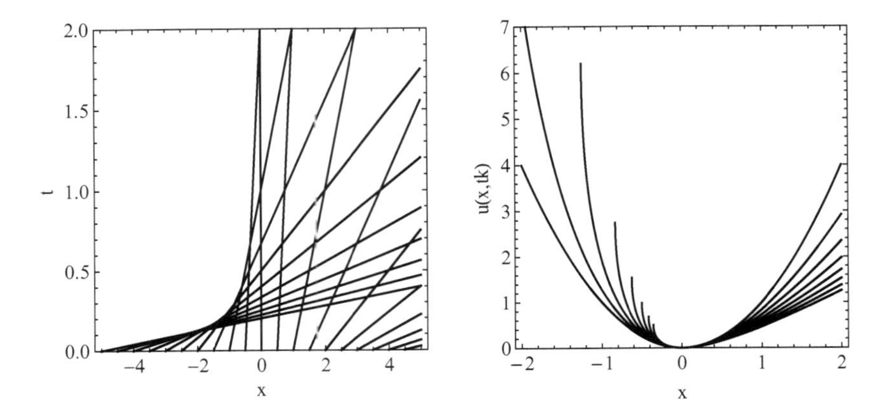

Fig. 3.2. The Cauchy problem $u_t+uu_x=0$, $u(x,0)=x^2$: characteristic lines and the solution $u(x,t)$ for $t_k = 0, 0.1, 0.2, \ldots, 0.7$

3. Visualize the characteristics, defined by the equation $\xi=x-\xi^2 t$ in the (x,t)-plane, and the solution $u(x,t)$ at various moments of time t (see Fig. 3.2).

Maple:

```
with(PDEtools): declare(u(x,t)); alias(u=u(x,t));
PDE:=u->diff(u(x,t),t)+c(u(x,t))*diff(u(x,t),x)=0; PDE(u);
IniCurve:=u(x,0)=f(x); IniData:=u(xi,0)=f(xi);
tr1:={X(t)=X,U(t)=U}; CharEqs:=[diff(X(t),t)=c(U),
 diff(U(t),t)=0]; Eq1:=dsolve(CharEqs[1],X(t));
Eq2:=isolate(Eq1,_C1); Eq3:=subs(tr1,Eq2); Eq4:=isolate(Eq3,X);
Eq5:=subs(_C1=xi,c(U)=c(rhs(IniData)),Eq4);
Eq6:=u(x,t)=(u(rhs(Eq5),t)=IniData);
Eq7:={u(x,t)=rhs(rhs(rhs(Eq6))),-Eq5+xi+X};
PDE1:=u->diff(u(x,t),t)+u(x,t)*diff(u(x,t),x)=0; PDE1(u);
IniCurve1:=(u(x,0)=x^2)=f(x);
u1:=unapply(subs(f(xi)=rhs(lhs(IniCurve1)),rhs(Eq7[2])),x,t);
Eq8:=u1=u1(xi,t);
Eq9:=subs(c(f(xi))=rhs(lhs(IniCurve1)),Eq7[1]);
Eq91:=subs(x=xi,Eq9); Eq10:=[solve(Eq91,xi)];
Eq11:=xi^2=expand(Eq10[1]^2); Eq12:=xi^2=expand(Eq10[2]^2);
SolFin1:=subs(Eq11,Eq8); SolFin2:=subs(Eq12,Eq8);
limit(rhs(SolFin1),t=0); limit(rhs(SolFin2),t=0);
CharsEq:=subs(xi=eta,Eq91); Chars1:=NULL:
```

```
for eta from -5 to 5 by 0.5 do Chars1:=Chars1,CharsEq; od:
Chars1; with(plots): setoptions(plot,numpoints=100):
implicitplot([Chars1],X=-5..5,t=0..2,color=blue);
U1:=rhs(SolFin1); GU0:=plot(rhs(op(1,IniCurve1)),x=-2..2):
GU3:=plot(subs(t=0.1,U1),X=-2..2,color=blue):
GU5:=plot(subs(t=0.2,U1),X=-2..2,color=blue):
display({GU0,GU3,GU5});
```

Mathematica:

```
p1=5; p2=2; chars1={}; gC={}; SetOptions[ContourPlot,Frame->True,
 ImageSize->500,ContourShading->False,ContourStyle->Blue];
SetOptions[Plot,Frame->True,ImageSize->500,PlotRange->{{-p2,p2},
 {0,7}}]; pde[u_]:=D[u[x,t],t]+c[u[x,t]]*D[u[x,t],x]==0;
{pde[u], iniCurve=u[x,0]->f[x], iniData=u[xi,0]->f[xi],
 tr1={xN[t]->xN,uN[t]->uN}, charEqs={xN'[t]==c[uN],uN'[t]==0}}
{eq1=DSolve[charEqs[[1]],xN[t],t]//First,
 eq2=Solve[eq1/.Rule->Equal,C[1]]//First, eq3=eq2/.tr1,
 eq4=Solve[eq3/.Rule->Equal,xN]//First,
 eq5=eq4/.C[1]->xi/.c[uN]->c[iniData[[2]]],
 eq6=u[x,t]->u[eq5[[1,2]],t]==iniData}
eq51=eq5/.Rule->Equal//First
eq52=Thread[Thread[eq51*(-1),Equal]+xi+xN,Equal]//Expand//ToRules
eq7={u[x,t]->eq6[[2,2,2]],eq52}//Flatten
pde1[u_]:=D[u[x,t],t]+u[x,t]*D[u[x,t],x]==0; pde1[u]
u1[xN_,tN_]:=eq7[[1,2]]/.f[xi]->iniCurve1[[1,2]]/.{x->xN,t->tN}
{iniCurve1=(u[x,0]==x^2)==f[x], u1[xi,t],
 eq9=eq7[[2]]/.c[f[xi]]->iniCurve1[[1,2]], eq91=eq9/.x->xi,
 eq10=Solve[eq91/.Rule->Equal,xi]//Flatten,
 eq11=xi^2->eq10[[1,2]]^2//Expand}
{eq12=xi^2->eq10[[2,2]]^2//Expand, solFin1=u1[xi,t]/.eq11}
{solFin2=u1[xi,t]/.eq12,Limit[solFin1,t->0],Limit[solFin2,t->0]}
{charsEq=eq91/.Rule->Equal/.xi->eta,
 charsEq1=charsEq[[1]]-charsEq[[2]], uN1=solFin2}
Do[chars1=Append[chars1,charsEq1],{eta,-p1,p1,0.5}]; chars1
Do[gC=Append[gC,ContourPlot[chars1[[i]]==0,{xN,-p1,p1},{t,0,2}]],
 {i,1,Length[chars1]}]; Show[gC]
gU0=Plot[iniCurve1[[1,2]],{x,-2,2},PlotStyle->Red]; gU3=Plot[
 uN1/.t->0.1,{xN,-2,2},PlotStyle->Blue]; gU5=Plot[uN1/.t->0.2,
 {xN,-2,2},PlotStyle->Blue]; Show[{gU0,gU3,gU5}]
```

□

Problem 3.7 *Kinematic wave equation. Characteristics. Cauchy problem.* Let us consider, as in the previous problem, the kinematic wave equation with the initial curve

$$u_t+c(u)u_x=0, \quad u(x,0)=f(x),$$

where $\{x \in \mathbb{R}, t \geq 0\}$.

1. Applying the method of characteristics, prove that the solution of the given Cauchy problem can be written in the *parametric form* $u(x,t)=f(\xi)$, $\xi=x-c(f(\xi))t$.

2. Solve the given Cauchy problem for $c(u)=u$, $g(x,t,u)=0$ (see the previous problem), and $u(x,0)=f(x)=\begin{cases}1-\xi^2, & |x|\leq 1,\\ 0, & |x|>1.\end{cases}$

3. Visualize the characteristics, defined by $\xi=x-(1-\xi^2)t$ in the (x,t)-plane and the solution $u(x,t)$ at various moments of time t.

Maple:

```
with(PDEtools): with(plots): setoptions(plot,numpoints=100):
declare(u(x,t)); alias(u=u(x,t)); BB:=color=blue: tR:=0..1:
PDE:=u->diff(u(x,t),t)+c(u(x,t))*diff(u(x,t),x)=0; PDE(u);
IniCurve:=u(x,0)=f(x); IniData:=u(xi,0)=f(xi);
tr1:={X(t)=X,U(t)=U}; Chs1:=NULL: Chs2:=NULL: Chs3:=NULL:
CharEqs:=[diff(X(t),t)=c(U),diff(U(t),t)=0];
Eq1:=dsolve(CharEqs[1],X(t)); Eq2:=isolate(Eq1,_C1);
Eq3:=subs(tr1,Eq2); Eq4:=isolate(Eq3,X); Eq5:=subs(_C1=xi,
 c(U)=c(rhs(IniData)),Eq4); Eq6:=u(x,t)=(u(rhs(Eq5),t)=IniData);
Eq7:={u(x,t)=rhs(rhs(rhs(Eq6))),-Eq5+xi+X};
PDE1:=u->diff(u(x,t),t)+u(x,t)*diff(u(x,t),x)=0; PDE1(u);
InCur1:=(u(x,0)=piecewise(abs(x)<=1,1^2-x^2,abs(x)>1,0))=f(x);
u1:=unapply(subs(f(xi)=rhs(lhs(InCur1)),rhs(Eq7[2])),x,t);
Eq8:=u1=u1(xi,t); Eq9:=subs(c(f(xi))=op(2,rhs(lhs(InCur1))),
 Eq7[1]); Eq91:=subs(x=xi,Eq9); Eq10:=[solve(Eq91,xi)];
Eq11:=xi^2=expand(Eq10[1]^2); Eq12:=xi^2=expand(Eq10[2]^2);
SolFin1:=subs(Eq11,Eq8); SolFin2:=subs(Eq12,Eq8);
ChsEq:=subs(xi=eta,Eq91);
for eta from -1 to 1 by 0.1 do Chs1:=Chs1,ChsEq;
od: for eta from -3 to -1 by 0.1 do Chs2:=Chs2,subs(t=0,ChsEq);
od: for eta from 1 to 3 by 0.1 do Chs3:=Chs3,subs(t=0,ChsEq);
od: Chs1; Chs2; Chs3;
G1:=implicitplot([Chs1],X=-1..1,t=tR,color=blue):
```

```
G2:=implicitplot([Chs2],X=-3..-1,t=tR):
G3:=implicitplot([Chs3],X=1..3,t=tR): display({G1,G2,G3});
U1:=[op(2,rhs(SolFin1)),op(2,rhs(SolFin2))];
GU0:=plot(rhs(op(1,InCur1)),x=-10..40): GU3:=plot(subs(t=3,U1),
 X=-10..40,0..3,BB): GU5:=plot(subs(t=30,U1),X=-10..40,0..3,BB):
display({GU0,GU3,GU5});
```

Mathematica:

```
p=10; tR=1; chs1={}; chs2={}; chs3={}; gC1={}; gC2={}; gC3={};
SetOptions[ContourPlot,Frame->True,ImageSize->500,
 ContourShading->False,ContourStyle->Red]; SetOptions[Plot,
 Frame->True,ImageSize->500,PlotRange->{{-p,p*4},{0,3}}];
pde[u_]:=D[u[x,t],t]+c[u[x,t]]*D[u[x,t],x]==0; pde[u]
{iniCurve=u[x,0]->f[x], iniData=u[xi,0]->f[xi],
 tr1={xN[t]->xN,uN[t]->uN}, charEqs={xN'[t]==c[uN],uN'[t]==0}}
{eq1=DSolve[charEqs[[1]],xN[t],t]//First, eq2=Solve[eq1/.
 Rule->Equal,C[1]]//First, eq3=eq2/.tr1, eq4=Solve[eq3/.
 Rule->Equal,xN]//First, eq5=eq4/.C[1]->xi/.c[uN]->
 c[iniData[[2]]], eq6=u[x,t]->u[eq5[[1,2]],t]==iniData}
{eq51=eq5/.Rule->Equal//First, eq52=Thread[Thread[eq51*(-1),
 Equal]+xi+xN,Equal]//Expand//ToRules}
eq7={u[x,t]->eq6[[2,2,2]],eq52}//Flatten
pde1[u_]:=D[u[x,t],t]+u[x,t]*D[u[x,t],x]==0; pde1[u]
u1[xN_,tN_]:=eq7[[1,2]]/.f[xi]->inCur1[[1]]/.{x->xN,t->tN}
{inCur1=(u[x,0]=Piecewise[{{1-x^2,Abs[x]<=1},
 {0,Abs[x]>1}}])==f[x], u1[xi,t], eq9=eq7[[2]]/.c[f[xi]]->
 inCur1[[1,1,1,1]], eq91=eq9/.x->xi, eq10=Solve[eq91/.
 Rule->Equal,xi]//Flatten, eq11=xi^2->eq10[[1,2]]^2//Expand}
{eq12=xi^2->eq10[[2,2]]^2//Expand, solFin1=u1[xi,t]/.eq11}
{solFin2=u1[xi,t]/.eq12, chsEq=eq91/.Rule->Equal/.xi->eta}
chsEq1=chsEq[[1]]-chsEq[[2]]
uN1={solFin1[[1,1,1]],solFin2[[1,1,1]]}
Do[chs1=Append[chs1,chsEq1],{eta,-1,1,0.1}]; chs1
Do[chs2=Append[chs2,chsEq1/.t->0],{eta,-3,-1,0.1}]; chs2
Do[chs3=Append[chs3,chsEq1/.t->0],{eta,1,3,0.1}]; chs3
Do[gC1=Append[gC1,ContourPlot[chs1[[i]]==0,{xN,-1,1},{t,0,tR},
 ContourStyle->Blue]],{i,1,Length[chs1]}]; Do[gC2=Append[gC2,
 ContourPlot[chs2[[i]]==0,{xN,-3,-1},{t,0,tR}]],{i,1,Length[
 chs2]}]; Do[gC3=Append[gC3,ContourPlot[chs3[[i]]==0,{xN,1,3},
 {t,0,tR}]],{i,1,Length[chs3]}];
Show[{gC1,gC2,gC3},PlotRange->{{-3,3},{0,1}}]
gU0=Plot[inCur1[[1]],{x,-10,40},PlotStyle->Red];
```

```
gU3=Plot[uN1/.t->3,{xN,-10,40},PlotStyle->Blue]; gU5=Plot[uN1/.
 t->30,{xN,-10,40},PlotStyle->Blue]; Show[{gU0,gU3,gU5}]
```
□

3.1.3 Solution Profile at Infinity

Let us consider initial value problems for some first-order equations arising in applied sciences and analyze the large time behavior of the solutions obtained. In particular, we show that some of the solutions tend to special exact solutions as time becomes large.

Problem 3.8 *Generalized inviscid Burgers equation. Large time behavior of the solution. Cauchy problem.* Let us consider the generalized inviscid Burgers equation with the initial curve

$$u_t+G(u)u_x=H(u), \quad u(x,0)=f(x),$$

where $\{x \in \mathbb{R}, t \geq 0\}$, and $G(u)$, $H(u)$ are nonnegative functions, e.g., $G(u)=1$, $H(u)=u(\lambda_1-\lambda_2 u)$.

1. Applying the transformation of the dependent variable $u(x,t) = 1/w(x,t)$ (`tr1`) and analytical methods (e.g., the method of characteristics), verify that the exact solution of the given Cauchy problem has the form $u(x,t) = \dfrac{f(x-t)\lambda_1}{-\lambda_2(e^{-\lambda_1 t}-1)f(x-t)+e^{-\lambda_1 t}\lambda_1}$ (`Solu1`).

2. Analyze the large time behavior of the solution and show that the solution (`Solu1`) as $t \to \infty$ $(x-t=O(1))$ tends to a constant $u \approx \lambda_1/\lambda_2$ (`ProfInf`).

Maple:

```
with(PDETools): declare(u(x,t),w(x,t)); U:=diff_table(u(x,t));
interface(showassumed=0);assume(f(x)>=0,lambda1>0,lambda2>0,t>0);
tr1:={u(x,t)=1/w(x,t)}; Eq1:=U[t]+U[x]-U[]*(lambda1-lambda2*U[]);
Eq2:=simplify(dchange(tr1,Eq1,[w(x,t)])*w(x,t)^2);
Solw1:=pdsolve({Eq2,w(x,0)=1/f(x)},w(x,t)); Solu1:=1/rhs(Solw1);
ProfInf:=limit(Solu1,t=infinity);
```

Mathematica:

```
eq1[u_]:=D[u,t]+D[u,x]-u*(lambda1-lambda2*u); inf=Infinity
{eq2=eq1[1/w[x,t]]//FullSimplify, eq21=eq2*w[x,t]^2//Expand}
solw1=DSolve[{eq2==0,w[x,0]==1/f[x]},w[x,t],{x,t}]//First
solu1=1/solw1[[1,2]]/.{f[-t+x]->f[z]}//FullSimplify
profInf=Limit[solu1,t->inf,Assumptions->{lambda1>0,lambda2>0}]
```
□

Problem 3.9 *Generalized inviscid Burgers equation. Large time behavior of the solution. Cauchy problem.* As in the previous problem, we consider the generalized inviscid Burgers equation with the initial curve,

$$u_t+G(u)u_x=H(u), \quad u(x,0)=x+1,$$

where $\{x \in \mathbb{R}, t \geq 0\}$, and $G(u)$, $H(u)$ are nonnegative functions, e.g., $G(u)=u$, $H(u)=-u$.

1. Solving this equation analytically, verify that the exact solution of this Cauchy problem has the form $u(x,t) = \dfrac{(x+1)e^{-t}}{2-e^{-t}}$ (SoluFin).

2. Analyze the large time behavior of the solution (SoluFin) as $t \to \infty$.

Maple:

```
with(PDETools): declare(u(x,t)); U:=diff_table(u(x,t));
interface(showassumed=0); assume(t>0);
Eq1:=U[t]+U[]*U[x]+U[]=0;
Solu:=pdsolve({Eq1,u(x,0)=x+1},u(x,t),HINT=f(x)*g(t),explicit);
Solu1:=rhs(Solu); SN:=numer(Solu1)*exp(-t);
SD:=combine(expand(denom(Solu1)*exp(-t))); SoluFin:=SN/SD;
Sol1:=expand(SoluFin); Sol11:=sort(op(1,Sol1));
Sol12:=sort(op(2,Sol1)); ProfInf1:=op(2,Sol11)*op(3,Sol11)
 *limit(op(1,Sol11),t=infinity);
ProfInf2:=op(2,Sol12)*limit(op(1,Sol12),t=infinity);
ProfInf:=simplify(ProfInf1+ProfInf2);
plot3d(ProfInf,x=-5..5,t=0..5,axes=boxed);
plots[contourplot](ProfInf,x=-5..5,t=0..5,filledregions=true,
      coloring=[magenta,blue]);
```

Mathematica:

```
eq1=D[u[x,t],t]+u[x,t]*D[u[x,t],x]+u[x,t]==0
solu=DSolve[{eq1,u[x,0]==x+1},u[x,t],{x,t}]//First
{solu1=solu[[1,2]], sN=Numerator[solu1]*Exp[-t]//Simplify}
{sD=Denominator[solu1]*Exp[-t]//Simplify, soluFin=sN/sD}
profInf=sN/Limit[sD,t->Infinity]
Plot3D[profInf,{x,-5,5},{t,0,5},PlotRange->All,Mesh->False]
ContourPlot[profInf,{x,-5,5},{t,0,5},PlotRange->All,
 ContourStyle->Hue[0.9],ContourShading->Automatic]
```

□

3.2 Generalized Method of Characteristics

In this section, we consider the *generalized method of characteristics* for solving first-order nonlinear PDEs of the form $\mathfrak{F}(x, y, u, u_x, u_y)=0$ or $\mathfrak{F}(x, y, u, p, q)=0$ (where $u_x=p$ and $u_y=q$). We will show how to obtain a *complete solution* or a *complete integral* of nonlinear equations and how to determine the general solution of a nonlinear PDE from its complete integral.

The generalized method of characteristics (as the method of characteristics considered in the previous section) allows us to reduce a nonlinear PDE to a system of ODEs along which the given PDE with some initial data (the Cauchy data) is integrable. But in the nonlinear case, the solution surface or *integral surface*, through which the complete integral have to pass, represents a cone, called the *Monge cone*. The *characteristic curves* depend on the *orientation of tangent planes* on the Monge cone at each point. So there exists a Monge cone of characteristics, and the *characteristic equations* or the *Charpit equations* for nonlinear equations have the form:

$$\frac{dx}{dt}=\mathfrak{F}_p, \quad \frac{dy}{dt}=\mathfrak{F}_q, \quad \frac{du}{dt}=p\mathfrak{F}_p+q\mathfrak{F}_q,$$
$$\frac{dp}{dt}=-(\mathfrak{F}_x+p\mathfrak{F}_u), \quad \frac{dq}{dt}=-(\mathfrak{F}_y+q\mathfrak{F}_u).$$

Once the system of ODEs is found, it can be solved along the characteristic curves and transformed into a complete integral for the original nonlinear PDE.

The generalized method of characteristics was proposed by J. L. Lagrange (who published the main results for PDEs in two independent variables in 1772 and 1779) and P. Charpit in 1784 (who has combined the methods for solving linear and nonlinear first-order PDEs by reducing an equation of this class to a system of ODEs). The method described in modern texts is referred to as the *Lagrange–Charpit* or *Charpit method*. Also this method is referred to as the *Cauchy method* due to the generalization (performed by Cauchy in 1819) to n independent variables of the method proposed by Lagrange. The Cauchy method of characteristics has been analyzed from the differential-geometric point of view by Monge and Cartan. During the last century, the method of characteristics was further developed by many mathematicians (P. Lax, E. F. F. Hopf, O. A. Oleinik, S. N. Kruzhkov, V. P. Maslov, L. C. Evans, and others).

3.2.1 Complete Integrals. General Solution

Applying the generalized method of characteristics, we will show how to obtain complete solutions or complete integrals of first-order nonlinear equations and how to derive the general solution from a complete integral.

Definition 3.6 A *complete integral* of the equation $\mathcal{F}(x,y,u,u_x,u_y)=0$ is a two-parameter family of surfaces of the form $f(x,y,u,a,b)=0$ (where a and b are parameters).

Problem 3.10 *Nonlinear first-order equation. Complete integral.* Let us consider the nonlinear first-order PDE

$$u^2u_t^2+u^2u_x^2+u^2=c^2,$$

where $\{x\in\mathbb{R}, t\geq 0\}$. Verify that the family of spheres, $(t-a)^2+(x-b)^2+u^2=c^2$, i.e., the complete integral of the equation, satisfies this nonlinear PDE.

Maple:

```
with(PDEtools): declare(u(x,t)); alias(u=u(x,t));
IntSurf:=(t-a)^2+(x-b)^2+u(x,t)^2=c^2;
PDE:=u->u(x,t)^2*(diff(u(x,t),t)^2+diff(u(x,t),x)^2+1)=c^2;
expand(PDE(u)); Eq1:=diff(IntSurf,t)/2;
Eq2:=diff(IntSurf,x)/2; Eq11:=isolate(Eq1,-a)+t;
Eq21:=isolate(Eq2,-b)+x;
Eq3:=lhs(Eq11)^2+lhs(Eq21)^2=rhs(Eq11)^2+rhs(Eq21)^2;
Eq4:=lhs(Eq3)=factor(rhs(Eq3)); Eq5:=Eq4+u^2;
Eq6:=expand(subs(lhs(Eq5)=rhs(IntSurf),Eq5));
expand(PDE(u)=Eq6); eliminate({IntSurf,Eq1,Eq2},{a,b});
```

Mathematica:

```
intSurf=(t-a)^2+(x-b)^2+u[x,t]^2==c^2
pde[u_]:=u[x,t]^2*(D[u[x,t],t]^2+D[u[x,t],x]^2+1)==c^2;
{pde[u]//Expand, eq1=Thread[D[intSurf,t]/2,Equal]//Expand}
{eq2=Thread[D[intSurf,x]/2,Equal]//Expand, eq11=Thread[
 eq1-eq1[[1,3]],Equal], eq21=Thread[eq2-eq2[[1,3]],Equal],
 eq3=Thread[Map[#^2&,eq11]+Map[#^2&,eq21],Equal]}
{eq4=Map[FullSimplify,eq3], eq5=Thread[eq4+u[x,t]^2,Equal]}
{eq6=eq5/.eq5[[1]]->intSurf[[2]], pde[u]==eq6//Expand}
Eliminate[{intSurf,eq1,eq2},{a,b}]//FullSimplify
```

□

Problem 3.11 *Nonlinear first-order equation. Complete integral.* Let us consider the nonlinear first-order PDE

$$(u_x)^2+yu_y=u,$$

where $\{x\in\mathbb{R}, y\in\mathbb{R}\}$. Applying the generalized method of characteristics, prove that the complete integral of this nonlinear equation has the form $u(x,t)=\frac{1}{4}(x+b)^2+ay$.

Maple:

```
with(Student[Precalculus]): F:=p^2+q*y-u;
CharEqs:=[diff(x(t),t)=diff(F,p),diff(y(t),t)=diff(F,q),
 diff(u(t),t)=p*diff(F,p)+q*diff(F,q), diff(p(t),t)=-(diff(F,x)+
 p*diff(F,u)),diff(q(t),t)=-(diff(F,y)+q*diff(F,u))];
tr1:=q(t)=q; tr2:=du=p*dx+q*dy; tr3:=u-a*y=v;
Eq1:=map(int,CharEqs[5],t); Eq2:=lhs(Eq1)=rhs(Eq1)+a;
Eq3:=[solve(subs(subs(tr1,Eq2),F),p)]; Eq4:=p=Eq3[1];
Eq5:=(dx/rhs(CharEqs[1])=dy/rhs(CharEqs[2]))=du/rhs(CharEqs[3]);
Eq6:=subs(tr2,Eq5); Eq7:=subs(p=rhs(Eq4),q=rhs(Eq2),tr2);
Eq8:=expand(isolate(Eq7,dx));
Eq81:=rhs(Eq8)=d(lhs(tr3))/denom(rhs(Eq8)); Eq82:=subs(tr3,Eq81);
Eq83:=int(1/denom(rhs(Eq82)),v)=int(lhs(Eq8)/dx,x)+b;
Eq84:=subs(v=lhs(tr3),lhs(Eq83))=rhs(Eq83);
Sol:=solve(Eq84,u); CompleteInt:=u(x,y)=CompleteSquare(Sol);
```

Mathematica:

```
trS1[eq_,var_]:=Select[eq,MemberQ[#,var,Infinity]&];
cS[a_.x_^2+b_.x_+c_.]:=a*((x+b/(2*a))^2-(b^2-4*a*c)/(4*a^2));
completeSquare[x_]:=If[TrueQ[x==Expand[x]],x,cS[Expand[x]]];
{fF=p^2+q*y-u, charEqs={x'[t]==D[fF,p],y'[t]==D[fF,q],
 u'[t]==p*D[fF,p]+q*D[fF,q], p'[t]==-(D[fF,x]+p*D[fF,u]),
 q'[t]==-(D[fF,y]+q*D[fF,u])}}
{tr1=q[t]->q, tr2=du->p*dx+q*dy, tr3=u-a*y->v}
eq1=Thread[Integrate[charEqs[[5]],t],Equal]
{eq2=eq1[[1]]->eq1[[2]]+a, eq3=Solve[fF==0/.{eq2/.tr1},p],
 eq4=p==eq3[[1,1,2]], eq5=(dx/charEqs[[1,2]]==
 dy/charEqs[[2,2]])==du/charEqs[[3,2]], eq6=eq5/.tr2}
eq7=du==(tr2/.{p->eq4[[2]],q->eq2[[2]]})[[2]]
{eq8=Solve[eq7,dx]//Expand//First, eq81=eq8[[1,2]]==
 d[tr3[[1]]]*trS1[eq8[[1,2,1]],a],eq82=eq81/.tr3,
 eq83=Integrate[eq82[[2,1]],v]==Integrate[eq8[[1,1]]/dx,x]+b}
{eq84=eq83[[1]]==eq83[[2]]/.v->tr3[[1]], sol=Solve[eq84,u]}
completeInt=u[x,y]==completeSquare[sol[[1,1,2]]]
```

□

Let us consider a *conservative dynamical system* and assume that $S{=}u{-}Et$, where S is the *Hamilton principle function*, E is a constant, $u(q_i)$ the unknown function, q_i are the generalized coordinates, and t is the time variable. We know that S is a solution of the *Hamilton–Jacobi equation*

$$S_t{+}H(q_i,p_i,t){=}0 \quad (p_i{=}S_{q_i},\ i{=}1,\dots,N)$$

if u satisfies the time-independent Hamilton–Jacobi equation

$$H(q_i,p_i){=}E \quad (p_i{=}u_{q_i},\ i{=}1,\dots,N).$$

Here H is called the *Hamiltonian* of the dynamical system and $u(q_i)$ is called *Hamilton's characteristic function.*

Let us consider the case of a single particle of mass m moving in a conservative force field ($N{=}3$). The Hamiltonian of this dynamical system is written as $H{=}T{+}V{=}\dfrac{1}{2m}\sum\limits_{i=1}^{N}p_i^2{+}V$, where $V{=}V(x,y,z)$ is the potential energy of the particle and $p_i{=}S_{q_i}{=}u_{q_i}$ $(i{=}1,\dots,N)$.

Setting $q_1{=}x$, $q_2{=}y$, and $q_3{=}z$, we obtain the following form of the Hamilton–Jacobi equation

$$(2m)^{-1}(u_x^2{+}u_y^2{+}u_z^2){+}V{=}E$$

or

$$(u_x^2{+}u_y^2{+}u_z^2){=}2m(E{-}V){=}f(x,y,z),$$

i.e., the nonlinear first-order PDE. Let us now solve this nonlinear equation which is very important in *nonlinear geometrical optics* and propagation of waves in continuous media (acoustics, elasticity, and electromagnetic theory), see Sect. 1.1.2.

Problem 3.12 *Eikonal equation. Complete integral.* Let us rewrite the above equations in the form

$$u_x^2{+}u_y^2{+}u_z^2{=}n^2(x,y,z),$$

where $\{x\in\mathbb{R}, y\in\mathbb{R}, z\in\mathbb{R}\}$.

In nonlinear optics, $n{=}c_0/c$ is the *refraction index*, $c{=}c(x,y,z)$ is the wave speed of light propagating in a medium along rays, and $u(x,y,z)$ is the unknown function. This equation is called the *eikonal equation* (see Sect. 1.1.2). Physically, the eikonal equation describes a surface of constant optical phase as *wave fronts* propagate orthogonal to the light rays. Let $N{=}2$ and n be a constant, e.g., $n{=}1$, then the eikonal equation takes the form $u_x^2{+}u_y^2{=}1$.

Applying the generalized method of characteristics, prove that the complete integral of the eikonal equation has the form

$$u(x,y)=ax+\sqrt{1-a^2}\,y+b,$$

where a and b are parameters. This solution coincides to the solution (Sol2) obtained via predefined functions in Problem 1.18.

Maple:

```
F:=p^2+q^2-1; CharEqs:=[diff(x(t),t)=diff(F,p),
 diff(y(t),t)=diff(F,q), diff(u(t),t)=p*diff(F,p)+q*diff(F,q),
 diff(p(t),t)=-(diff(F,x)+p*diff(F,u)),
 diff(q(t),t)=-(diff(F,y)+q*diff(F,u))];
tr1:={q(t)=q,p(t)=p}; tr2:=du=p*dx+q*dy;
Eq1:=dsolve(CharEqs[4],p(t)); Eq2:=subs(tr1,_C1=a,Eq1);
Eq3:=dsolve(CharEqs[5],q(t)); Eq4:=subs(tr1,_C1=c,Eq3);
Eq5:=subs(Eq2,Eq4,F); Eq51:=[solve(Eq5,c)]; Eq52:=c=Eq51[1];
Eq6:=subs(p=rhs(Eq2),q=rhs(Eq4),Eq52,tr2);
CompleteInt:=int(lhs(Eq6)/du,u)=int(op(1,rhs(Eq6))/dx,x)
           +int(op(2,rhs(Eq6))/dy,y)+b;
```

Mathematica:

```
{fF=p^2+q^2-1, charEqs={x'[t]==D[fF,p], y'[t]==D[fF,q],
 u'[t]==p*D[fF,p]+q*D[fF,q], p'[t]==-(D[fF,x]+p*D[fF,u]),
 q'[t]==-(D[fF,y]+q*D[fF,u])}}
{tr1={q[t]->q, p[t]->p}, tr2=du->p*dx+q*dy}
{eq1=DSolve[charEqs[[4]],p[t],t], eq2=eq1/.tr1/.C[1]->a//First}
{eq3=DSolve[charEqs[[5]],q[t],t], eq4=eq3/.tr1/.C[1]->c//First}
{eq5=fF/.eq2/.eq4, eq51=Solve[eq5==0,c]//Flatten, eq52=c->
 eq51[[2,2]], eq6=tr2/.p->eq2[[1,2]]/.q->eq4[[1,2]]/.eq52}
completeInt=Integrate[eq6[[1]]/du,u]==
 Integrate[eq6[[2,1]]/dx,x]+Integrate[eq6[[2,2]]/dy,y]+b
```

□

Now, considering a nonlinear first-order equation of general form in N independent variables, let us show how to build from a complete integral (which depends upon N arbitrary constants) more complicated solutions, i.e., solutions depending on an arbitrary function G of $N-1$ variables. As an example, we will consider nonlinear equations of the form $\mathcal{F}(x,y,u,u_x,u_y)=0$, i.e., for $N=2$.

Problem 3.13 *Eikonal equation. General solution.* As in the previous problem, we consider the eikonal equation

$$u_x^2+u_y^2=1,$$

where $\{x\in\mathbb{R}, y\in\mathbb{R}\}$ and $u(x,y)=ax+\sqrt{1-a^2}y+b$ (`CInt`) is the complete integral determined in the previous problem.

1. First, we visualize particular solutions (`CI1`, `CI2`) described by the above expression (the family of planes).

2. Substituting $a=\cos(\alpha)$ into the complete integral, we obtain an alternative form for a complete integral of the eikonal equation (for $N=2$): $u(x,y;\alpha,b)=\cos(\alpha)x+\sin(\alpha)y+b$ (`CInt1`).

3. Applying *Maple* predefined functions, we find the general solution of the eikonal equation. First, we replace the second parameter b by an arbitrary function of the first parameter α, i.e $b=G(\alpha)$ (`Eq1`). We consider the *envelope* of the one-parameter family of solutions (family of planes), which is described by the system of equations: $u=\phi(x,y;\alpha,G(\alpha))$, $\phi_\alpha(x,y;\alpha,G(\alpha))+\phi_G(x,y;\alpha,G(\alpha))G'(\alpha)=0$*. In our case, the system of two equations, $u=\cos(\alpha)x+\sin(\alpha)y+G(\alpha)$, $0=-\sin(\alpha)x+\cos(\alpha)y+G_\alpha(\alpha)$ (`Eq1`, `Eq2`), define the general solution. Then, performing elimination of the parameter α between these two equations, we can obtain an expression for the envelope (which involves the arbitrary function $G(\alpha)$ and represents a solution of the eikonal equation), i.e., a general solution of the given PDE (`GenSol`).

4. Finally, considering some explicit choice of the function $G(\alpha)$, i.e., $G(\alpha)=0$, we obtain a subfamily of solutions of the eikonal equation and compute the envelope $u=\cos(\alpha)x+\sin(\alpha)y$, $0=-\sin(\alpha)x+\cos(\alpha)y$ (`sys1`). Since $\alpha=\arctan(y/x)$, we obtain the solutions of the eikonal equation $v=\pm\sqrt{x^2+y^2}$ (for $x\neq 0$).

Maple:

```
CInt:=u=a*x+sqrt(-a^2+1)*y+b; CI1:=subs(b=1,a=0.1,CInt);
CI2:=subs(b=-1,a=0.5,CInt);
plot3d({rhs(CI1),rhs(CI2)},x=-10..10,y=-10..10,axes=boxed);
CInt1:=simplify(subs(a=cos(alpha),CInt)) assuming sin(alpha)>0;
Eq1:=subs(b=G(alpha),CInt1); Eq2:=diff(Eq1,alpha);
Eq3:=eliminate([Eq1,Eq2],alpha); tralpha:=op(Eq3[1]);
```

*The second equation is obtained (from the first equation) by performing a partial differentiation with respect to α.

```
GenSol:=subs(tralpha,u=v,Eq1);
sys1:=eval([Eq1,Eq2], G(alpha)=0);
Eq31:=eliminate(sys1,alpha); tralpha1:=op(Eq31[1]);
Sol1:=simplify(subs(tralpha1,u=v,sys1[1])) assuming x>0, y>0;
Sol2:=simplify(subs(tralpha1,u=v,sys1[1])) assuming x<0, y<0;
```
□

3.2.2 The Monge Cone. Characteristic Directions

Again, we consider the nonlinear first-order equation in two independent variables $\mathcal{F}(x,y,u,p,q)=0$, where $p=u_x$, $q=u_y$, $u=u(x,y)$ is the dependent variable, and $\mathcal{F}$ is a C^1 function of its arguments, $\mathcal{F}_p^2+\mathcal{F}_q^2\neq 0$. From the geometrical point of view, this PDE represents a relation between the coordinates of the point (x,y,u) on an integral surface $u=u(x,y)$ and the direction of the normal vector $(p,q,-1)$ at this point. The associated tangent plane to a solution through (x_0,y_0,u_0) is given by the formula $u-u_0=p(x-x_0)+q(y-y_0)$. For a given value (x_0,y_0,u_0), we can obtain different values of p and q, and hence have a one-parameter family of tangent planes (parameterized by p), which have as their envelope a cone, called the *Monge cone* for the given first-order PDE at that point.

Problem 3.14 *Eikonal equation. Monge cone.* Let consider the eikonal equation (see Problem 3.13)

$$u_x^2+u_y^2=1,$$

where $\{x\in\mathbb{R}, y\in\mathbb{R}\}$. Find the equation of the Monge cone through a point (x_0,y_0,u_0) for the eikonal equation.

1. In the case of the Monge cone, we assume $q=q(p)$ and consider the envelope of the one-parameter family of solutions (family of planes), which is described by the system of equations (see Problem 3.13): $u-u_0=p(x-x_0)+q(p)(y-y_0)$, $0=(x-x_0)+q'(p)(y-y_0)$.

In our case, we obtain $q=\pm\sqrt{1-p^2}$ (`Solsq`) and the corresponding system of equations takes the form $u-u_0=p(x-x_0)\pm\sqrt{1-p^2}(y-y_0)$ (`Eq11`, `Eq12`), $0=x-x_0\pm p/\sqrt{1-p^2}(y-y_0)$ (`Eq21`, `Eq22`).

Squaring both sets of equations (`Eq11`, `Eq21`) and adding, we obtain the equation of the Monge cone through (x_0,y_0,u_0) for the eikonal equation: $(u-u_0)^2=(y-y0)^2+(x-x0)^2$ (`EqMC1`). The same result can be obtained for the equations `Eq12` and `Eq22`.

2. Finally, considering a particular point $x_0=y_0=u_0=0$, we visualize a family of tangent planes to a solution through the point $(0,0,0)$ for which $p=0$, $p=1$, $p=-1$, and the Monge cone through this point, represents an envelope of these planes (see Fig. 3.3).

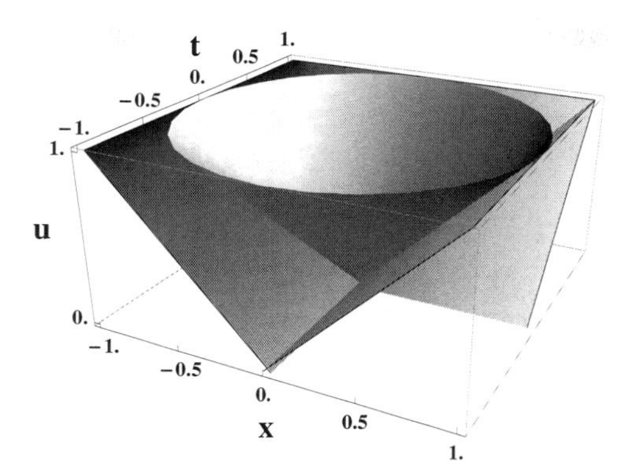

Fig. 3.3. The Monge cone and a family of tangent planes to a solution of the eikonal equation through the point $(0, 0, 0)$

Maple:

```
with(plots): TPlane:=u-u0=p*(x-x0)+q*(y-y0); PDE1:=p^2+q^2-1=0;
Solsq:=[solve(PDE1,q)]; uR:=0..1; vR:=0..2*Pi; Op1:=color=blue;
Eq11:=subs(q=Solsq[1],TPlane); Eq12:=subs(q=Solsq[2],TPlane);
Eq21:=diff(Eq11,p); Eq22:=diff(Eq12,p);
tr1:={x-x0=X,y-y0=Y,u-u0=U,-y+y0=-Y};
tr2:={X=x-x0,Y=y-y0,U=u-u0};
Eq3:=subs(tr1,Eq11); Eq4:=subs(tr1,collect(Eq21*sqrt(1-p^2),p));
Eq31:=map(`^`, Eq3, 2); Eq41:=map(`^`, Eq4, 2);
EqMC1:=subs(tr2,simplify(expand(Eq31+Eq41))); sys1:=[Eq11,Eq21];
sys2:=[Eq12,Eq22]; params:={x0=0,y0=0,u0=0};
TP1:=subs(p=0,params,sys1[1]); TP2:=subs(p=1,params,sys1[1]);
TP3:=subs(p=-1,params,sys1[1]);
GMC:=plot3d([u*cos(v),u*sin(v),u],u=uR,v=vR,axes=boxed,Op1):
for i from 1 to 3 do G||i:=plot3d(rhs(TP||i),x=-1..1,y=-1..1,
 color=x+y): od: display({GMC,G1,G2,G3});
```

Mathematica:

```
{tPlane=u-u0==p*(x-x0)+q*(y-y0), pde1=p^2+q^2-1==0,
 solsq=Solve[pde1,q]//Flatten, eq11=tPlane/.solsq[[2]],
 eq12=tPlane/.solsq[[1]], eq21=Thread[D[eq11,p],Equal],
 eq22=Thread[D[eq12,p],Equal]}
{tr1={x-x0->xN,y-y0->yN,u-u0->uN,-y+y0->-yN}, tr2={xN->x-x0,
 yN->y-y0,uN->u-u0}, eq3=eq11/.tr1}
```

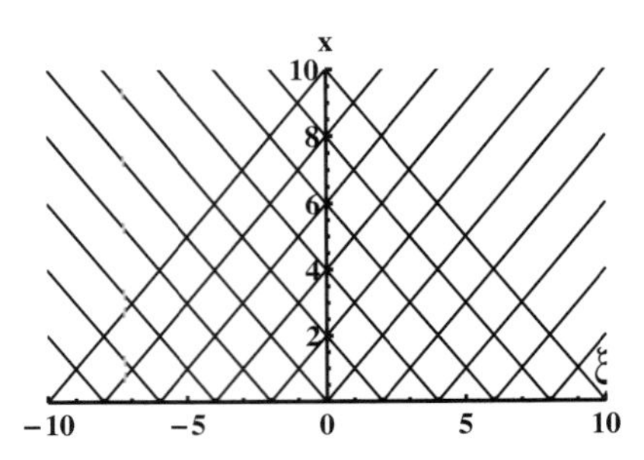

Fig. 3.4. Characteristic directions for the sine–Gordon equation

```
eq4=FullSimplify[Thread[eq21*Sqrt[1-p^2],Equal]]/.tr1
{eq31=Map[Power[#,2]&,eq3], eq41=Map[Power[#,2]&,eq4]}
{eq42=0==eq41[[1]]-eq41[[2]], eqMC1=FullSimplify[Expand[Thread[
 eq31+eq42,Equal]]/.tr2], sys1={eq11,eq21}, sys2={eq12,eq22}}
{params={x0->0,y0->0,u0->0}, tP[1]=sys1[[1]]/.p->0/.params,
 tP[2]=sys1[[1]]/.p->1/.params, tP[3]=sys1[[1]]/.p->-1/.params}
gMC=ParametricPlot3D[{u*Cos[v],u*Sin[v],u},{u,0,1},{v,0,2*Pi},
 PlotStyle->Blue]; Do[g[i]=Plot3D[tP[i][[2]],{x,-1,1},{y,-1,1},
 ColorFunction->Function[t,Hue[.95*(0.01+t)]]],{i,1,3}];
Show[{gMC,g[1],g[2],g[3]}]
```
□

Problem 3.15 *Sine–Gordon equation. Slopes of characteristic directions.* Let us consider the sine–Gordon equation

$$u_{xx}-u_{tt}=\sin u,$$

where $\{x\in\mathbb{R}, t\geq 0\}$. Applying the generalized method of characteristics, prove that the sine–Gordon equation has the two characteristic directions (see Fig. 3.4) whose slopes are given by the equations $dt/dx=-1$ and $dt/dx=1$.

Maple:

```
with(plots): G:=NULL: F:=p^2-q^2-sin(u);
CharEqs:=[dx/dr=subs(p=p(r),diff(F,p)), dt/dr=subs(q=q(r),
 diff(F,q)),diff(u(r),r)=p(r)*subs(p=p(r),diff(F,p))+
 q(r)*subs(q=q(r),diff(F,q)),diff(p(r),r)=-(diff(F,x)+
 p(r)*diff(F,u)),diff(q(r),r)=-(diff(F,t)+q(r)*diff(F,u))];
Sol45:=subs(u=u(r),dsolve({CharEqs[4],CharEqs[5]},
 {p(r),q(r)}));
```

```
Eq21:=combine(lhs(CharEqs[2])/lhs(CharEqs[1])=
 rhs(CharEqs[2])/rhs(CharEqs[1])); Eq45:=subs(Sol45,Eq21);
Eq451:=simplify(2*map(`^`,Sol45[1],2)-2*map(`^`,Sol45[2],2));
Eq3:=subs(Eq451,CharEqs[3]); dsolve(Eq3); tr1:={_C1=1,_C2=1};
tr2:={_C1=-1,_C2=1}; Eq31:=subs(tr1,Eq3); Eq31:=subs(tr2,Eq3);
Chars:=[subs(tr1,Eq45),subs(tr2,Eq45)];
for x from -10 to 10 by 2 do
 G:=G,plot({-xi-x,xi+x},xi=-10..10,color=[blue,magenta]); od:
display([G], view=[-10..10,0..10]);
```

Mathematica:

```
{fF=p^2-q^2-Sin[u], charEqs={dx/dr==D[fF,p]/.p->p[r],
 dt/dr==D[fF,q]/.q->q[r], u'[r]==p[r]*(D[fF,p]/.p->p[r])+
 q[r]*(D[fF,q]/.q->q[r]), p'[r]==-(D[fF,x]+p[r]*D[fF,u]),
 q'[r]==-(D[fF,t]+q[r]*D[fF,u])}}
{sol45=DSolve[{charEqs[[4]],charEqs[[5]]},{p[r],q[r]},r]/.
 u->u[r], eq21=charEqs[[2,1]]/charEqs[[1,1]]==
 charEqs[[2,2]]/charEqs[[1,2]], eq45=eq21/.sol45[[1]]}
{eq451=Apply[Subtract,Map[2*#^2&,{sol45[[1,1,1]],
 sol45[[1,2,1]]}]]->Apply[Subtract,Map[2*#^2&,{sol45[[1,1,2]],
 sol45[[1,2,2]]}]]//Simplify, eq3=charEqs[[3]]/.eq451}
{tr1={C[1]->1,C[2]->1}, tr2={C[1]->-1,C[2]->1}}
{eq31=eq3/.tr1, eq31=eq3/.tr2, chars={eq45/.tr1,eq45/.tr2}}
g=Table[Plot[{-xi1-x1,xi1+x1},{xi1,-10,10}],{x1,-10,10,2}];
Show[g,PlotRange->{{-10,10},{0,10}}]
```

□

3.2.3 Integral Surfaces. Cauchy Problem

In this section, applying the generalized method of characteristics, we will determine exact solutions of classical Cauchy problems and generalized Cauchy problems.

Problem 3.16 *Nonlinear first-order equation. Integral surfaces. Cauchy problem.* Let us consider the nonlinear first-order PDE with the initial data

$$(u_x)^2 u_y = 1, \quad u(x,0) = x,$$

where $\{x \in \mathbb{R}, y \geq 0\}$. Applying the generalized method of characteristics, prove that the solution of the given Cauchy problem has the form $u(x,y) = x + y$.

Maple:

```
F:=p^2*q-1; IniData:={u=x,y=0}; CharEqs:=[diff(x(t),t)=
 diff(F,p),diff(y(t),t)=diff(F,q), diff(u(t),t)=p*diff(F,p)+
 q*diff(F,q), diff(p(t),t)=-(diff(F,x)+p*diff(F,u)),
 diff(q(t),t)=-(diff(F,y)+q*diff(F,u))]; tr1:={q(t)=q,p(t)=p};
tr2:=du=p*dx+q*dy; Eq1:=dsolve(CharEqs[4],p(t));
Eq2:=subs(tr1,_C1=a,Eq1); Eq3:=dsolve(CharEqs[5],q(t));
Eq4:=subs(tr1,_C1=c,Eq3); Eq5:=subs(Eq2,Eq4,F);
Eq51:=isolate(Eq5,c); Eq6:=subs(Eq51,tr2); Eq7:=subs(p=rhs(Eq2),
 q=rhs(Eq4),Eq51,tr2); Eq8:=expand(isolate(Eq7,du));
Sol:=int(lhs(Eq8)/du,u)=int(op(1,rhs(Eq8))/dx,x)
 +int(op(2,rhs(Eq8))/dy,y)+b; Sol1:=subs(IniData,Sol);
Consts:={a=coeff(lhs(Sol1),x),b=coeff(lhs(Sol1),x,0)};
SolCauchy:=subs(Consts,Sol);
```

Mathematica:

```
{fF=p^2*q-1, iniData={u->x,y->0},charEqs={x'[t]==D[fF,p],
 y'[t]==D[fF,q], u'[t]==p*D[fF,p]+q*D[fF,q], p'[t]==
 -(D[fF,x]+p*D[fF,u]), q'[t]==-(D[fF,y]+q*D[fF,u])}}
{tr1={q[t]->q, p[t]->p}, tr2=du->p*dx+q*dy}
{eq1=DSolve[charEqs[[4]],p[t],t], eq2=eq1/.tr1/.C[1]->a}
{eq3=DSolve[charEqs[[5]],q[t],t], eq4=eq3/.tr1/.C[1]->c}
{eq5=Flatten[fF==0/.eq2/.eq4], eq51=Solve[eq5,c], eq6=tr2/.eq51}
eq7=tr2/.p->eq2[[1,1,2]]/.q->eq4[[1,1,2]]/.eq51/.Rule->Equal
{sol=Integrate[eq7[[1,1]]/du,u]==Integrate[eq7[[1,2,1]]/dx,x]+
 Integrate[eq7[[1,2,2]]/dy,y]+b, sol1=sol/.iniData}
{consts={a->Coefficient[sol1[[1]],x],b->Coefficient[
 sol1[[1]],x,0]}, solCauchy=sol/.consts}
```

□

Problem 3.17 *Nonlinear first-order equation. Integral surfaces. Cauchy problem.* Let us consider the nonlinear first-order PDE with the initial condition

$$(u_x)^2+u_y+u=0, \quad u(x,0)=x,$$

where $\{x\in\mathbb{R}, y\geq 0\}$. Let us rewrite the initial condition in the equivalent parametric form, i.e., find a solution of this equation that passes through the initial curve represented parametrically by the equations $x(r,0)=r$, $y(r,0)=0$, $u(r,0)=r$ (r is a parameter). Applying the generalized method of characteristics, prove that the solution of the given Cauchy problem has the form $u(x,y)=(x-1)e^{-y}+e^{-2y}$.

Maple:

```
interface(showassumed=0); assume(t>0); F:=p^2+q+u;
F1:=p(r,t)^2+q(r,t)+u(r,t); IniData:=[x(r,0)=r,y(r,0)=0,
 u(r,0)=r]; IniData1:=[x(0)=r,y(0)=0,u(0)=r];
Eq1:=p(r,0)=rhs(diff(IniData[1],r));
Eq2:=subs(Eq1,IniData[3],subs(t=0,F1)); Eq3:=isolate(Eq2,q(r,0));
CharEqs:=[diff(x(t),t)=diff(F,p),diff(y(t),t)=diff(F,q),
          diff(u(t),t)=p*diff(F,p)+q*diff(F,q),
          diff(p(t),t)=-(diff(F,x)+p*diff(F,u )),
          diff(q(t),t)=-(diff(F,y)+q*diff(F,u))];
tr1:={p(t)=p(r,t),q(t)=q(r,t)}; tr2:={p=p(r,t),q=q(r,t)};
Eq4:=dsolve({subs(tr1,lhs(CharEqs[4]))=
             subs(tr2,rhs(CharEqs[4])),Eq1},p(r,t));
Eq5:=dsolve({subs(tr1,lhs(CharEqs[5]))=
             subs(tr2,rhs(CharEqs[5])),Eq3},q(r,t));
tr4:={p=rhs(Eq4),q=rhs(Eq5)}; tr5:={x(t)=x,y(t)=y,u(t)=u};
Eq6:=combine(subs(tr4,[CharEqs[i] $ i=1..3]));
Eq7:=dsolve({op(Eq6),op(IniData1)},{x(t),y(t),u(t)});
Eq8:=eliminate(subs(tr5,Eq7),{t,r}); SolFin:=collect(combine(
 expand(isolate(op(Eq8[2]),u))),exp); pdetest(u(x,y)=rhs(SolFin),
 diff(u(x,y),x)^2+diff(u(x,y),y)+u(x,y)=0);
```

Mathematica:

```
{fF=p^2+q+u, f1=p[r,t]^2+q[r,t]+u[r,t], iniData={x[r,0]->r,
 y[r,0]->0,u[r,0]->r}, iniData1={x[0]==r,y[0]==0,u[0]==r}}
{eq1=p[r,0]->D[iniData[[1]],r][[2]], eq2=(f1/.t->0)/.eq1/.
 iniData[[3]], eq3=Solve[eq2==0,q[r,0]]}
{charEqs={x'[t]==D[fF,p], y'[t]==D[fF,q], u'[t]==p*D[fF,p]+
 q*D[fF,q], p'[t]==-(D[fF,x]+p*D[fF,u]),
 q'[t]==-(D[fF,y]+q*D[fF,u])}, tr1={p->p[r,t],q->q[r,t]}}
tr2={p'[t]->D[p[r,t],t], q'[t]->D[q[r,t],t]}
eq4=DSolve[{(charEqs[[4,1]]/.tr2)==(charEqs[[4,2]]/.tr1),
 (eq1/.Rule->Equal)},p[r,t],{r,t}]
eq5=DSolve[{(charEqs[[5,1]]/.tr2)==(charEqs[[5,2]]/.tr1),
 (eq3/.Rule->Equal)},q[r,t],{r,t}]
{tr4={p->eq4[[1,1,2]],q->eq5[[1,1,2]]},tr5={x[t]->x,y[t]->y,
 u[t]->u}, eq6=Table[charEqs[[i]],{i,1,3}]/.tr4}
eq7=DSolve[{eq6,iniData1},{x[t],y[t],u[t]},t]
{sys7=(eq7/.tr5/.Rule->Equal)[[1]]//Expand, sys8=Eliminate[
 sys7,r], eq8=sys8[[2]]/.(sys8[[1]]/.Equal->Rule),
 solFin=(Solve[eq8,u]//Simplify)/.Rule->Equal}
```

□

Problem 3.18 *Nonlinear first-order equation. Generalized Cauchy problem.* We consider the same nonlinear first-order PDE as in Problem 3.17:

$$(u_x)^2+u_y+u=0,$$

where $\{x \in \mathbb{R}, y \geq 0\}$, and find a solution of this equation that passes through a more complicated initial curve (compared to the previous problem), which is given by the equations $x(r,0)=r$, $y(r,0)=r$, and $u(r,0)=2r-1$ (r is a parameter).

1. Applying the generalized method of characteristics, let us formulate the Cauchy problem for this nonlinear equation to determine the family of curves in the parametric form, $u=u(r,t)$, $x=x(r,t)$, $y=y(r,t)$, and the values $p=p(r,t)$, $q=q(r,t)$ (r and t are parameters). Looking for a solution surface $u=u(x,y)$ with the given initial data at $t=0$, we first determine the initial conditions for $p(r,t)$ and $q(r,t)$ at $t=0$ (`sols`): $p(r,0)=\frac{1}{2} \mp \frac{1}{2}\sqrt{-3-8r}$ and $q(r,0)=\frac{3}{2} \pm \frac{1}{2}\sqrt{-3-8r}$, which must satisfy the original equation and the *strip condition* (`StripCond`)

$$u_r(r,0)=p(r,0)x_r(r,0)+q(r,0)y_r(r,0).$$

We take the first pair of the values of $p(r,0)$ and $q(r,0)$. Then, we can determine $p(r,t)=-\frac{1}{2}(-1+S)e^{-t}$ and $q(r,t)=\frac{1}{2}(3+S)e^{-t}$ (`Solp`, `Solq`), where $S=\sqrt{-3-8r}$, and the solution of the given Cauchy problem as the family of curves in the parametric form, $u(r,t)=-\left(\frac{3}{2}+\frac{1}{2}S\right)e^{-t}+\left(\frac{1}{2}(1+S)+2r\right)e^{-2t}$, $x(r,t)=(1-S)(1-e^{-t})+r$, $y(r,t)=t+r$ (`SolFin`). We verify that this solution is exact solution of the given nonlinear PDE (`test1`).

Maple:

```
interface(showassumed=0); assume(t>0); F:=p^2+q+u;
F1:=p(r,t)^2+q(r,t)+u(r,t); IniData:=[x(r,0)=r,y(r,0)=r,
 u(r,0)=2*r-1]; IniData1:=[x(0)=r,y(0)=r,u(0)=2*r-1];
CharEqs:=[diff(x(t),t)=diff(F,p), diff(y(t),t)=diff(F,q),
 diff(u(t),t)=p*diff(F,p)+q*diff(F,q),
 diff(p(t),t)=-(diff(F,x)+p*diff(F,u)),
 diff(q(t),t)=-(diff(F,y)+q*diff(F,u))];
tr1:={p(t)=p(r,t),q(t)=q(r,t)}; tr2:={p=p(r,t),q=q(r,t)};
Eq1:=subs(IniData,subs(t=0,F1))=0; StripCond:=diff(u(r,0),r)=
 p(r,0)*diff(x(r,0),r)+q(r,0)*diff(y(r,0),r);
IniData2:=diff(IniData,r); Eq2:=subs(IniData2,StripCond);
sys1:={Eq1,Eq2}; vars:=indets(sys1) minus {r};
sols:=[allvalues(solve(sys1,vars))];
```

```
for k from 4 to 5 do  Eq||k||1:=subs(tr1,lhs(CharEqs[k]))=
 subs(tr2,rhs(CharEqs[k])); od;
Solp:=dsolve({Eq41} union {op(1,sols[1])},p(r,t));
Solq:=dsolve({Eq51} union {op(2,sols[1])},q(r,t));
tr4:={p=rhs(Solp),q=rhs(Solq)}; tr5:={x(t)=x,y(t)=y,u(t)=u};
Eq6:=combine(subs(tr4,[CharEqs[i] $ i=1..3]));
Eq7:=dsolve({op(Eq6),op(IniData1)},{x(t),y(t),u(t)});
Eq71:= Eq7 union {Solp} union {Solq};
SolFin:=subs({u(t)=u(r,t),x(t)=x(r,t),y(t)=y(r,t)},Eq71);
test1:=simplify(subs(SolFin,F1));
```

Mathematica:

```
{fF=p^2+q+u, f1=p[r,t]^2+q[r,t]+u[r,t], iniData={x[r,0]->r,
 y[r,0]->r,u[r,0]->2*r-1}, iniData1={x[0]==r,y[0]==r,
 u[0]==2*r-1}, charEqs={x'[t]==D[fF,p], y'[t]==D[fF,q],
 u'[t]==p*D[fF,p]+q*D[fF,q], p'[t]==-(D[fF,x]+p*D[fF,u]),
 q'[t]==-(D[fF,y]+q*D[fF,u])}}
{tr1={p->p[r,t], q->q[r,t]}, tr12={p'[t]->D[p[r,t],t],
 q'[t]->D[q[r,t],t]}, tr2={p->p[r,t],q->q[r,t]}}
{eq1=(f1/.t->0/.iniData)==0, stripCond=D[u[r,0],r]==p[r,0]*
 D[x[r,0],r]+q[r,0]*D[y[r,0],r], iniData2=D[iniData,r],
 eq2=stripCond/.iniData2, sys1={eq1,eq2}}
{vars=Complement[Variables[{eq1[[1]],eq2[[1]]}],{r}],
 sols=Solve[sys1,vars], sols1=sols//First//Sort}
{Table[eq[k]=(charEqs[[k]][[1]]/.tr12)==(charEqs[[k]][[2]]/.tr2),
 {k,4,5}], sols2={vars[[1]]==sols1[[1,2]],
 vars[[2]]==sols1[[2,2]]}, solp=DSolve[{eq[4],sols2[[1]]},p[r,t],
 {r,t}], solq=DSolve[{eq[5],sols2[[2]]},q[r,t],{r,t}]}
{tr4={p->solp[[1,1,2]], q->solq[[1,1,2]]}, tr5={x[t]->x,y[t]->y,
 u[t]->u}, tr6={u[t]->u[r,t], x[t]->x[r,t], y[t]->y[r,t]}}
eq6=Table[charEqs[[i]],{i,1,3}]/.tr4//Together
{eq7=DSolve[{eq6,iniData1},{x[t],y[t],u[t]},{t}],
 eq71={eq7,solp,solq}, solFin=eq71/.tr6//Flatten,
 test1=f1/.solFin//Simplify}
```

2. Now let us apply these programs for solving Problem 3.17, i.e., we find the exact solution of the generalized Cauchy problem and compare it with the exact solution of the Cauchy problem, obtained in the previous problem. We can observe that the exact solutions are identical. For example, for the *Maple* program, we only have to change the lines:
`IniData:=[x(r,0)=r,y(r,0)=0,u(r,0)=r];`
`IniData1:=[x(0)=r,y(0)=0,u(0)=r];` and to add the following new lines:

```
Sol1:=[op(SolFin)];
tr6:={p(r,t)=p,q(r,t)=q,u(r,t)=u,x(r,t)=x,y(r,t)=y};
Sol2:=eliminate(subs(tr6,{Sol1[3],Sol1[4],Sol1[5]}),{r,t});
Sol3:=combine(op(Sol2[2]))=0;
SolFin1:=map(combine,collect(expand(isolate(Sol3,u)),exp(y))); □
```

Problem 3.19 *Eikonal equation. Integral surfaces. Generalized Cauchy problem.* Let us solve the eikonal equation (considered in Problems 3.12—3.14)

$$u_x^2 + u_y^2 = n^2,$$

where $\{x \in \mathbb{R}, y \geq 0\}$, with the initial data (for $r=0$) $x(0,s)=x_0(s)$, $y(0,s)=y_0(s)$, $u(0,s)=u_0(s)$, $p(0,s)=p_0(s)$, $q(0,s)=q_0(s)$ (r and s are parameters).

1. Applying the generalized method of characteristics, we obtain that the solution of the eikonal equation with the given Cauchy data takes the form $u(x,t)=u_0+n\sqrt{(x-x_0)^2+(y-y_0)^2}$.

Maple:

```
with(Student[Precalculus]): interface(showassumed=0);
assume(t>0); assume(x>x0); assume(y>y0); assume(u>u0);
F:=p^2+q^2-n^2; F1:=p(r,t)^2+q(r,t)^2-n^2;
IniData:=[x(0,t)=x0(t),y(0,t)=y0(t),u(0,t)=u0(t)];
IniData1:=[x(0)=x0(t),y(0)=y0(t),u(0)=u0(t)];
tr1:={p(t)=p(r,t),q(t)=q(r,t)}; tr2:={p=p(r,t),q=q(r,t)};
CharEqs:=[diff(x(r),r)=diff(F,p),diff(y(r),r)=diff(F,q),
 diff(u(r),r)=p*diff(F,p)+q*diff(F,q),
 diff(p(r),r)=-(diff(F,x)+p*diff(F,u )), diff(q(r),r)=
 -(diff(F,y)+q*diff(F,u))];
Eq1:=subs(_C1=p0(t),p(r,t)=rhs(dsolve(CharEqs[4],p(r))));
Eq2:=subs(_C1=q0(t),q(r,t)=rhs(dsolve(CharEqs[5],q(r))));
tr4:={p=rhs(Eq1),q=rhs(Eq2)};
tr5:={x(r)=x,y(r)=y,u(r)=u,p0(t)=p0,q0(t)=q0,x0(t)=x0,
 y0(t)=y0,u0(t)=u0}; tr6:=p0(t)^2+q0(t)^2=n^2;
Eq3:=combine(subs(tr4,[CharEqs[i] $ i=1..3]));
Eq4:=dsolve({op(Eq3),op(IniData1)},{x(r),y(r),u(r)});
Eq41:=op(1,Eq4); Eq42:=map(factor,collect(rhs(Eq41),r));
Eq43:=lhs(Eq41)=subs(tr6,Eq42); Eq5:=eliminate(subs(tr5,
 {Eq43,Eq4[2],Eq4[3],tr6}),{p0,q0,r});
Eq6:=CompleteSquare(op(Eq5[2])); Eq7:=collect(isolate(Eq6,
 (u-u0)^2),n^2); SolFin:=factor([solve(Eq7,u) assuming u>u0]);
```

Mathematica:

```
{fF=p^2+q^2-n^2, f1=p[r,t]^2+q[r,t]^2-n^2}
iniData={x[0,t]->x0[t],y[0,t]->y0[t],u[0,t]->u0[t]}
iniData1={x[0]->x0[t],y[0]->y0[t],u[0]->u0[t]}
{tr1={p[t]->p[r,t],q[t]->q[r,t]}, tr2={p->p[r,t],q->q[r,t]}}
charEqs={x'[r]==D[fF,p], y'[r]==D[fF,q], u'[r]==p*D[fF,p]+
 q*D[fF,q],p'[r]==-(D[fF,x]+p*D[fF,u]),
 q'[r]==-(D[fF,y]+q*D[fF,u])}
{eq1=p[r,t]==(First[DSolve[charEqs[[4]],p[r],r]/.
 C[1]->p0[t]])[[1,2]], eq2=q[r,t]==(First[DSolve[charEqs[[5]],
 q[r],r]/.C[1]->q0[t]])[[1,2]]}
{tr4={p->eq1[[2]],q->eq2[[2]]}, tr5={x[r]->x,y[r]->y,u[r]->u,
 p0[t]->p0,q0[t]->q0,x0[t]->x0,y0[t]->y0,u0[t]->u0},
 tr6=p0[t]^2+q0[t]^2->n^2}
{eq3=Table[charEqs[[i]],{i,1,3}]/.tr4, eq4=DSolve[{eq3,
 iniData1/.Rule->Equal}//Flatten,{x[r],y[r],u[r]},r]//Flatten}
{eq41=eq4[[3]], eq42=eq41[[2]]//FullSimplify,
 eq43=eq41[[1]]==eq42/.tr6, eq5=Eliminate[{eq43,eq4[[1]],
 eq4[[2]],tr6}/.Rule->Equal/.tr5,{p0,q0,r}]}
{eq51=eq5[[1]]-eq5[[2]], eq6=FullSimplify[eq51]==0,
 eq7=Thread[eq6+(u-u0)^2,Equal]}
solFin=Solve[eq7,u]//FullSimplify//Flatten
```

2. We visualize the solution $u(x,t)=u_0+n\sqrt{(x-x_0)^2+(y-y_0)^2}$ of the eikonal equation obtained for the Cauchy data $x_0(s)=0$, $y_0(s)=0$, and $u_0(s)=0$. We show that the solution obtained determines a cone with the vertex at (x_0, y_0, u_0) and identical to the corresponding *Monge cone* (see Problem 3.14). Verify that this solution is *singular*, i.e., satisfies the eikonal equation everywhere except for the initial point (x_0, y_0), and u_x and u_y are singular. Visualize the solution of the eikonal equation obtained for the Cauchy data $x_0(t)=t$, $y_0(t)=t$, $u_0(t)=t^2$.

Maple:

```
SolFin:=u=u0+n*sqrt((x-x0)^2+(y-y0)^2); trn:={n=1};
U:=unapply(rhs(SolFin),x,y); tr0:={p0(t)=n*cos(alpha),
 q0(t)=n*sin(alpha)}; combine(subs(tr0,p0(t)^2+q0(t)^2=n^2));
tr1:={x0=0,y0=0,u0=0}; tr2:={x0(t)=t,y0(t)=t,u0(t)=t^2};
plot3d(subs(tr1,trn,U(x,y)),x=-10..10,y=-10..10);
EikonalEq:=diff(U(x,y),x)^2+diff(U(x,y),y)^2-n^2;
sys1:={u=2*n^2*r+u0(t),x=2*p0(t)*r+x0(t),
       y=2*q0(t)*r+y0(t)}; sys2:=subs(tr2,tr0,sys1);
```

```
Sol2:=eliminate(sys2,{r,t}); Sol21:=isolate(op(Sol2[2]),u);
U2:=unapply(collect(rhs(Sol21),[n,x,y]),x,y);
plot3d(subs(trn,alpha=Pi,U2(x,y)),x=-10..10,y=-10..10);
```

Mathematica:

```
SetOptions[Plot3D,PlotRange->All,Mesh->False,BoxRatios->{1,1,1}];
{solFin=u==u0+n*Sqrt[(x-x0)^2+(y-y0)^2], trn=n->1}
uN[xN_,yN_]:=solFin[[2]]/.x->xN/.y->yN; uN[x,y]
{tr0={p0[t]->n*Cos[alpha],q0[t]->n*Sin[alpha]},
 p0[t]^2+q0[t]^2==n^2/.tr0, tr1={x0->0,y0->0,u0->0},
 tr2={x0[t]->t,y0[t]->t,u0[t]->t^2}}
Plot3D[Evaluate[uN[x,y]/.tr1/.trn],{x,-10,10},{y,-10,10}]
eikonalEq=D[uN[x,y],x]^2+D[uN[x,y],y]^2-n^2
sys1={u==2*n^2*r+u0[t], x==2*p0[t]*r+x0[t], y==2*q0[t]*r+y0[t]}
{sys2=sys1/.tr0/.tr2, sol2=Eliminate[sys2,{r,t}],
 sol21=Solve[sol2,u]//First}
uN2[xN_,yN_]:=sol21[[1,2]]//FullSimplify/.x->Xn/.y->yN;
Plot3D[Evaluate[uN2[x,y]/.alpha->Pi/.trn],{x,-10,10},
 {y,-10,10}]
```

□

3.3 Qualitative Analysis

It is known that only some nonlinear PDEs can be solved exactly and therefore we have to deal with another approach for finding approximate solutions. One of these approaches, the *qualitative analysis* or the phase-space analysis (or phase-plane analysis), can be applied for studying qualitatively the nature or types of solutions for a given nonlinear system of ODEs with specific initial conditions. This approach can be applied together with other approaches, numerical analysis or perturbation analysis (that we consider in Sect. 5.2, 5.2.2, and Chap. 6, 7).

In general, a dynamical system geometrically represents a vector-field (or a tensor-field) in the phase-space manifold M, which (by integration) defines a phase-flow in M. The phase-flow (that describes the complete behavior of a dynamical system) can be linear, nonlinear, or chaotic.

The theory of dynamical systems goes back to the qualitative study of differential equations (works by Newton, Euler, Hamilton, Maxwell). At the end of the 19th century, H. Poincaré was first to shift investigations from finding explicit forms of solutions to discovering their geometric properties. Then G. D. Birkhoff developed the methods of Poincaré and proposed the term "dynamical systems". In the last century, many mathematicians have made important contributions to the theory of dynamical systems (works by A. M. Lyapunov, L. S. Pontryagin, A. A. Andronov, S. Smale, A. N. Kolmogorov, V. I. Arnold, Y. G. Sinai, E. N. Lorenz, F. Takens). Nowadays, nonlinear dynamical systems can describe various scientific and engineering phenomena and have been applied to problems arising in mathematics, physics, chemistry, biology, medicine, economics, and engineering. In modern day mathematics there exist computers, supercomputers, and computer algebra systems that can aid to implement sophisticated mathematical methods for solving a wide class of dynamical systems and determining its orbits [92].

3.3.1 Nonlinear PDEs

In this section, we follow the standard stability analysis, based on the the Poincaré-Bendixson theory, for studying a wide class of nonlinear problems described by the system of first-order autonomous ordinary differential equations: $\dfrac{du}{d\xi} = P(u, v)$, $\dfrac{dv}{d\xi} = Q(u, v)$, where P and Q are nonlinear functions of the two dependent variables u and v, and the independent variable ξ can be eliminated obtaining $\dfrac{du}{dv} = \dfrac{P(u, v)}{Q(u, v)}$. The solution of this equation can be represented as a *phase diagram* on the plane (u, v) with some *phase trajectories* along which the system will evolve as ξ increases. Since the integration for finding $u(v)$ may not be possible, the phase plane analysis can be useful to determine *stationary* or *singular points* u_0, v_0 in some regions of the *Poincaré phase plane* and to analyze the nature of solutions. Then the entire phase plane will consists of all these regions (see Problem 3.22). In practice, since the phase plane portraits may be complicated it is convenient to obtain them with *Maple* and *Mathematica*.

Problem 3.20 *Fisher equation. Qualitative analysis.* Let us consider the Fisher equation

$$u_t - u_{xx} = F(u),$$

where $F(u) = u(1-u)$, $\{x \in \mathbb{R}, t \geq 0\}$. This equation admits a traveling wave solution of the form $u(x,t) = W(z)$, $z = x - ct$ $(0 \leq u(x,t) \leq 1)$, where c is the wave speed and the waveform $W(z)$ satisfies the boundary conditions $\lim_{z \to -\infty} W(z) = 1$, $\lim_{z \to \infty} W(z) = 0$.

1. We show that the Fisher equation can be reduced to the nonlinear ODE $-W''(z) - W'(z)c - W(z) + W(z)^2 = 0$, or to the system of two first-order ODEs $W' = V$, $V' = -cV - W + W^2$, or equivalently to the equation

$$\frac{dV}{dW}=\frac{-cV-W+W^2}{V}.$$

2. We show that the singular points (W_0, V_0) of the dynamical system obtained are $(0,0)$ and $(1,0)$.

3. Following the phase plane analysis, we study the nature of the given nonlinear system. We prove that the matrix associated with this system at singular points, $(0,0)$ and $(1,0)$, has, respectively, the eigenvalues $-\frac{1}{2}(c\mp\sqrt{c^2-4})$ and $\frac{1}{2}(-c\pm\sqrt{c^2+4})$.

4. We show, according to the theory of dynamical systems, that the point $(0,0)$ is a stable node for $c\geq 2$ and the point $(1,0)$ is a saddle point. We construct the phase portrait on the plane (W,V), showing that there exists a unique separatrix joining the stable node $(0,0)$ with the saddle point $(1,0)$ (for $c\geq 2$).

Maple:

```
with(PDEtools): with(plots): with(DEtools): with(LinearAlgebra):
Ops:=arrows=medium,dirgrid=[20,20],stepsize=0.1,thickness=2,
 linecolour=blue,color=green;
tr1:=x-c*t=z; tr2:={V(z)=V,W(z)=W}; vars:=[W(z),V(z)];
Eq1:=u->diff(u,t)-diff(u,x$2)-u*(1-u);
Eq2:=expand(Eq1(W(lhs(tr1)))); Eq3:=algsubs(tr1,Eq2)=0;
Eq4:=map(convert,Eq3,diff); Eq5:=diff(W(z),z)=V(z);
Eq6:=isolate(subs(Eq5,Eq4),diff(V(z),z)); Eq7:=subs(tr2,
 Diff(V,W)=rhs(Eq6)/rhs(Eq5)); Eq81:=denom(rhs(Eq7))=0;
Eq82:=[solve(subs(Eq81,numer(rhs(Eq7))),W)];
SingularPoints:=[[Eq82[1],rhs(Eq81)],[Eq82[2],rhs(Eq81)]];
P:=subs(tr2,rhs(Eq5)); Q:=subs(tr2,rhs(Eq6));
A:=<diff(P,W),diff(Q,W)|diff(P,V),diff(Q,V)>;
A1:=subs({W=0,V=0},A); A2:=subs({W=1,V=0},A);
map(Eigenvalues,{A1,A2}); c>=2; Eqs:=subs(c=3,[Eq5,Eq6]);
IC:=[[W(0)=-0.5,V(0)=-1.],[W(0)=0.1,V(0)=1.],
 [W(0)=0.96968212,V(0)=0.1],[W(0)=0.35,V(0)=1.],
 [W(0)=0.5,V(0)=1.],[W(0)=0.8,V(0)=-1.],[W(0)=0.4,V(0)=-1.],
 [W(0)=0.1,V(0)=-1.],[W(0)=0.999,V(0)=-1.],[W(0)=1.1,V(0)=-1.],
 [W(0)=1.29889,V(0)=-1.],[W(0)=0.65999,V(0)=1.1]];
phaseportrait(Eqs,vars,z=0..60,IC,Ops,view=[-0.7..1,-1..1]);
```

Mathematica:

```
eq1[u_]:=D[u,t]-D[u,{x,2}]-u*(1-u);
{tr1=x-c*t->z, tr2={vN[z]->vN,uN[z]->uN}}
```

```
tr3={uN'[z]->vN[z],uN''[z]->vN'[z]}
{eq2=eq1[uN[tr1[[1]]]]//Expand, eq3=(eq2/.tr1)==0,
 eq4=eq3//TraditionalForm, eq5=eq3/.tr3}
{eq6=Solve[eq5,vN'[z]], eq7=Hold[D[vN,uN]]==eq6[[1,1,2]]/vN[z]}
{eq81=eq7[[2,1,1]]->0, eq82=Solve[eq7[[2,2]]==0/.eq81,uN[z]]}
singPoints={{eq82[[1,1,2]],eq81[[2]]},{eq82[[2,1,2]],eq81[[2]]}}
{p1=tr3[[1,2]]/.tr2, q1=eq6[[1,1,2]]/.tr2}
ma={{D[p1,uN],D[p1,vN]},{D[q1,uN],D[q1,vN]}}
{a1=ma/.{uN->0,vN->0}, a2=ma/.{uN->1,vN->0}}
{Eigenvalues[a1], Eigenvalues[a2], c>=2}
{eqs=({tr3[[1]],eq6[[1,1]]}/.c->3)/.Rule->Equal, vars={uN,vN}}
ic={{-0.5,-1.},{0.1,1.},{0.96968212,0.1},{0.35,1.},{0.5,1.},
    {0.8,-1.},{0.4,-1.},{0.1,-1.},{0.999,-1.},{1.1,-1.},
    {1.29889,-1.},{0.65999,1.1}}; n=Length[ic]; zF=60;
Do[{sys[i]={eqs[[1]],eqs[[2]],uN[0]==ic[[i,1]],vN[0]==ic[[i,2]]};
  sols=NDSolve[sys[i],vars,{z,0,zF}]; cu=uN/.sols[[1]];
  cv=vN/.sols[[1]]; c[i]=ParametricPlot[Evaluate[{cu[z],cv[z]}],
  {z,0,zF},PlotStyle->{Hue[0.1*i+0.2],Thickness[.01]}];},{i,1,n}]
fu=eqs[[1,2]]/.tr2; fv=eqs[[2,2]]/.tr2;
fd=VectorPlot[{fu,fv},{uN,-0.7,1.},{vN,-1.,1.},
   VectorColorFunction->Hue]; Show[fd,Table[c[i],{i,1,n}]]
```
□

Problem 3.21 *Sine–Gordon equation. Qualitative analysis.* Let us consider the sine–Gordon equation

$$u_{xx} - u_{tt} = \sin(u),$$

where $\{x \in \mathbb{R}, t \geq 0\}$. This equation also admits traveling wave solution of the form $u(x,t) = W(z)$, $z = x - ct$, where c is a wave speed.

1. We show that the sine–Gordon equation can be reduced to the ODE $W''(z)(1-c^2) - \sin(W(z)) = 0$ or to the system of two first-order ODEs $W' = V$, $V' = \sin(W)/(1-c^2)$ or equivalently $\dfrac{dV}{dW} = \dfrac{\sin(W)}{(1-c^2)V}$. Note that $c \neq 1$, i.e., the nature of the given nonlinear system depends on the parameter c. Let us choose $c < 1$.

2. We show, according to the theory of dynamical systems, that the points $(-\pi, 0)$ and $(\pi, 0)$ are stable nodes for $c<1$ and the points $(-2\pi, 0)$, $(0,0)$, and $(2\pi, 0)$, are saddle points.

3. We construct the phase portrait on the plane (W, V) (see Fig. 3.5), showing that there exist two separatrixes between the saddle points

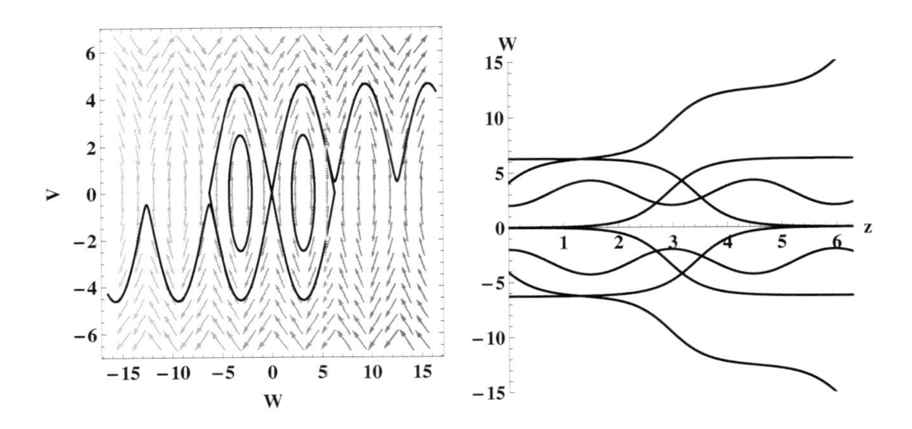

Fig. 3.5. Traveling wave solution of the sine–Gordon equation: phase portrait and kink, antikink, and periodic solutions

$(-2\pi, 0)$ and $(0, 0)$, and two separatrixes between the saddle points $(0, 0)$ and $(2\pi, 0)$ (for $c < 1$).

Note that these separatixes correspond to a *kink* and *antikink* solutions of the sine–Gordon equation (see Problems 2.11, 2.20, 2.32). The solutions inside and outside the separatrixes are periodic. The plots of kink and antikink solutions, i.e., the function $W(z)$, are shown in Fig. 3.5.

Maple:

```
with(PDEtools): with(plots): with(DEtools): declare(u(x,t));
Ops:=arrows=medium,dirgrid=[20,20],stepsize=0.1,thickness=2,
     linecolour=blue,color=green;
tr1:=x-c*t=z; tr2:={V(z)=V,W(z)=W};
Eq1:=u->diff(u,x$2)-diff(u,t$2)-sin(u);
Eq2:=expand(Eq1(W(lhs(tr1)))); Eq3:=algsubs(tr1,Eq2)=0;
Eq4:=map(convert,Eq3,diff); Eq5:=diff(W(z),z)=V(z);
Eq6:=isolate(collect(subs(Eq5,Eq4),diff),V(z));
Eq7:=subs(tr2,Diff(V,W)=rhs(Eq6)/rhs(Eq5));
Eqs:=subs(c=0.9,[Eq5,Eq6]); vars:=[W(z),V(z)];
IC:=[[W(0)=0.01,V(0)=0.],[W(0)=-0.01,V(0)=0.],[W(0)=2,V(0)=0.],
 [W(0)=-2,V(0)=0.],[W(0)=4.1,V(0)=4.1],[W(0)=-4.1,V(0)=-4.1],
 [W(0)=6.28,V(0)=0.],[W(0)=-6.28,V(0)=0.]];
P:=(X,Y)->phaseportrait(Eqs,vars,z=0..2*Pi,IC,scene=[X,Y],
          Ops,view=[-7..7,-7..7]); P(W,V); P(z,W);
```

Mathematica:

```
n1=20; p=15; zF=2*Pi//N;
trS1[eq_,var_]:=Select[eq,MemberQ[#,var,Infinity]&];
trS3[eq_,var_]:=Select[eq,FreeQ[#,var]&];
SetOptions[ParametricPlot,AspectRatio->1,
 PlotStyle->{Blue,Thickness[.01]}];
SetOptions[VectorPlot,VectorColorFunction->Hue,
 VectorPoints->{n1,n1},VectorStyle->Arrowheads[0.02],
 PlotRange->All]; {tr1=x-c*t->z, tr2={v[z]->v,w[z]->w}}
eq1[u_]:=D[u,{x,2}]-D[u,{t,2}]-Sin[u];
{eq2=eq1[w[tr1[[1]]]]//Expand, eq3=(eq2/.tr1)==0,
 eq4=eq3//TraditionalForm, eq5={w'[z]->v[z],w''[z]->v'[z]}}
{eq6=eq3/.eq5//FullSimplify, termc=trS3[trS1[eq6[[1]],c],z]}
eq61=Thread[eq6/termc,Equal]//FullSimplify
eq62=Thread[Thread[(-1)*eq61,Equal]+v'[z],Equal]
eq7=HoldForm[D[v,w]]==(eq62[[1]]/eq5[[1,2]])/.tr2
eqs={eq5[[1]]/.Rule->Equal, eq62[[2]]==eq62[[1]]}/.c->0.9
vars={w,v}; ic={{0.01,0.},{-0.01,0.},{2,0.},{-2,0.},
 {4.1,4.1},{-4.1,-4.1},{6.28,0.},{-6.28,0.}}
n=Length[ic];
Do[{sys[i]={eqs[[1]],eqs[[2]], w[0]==ic[[i,1]],
 v[0]==ic[[i,2]]}; sols=NDSolve[sys[i],vars,{z,0,zF}];
 cw=w/.sols[[1]]; cv=v/.sols[[1]];
 wv[i]=ParametricPlot[Evaluate[{cw[z],cv[z]}],{z,0,zF},
  PlotRange->{{-p,p},{-zF,zF}}];
 zw[i]=ParametricPlot[Evaluate[{z,cw[z]}],{z,0,zF},
  PlotRange->{{0,zF},{-p,p}}];},{i,1,n}];
fw=eqs[[1,2]]/.tr2; fv=eqs[[2,2]]/.tr2;
fd=VectorPlot[{fw,fv},{w,-p,p},{v,-zF,zF}];
Show[fd,Table[wv[i],{i,1,n}]]
Show[Table[zw[i],{i,1,n}]]
```

□

3.3.2 Nonlinear Systems

In this section, we will consider the problem of subharmonic excitation of surface standing nonlinear water waves in a vertically oscillating container (see Problem 5.10). The Lagrangian formulation is introduced to write out the exact nonlinear equations of hydrodynamics and the boundary conditions. The asymptotic procedure, based on the Krylov–Bogolyubov averaging method is described in Sect. 5.2.2.

Problem 3.22 *Nonlinear standing waves. Qualitative analysis.* Let us consider the system describing the evolution of the amplitude $\mathcal{C}$ and the slow phase θ (obtained by the averaging transformations)

$$\frac{d\mathcal{C}}{dt}=\frac{\mathcal{C}}{4}\varepsilon\omega\sin(2\theta),$$
$$\frac{d\theta}{dt}=\Delta-\sqrt{\varepsilon}\,\omega\phi_2\mathcal{C}^2-\varepsilon\omega\phi_4\mathcal{C}^4+\frac{1}{4}\varepsilon\omega\cos(2\theta),$$

where ϕ_2 and ϕ_4 are the second- and the fourth-order nonlinear corrections to the wave frequency, respectively,

$$\phi_2=\frac{P_2(\lambda)}{64\lambda^4},\quad \phi_4=\frac{P_4(\lambda)}{16384\lambda^{10}},$$

and the corresponding polynomials take the form:

$$P_2(\lambda)=2\lambda^6+3\lambda^4+12\lambda^2-9,$$
$$P_4(\lambda)=36\lambda^{16}-54\lambda^{14}-883\lambda^{12}+691\lambda^{10}-1611\lambda^8+6138\lambda^6$$
$$-4077\lambda^4+1323\lambda^2-81.$$

1. We consider a special case where the fluid depth h is equal or close to the critical value χ_c, i.e., the depth at which the third-order nonlinear correction to the wave frequency vanishes. The critical depth can be determined from the equation $\phi_2(\chi_c)=0$, $h_c=\chi_c/\kappa\approx 1.06/\kappa$.

Introducing the new variables u, v, we rewrite the system in the form

$$\frac{du}{d\tau}=\varepsilon v\left[\tfrac{1}{4}+\tilde{\Delta}-\tfrac{1}{2}(u^2+v^2)\tilde{\phi}_1+\tfrac{1}{4}(u^2+v^2)^2\phi_2\right],$$
$$\frac{dv}{d\tau}=\varepsilon u\left[\tfrac{1}{4}-\tilde{\Delta}+\tfrac{1}{2}(u^2+v^2)\tilde{\phi}_1-\tfrac{1}{4}(u^2+v^2)^2\phi_2\right],$$

where $u=\sqrt{2}\,\mathcal{C}\sin\theta$, $v=\sqrt{2}\,\mathcal{C}\cos\theta$, $\tilde{\Delta}=\Delta/(\varepsilon\omega)$, $\tilde{\phi}_1=\phi_1/\varepsilon^{1/2}$, and $\tau=\omega t$. This is the Hamiltonian system with Hamiltonian

$$\mathcal{H}=\varepsilon\left[\tfrac{1}{2}(u^2+v^2)\tilde{\Delta}+\tfrac{1}{8}(v^2-u^2)-\tfrac{1}{8}(u^2+v^2)^2\tilde{\phi}_1+\tfrac{1}{24}(u^2+v^2)^3\phi_2\right].$$

2. Analyzing the fixed points of this system, we obtain that apart from the zero (or trivial) fixed point there exist two classes of nonzero fixed points, (u^-,v^-) and (u^+,v^+):

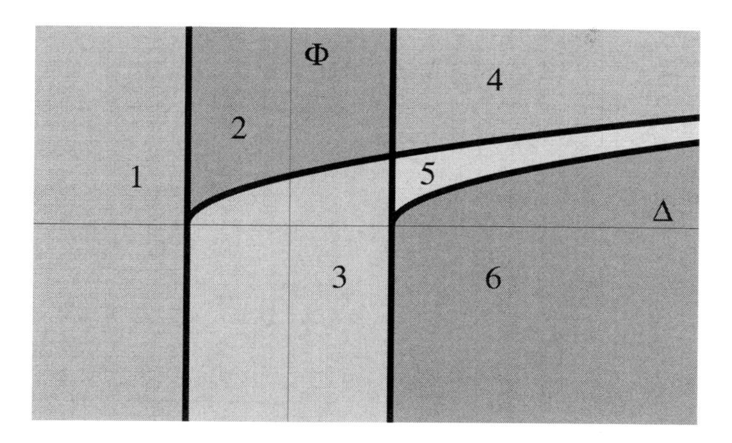

Fig. 3.6. The regions where the fixed points of all classes exist without bifurcations (for $\phi_2 > 0$)

Class (u^-, v^-): $\theta = \frac{\pi}{2}, \frac{3\pi}{2}$;

$$\mathcal{C}^-_+ = \sqrt{\frac{\tilde{\phi}_1 + \sqrt{\tilde{\phi}_1^2 - 4\phi_2(\tilde{\Delta} - \frac{1}{4})}}{2\phi_2}}, \ u^-_{+1} = \mathcal{C}^-_+\sqrt{2}, \ v^-_{+1} = 0, \ u^-_{+2} = -\mathcal{C}^-_+\sqrt{2}, \ v^-_{+2} = 0;$$

$$\mathcal{C}^-_- = \sqrt{\frac{\tilde{\phi}_1 - \sqrt{\tilde{\phi}_1^2 - 4\phi_2(\tilde{\Delta} - \frac{1}{4})}}{2\phi_2}}, \ u^-_{-1} = \mathcal{C}^-_-\sqrt{2}, \ v^-_{-1} = 0, \ u^-_{-2} = -\mathcal{C}^-_-\sqrt{2}, \ v^-_{-2} = 0;$$

Class (u^+, v^+): $\theta = 0, \pi$;

$$\mathcal{C}^+_+ = \sqrt{\frac{\tilde{\phi}_1 + \sqrt{\tilde{\phi}_1^2 - 4\phi_2(\tilde{\Delta} + \frac{1}{4})}}{2\phi_2}}, \ u^+_{+1} = 0, \ v^+_{+1} = \mathcal{C}^+_+\sqrt{2}, \ u^+_{+2} = 0, \ v^+_{+2} = -\mathcal{C}^+_+\sqrt{2};$$

$$\mathcal{C}^+_- = \sqrt{\frac{\tilde{\phi}_1 - \sqrt{\tilde{\phi}_1^2 - 4\phi_2(\tilde{\Delta} + \frac{1}{4})}}{2\phi_2}}, \ u^+_{-1} = 0, \ v^+_{-1} = \mathcal{C}^+_-\sqrt{2}, \ u^+_{-2} = 0, \ v^+_{-2} = -\mathcal{C}^+_-\sqrt{2}.$$

The regions (where the fixed points of all classes exist without bifurcations) and their characteristics for $\phi_2 > 0$ and $\phi_2 < 0$ are represented, respectively, in Fig. 3.6, Tab. 3.1 and in Fig. 3.7, Tab. 3.2.

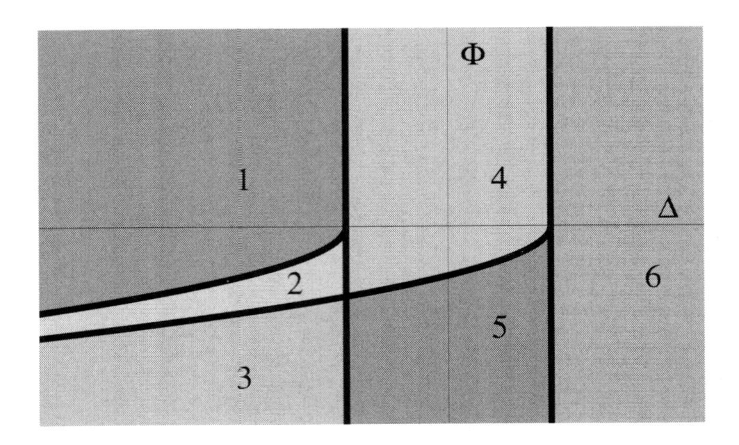

Fig. 3.7. The regions where the fixed points of all classes exist without bifurcations (for $\phi_2 < 0$)

Table 3.1.* Characteristics of the regions on the (Δ,Φ) plane for $\phi_2 > 0$

	Tr.	$u^+_{-1}\, v^+_{-1}$	$u^+_{-2}\, v^+_{-2}$	$u^+_{+1}\, v^+_{+1}$	$u^+_{+2}\, v^+_{+2}$	$u^-_{+1}\, v^-_{+1}$	$u^-_{+2}\, v^-_{+2}$	$u^-_{-1}\, v^-_{-1}$	$u^-_{-2}\, v^-_{-2}$
1	S ∃			US ∃	US ∃	S ∃	S ∃		
2	US ∃	S ∃	S ∃	US ∃	US ∃	S ∃	S ∃		
3	US ∃					S ∃	S ∃		
4	S ∃	S ∃	S ∃	US ∃	US ∃	S ∃	S ∃	US ∃	US ∃
5	S ∃					S ∃	S ∃	US ∃	US ∃
6	S ∃								

For example, the fixed points $\{u^-_{+1}, v^-_{+1}\}$, $\{u^-_{+2}, v^-_{+2}\}$ exist in the regions 1–5 for $\phi_2 > 0$, and in the regions 3, 5 for $\phi_2 < 0$. Nonzero fixed points of each class are always either stable or unstable (in the regions where they exist without bifurcations).

The fixed points $\{u^-_{+1}, v^-_{+1}\}$, $\{u^-_{+2}, v^-_{+2}\}$ are always stable and the fixed points $\{u^-_{-1}, v^-_{-1}\}$, $\{u^-_{-2}, v^-_{-2}\}$ are always unstable.

Table 3.2. Characteristics of the regions on the (Δ,Φ) plane for $\phi_2<0$

	Tr.	$u^+_{-1} v^+_{-1}$	$u^+_{-2} v^+_{-2}$	$u^+_{+1} v^+_{+1}$	$u^+_{+2} v^+_{+2}$	$u^-_{+1} v^-_{+1}$	$u^-_{+2} v^-_{+2}$	$u^-_{-1} v^-_{-1}$	$u^-_{-2} v^-_{-2}$
1	S ∃								
2	S ∃	S ∃	S ∃	US ∃	US ∃				
3	S ∃	S ∃	S ∃	US ∃	US ∃	S ∃	S ∃	US ∃	US ∃
4	US ∃	S ∃	S ∃						
5	US ∃	S ∃	S ∃			S ∃	S ∃	US ∃	US ∃
6	S ∃	S ∃	S ∃					US ∃	US ∃

3. By varying the parameters of the system, we can construct phase portraits that correspond to the six regions, where the solution exists.

Let us choose one special case, e.g., the region 1 (for $\phi_2 > 0$), and construct the phase portrait (see Fig. 3.8). The other phase portraits can be obtained in a similar manner.

Maple:

```
with(plots): with(DEtools):
Ops:=arrows=medium,dirgrid=[20,20],stepsize=0.1,thickness=2,
 linecolour=blue,color=magenta;
phi_1:=1; phi_2:=1; epsilon:=0.1; delta:=-1/2;
Eq1:=D(v)(t)=epsilon*u(t)*(-delta+1/4+phi_1/2*(u(t)^2+v(t)^2)
 -phi_2/4*(u(t)^2+v(t)^2)^2);
Eq2:=D(u)(t)=epsilon*v(t)*(delta+1/4-phi_1/2*(u(t)^2+v(t)^2)
 +phi_2/4*(u(t)^2+v(t)^2)^2);
Eqs:=[Eq1,Eq2]; vars:=[v(t),u(t)];
IC:=[[u(0)=0,v(0)=1.1033],[u(0)=0,v(0)=-1.1033],
 [u(0)=1.1055,v(0)=0],[u(0)=-1.1055,v(0)=0],
 [u(0)=0,v(0)=1.613],[u(0)=0,v(0)=-1.613],
 [u(0)=0.2,v(0)=0],[u(0)=0.4,v(0)=0],
 [u(0)=1.3,v(0)=1.2],[u(0)=-1.3,v(0)=-1.2],
 [u(0)=0,v(0)=1.5]];
phaseportrait(Eqs,vars,t=-48..400,IC,Ops);
```

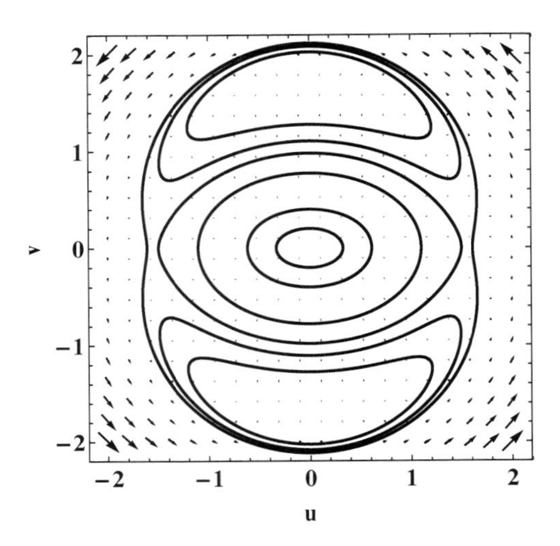

Fig. 3.8. Phase portrait for the region 1 ($\phi_2 > 0$)

Mathematica:

```
n1=20; tI=-48; tF=400;
{delta=-1/2,phi1=1,phi2=1,epsilon=0.1}
{eq1=epsilon*u[t]*(-delta+1/4+phi1/2*(u[t]^2+v[t]^2)-
 phi2/4*(u[t]^2+v[t]^2)^2),
 eq2=epsilon*v[t]*(delta+1/4-phi1/2*(u[t]^2+v[t]^2)+
 phi2/4*(u[t]^2+v[t]^2)^2)}
ic={{0,1.1033},{0,-1.1033},{1.1055,0},{-1.1055,0},{0,1.613},
 {0,-1.613},{0.2,0},{0.4,0},{1.3,1.2},{-1.3,-1.2},{0,1.5}}
n=Length[ic];
Do[{sys[i]={v'[t]==eq1,u'[t]==eq2,v[0]==ic[[i,2]],
 u[0]==ic[[i,1]]}; sols=NDSolve[sys[i],{v,u},{t,tI,tF}];
 cv=v/.sols[[1]]; cu=u/.sols[[1]];
 c[i]=ParametricPlot[Evaluate[{cv[t],cu[t]}],{t,tI,tF},
  AspectRatio->1,PlotStyle->{Blue,Thickness[.01]}];},
{i,1,n}];
fv=eq1/.{v[t]->v,u[t]->u}; fu=eq2/.{v[t]->v,u[t]->u};
fd=VectorPlot[{fv,fu},{v,-2,2},{u,-2,2},Frame->True,
 VectorColorFunction->Function[{x},Hue[0.8-Log[x]/30]],
 VectorPoints->{n1,n1},VectorStyle->Arrowheads[0.02]];
Show[fd,Table[c[i],{i,1,n}]]
```

□

Chapter 4
General Analytical Approach. Integrability

In this chapter, following the general analytical approach, we consider the basic concepts, ideas, and the most important methods for solving analytically nonlinear partial differential equations with the aid of *Maple* and *Mathematica*. In particular, we will consider the concepts of integrability, the Painlevé integrability, complete integrability for evolution equations, the Lax pairs, the variational principle. We will construct conservation laws of finite arbitrary order, nonlinear superposition formulas (which relate four different solutions of a nonlinear equation), exact multiple-soliton solutions of integrable nonlinear evolution equations by applying the Hirota method. Finally, we will consider various analytical methods for solving nonlinear systems of PDEs.

Let us consider a PDE with the independent variables x_i $(i{=}1,\dots,n)$:

$$\mathcal{F}(x_i,\dots,u,u_{x_i},\dots)=0. \tag{4.1}$$

Definition 4.1 Equation (4.1) is called *integrable* if at least one of the following properties holds:
it is possible *to linearize* the PDE;
it is possible to obtain a *general solution*;
it is possible to construct an *auto-Bäcklund transformation* for $n>1$;
it is possible to construct a *Bäcklund transformation* (BT) to an integrable PDE.

For example, it is possible to find the general solution of the form $u(x,t)=\big(\int f(x)\,dx+F(t-x)\big)^{-1}$ to the nonlinear PDE $u_t+u_x+f(x)u^2=0$, where $f(x)$ and $F(t-x)$ are arbitrary functions (see Problem 1.8); there exists a connection between the Burgers equation $u_t+uu_x=\nu u_{xx}$ and the linear heat equation $v_t=\nu v_{xx}$ by applying the Hopf–Cole transformation (see Problem 2.14); there exists the auto-Bäcklund transformation for the sine–Gordon equation $u_{xt}=\sin u$ (see Problem 2.11); there exists a connection between the modified KdV equation $v_t - 6v^2 v_x + v_{xxx} = 0$ and the KdV equation $u_t - 6uu_x + u_{xxx} = 0$ by applying the BT or the Miura transformation (as a particular case of BT) (see Problem 2.12).

Partially integrable and nonintegrable equations (the majority of physical equations) do not admit Bäcklund transformations.

4.1 Painlevé Test and Integrability

In this section, we will discuss the extension of the *Painlevé property* and *Painlevé test* (initially introduced for ODEs) to nonlinear PDEs. We will examine the *Painlevé integrability*, i.e., we will determine if a given PDE admits a single-valued general solution and find it. We will consider two kinds of nonlinear PDEs: equations that pass the Painlevé test (where we can find a general solution), and equations that fail the Painlevé test (where we can find a particular single-valued solution).

4.1.1 Painlevé Property and Test

It is known [32] that singularities of a given nonlinear PDE are not isolated in the space of the independent variables $(x_1, \ldots, x_n)$, i.e., they are not located in the neighborhood of a point $(x_{10}, \ldots, x_{n0})$, they lie on a manifold $\phi(x_1, \ldots, x_n) - \phi_0 = 0$, where ϕ and ϕ_0, are, respectively, the arbitrary function of the independent variables and the arbitrary movable constant.

The *Painlevé property* (PP) of a given nonlinear PDE is its integrability and the absence of movable critical singularities about arbitrary noncharacteristic manifold.

A singular point of a solution is *critical* if the solution is multi-valued in its neighborhood and *movable* if its location depends on initial conditions. A manifold is *noncharacteristic* if it is possible to apply the Cauchy–Kowalevski* existence theorem.

The *Painlevè analysis* consists in the application of two methods: to determine necessary conditions for a given nonlinear PDE to have the PP (the *Painlevè test*); to find the integrable properties discussed in Definition 4.1 (if these conditions are satisfied or at least some of them).

For example, in the integrable case, it is possible to find an auto-BT or a BT to another PDE with the PP; in the partially integrable case, it is possible to find degenerate forms.

The Painlevé test is an algorithm, which determines necessary conditions for a nonlinear PDE to have the Painlevé property. The first

*Kowalevski is an alternative transliteration for academic publications of the great mathematician Sofia Vasilyevna Kovalevskaya.

algorithm was developed by S. V. Kovalevskaya in 1889 [84]. A more general algorithm, developed by P. Painlevé in 1895 [117], is known as the α-method. Then, the Ablowitz–Ramani–Segur (ARS) algorithm of the Painlevé test for ODEs has been developed in 1980 [2]. For nonlinear PDEs, there exist different methods, e.g., the Weiss–Tabor–Carnevale (WTC) method [160], the Kruskal simplification method [80], the invariant and perturbation methods (see [31], [32]) developed by Conte and his colleagues.

The most widely applied methods for verifying the Painlevé property are the WTC method and the Kruskal simplification method. In some cases it is sufficient to apply the Kruskal simplification method or WTC method (that gives more information about integrability).

It should be noted that the application of computer algebra systems, *Maple*, *Mathematica*, *Macsyma*, can be very helpful for performing the Painlevé test. Several programs for performing the Painlevé test of nonlinear ODEs and PDEs have been developed (see, e.g., [50], [170], [15], [171]).

Problem 4.1 *Burgers equation. Painlevé test. Integrable case.* Let us consider the Burgers equation

$$u_t+auu_x+bu_{xx}=0,$$

where $\{x \in \mathbb{R}, t \geq 0\}$. Applying the Kruskal method, verify that the Burgers equation passes the Painlevé test and its solution has the form:

$$u(x,t)=\frac{2b}{a(x-\psi)}+\frac{2}{a}\psi_t+(x-\psi)u_2.$$

Verify that if $a=1$ and $b=-\nu$, the solution of the Burgers equation $u_t+uu_x-\nu u_{xx}=0$ coincides with the solution described in [125], i.e., has the form $u(x,t)=-2\nu/(x-\psi)+2\psi_t+(x-\psi)u_2$.

According to the Kruskal simplification method, the solutions for the Burgers equation can be expanded in terms of the Laurent series:

$$u(x,t)=\phi^{-\rho}\sum_{j=0}^{\infty}\phi^j u_j(t), \tag{4.2}$$

where $\phi=x-\psi(t)$ and $u_j=u_j(t)$ are analytic functions in a neighborhood of the noncharacteristic singular manifold $x-\psi(t)=0$, $u_0\neq 0$, and ρ is a positive integer to be determined. The algorithm of the Painlevé analysis consists of the following three steps:

1. Leading order term analysis. The leading order term of the solution for the Burgers equation is assumed to be $u \approx \phi^{-\rho} u_0$, where $u_0(t)$ is an analytic function. Substituting this expression into the original equation, we obtain $-\rho(-b\rho+\psi\psi_t-b-x\psi_t)u_0-a\rho\phi^{1-\rho}u_0^2+\phi^2 u_{0t}=0$ (`Eq5`). Then, balancing the highest-order derivative and nonlinear terms, we obtain $\rho=1$ and $u_0=-(-2b+\psi\psi_t-x\psi_t)/a$ (`tr3`, `tr4`). Since ρ is a positive integer, the first necessary condition of the Painlevé test is satisfied.

2. Finding the resonances (or the Fuchs indices). We find the resonances, i.e., the powers of ϕ (at which the arbitrary coefficients appear in the series) are determined by performing a resonance analysis. Substituting $u_0(t)\phi^{-\rho}+u_j(t)\phi^{j-\rho}$ (`tr5`, `tr51`) into the original PDE and collecting the terms with the lowest power of ϕ, we obtain the equation $b\phi^{j-3}(j+1)(j-2)u_j+\ldots=0$ (`Eq9`, `Eq10`), where u_j turns to be arbitrary when $j=-1,2$. The resonance $j=-1$ usually corresponds to the arbitrariness of the function ϕ.

3. Verifying the compatibility condition. We consider the truncated expansion $u(x,t)=\phi^{-1}\sum_{j=0}^{2}\phi^j u_j(t)$, where the upper limit of the sum is equal to the largest resonance $j=2$ and $\rho=1$. Substituting this expansion into the original equation, collecting terms of like powers of ϕ, and equating the coefficient powers to zero, we obtain the following system of equations for u_j (`sys1`): $u_{0t}=0$, $-u_0(au_1-\psi_t)=0$, and $-u_0(au_0-2b)=0$. If the Burgers equation passes the Painlevé test, ϕ and u_2 should be arbitrary functions in truncated expansion. The detailed calculations show that $u_0=-2\nu$, $u_1=\psi_t/a$ (`tru1`), u_2 and $\psi=\psi(t)$ are arbitrary functions, and the solution to the Burgers equation has the following form: $u(x,t)=2b/(a(x-\psi))+(2/a)\psi_t+(x-\psi)u_2$ (`SolFin`). Therefore, through the above three steps, we have confirmed that the Burgers equation is Painlevé integrable.

If we set $a=1$ and $b=-\nu$, we obtain the solution of the Burgers equation $u_t+uu_x-\nu u_{xx}=0$ that coincides with the solution described in [125], i.e., has the form $u(x,t)=-2\nu/(x-\psi)+2\psi_t+(x-\psi)u_2$ (`Sol2`).

Maple:

```
interface(showassumed=0): assume(j>0): with(PDETools):
declare(u(x,t),u[0](t),psi(t)); U:=diff_table(u(x,t));
tr11:=N->1/phi^(rho)*sum(u[j](t)*phi^j,j=0..N);
tr21:=phi=x-psi(t); tr22:=rhs(tr21)=phi;
tr23:=-rhs(tr21)=-phi; Eq1:=U[t]+a*U[]*U[x]+b*U[x,x]=0;
Eq2:=algsubs(u(x,t)=subs(tr21,tr11(0)),Eq1)*(rhs(tr21))^(rho+2);
```

```
Eq3:=collect(subs(tr22,simplify(Eq2)),[u[0](t),diff]);
Eq4:=collect(map(factor,lhs(Eq3)),diff); Eq5:=combine(subs(
 tr23,Eq4)); Eq51:=remove(has,Eq5,diff);
termH:=select(has,select(has,indets(Eq51),phi),rho);
tr3:=isolate(op(2,op(termH)),rho);
tr4:=u[0](t)=[solve(remove(has,subs(tr3,Eq5),phi),u[0](t))][2];
tr5:=subs(tr21,u[0](t)*phi^(-rho)+u[j](t)*phi^(j-rho));
tr51:=u(x,t)=subs(tr3,tr4,tr5);
Eq6:=expand(algsubs(tr51,a*U[]*U[x]+b*U[x,x]));
Eq7:=subs(tr23,subs(tr22,combine(Eq6))); Eq9:=collect(Eq7,
 [u[j](t)]); Eq10:=factor(combine(coeff(Eq9,u[j](t),1)));
FI:=[solve(remove(has,Eq10,phi),j) assuming j<>0]; FI[2];
tr8:=u(x,t)=subs(tr21,tr3,expand(tr11(FI[2])));
Eq11:=simplify(subs(tr23,expand(algsubs(tr8,Eq1)))*phi^3);
Eq12:=collect(Eq11,[diff,u[1](t),u[2](t)]);
Eq13:=map(factor,lhs(Eq12)); Eq14:=subs(tr23,Eq13);
Eq15:=collect(Eq14,[phi,u[1](t),u[2](t)]);
Eq16:=map(factor,Eq15); Eq17:=collect(Eq16,[u[2](t)]);
Eq18:=map(factor,Eq17); Eq19:=subs(tr23,Eq18);
sys1:=factor([coeff(Eq19,phi,2)=0,coeff(Eq19,phi,1)=0,
 coeff(Eq19,phi,0)=0]); tru1:=isolate(sys1[2],u[1](t));
SolFin:=u(x,t)=expand(subs(tr4,subs(tr3,tr21,tru1,tr11(FI[2]))),
 rhs(tr21)); Sol2:=subs(a=1,b=-nu,SolFin);
```

Mathematica:

```
trS1[eq_,var_]:=Select[eq,MemberQ[#,var,Infinity]&];
trS3[eq_,var_]:=Select[eq,FreeQ[#,var]&]; tr11[n_]:=1/phi^(rho)*
 Sum[u[t][j]*phi^j,{j,0,n}]; trD[u_,var_]:=Table[D[u,{var,i}],
 {i,1,2}]//Flatten; {tr21=phi->x-psi[t], tr22=tr21[[2]]->phi,
 tr23=-tr21[[2]]->-phi}
eq1=D[u[x,t],t]+a*u[x,t]*D[u[x,t],x]+b*D[u[x,t],{x,2}]==0
{tr24={u[x,t]->tr11[0]/.tr21}, eq11=eq1/.tr24/.trD[tr24,t]/.
 trD[tr24,x], eq2=Thread[eq11*(tr21[[2]])^(rho+2),Equal]//
 Simplify, eq3=Collect[eq2/.tr22,{u[t][0],u'[t][0]}]}
{eq4=Map[Factor,eq3[[1]]], eq51=trS1[eq4,a],
 termH=trS1[eq51,phi], tr3=Solve[termH[[2]]==0,rho]//First}
tr4=Solve[trS3[Expand[eq4/.tr3],phi]==0,u[t][0]]//Flatten
{tr5=u[t][0]*phi^(-rho)+u[t][j]*phi^(j-rho)/.tr21, tr51=u[x,t]->
 tr5/.tr4[[2]]/.tr3, eq6=a*u[x,t]*D[u[x,t],x]+b*D[u[x,t],
 {x,2}]/.tr51/.trD[tr51,x]//Expand//FullSimplify}
{eq7=eq6/.tr22/.tr23, eq9=Collect[eq7//Expand,u[t][j]]}
eq10=Coefficient[eq9,u[t][j],1]//FullSimplify
```

```
{fI=Solve[trS3[eq10,phi]==0,j]//Flatten, fI[[2]]}
tr8=u[x,t]->(tr11[fI[[2,2]]]//Expand)/.tr3/.tr21
eq11=(((eq1/.tr8/.trD[tr8,x]/.trD[tr8,t])//Expand)/.tr22)
eq111=Thread[eq11*phi^3,Equal]//Expand
{eq12=(eq111/.x->phi+psi[t])//Factor, eq13=eq12[[1]]}
{sys1={Coefficient[eq13,phi,2]==0,Coefficient[eq13,phi,1]==0,
 Coefficient[eq13,phi,0]==0}//Factor, tru1=Solve[sys1[[2]],
 u[t][1]]//First, sol1=Collect[tr11[fI[[2,2]]]/.tru1/.tr21/.
 tr3/.tr4[[2]]//Expand,{b,u[t][2],a}]}
{solFin=u[x,t]->Map[Factor,sol1], sol2=solFin/.a->1/.b->-nu}
```
□

Problem 4.2 *Generalized Kawahara equation. Painlevè test. Nonintegrable case.* Let us consider the generalized Kawahara equation [31]

$$u_t+u_x+u^2u_x+au_{xxx}+bu_{xxxxx}=0,$$

where $\{x \in \mathbb{R}, t \geq 0\}$. Applying the Kruskal method, verify that the generalized Kawahara equation fails the Painlevé test.

1. Leading order term analysis. The leading order term of the solution for the generalized Kawahara equation is assumed to be $u \approx \phi^{-\rho} u_0$, where $u_0(t)$ is an analytic function. Substituting this expression into the original equation, we obtain $-\rho\phi^{4-2\rho}u_0^3+\phi^5 u_{0t}+Au_0=0$ (Eq5). Next, balancing the highest-order derivative and nonlinear terms, we obtain $\rho=2$ and $u_0=B$ (tr3, tr4). Since ρ is a positive integer, the first necessary condition of the Painlevé test is satisfied. We note that the expressions for the coefficients A and B are too long, and we don't present them here.

2. Finding the resonances (or the Fuchs indices). We find the resonances, i.e., the powers of ϕ (at which the arbitrary coefficients appear in the series) are determined by performing a resonance analysis. Substituting $u_0(t)\phi^{-\rho}+u_j(t)\phi^{j-\rho}$ (tr5, tr51) into the original PDE and collecting the terms with the lowest power of ϕ, we obtain the equation $b(j-6)(j-8)(j+1)(j^2-7j+30)+\ldots=0$ (Eq9, Eq10). The Fuchs indices or resonances are $\{-1, 6, 8, (7 \pm i\sqrt{71})/2\}$. Since there exist complex resonance points, we conclude that the generalized Kawahara equation fails the Painlevé test, that coincides with the result obtained by Conte [31].

Maple:

```
interface(showassumed=0): assume(j>0,rho>0): with(PDETools):
declare(u(x,t),u[0](t),psi(t)); U:=diff_table(u(x,t));
```

```
tr11:=N->1/phi^(rho)*sum(u[j](t)*phi^j,j=0..N);
tr21:=phi=x-psi(t); tr22:=rhs(tr21)=phi; tr23:=-rhs(tr21)=-phi;
Eq1:=U[t]+U[x]+U[]^2*U[x]+a*U[x,x,x]+b*U[x,x,x,x,x]=0;
Eq2:=algsubs(u(x,t)=subs(tr21,tr11(0)),Eq1)*(rhs(tr21))^(rho+5);
Eq3:=collect(subs(tr22,simplify(Eq2)),[u[0](t),diff]);
Eq4:=map(factor,lhs(Eq3)); Eq5:=combine(subs(tr23,Eq4));
Eq51:=remove(has,Eq5,diff);
termH:=select(has,select(has,indets(Eq51),phi),rho);
tr3:=isolate(op(2,op(termH)),rho);
tr4:=u[0](t)=[solve(remove(has,subs(tr3,Eq5),phi),u[0](t))][2];
tr5:=subs(tr21,u[0](t)*phi^(-rho)+u[j](t)*phi^(j-rho));
tr51:=u(x,t)=subs(tr3,tr4,tr5);
Eq6:=expand(algsubs(tr51,U[]^2*U[x]+b*U[x,x,x,x,x])):
Eq7:=subs(tr23,subs(tr22,combine(Eq6))):
Eq8:=remove(has,Eq7,diff); Eq9:=collect(Eq8,[u[j](t)]);
Eq10:=factor(combine(coeff(Eq9,u[j](t),1)));
Eq11:=remove(has,subs(tr23,Eq10),phi);
Eq12:=map(factor,collect(Eq11,[x,psi,a,b]));
Eq13:=select(has,Eq12,[b]); FI:=[solve(Eq13,j) assuming j<>0];
```

Mathematica:

```
trS1[eq_,var_]:=Select[eq,MemberQ[#,var,Infinity]&];
trS3[eq_,var_]:=Select[eq,FreeQ[#,var]&]; tr11[n_]:=1/phi^(rho)*
 Sum[u[t][j]*phi^j,{j,0,n}]; trD[u_,var_]:=Table[D[u,{var,i}],
 {i,1,5}]//Flatten; {tr21=phi->x-psi[t], tr22=tr21[[2]]->phi,
 tr23=-tr21[[2]]->-phi}
{eq1=D[u[x,t],t]+D[u[x,t],x]+u[x,t]^2*D[u[x,t],x]+a*D[u[x,t],
 {x,3}]+b*D[u[x,t],{x,5}]==0, tr24={u[x,t]->tr11[0]/.tr21}}
eq11=eq1/.tr24/.trD[tr24,t]/.trD[tr24,x]
eq2=Thread[eq11*(tr21[[2]])^(rho+5),Equal]//Simplify
eq3=Collect[eq2/.tr22,{u[t][0],u'[t][0]}]
{eq4=Map[Factor,eq3[[1]]], eq51=trS1[eq4,u[t][0]^3],
 termH=trS1[eq51,phi], tr3=Solve[termH[[2]]==0,rho]//First}
tr4=Solve[trS3[Expand[eq4/.tr3],phi^5]==0,u[t][0]]//Flatten
tr5=u[t][0]*phi^(-rho)+u[t][j]*phi^(j-rho)/.tr21
{tr51=u[x,t]->tr5/.tr4[[3]]/.tr3, eq6=u[x,t]^2*D[u[x,t],x]+
 b*D[u[x,t],{x,5}]/.tr51/.trD[tr51,x]//Expand//FullSimplify}
{eq7=eq6/.tr22/.tr23, eq9=Collect[eq7//Expand,u[t][j]]}
eq10=Coefficient[eq9,u[t][j],1]//FullSimplify
termphi=trS1[trS3[eq10//Factor,psi],phi]
{eq11=trS3[trS3[Thread[eq10/termphi,Equal]//Expand,psi],phi]//
 Factor, fI=Solve[eq11==0,j]//Flatten}
```

□

4.1.2 Truncated expansions

It is possible to find integrability properties by means of *truncated expansions* (that can be considered as an extension of various Painlevé tests), or the *singular manifold method* developed by J. Weiss [161]. The main idea of the method consists in truncation of an expansion, e.g Eq. (4.2), usually at the zero-order terms ϕ_0. This truncation allows us to construct exact solutions, obtain a Bäcklund transformation of a given PDE. In this case the singular manifold function ϕ is not arbitrary, it satisfies some conditions. For integrable equations (that are solvable by the inverse scattering transform), the singular manifold method can be used to derive the associated Lax pair (see Sect. 4.2.4). For linearizable equations (e.g., the Burgers equation), the method allows us to construct the linearization. For nonintegrable PDEs, the singular manifold method can be used to obtain exact solutions.

Let us consider the two equations (integrable and nonintegrable) analyzed in Problems 4.1, 4.2.

Problem 4.3 *Burgers equation. Truncated expansion. Integrable case.* Let us consider the Burgers equation

$$u_t+auu_x+bu_{xx}=0,$$

where $\{x\in\mathbb{R}, t\geq 0\}$ and a, b are real constants. Applying the truncated expansion, construct the linearization of the Burgers equation, i.e., deduce the Hopf–Cole transformation (see Sect. 2.4.1) and the linear heat equation.

1. Generating the system of equations. Applying the truncated expansion of the form $u(x,t)=\sum_{j=0}^{N} u_j(x,t)\phi(x,t)^{-(N-j)}$ (`tr1`), with $N=1$, we obtain $u=u_0/\phi+u_1$ (`tr1(1)`), where $u=u(x,t)$, $u_i=u_i(x,t)$ $(i=0,1)$, and $\phi=\phi(x,t)$. Substituting it into the Burgers equation and collecting the terms of equal powers in ϕ, we obtain an equation (`Eq3`), from which we generate the system of four equations (`E0`–`E3`) by equating the coefficients of like powers of ϕ to zero.

2. Solving the system of equations. From the first equation we find $u_0=2b\phi_x/a$ (`tru0`). Substituting u_0 into the second and the third equations of the system, we obtain the two equations (`EE1`, `EE2`), which are consistent. Therefore, the truncated expansion (with $N=1$) takes the form $u=2b\phi_x/(a\phi)+u_1$ (`tr11`), where the functions u_1 and u satisfy the Burgers equation (since the equation `E0` is the Burgers equation) and the function ϕ is described by the linear equation `EE1`.

3. Constructing the linearization. Considering a particular solution of the Burgers equation $u_1=0$ and substituting it into the final expression of truncated expansion (`tr11`) and system of two equations (`EE1`, `EE2`), we obtain, respectively, the Hopf–Cole transformation, `trHC`, and the linear heat equation, `EE21` (assuming that $\phi_x \neq 0$, $a \neq 0$).

Maple:

```
interface(showassumed=0): assume(j>0): with(PDETools):
declare(u(x,t),u[0](x,t),u[1](x,t),phi(x,t));
U:=diff_table(u(x,t)); Eq1:=U[t]+a*U[]*U[x]+b*U[x,x]=0;
tr1:=N->u(x,t)=sum(u[j](x,t)/phi(x,t)^(N-j),j=0..N); tr1(1);
Eq2:=algsubs(tr1(1),Eq1); Eq3:=collect(lhs(Eq2),[phi(x,t)]);
for i from 1 to 3 do
 E||i:=select(has,select(has,Eq3,phi(x,t)^(-i)),diff)=0; od;
Eq0:=remove(has,Eq3,phi(x,t))=0;
tru0:=u[0](x,t)=[solve(factor(E3),u[0](x,t))][2];
EE1:=factor(algsubs(tru0,E1)); EE2:=factor(algsubs(tru0,E2));
tr11:=subs(tru0,tr1(1)); tr2:=u[1](x,t)=0;
trHC:=subs(tr2,tr11); EE21:=subs(tr2,EE2);
  subs(a=1,b=-nu,trHC); subs(a=1,b=-nu,EE21);
```

Mathematica:

```
trS1[eq_,var_]:=Select[eq,MemberQ[#,var,Infinity]&];
trS3[eq_,var_]:=Select[eq,FreeQ[#,var]&];
trD[u_,var_]:=Table[D[u,{var,i}],{i,1,5}]//Flatten;
tr1[n_]:=u[x,t]->Sum[u[x,t][j]/phi[x,t]^(n-j),{j,0,n}];
{tr1[1], eq1=D[u[x,t],t]+a*u[x,t]*D[u[x,t],x]+b*D[u[x,t],
 {x,2}]==0, eq2=eq1/.tr1[1]/.trD[tr1[1],t]/.trD[tr1[1],x]}
{eq3=Collect[eq2,phi[x,t]], Table[e[i]=trS3[trS1[
 eq3[[1]],phi[x,t]^(-i)],phi[x,t]]==0,{i,1,3}],
 eq0=trS3[eq3[[1]],phi[x,t]]==0}
tru0=Solve[Factor[e[3]],u[x,t][0]]//Flatten
{tru01=tru0[[2]], ee1=e[1]/.tru01/.trD[tru01,x]/.trD[tru01,t],
 ee2=e[2]/.tru01/.trD[tru01,x]/.trD[tru01,t]}
{tr11=tr1[1]/.tru01, tr2=u[x,t][1]->0, tr3={a->1,b->-nu}}
{trHC=tr11/.tr2, ee21=ee2/.tr2, trHC/.tr3, ee21/.tr3//Factor}
```
□

Problem 4.4 *Generalized Kawahara equation. Truncated expansion. Nonintegrable case.* Let us consider the generalized Kawahara equation

$$u_t+u_x+u^2u_x+au_{xxx}+bu_{xxxxx}=0,$$

where $\{x \in \mathbb{R}, t \geq 0\}$. It is known that the generalized Kawahara equation fails the Painlevé test (see Problem 4.2). Construct the truncated expansion of this equation and an exact solution.

1. Generating the system of equations. As in the previous problem, applying the truncated expansion (tr1), with N=2, we obtain $u=u_0/\phi^2+u_1/\phi+u_2$ (tr1(2)), where $u=u(x,t)$, $u_i=u_i(x,t)$ $(i=0,1,2)$, and $\phi=\phi(x,t)$. Substituting it into the generalized Kawahara equation and collecting the terms of equal powers in ϕ, we obtain an equation (Eq3), from which we generate the system of eight equations (E0–E7) by equating the coefficients of like powers of ϕ to zero.

2. Constructing truncated expansion. From equation E7 we find $u_0 = \pm 6\sqrt{-10b}\,\phi_x^2$ (tru01, tru02). Substituting u_0 into equation E6, we obtain the equation, from which we find $u_1 = \mp 6\sqrt{-10b}\,\phi_{xx}$ (tru11, tru12). Then, substituting u_0, u_1 into equation E5, we obtain the equation, from which we find u_2 (tru21, tru22). Hence, the truncated expansion (with N=2) takes the form (tr11, tr12):

$$u=\pm 6\sqrt{-10b}\frac{\phi_x^2}{\phi^2}\mp 6\sqrt{-10b}\frac{\phi_{xx}}{\phi}\pm\frac{\sqrt{-10b}}{10}\Big(\frac{-15b\phi_{xx}^2+20b\phi_x\phi_{xxx}+a\phi_x^2}{b\phi_x^2}\Big). \quad (4.3)$$

Maple:

```
with(PDETools): declare(u(x,t),u[0](x,t),u[1](x,t),u[2](x,t),
 phi(x,t)); U:=diff_table(u(x,t));
Eq1:=U[t]+U[x]+U[x]*U[]^2+a*U[x,x,x]+b*U[x,x,x,x,x]=0;
tr1:=N->u(x,t)=sum(u[j](x,t)/phi(x,t)^(N-j),j=0..N); tr1(2);
Eq2:=algsubs(tr1(2),Eq1); Eq3:=collect(lhs(Eq2),[phi(x,t)]);
for i from 1 to 7 do
 E||i:=select(has,select(has,Eq3,phi(x,t)^(-i)),diff)=0; od;
Eq0:=remove(has,Eq3,phi(x,t))=0;
tru01:=u[0](x,t)=[solve(E7,u[0](x,t))][2];
tru02:=u[0](x,t)=[solve(E7,u[0](x,t))][3];
for i from 1 to 2 do
 EE6||i:=algsubs(tru0||i,E6);
 tru1||i:=u[1](x,t)=solve(EE6||i,u[1](x,t));
 EE5||i:=algsubs(tru1||i,algsubs(tru0||i,E5));
 tru2||i:=u[2](x,t)=solve(EE5||i,u[2](x,t));
 tr1||i:=subs(tru0||i,tru1||i,tru2||i,tr1(2)); od;
```

Mathematica:

```
trS1[eq_,var_]:=Select[eq,MemberQ[#,var,Infinity]&];
trS3[eq_,var_]:=Select[eq,FreeQ[#,var]&]; trD[u_,var_]:=Table[
 D[u,{var,i}],{i,1,5}]//Flatten;
tr1[n_]:=u[x,t]->Sum[u[x,t][j]/phi[x,t]^(n-j),{j,0,n}]; tr1[2]
{eq1=D[u[x,t],t]+D[u[x,t],x]+u[x,t]^2*D[u[x,t],x]+a*D[u[x,t],
 {x,3}]+b*D[u[x,t],{x,5}]==0, eq2=eq1/.tr1[2]/.trD[tr1[2],t]/.
 trD[tr1[2],x], eq3=Collect[eq2,phi[x,t]]}
{Table[e[i]=trS3[trS1[eq3[[1]],phi[x,t]^(-i)],phi[x,t]]==0,
 {i,1,7}], Do[Print["e[",i,"]=",e[i]],{i,1,7}];}
{eq0=trS3[eq3[[1]],phi[x,t]]==0, tru0=Solve[Factor[e[7]],
 u[x,t][0]]//Flatten, tru0N[1]=tru0[[3]], tru0N[2]=tru0[[2]]}
Do[ee6[i]=e[6]/.tru0N[i]/.trD[tru0N[i],x]/.trD[tru0N[i],t];
 tru1[i]=Solve[ee6[i],u[x,t][1]]//Flatten; ee5[i]=e[5]/.
 tru0N[i]/.trD[tru0N[i],x]/.trD[tru0N[i],t]/.tru1[i]/.
 trD[tru1[i],x]/.trD[tru1[i],t]; tru2[i]=Solve[ee5[i],
 u[x,t][2]]//Flatten; tr1N[i]=tr1[2]/.tru0N[i]/.tru1[i]/.tru2[i];
Map[Print,{ee6[i],tru1[i],ee5[i],tru2[i],tr1N[i]}],{i,1,2}];
```

3. Constructing exact solution. Applying *Maple* predefined functions, let us construct the exact solution. Substituting the truncated expansion (4.3) into the given equation, we obtain the overdetermined system of differential equations for ϕ. There exist special solutions. We find the solution $\phi=\Big(-2F_2(t)\tanh[\frac{1}{2}[x+F_3(t)]/F_1(t)]+F_4(t)F_1(t)\Big)[F_1(t)]^{-1}$ (`Solphi1`), where $F_i(t)$ $(i=1,\dots,4)$ are arbitrary functions. Then we build the exact solution $u=-\dfrac{\sqrt{10}}{10}\dfrac{-4\cosh(\frac{1}{2}x-\frac{13}{10}t)^2-11}{\cosh(\frac{1}{2}x-\frac{13}{10}t)^2-1}$ (`SolFin2`) and verify that this solution is exact solution of the given nonlinear PDE.

Maple:

```
S1:=algsubs(tr11,Eq1); S2:=collect(simplify(S1),phi(x,t));
S3:=op(1,S2); Solphi1:=convert(pdsolve(expand(op(5,S3)),
 phi(x,t)),tanh); Solphi2:=subs(_F1(t)=1,_F4(t)=0,Solphi1);
for i from 1 to 4 do
Q||i:=simplify(subs(Solphi2,algsubs(Solphi2,op(i,S3)))); od;
SolF3:=[dsolve(Q1,_F3(t))][2]; SolF2:=dsolve(algsubs(
 SolF3,Q4),_F2(t)); Solphi3:=subs(SolF3,SolF2,Solphi2);
SolFin1:=simplify(subs(Solphi3,algsubs(Solphi3,tr11)));
params:={a=1,b=-1,_C1=0}; SolFin2:=subs(params,SolFin1);
Test1:=pdetest(SolFin2,subs(params,Eq1));
```

We note that we cannot build this exact solution in *Mathematica* (since we cannot obtain the analytical solution `Solphi1` via the predefined function `DSolve`). □

4.2 Complete Integrability. Evolution Equations

It is known that some nonlinear evolution equations admit one-soliton solution (e.g., the KdV equation with time-dependent coefficients) or two-soliton solutions (e.g., the ninth-order KdV equation, the sixth-order Boussinesq equation), but do not admit N-soliton solutions (for $N>2$). If an equation has two-soliton solutions, it does not imply it has three-soliton solutions, but if an equation has three-soliton solutions, it also has N-soliton solutions (for $N>3$), and is called completely integrable (see e.g., [55]–[59], [53], [95], [96], [51], [52]).

There are several definitions or criteria for *complete integrability* of evolution equations (e.g., see [146], [99], [114]): the existence of an infinite number of conservation laws, the existence of N-soliton solutions of any order, the existence of higher-order symmetries. However, the nonexistence of a sequence of conservation laws does not prevent the complete integrability, e.g., the Burgers equation (that has only one conserved density and is completely integrable).

F. Calogero proposed to call the equations that can be related to linear equations by a transformation, C-integrable, while the equations that are integrable by means of the inverse scattering transform (IST), to call S-integrable. The IST is a method for finding solitons for a PDE of all orders. If it can be done, the PDE is completely integrable or completely solved. However the application of the IST to a PDE can be very complicated task, so the existence of an infinite number of conservation laws can predict complete integrability of a PDE. For example, the Burgers equation is C-integrable equation, that can be related to the linear heat equation by the Hopf–Cole transformation (see Problem 2.14), that has only one conserved density, and have infinitely many symmetries.

4.2.1 Conservation Laws

Conservation laws, arising in various branches of science and engineering, describe the conservation of basic quantities of a physical system. The existence of a sequence of conserved densities predicts complete integrability.

Definition 4.2 Let $\mathcal{F}(x, t, u, u_x, u_t, \ldots)=0$ be a PDE, where t is interpreted as a time-like variable. A *conservation law* for this equation is the equation

$$T_t + X_x = 0, \tag{4.4}$$

where T is the density, X is the flux, and both T, X do not involve derivatives with respect to t.

Definition 4.3 Let T, X be integrable on $(-\infty, \infty)$ so that $X \to \text{const}$ as $|x| \to \infty$. The integral $\int_{-\infty}^{\infty} T\, dx = \text{const}$, which is obtained by integrating Eq. (4.4),* is called a *constant of motion.*

For most equations, the density-flux pairs are polynomials in u and derivatives of u with respect to x. The lower-order polynomial-type conservation laws have a physical interpretation and can be find by hand, whereas the higher-order conservation laws require a significant amount of computation. There are several methods for computing explicitly conservation laws [110], some of them are suitable for computer algebra systems. Therefore several symbolic programs have been developed with the aid of different CAS (for details see [130] [167], and [67]). In this section, we will construct polynomial-type conservation laws and constants of motion of finite arbitrary order for the KdV equation applying the Miura and Gardner transformations and following the perturbation approach.

Problem 4.5 *KdV equation. Conservation laws.* For the Korteweg–de Vries equation

$$u_t - 6uu_x + u_{xxx} = 0,$$

where $\{x \in \mathbb{R}, t \geq 0\}$. Determine and verify the first three conservation laws and constants of motion.

1. The first conservation law and constant of motion. The KdV equation can be rewritten in conservation form with $T_1 = u$, $X_1 = u_{xx} - 3u^2$ (`Eq11`) and the constant of motion is $\int_{-\infty}^{\infty} u\, dx = \text{const}$ (`CM1`).

2. The second conservation law and constant of motion. Multiplying the KdV equation by u, we obtain the second density-flux pair $T_2 = \frac{1}{2}u^2$, $X_2 = uu_{xx} - \frac{1}{2}u_x^2 - 2u^3$ (`Eq21`), and the constant of motion is $\int_{-\infty}^{\infty} u^2\, dx = \text{const}$ (`CM2`).

---

*Integrating Eq. (4.4), we have $\frac{d}{dt}\Big(\int_{-\infty}^{\infty} T\, dx\Big) = -X\Big|_{-\infty}^{\infty} = 0.$

3. The third conservation law and constant of motion. Multiplying the KdV equation by $3u^2$, multiplying the partial derivative of the KdV equation with respect to x by u_x, and adding these results, we obtain $T_3{=}u^3{+}\frac{1}{2}u_x^2$, $X_3{=}{-}\frac{9}{2}u^4{+}3u^2u_{xx}{-}6u(u_x)^2{+}u_xu_{xxx}{-}\frac{1}{2}(u_{xx})^2$ (`Eq31`), the constant of motion is $\int_{-\infty}^{\infty}(u^3{+}\frac{1}{2}u_x^2)\,dx{=}\text{const}$ (`CM3`), and we verify the results obtained (`test1`, `test2`, `test3`).

Maple:

```
with(PDEtools): declare(T(x,t),X(x,t),u(x,t));
alias(u=u(x,t),T=T(x,t),X=X(x,t)); U:=diff_table(u(x,t));
EqC:=(T,X)->diff(T,t)+diff(X,x)=0; ab:=-infinity..infinity;
CM:=T->int(T,x=ab)=C; PDE1:=U[t]-6*U[]*U[x]+U[x,x,x]=0;
T1:=U[]; X1:=U[x,x]-3*U[]^2; Eq11:=expand(EqC(T1,X1));
CM1:=CM(T1); Eq20:=expand(PDE1*u); T2:=U[]^2/2;
X2:=U[]*U[x,x]-U[x]^2/2-2*U[]^3; Eq21:=expand(EqC(T2,X2));
CM2:=subs(2*C=C,expand(CM(T2)*2)); Eq301:=PDE1*3*U[]^2;
Eq302:=diff(PDE1,x)*U[x]; Eq303:=expand(Eq301+Eq302);
T3:=U[]^3+U[x]^2/2; X3:=-9/2*U[]^4+3*U[]^2*U[x,x]
    -6*U[]*U[x]^2+U[x]*U[x,x,x]-U[x,x]^2/2;
Eq31:=expand(EqC(T3,X3)); CM3:=simplify(CM(T3));
Test1:=Eq11-PDE1; Test2:=Eq21-Eq20; Test3:=Eq303-Eq31;
```

Mathematica:

```
var=Sequence[x,t]; eqC[tN_,xN_]:=D[tN,t]+D[xN,x]==0;
inf=Infinity; cM[tN_]:=Integrate[tN,{x,-inf,inf}]==c;
pde1=D[u[var],t]-6*u[var]*D[u[var],x]+D[u[var],{x,3}]==0
{tN1=u[var], xN1=D[u[var],{x,2}]-3*u[var]^2}
{eq11=eqC[tN1,xN1]//Expand, cM1=cM[tN1]}
{eq20=Thread[pde1*u[var],Equal]//Expand}
{tN2=1/2*u[var]^2, xN2=u[var]*D[u[var],{x,2}]-
 D[u[var],x]^2/2-2*u[var]^3}
{eq21=eqC[tN2, xN2]//Expand, cM2=(Thread[cM[tN2]*2,
 Equal]//Expand)/.{2*c->c}//Expand, eq301=Thread[
 pde1*3*u[var]^2,Equal], eq302=Thread[D[pde1,x]*D[u[var],x],
 Equal], eq303=Thread[eq301+eq302,Equal]//Expand}
{tN3=u[var]^3+D[u[var],x]^2/2, xN3=-9/2*u[var]^4+3*u[var]^2*
 D[u[var],{x,2}]-6*u[var]*D[u[var],x]^2+D[u[var],x]*D[u[var],
 {x,3}]-D[u[var],{x,2}]^2/2}
{eq31=eqC[tN3,xN3]//Expand, cM3=cM[tN3]//Simplify}
{eq11==pde1,eq21==eq20,eq303==eq31}
```

The first three constants of motion describe, respectively, the conservation of mass, the conservation of horizontal momentum for water waves (in the framework of the shallow water approximation theory), and the conservation of energy of water waves. □

Problem 4.6 *KdV equation. Constructing conservation laws of finite arbitrary order.* Let us consider the Korteweg–de Vries equation

$$u_t - 6uu_x + u_{xxx} = 0,$$

where $\{x \in \mathbb{R}, t \geq 0\}$. Applying the Miura and Gardner transformations and following the perturbation approach, construct conservation laws of finite arbitrary order.

1. The Miura and Gardner transformations, the Gardner equation. Applying the Miura transformation (see Problem 2.12), $u=v^2+v_x$ (`trMi`), that relates one nonlinear equation to another nonlinear equation (in our case, the KdV equation to the mKdV equation), introducing the new dependent variable $w(x,t)$ defined by $v=\frac{1}{2}\varepsilon^{-1}+\varepsilon w$ (`trv`), we obtain the Gardner transformation (see Problem 2.13), $u=w+\varepsilon w_x+\varepsilon^2 w^2$ (`trGar`) and the Gardner equation $w_t+(-6\varepsilon^2 w^2-6w)w_x+w_{xxx}=0$ (`EqGar`). We verify that if $\varepsilon=0$ and $w=u$, the Gardner equation becomes the KdV equation (`test2`).

Maple:

```
with(PDEtools): declare((u,v,w,T,X)(x,t));
alias(u=u(x,t),v=v(x,t),w=w(x,t),T=T(x,t),X=X(x,t));
W:=diff_table(w(x,t));  trMi:=u=v^2+diff(v,x);
KdV:=diff(u,t)-6*u*diff(u,x)+diff(u,x$3)=0; Eq1:=algsubs(trMi,
 KdV); mKdV:=diff(v,t)-6*v^2*diff(v,x)+diff(v,x$3)=Mv;
subs(mKdV,Eq1); MvL:=lhs(mKdV); Eq2:=2*v*Mv+Diff(Mv,x)=0;
Eq3:=expand(2*v*MvL+diff(MvL,x))=0; evalb(Eq1=Eq3);
trv:=v=1/(2*epsilon)+epsilon*w; Eq4:=expand(algsubs(trv,trMi));
trGar:=subs(1/epsilon^2=0,Eq4);
Eq5:=collect(expand(subs(trGar,KdV)),diff);
Eq51:=map(factor,lhs(Eq5)); tr1:=1+2*epsilon^2*w=G;
Eq52:=collect(subs(tr1,Eq51),G);
Eq53:=collect(expand((Eq52-op(1,Eq52))/epsilon),diff);
term1:=op(1,op(2,Eq53))=G1;
EqGar:=subs(G1=lhs(term1),map(int,subs(term1,Eq53),x))=0;
Eq54:=expand(subs(G=lhs(tr1),G*EqGar+epsilon*diff(EqGar,x)));
Test1:=expand(Eq54-Eq5);
Test2:=evalb(KdV=algsubs(w=u,subs(epsilon=0,EqGar)));
```

Mathematica:

```
trD[u_,var_]:=Table[D[u,{var,i}],{i,1,6}]//Flatten;
trS1[eq_,var_]:=Select[eq,MemberQ[#,var,Infinity]&];
trS3[eq_,var_]:=Select[eq,FreeQ[#,var]&]; var=Sequence[x,t];
{trMi=u[var]->v[var]^2+D[v[var],x], kdV=D[u[var],t]-6*u[var]*
 D[u[var],x]+D[u[var],{x,3}]==0}
eq1=kdV/.trMi/.trD[trMi,x]/.trD[trMi,t]
mKdV=D[v[var],t]-6*v[var]^2*D[v[var],x]+D[v[var],{x,3}]->mv
{eq1/.mKdV, mvL=mKdV[[1]], eq2=2*v[var]*mv+Hold[D[mv,x]]==0}
{eq3=Expand[2*v[var]*mvL+D[mvL,x]]==0, Map[Expand,eq1==eq3]}
{trv=v[var]->1/(2*epsilon)+epsilon*w[var], eq4=trMi/.trv/.
 trD[trv,x]//Expand, trGar=eq4/.{1/epsilon^2->0}}
l1={trD[w[x,t],x],trD[w[x,t],t]}//Flatten
eq5=Collect[kdV/.trGar/.trD[trGar,x]/.trD[trGar,t]//Expand,l1]
{eq51=Map[Factor,eq5[[1]]], tr1=1+2*epsilon^2*w[var]->g}
{eq52=Collect[eq51/.tr1,g], termg=trS1[eq52,g]}
eq53=Collect[Thread[(eq52-termg)/epsilon,Equal]//Expand,l1]
termepsilon=trS3[trS1[eq53,epsilon],D[w[var],{x,2}]]
{term1=termepsilon->g1, eqGar=(Map[Integrate[#,x]&,eq53/.term1]/.
 g1->term1[[1]])==0, eq541=Thread[g*eqGar,Equal]//Expand}
eq542=Thread[epsilon*D[eqGar,x],Equal]//Expand
eq54=Thread[eq541+eq542,Equal]/.g->tr1[[1]]//Expand
test1=Thread[eq54-Expand[eq5],Equal]//Expand
test2=kdV==(eqGar/.epsilon->0/.w->u)
```

2. Conservation laws and constants of motion of finite arbitrary order. The Gardner equation can be written in the conservation form with $T_1{=}w$, $X_1{=}w_{xx}{-}3w^2{-}2\varepsilon^2 w^3$ (`EqC1`), and the constant of motion is $\displaystyle\int_{-\infty}^{\infty} u\,dx{=}\text{const}$ (`CM1`).

Then, introducing the asymptotic expansion of $w(x,t;\varepsilon)$ in ε by the formula (`SolSer`) $w(x,t;\varepsilon){\sim}\displaystyle\sum_{i=0}^{N}\varepsilon^i w_i(x,t)$ as $\varepsilon{\to}0$, we can obtain conservation laws and constants of motion of finite arbitrary order for the KdV equation (since $w{\to}u$ as $\varepsilon{\to}0$).

Substituting `SolSer` into the Gardner transformation and equating coefficients of ε^i (for $i{=}0,1,\ldots,N$), we can obtain constants of motion of finite arbitrary order. It can be observed that if i is even, the constant of motion $\displaystyle\int_{-\infty}^{\infty} w_i\,dx{=}\text{const}$ will generate conservation laws, and if i is odd, w_i is an exact differential (i.e., will not generate any conservation law).

Maple:

```
EqC:=(T,X)->diff(T,t)+diff(X,x)=0; ab:=-infinity..infinity;
CM:=T->int(T,x=ab)=C; T1:=w;X1:=W[x,x]-3*W[]^2-2*epsilon^2*W[]^3;
EqC1:=expand(EqC(T1,X1)); CM1:=CM(T1); test1:=expand(EqC1-EqGar);
SolSer:=(j,K)->w=Sum(epsilon^j*V[j](x,t),j=0..K);
PEq0:=value(algsubs(SolSer(i1,0),trGar)); S0:=coeff(lhs(PEq0),
 epsilon,0)=coeff(rhs(PEq0),epsilon,0);
S01:=isolate(S0,V[0](x,t)); SS:={S01}: for k from 1 to 6 do
 PEq||k:=expand(value(algsubs(SolSer(i1,k),trGar)));
 S||k:=coeff(lhs(PEq||k),epsilon,k)=coeff(rhs(PEq||k),epsilon,k);
 S||k||1:=isolate(S||k,V[k](x,t)); SS:=SS union
  {seq(S||j||1,j=1..k)}; S||k||2:=expand(subs(SS,subs(SS,subs(SS,
  subs(SS,subs(SS,subs(SS,subs(SS,S||k||1))))))));
  print(V[k]=S||k||2); od: CM0:=int(rhs(S01),x);
for i from 1 to 6 do CM||i:=map(int,rhs(S||i||2),x); od;
```

Mathematica:

```
eqC[tN_,xN_]:=D[tN,t]+D[xN,x]==0; inf=Infinity;
cM[tN_]:=Integrate[tN,{x,-inf,inf}]==c; {tN1=w[var],
 xN1=D[w[var],{x,2}]-3*w[var]^2-2*epsilon^2*w[var]^3}
{eqC1=eqC[tN1,xN1]//Expand, cM1=cM[tN1]}
test1=Thread[eqC1-Expand[eqGar],Equal]//Simplify
solSer[j_,k_]:=w[var]->Sum[epsilon^j*v[var][j],{j,0,k}];
pEq0=trGar/.solSer[i1,0]/.trD[solSer[i1,0],x]
{s0=Coefficient[pEq0[[1]],epsilon,0]==Coefficient[pEq0[[2]],
 epsilon,0], s01=Solve[s0,v[var][0]]//First, sS={s01}}
Do[pEq[k]=trGar/.solSer[i1,k]/.trD[solSer[i1,k],x]//Expand;
 s[k]=Coefficient[pEq[k][[1]],epsilon,k]==Coefficient[
  pEq[k][[2]],epsilon,k]; s1[k]=(Solve[s[k],v[var][k]]//First);
 sS=Union[sS,Table[s1[j],{j,1,k}]]//Flatten;
 s2[k]=s1[k]//.sS//.trD[sS,x]//.sS//.trD[sS,x]//.sS//Expand;
 Print[v[var][k],"=",s2[k]],{k,1,6}];
cM0=Integrate[s01[[1,2]],x]
Do[cM[k]=Map[Integrate[#,x]&,s2[k][[1,2]]]//Simplify;
 Print["cM[",k,"]=",cM[k]],{k,1,6}];
```

□

4.2.2 Nonlinear Superposition Formulas

It is known that almost all of the methods for solving linear PDEs are based on the *superposition principle*. However, for nonlinear differential equations the superposition principle in general does not apply. For some classes of integrable nonlinear evolution equations, it is possible to construct N-soliton solutions by establishing from the Bäcklund transformation (BT) a nonlinear superposition formula (NLSF).

Definition 4.4 The *nonlinear superposition formula* (NLSF) establishes a relation between four different solutions of an integrable PDE.

It requires both the existence of a BT (with spectral parameter λ relating two solutions u and $\tilde{u}$) and the validity of the *permutability theorem* for the BT (which for the first time has been proved by G. Darboux [35]). This theorem states that, if u_n, $\tilde{u}_n$ are the transforms of u_{n-1} under the BT with the respective parameters λ_n and λ_{n+1}, then the solution u_{n+1} is the transform of u_n and $\tilde{u}_n$ with the respective parameters λ_{n+1} and λ_n.

Problem 4.7 For the nonlinear PDE

$$u_x^2+u_t^2-u^4=0,$$

where $\{x \in \mathbb{R}, t \geq 0\}$, verify that the $u_1=-1/x$, $u_2=1/(1-x)$ are solutions of the nonlinear equation, but their sum u_1+u_2 is not a solution of this equation.

Maple:

```
with(PDEtools): declare(u(x,t)); U:=diff_table(u(x,t));
PDE1:=U[x]^2+U[t]^2-U[]^4=0; u1:=-1/x; u2:=1/(1-x);
uN:=u1+u2; pdetest(u(x,t)=u1,PDE1);
pdetest(u(x,t)=u2,PDE1); pdetest(u(x,t)=u1+u2,PDE1);
```

Mathematica:

```
trD[u_,var_]:=Table[D[u,{var,i}],{i,1,2}]//Flatten;
pde1=D[u[x,t],x]^2+D[u[x,t],t]^2-u[x,t]^4==0
{u1=u[x,t]->-1/x, u2=u[x,t]->1/(1-x), uN=u[x,t]->
 u1[[2]]+u2[[2]]}
{pde1/.u1/.trD[u1,x]/.trD[u1,t], pde1/.u2/.trD[u2,x]/.
 trD[u2,t], pde1/.uN/.trD[uN,x]/.trD[uN,t]//Factor}
```

Therefore, most solution methods for linear equations cannot be applied to nonlinear equations. Moreover, since there is no general method for finding analytical solutions of nonlinear PDEs, most nonlinear PDEs requires new methods (analytical, numerical, analytical-numerical). □

Problem 4.8 *Sine–Gordon equation. Nonlinear superposition formula.* For the sine–Gordon equation

$$u_{xt} = \sin u = 0,$$

where $\{x \in \mathbb{R}, t \geq 0\}$, construct the nonlinear superposition formula.

We consider the four expressions of the x-part of the auto-Bäcklund transformation $\phi_1(u,v;\lambda)=(u+v)_x-2\lambda\sin\left(\frac{1}{2}(u-v)\right)=0$ (see Problem 2.11), namely,

$$\phi_1(u_n, u_{n-1}, \lambda_n),\ \phi_1(\tilde{u}_n, u_{n-1}, \lambda_{n+1}),\ \phi_1(u_{n+1}, u_n, \lambda_{n+1}),\ \phi_1(u_{n+1}, \tilde{u}_n, \lambda_n)$$

(`Eq1`, `Eq2`, `Eq3`, `Eq4`). Eliminating the first derivatives of u_{n+1}, u_n, $\tilde{u}_n$, u_{n-1} between these four relations, we obtain the following algebraic relation (`Eq10`):

$$\left[\sin(\tfrac{u_{n+1}-\tilde{u}_n}{2})+\sin(\tfrac{u_n-u_{n-1}}{2})\right]\lambda_n+\left[\sin(\tfrac{u_n-u_{n+1}}{2})+\sin(\tfrac{u_{n-1}-\tilde{u}_n}{2})\right]\lambda_{n+1}=0.$$

Applying the trigonometric identity $\sin x+\sin y=2\sin[\frac{1}{2}(x+y)]\cos[\frac{1}{2}(x-y)]$ (`tr1`), this equation can be reduced to the expression (`Eq24`):

$$\lambda_{n+1}\sin(z_1)\cos(z_2)+\lambda_n\sin(z_3)\cos(z_2),$$

where $z_1=\frac{1}{4}(-u_{n+1}+u_n-\tilde{u}_n+u_{n-1})$, $z_2=\frac{1}{4}(-u_{n+1}+u_n+\tilde{u}_n-u_{n-1})$, and $z_3=\frac{1}{4}(u_{n+1}+u_n-\tilde{u}_n-u_{n-1})$. It can be shown (`test1`) that this expression is equivalent to the following trigonometric expression (`tr2`):

$$-(\lambda_{n+1}-\lambda_n)\sin(A)\cos(B)-(\lambda_{n+1}+\lambda_n)\cos(A)\sin(B),$$

where $A=\frac{1}{4}(-u_{n+1}+u_{n-1})$ and $B=\frac{1}{4}(-\tilde{u}_n+u_n)$. Dividing this equation by $\cos(B)$ (`Eq25`) and by $\cos(A)$ (`Eq26`) and then simplifying the resulting expression, we can obtain the final result, i.e., the nonlinear superposition formula for the sine–Gordon equation:

$$u_{n+1}-u_{n-1}-4\arctan\left(\frac{\lambda_{n+1}+\lambda_n}{\lambda_{n+1}-\lambda_n}\tan\left(\frac{-\tilde{u}_n+u_n}{4}\right)\right)=0.$$

Maple:

```
alias(u[n]=u[n](x,t),uT[n]=uT[n](x,t),u[n-1]=u[n-1](x,t),
      u[n+1]=u[n+1](x,t)); L1:=[lambda[n],lambda[n+1]];
phi1:=(u,v,lambda)->diff((u+v),x)-2*lambda*sin((u-v)/2)=0;
Eq1:=phi1(u[n],u[n-1],L1[1]); Eq2:=phi1(uT[n],u[n-1],L1[2]);
```

```
Eq3:=phi1(u[n+1],u[n],L1[2]); Eq4:=phi1(u[n+1],uT[n],L1[1]);
Eq34:=Eq3-Eq4; Eq12:=Eq1-Eq2; Eq1234:=factor((Eq34-Eq12)/2);
Eq10:=collect(Eq1234,L1); k1:=u[n+1]; k2:=u[n-1];
A:=(u[n+1]-u[n-1])/4; B:=(uT[n]-u[n])/4;
tr1:=sin(X)+sin(Y)=2*sin((X+Y)/2)*cos((X-Y)/2);
Eq11:=select(has,lhs(Eq10),L1[1])/L1[1];
Eq21:=select(has,lhs(Eq10),L1[2])/L1[2];
A1:=op(1,select(has,Eq11,k1)); A2:=op(1,select(has,Eq11,k2));
A3:=op(1,select(has,Eq21,k1)); A4:=op(1,select(has,Eq21,k2));
Eq13:=subs(X=A1,Y=A2,tr1); Eq23:=subs(X=A3,Y=A4,tr1);
Eq24:=rhs(L1[2]*Eq23+L1[1]*Eq13)/2; tr2:=(L1[2]-L1[1])*sin(A)*
 cos(B)+(L1[2]+L1[1])*cos(A)*sin(B); T1:=factor(Eq24/op(3,
 op(1,Eq24))); T2:=combine(tr2); test1:=factor(T1+T2);
Eq25:=map(`/`,tr2,cos(B)); Eq26:=map(`/`,Eq25,cos(A));
Eq27:=convert(Eq26,tan); tlam:=select(has,op(1,Eq27),
 lambda[n+1]); Eq28:=map(`/`,Eq27,tlam);
Eq29:=(-op(1,op(2,op(1,Eq28)))+arctan(op(2,Eq28)))*4=0;
```

Mathematica:

```
trD[u_,var_]:=Table[D[u,{var,i}],{i,1,6}]//Flatten;
trS1[eq_,var_]:=Select[eq,MemberQ[#,var,Infinity]&];
trS3[eq_,var_]:=Select[eq,FreeQ[#,var]&]; var=Sequence[x,t];
var=Sequence[x,t]; l1={lambda[n],lambda[n+1]}
phi1[u_,v_,lambda_]:=D[(u+v),x]-2*lambda*Sin[(u-v)/2]==0;
{eq1=phi1[u[var][n],u[var][n-1],l1[[1]]], eq2=phi1[uT[var][n],
 u[var][n-1],l1[[2]]], eq3=phi1[u[var][n+1],u[var][n],l1[[2]]],
 eq4=phi1[u[var][n+1],uT[var][n],l1[[1]]]}
eq34=Thread[eq3+Thread[(-1)*eq4,Equal],Equal]
eq12=Thread[eq1+Thread[(-1)*eq2,Equal],Equal]
{eq1234=Thread[Thread[eq34+Thread[(-1)*eq12,Equal],Equal]/2,
 Equal]//Expand, eq10=Collect[eq1234,l1]}
{k1=u[var][n+1], k2=u[var][n-1], a=(u[var][n+1]-u[var][n-1])/4,
 b=(uT[var][n]-u[var][n])/4, tr1=Sin[xN]+Sin[yN]->2*Sin[
 (xN+yN)/2]*Cos[(xN-yN)/2],eq11=trS1[eq10[[1]],l1[[1]]]/l1[[1]],
 eq21=trS1[eq10[[1]],l1[[2]]]/l1[[2]]}
aF[eq_,var_]:=(trS1[eq,var]//Simplify)[[1]]//Expand;
{a1=aF[eq11,k1],a2=aF[eq11,k2],a3=aF[eq21,k1],a4=aF[eq21,k2]}
eq13=tr1/.xN->a1/.yN->a2/.Rule->Equal//ExpandAll
eq23=tr1/.xN->a3/.yN->a4/.Rule->Equal//ExpandAll
{eq24=(Thread[Thread[l1[[2]]*eq23,Equal]+Thread[l1[[1]]*eq13,
 Equal],Equal])[[2]]/2//Expand, tr2=(l1[[2]]-l1[[1]])*Sin[a]*
 Cos[b]+(l1[[2]]+l1[[1]])*Cos[a]*Sin[b]}
```

```
termCos=trS3[trS3[trS1[eq24,l1[[2]]],l1[[2]]],Sin[_]]
{t1=eq24/termCos//Factor, t2=Together[tr2//ExpandAll,
 Trig->True], test1=(t1+t2)//FullSimplify}
{eq25=Map[Divide[#,Cos[b]]&,tr2], eq26=Map[Divide[#,Cos[a]]&,
 eq25], tlam=trS3[trS1[eq26,l1[[2]]][[1]],Tan[_]]}
{eq28=Map[Divide[#,tlam]&,eq26], eq29=Simplify[(eq29=Map[
 ArcTan[#]&,eq28]*4)//PowerExpand//ExpandAll]}
```

□

Problem 4.9 *KdV equation. Nonlinear superposition formula. Two-soliton solution.* For the Korteweg–de Vries equation

$$u_t - 6uu_x + u_{xxx} = 0,$$

where $\{x \in \mathbb{R}, t \geq 0\}$, construct the nonlinear superposition formula and the two-soliton solution.

1. Nonlinear superposition formula. As in the previous problem, we consider four copies of the x-part of the auto-Bäcklund transformation $\phi_1(u, v; \lambda) = (u+v)_x - 2\lambda - \frac{1}{2}(u-v)^2 = 0$, namely,

$$\phi_1(u_n, u_{n-1}, \lambda_n), \; \phi_1(\tilde{u}_n, u_{n-1}, \lambda_{n+1}), \; \phi_1(u_{n+1}, u_n, \lambda_{n+1}), \; \phi_1(u_{n+1}, \tilde{u}_n, \lambda_n)$$

(`Eq1`, `Eq2`, `Eq3`, `Eq4`). Eliminating the first derivatives of u_{n+1}, u_n, $\tilde{u}_n$, u_{n-1} between these four relations, we obtain the following nonlinear superposition formula for the KdV equation (`Eq5`):

$$(u_{n+1} - u_{n-1})(u_n - \tilde{u}_n) - 4(\lambda_{n+1} - \lambda_n) = 0.$$

Maple:

```
with(PDEtools): with(plots): declare(w1(x,t),w2(x,t));
alias(w1=w1(x,t),w2=w2(x,t),u[n]=u[n](x,t),uT[n]=uT[n](x,t),
 u[n-1]=u[n-1](x,t),u[n+1]=u[n+1](x,t));
L1:=[lambda[n],lambda[n+1]];
phi1:=(u,v,lambda)->diff((u+v),x)-2*lambda-(u-v)^2/2=0;
Eq1:=phi1(u[n],u[n-1],L1[1]); Eq2:=phi1(uT[n],u[n-1],L1[2]);
Eq3:=phi1(u[n+1],u[n],L1[2]); Eq4:=phi1(u[n+1],uT[n],L1[1]);
Eq34:=Eq3-Eq4; Eq12:=Eq1-Eq2; Eq1234:=simplify(Eq34-Eq12);
Eq5:=collect(Eq1234,[u[n],uT[n]]);
```

Mathematica:

```
var=Sequence[x,t]; l1={lambda[n],lambda[n+1]}
phi1[u_,v_,lambda_]:=D[(u+v),x]-2*lambda-(u-v)^2/2==0;
eq1=phi1[u[var][n],u[var][n-1],l1[[1]]]
eq2=phi1[uT[var][n],u[var][n-1],l1[[2]]]
eq3=phi1[u[var][n+1],u[var][n],l1[[2]]]
eq4=phi1[u[var][n+1],uT[var][n],l1[[1]]]
eq34=Thread[eq3+Thread[(-1)*eq4,Equal],Equal]
eq12=Thread[eq1+Thread[(-1)*eq2,Equal],Equal]
eq1234=Thread[eq34+Thread[(-1)*eq12,Equal],Equal]//Expand
eq5=Collect[eq1234,{u[var][n],uT[var][n]}]
```

2. Two-soliton solution. Applying *Maple* predefined functions, we will construct the two-soliton solution `U12(x,t)` from the vacuum solution $u_0{=}0$. In this case, $n{=}1$, $u_{n+1}{=}u_{12}$, $u_n{=}u_1$, $\tilde{u}_n{=}u_2$, $u_{n-1}{=}u_0$, and $u_{12}{=}4(\lambda_2{-}\lambda_1)/(u_1{-}u_2)$ (`Eq8`). Solving the auto-BT (`BTx`, `BTt`)

$$(w_1+w_2)_x-2\lambda-\tfrac{1}{2}(w_1-w_2)^2, \quad (w_1-w_2)_t-3[(w_{1x})^2-(w_{2x})^2]+(w_1-w_2)_{xxx}$$

for w_1 with $w_2{=}0$ and for w_2 with $w_1{=}0$, we can obtain the following solutions (`sol1`, `sol2`):

$$w_1=-2\kappa\tanh(C_1+\kappa x+4\lambda\kappa t), \quad w_2=-2\kappa\coth(C_1+\kappa x+4\lambda\kappa t),$$

where $\kappa{=}\sqrt{-\lambda}$. We take $\lambda_1{=}-1$, $\lambda_2{=}-4$ and obtain $u_1{=}-2\tanh(x{-}4t)$, $u_2{=}-4\coth(2x{-}32t)$, $u_{12}{=}-6/[\tanh(-x{+}4t){-}2\coth(-2x{+}32t)]$. The corresponding solution of the KdV equation follows from this result

$$U_{12}(x,t)=\frac{6(3+\tanh(-x+4t)^2-4\coth(-2x+32t)^2)}{(\tanh(-x+4t)-2\coth(-2x+32t))^2}.$$

Finally, we verify that this solution (that represents the two-soliton solution) is exact solution of the KdV equation and visualize the solution at different times.

```
Eq6:=collect(subs(n=1,u[0](x,t)=0,Eq5),[u[2](x,t)]);
Eq7:=subs(u[2](x,t)=u[12],u[1](x,t)=u[1],uT[1](x,t)=u[2],Eq6);
Eq8:=isolate(Eq7,u[12]); params:={lambda[1]=-1,lambda[2]=-4};
BTx:=diff(w1+w2,x)-2*lambda-1/2*(w1-w2)^2;
BTt:=diff(w1-w2,t)-3*(diff(w1,x)^2-diff(w2,x)^2)+diff(w1-w2,x$3);
Eq11:=subs(w2=0,BTx); Eq12:=subs(w2=0,BTt);
sol1:=pdsolve({Eq11,Eq12},w1,HINT='TWS(tanh)');
Eq21:=subs(w1=0,BTx); Eq22:=subs(w1=0,BTt);
```

```
sol2:=pdsolve({Eq21,Eq22},w2,HINT='TWS(coth)');
sols:=simplify({u[1]=subs(lambda=-1,_C1=0,
 rhs(op(sol1[4]))),u[2]=subs(lambda=-4,_C1=0,rhs(op(sol2[4])))});
Eq9:=factor(subs(params,sols,Eq8));
U12:=unapply(rhs(diff(Eq9,x)),x,t);
EqKdV:=diff(u(x,t),t)-6*u(x,t)*diff(u(x,t),x)+diff(u(x,t),x$3)=0;
test1:=simplify(algsubs(-u(x,t)=-U12(x,t),EqKdV));
animate(-U12(x,t),x=-50..50,t=-10..10,numpoints=400,frames=200);
```

We note that the *Mathematica* predefined function `DSolve` does not allow us to find traveling wave solutions `sol1` and `sol2`, and hence to construct the final two-soliton solution of the KdV equation. □

4.2.3 Hirota Method

In 1971 R. Hirota proposed a new method, called *the Hirota direct method* or the *Hirota bilinear formalism*, for constructing exact *multiple-soliton solutions* of integrable nonlinear evolution equations [55]. The method is based on a transformation into new variables, so that in these new variables integrable equations can be written in a special *Hirota bilinear form*, all derivatives appear as the *Hirota bilinear derivatives*, and multiple-soliton solutions appear in a particularly simple form [57], [53].

The existence of multiple-soliton solutions can be used as an integrability condition and as a method of finding new integrable equations. It was shown that the Hirota method gives N-soliton solutions to the Korteweg–de Vries, modified Korteweg–de Vries, sine–Gordon, nonlinear Schrödinger equations. For nonintegrable equations, only exact two-soliton solutions can be obtained. Moreover, the Hirota method, reformulated in terms of τ-functions, can be applied for further mathematical developments (e.g., the Sato theory [76]).

Problem 4.10 *KdV equation. Multiple-soliton solutions.* Let us consider the Korteweg–de Vries equation

$$u_t+6uu_x+u_{xxx}=0,$$

where $\{x \in \mathbb{R}, t \geq 0\}$. Following the ideas of the Hirota method and a simplified version of the Hirota method proposed by Hereman [52], we propose a new version of the Hirota method and construct multiple-soliton solutions (one-soliton solution, `SS1`, two-soliton solution, `SS2`, three-soliton solution `SS3`) of the KdV equation.

1. Bilinear form. We have to determine a bilinear form of the KdV equation (see Problem 2.15). Introducing a new dependent variable w

by $u=w_{xx}$ (tr1), we rewrite the KdV equation in the following form: $w_{txx}+6w_{xx}w_{xxx}+w_{xxxxx}=0$ (Eq1). Integrating this equation with respect to x, we obtain the *potential form* of KdV equation $w_{tx}+3w_{xx}^2+w_{xxxx}=0$ (Eq2)*. Introducing a new dependent variable $F(x,t)$ defined by the equation $w=2\log F$ (tr2), we transform the potential form of KdV equation into the bilinear equation $FF_{xt}+FF_{xxxx}-F_xF_t-4F_xF_{xxx}+3F_{xx}^2=0$ (Eq4). According to the Hirota method, we have to rewrite this bilinear equation in operator form by using the *Hirota bilinear operator* or *Hirota D operator*:

$$D_t^n D_x^m(f\cdot g)=\left(\frac{\partial}{\partial t_1}-\frac{\partial}{\partial t_2}\right)^n\left(\frac{\partial}{\partial x_1}-\frac{\partial}{\partial x_2}\right)^m f(x_1,t_1)g(x_2,t_2)\Big|_{x_2=x_1,t_2=t_1}.$$

For our problem, we define two differential operators Dx^n and D_xD_t (Dxn, DxDt). Then we verify that KdV equation can be written using the Hirota D operator (test1), i.e., the bilinear Hirota form is

$$B(F,F)=(D_x^4+D_xD_t)(F\cdot F)=0.$$

Maple:

```
with(PDEtools): with(plots):  declare((u,w,F,F1,F2,f,g)(x,t));
alias(u=u(x,t),w=w(x,t),F=F(x,t),F1=F1(x,t),F2=F2(x,t),f=f(x,t),
 g=g(x,t)); tr1:=u=diff(w,x$2);
PDE1:=u->diff(u,t)+6*u*diff(u,x)+diff(u,x$3)=0;
Eq1:=expand(PDE1(rhs(tr1))); Eq2:=map(int,Eq1,x);
tr2:=w=2*log(F); Eq4:=simplify(algsubs(tr2,Eq2)/2*F^2);
Dxn:=(f,g,n)->sum((-1)^k*binomial(n,k)*diff(f,x$(n-k))
     *diff(g,x$k),k=0..n); Dxn(f,g,1); Dxn(f,g,2); Dxn(f,g,4);
DxDt:=(f,g)->f*diff(g,x,t)-diff(f,x)*diff(g,t)
      -diff(f,t)*diff(g,x)+diff(f,x,t)*g;
Eq5:=(Dxn(F,F,4)+DxDt(F,F)=0)/2; test1:=Eq4-Eq5;
```

Mathematica:

```
trD[u_,var_]:=Table[D[u,{var,i}],{i,1,6}]//Flatten;
var=Sequence[x,t]; tr1=u[var]->D[w[var],{x,2}]
pde1[u_]:=D[u,t]+6*u*D[u,x]+D[u,{x,3}]==0
{eq1=pde1[tr1[[2]]]//Expand, eq2=Map[Integrate[#,x]&,eq1]}
tr2=w[var]->2*Log[f[var]]
eq41=(eq2/.tr2/.trD[tr2,x]/.D[tr2,x,t])//Expand
eq4=Thread[eq41/2*f[var]^2,Equal]//Expand
```

*In this case we set an arbitrary integration function of t to be equal to 0.

```
dxn[f_,g_,n_]:=Sum[(-1)^k*Binomial[n,k]*D[f[var],{x,n-k}]*
 D[g[var],{x,k}],{k,0,n}]; {dxn[f,g,1], dxn[f,g,2], dxn[f,g,4]}
dxdt[f_,g_]:=f[var]*D[g[var],x,t]-D[f[var],x]*D[g[var],t]-
 D[f[var],t]*D[g[var],x]+D[f[var],x,t]*g[var];
{eq50=dxn[f,f,4]+dxdt[f,f]==0, eq5=Thread[eq50/2,Equal]//Expand}
test1=Thread[eq4-eq5,Equal]//Expand
```

2. One-soliton solution. Let us construct the one-soliton solution of the KdV equation. Following a simplified version of the Hirota method proposed by Hereman (see [49], [51]) and applying the transformation $u{=}2\ln(F)_{xx}$ (`Eq6`), we rewrite the KdV equation in the following form: $(F_{xt}{+}F_{xxxx})F{-}4F_xF_{xxx}{-}F_xF_t{+}3F_{xx}^2{=}0$ (`Eq41`). This equation can be decomposed into the linear operator L_D and nonlinear operator N_D defined by $L_D(F){=}F_{xt}{+}F_{xxxx}$, $N_D(F\cdot F){=}{-}4F_xF_{xxx}{-}F_xF_t{+}3F_{xx}^2$ (`LD, ND`).

First, we find the zero-soliton solution or the vacuum $V_0{=}1$ (`SolVac`). Then, multiple-soliton solutions are obtained by the perturbation expansion around the vacuum solution $V_0{=}1$: $F(x,t){=}V_0{+}\sum_{n=1}^{N}\varepsilon^nV_n$ (`SolSer`), where ε is a formal expansion parameter. Substituting `SolSer` into the bilinear equation and equating to zero the powers of ε, we obtain the following equations $O(\varepsilon^n):B\Big(\sum_{i=0}^{n}F_j{\cdot}F_{n-j}\Big)$, or in terms of the operators L_D and N_D up to $O(\varepsilon^3)$ we obtain:*

$$O(\varepsilon^1):L_D(F_1){=}0, \qquad O(\varepsilon^2):L_D(F_2){=}{-}N_D(F_1,F_1), \tag{4.5}$$
$$O(\varepsilon^3):L_D(F_3){=}{-}F_1L_D(F_2){-}F_2L_D(F_1){-}N_D(F_1,F_2). \tag{4.6}$$

The N-soliton solution is obtained from $V_1{=}\sum_{i=1}^{K}e^{\theta_i}$ (`tr4`), where k_i, c_i are arbitrary constants, and $\theta_i{=}k_ix{-}c_it$ (`tr3`). Substituting $u(x,t){=}e^{\theta_i}$ (`tr5`) into the linear terms of the KdV equation (or into the first relation of Eq. (4.5), in terms of operators), we obtain the dispersion relation $c_i{=}k_i^3$ (`tr6`). We therefore find $\theta_i{=}k_ix{-}k_i^3t$ (`tr31`) and the first correction $V_1{=}e^{k_1x-k_1^3t}$ (`Corr11`). For the one-soliton solution we have $F{=}1{+}e^{k_1x-k_1^3t}$ (`SS1`), where we set $\varepsilon{=}1$. In terms of the original dependent variable $u(x,t)$, we obtain $u(x,t){=}\dfrac{2k_1^2e^{-k_1(k_1^2t-x)}}{(1{+}e^{-k_1(k_1^2t-x)})^2}$ (`SS1u`) and visualize it at different times.

*The term of order ε^0 vanishes.

Maple:

```
Eq41:=lhs(collect(Eq4,F)); Eq6:=subs(tr2,tr1);
LD:=op(1,Eq41)/F; ND:=Eq41-op(1,Eq41);
M:=1; Eq7:=simplify(exp(lhs(Eq6))=exp(rhs(Eq6)));
Eq71:=op(1,lhs(Eq7))=op(1,rhs(Eq7)); Eq72:=subs(u=0,Eq71);
Eq73:=subs(F=V[0],pdsolve(Eq72,F,explicit));
arbFun:={_F1(t)=0,_F2(t)=1}; SolVac:=eval(Eq73,arbFun);
SolSer:=N->F=V[0]+Sum(epsilon^n*V[n],n=1..N);
tr3:=i->theta[i]=k[i]*x-c[i]*t;
tr4:=(j,K)->subs(tr3(j),V[1]='Sum(exp(theta[j]),j=1..K)');
tr5:=subs(tr3(i),u=exp(theta[i])); Eq8:=PDE1(rhs(tr5));
Eq81:=op(1,lhs(Eq8))+op(3,lhs(Eq8))=0;
tr6:=unapply(isolate(Eq81,c[i]),i); tr31:=subs(tr6(i),tr3(i));
Corr11:=value(subs(tr6(i),tr4(i,M)));
SS1:=subs(epsilon=1,Corr11,value(subs(SolVac,SolSer(M))));
SS1u:=simplify(subs(SS1,Eq6)); SS1uG:=subs(k[1]=1,rhs(SS1u));
animate(SS1uG,x=-50..50,t=-10..10,numpoints=200,frames=50);
```

Mathematica:

```
SetOptions[Plot,PlotPoints->300,ImageSize->500,PlotStyle->
 {Hue[0.9],Thickness[0.01]}]; eq41=Collect[eq4,f[var]][[1]]
trS1[eq_,var_]:=Select[eq,MemberQ[#,var,Infinity]&];
trS3[eq_,var_]:=Select[eq,FreeQ[#,var]&]; var=Sequence[x,t];
{eq6=tr1/.tr2/.trD[tr2,x]/.Rule->Equal,
 termf=trS1[eq41,f[var]], lD=termf/f[var]//Simplify,
 nD=eq41-termf//Expand, m=1, eq7=Map[Exp[#]&,eq6],
 eq71=eq7[[1,2]]==eq7[[2,2]], eq72=eq71/.u[var]->0}
eq73=DSolve[eq72,f[var],{x,t}]/.f[var]->v[var][0]//First
{arbFun={C[1][t]->0,C[2][t]->1}, solVac=eq73/.arbFun}
solSer[nN_]:=f[var]->v[var][0]+Sum[epsilon^n*v[var][n],{n,1,nN}];
tr3[i_]:=theta[i]->k[i]*x-c[i]*t; tr4[j_,k_]:=v[var][1]->
 Sum[Exp[theta[j]],{j,1,k}]/.tr3[j]; {tr5=u[var]->Exp[theta[i]]/.
 tr3[i], eq8=pde1[tr5[[2]]], eq81=trS3[eq8[[1]],6]==0}
tr6[i1_]:=((Solve[eq81,c[i]][[1,1]]))/.i->i1; tr31[i1_]:=tr3[i]/.
 tr6[i]/.i->i1; corr11=tr4[i,m]/.tr5/.tr31[1]
sS1=solSer[m]/.solVac/.corr11/.epsilon->1
sS1u=eq6/.sS1/.trD[sS1,x]/.trD[sS1,t]//Simplify
sS1uG[xN_,tN_]:=(sS1u[[2]]/.k[1]->1)/.{x->xN,t->tN}; sS1uG[x,t]
Animate[Plot[Evaluate[sS1uG[x,t]],{x,-50,50},
 PlotRange->{{-50,50},{0,0.6}}],{t,-10,10},AnimationRate->0.9]
```

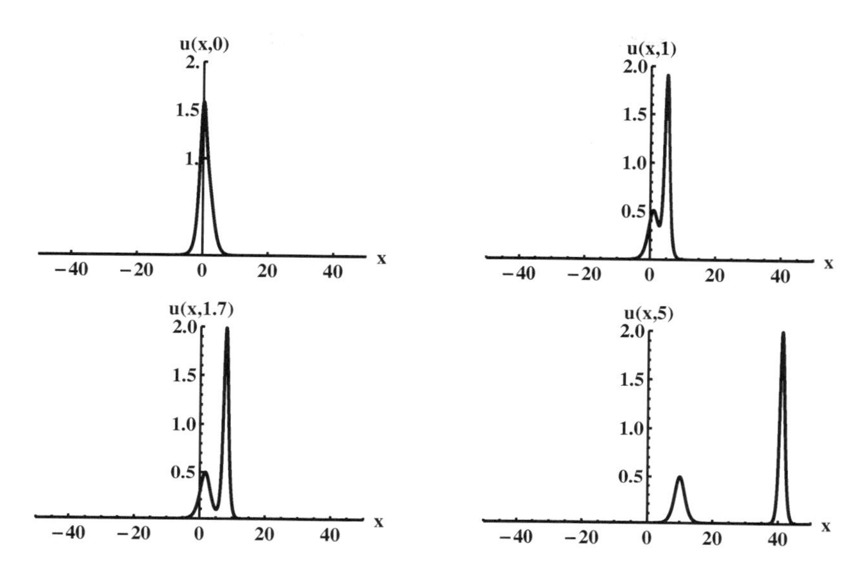

Fig. 4.1. Hirota method: two-soliton solution of the KdV equation at different times $t_k = 0, 1, 1.7, 10$

3. Two-soliton solution. Let us construct the two-soliton solution of the KdV equation. Setting $K=2$ (`tr4`), we obtain $V_1 = e^{k_1 x - c_1 t} + e^{k_2 x - c_2 t}$ (`tr2V1`). Then, we set $V_2 = a_{12} e^{k_1 x - c_1 t + k_2 x - c_2 t}$ (`tr2V2`). To determine a_{12}, we substitute V_1, V_2 into the second relation of Eq. (4.5). Evaluating the left hand side (`Eq9L`) and equate it with the right hand side (`Eq9R`), we obtain $a_{12} = (k_1 - k_2)^2/(k_1 + k_2)^2$ (`a12`). For the two-soliton solution, calculating the first and second corrections (`Corr21`, `Corr22`), we obtain (`SS2`)

$$F = 1 + e^{k_1 x - k_1^3 t} + e^{k_2 x - k_2^3 t} + \frac{(k_1 - k_2)^2}{(k_1 + k_2)^2} e^{k_1 x - k_1^3 t + k_2 x - k_2^3 t}.$$

Then we find the solution in terms of the original dependent variable $u(x,t)$ (`SS2u`) and visualize it at different times (see Fig. 4.1).

```
M:=2;  tr32:=seq(tr3(i),i=1..M); tr2V1:=value(tr4(i,M));
tr2V2:=subs(tr32,V[2]=a[1,2]*exp(theta[1]+theta[2]));
Eq9L:=algsubs(F=rhs(tr2V2),LD); Eq9R:=-algsubs(F=rhs(tr2V1),ND);
Eq91:=collect(simplify(Eq9L-Eq9R),exp);
a12:=factor(isolate(subs(seq(tr6(i),i=1..M),Eq91),a[1,2]));
Corr21:=value(subs(tr6(i),tr4(i,M))); trk:={k[1]=1,k[2]=2};
Corr22:=subs(a12,seq(tr6(i),i=1..M),tr2V2);
```

```
SS2:=subs(epsilon=1,Corr21,Corr22,value(subs(SolVac,SolSer(M))));
SS2u:=simplify(subs(SS2,Eq6)); SS2uG:=subs(trk,rhs(SS2u));
animate(SS2uG,x=-50..50,t=-10..10,numpoints=200,frames=50);
```

Mathematica:

```
{m=2, tr32=Table[tr3[i],{i,1,m}], tr2v1=tr4[i,m]/.tr5/.tr31[1]/.
 tr31[2], tr2v2=v[var][2]->a[1,2]*Exp[theta[1]+theta[2]]/.tr32,
 trf2=f[var]->tr2v2[[2]],trf1=f[var]->tr2v1[[2]]}
eq9L=lD/.trf2/.trD[trf2,x]/.D[trf2,x,t]
eq9R=-nD/.trf1/.trD[trf1,x]/.trD[trf1,t]/.D[trf1,x,t]
{eq91=eq9L-eq9R//FullSimplify, tr6seq=Table[tr6[i],{i,1,m}]}
a12=Solve[(eq91/.tr6seq)==0,a[1,2]]//First
corr21=tr4[i,m]/.tr6[i]/.tr31[1]/.tr31[2]
{trk={k[1]->1,k[2]->2}, corr22=tr2v2/.a12/.tr6seq}
sS2=solSer[m]/.solVac/.corr22/.corr21/.epsilon->1
sS2u=eq6/.sS2/.trD[sS2,x]/.trD[sS2,t]//Simplify
sS2uG[xN_,tN_]:=(sS2u[[2]]/.trk)/.{x->xN,t->tN}; sS2uG[x,t]
Animate[Plot[Evaluate[sS2uG[x,t]],{x,-50,50},
 PlotRange->{{-50,50},{0,2}}],{t,-10,10},AnimationRate->0.9]
```

4. Three-soliton solution. Let us construct the three-soliton solution of the KdV equation. Setting $K{=}3$ in `tr4`, we obtain $V_1{=}e^{\theta_1}{+}e^{\theta_2}$ (`tr3V1`), where $\theta_i{=}k_i x{-}c_i t$ (`tr3`). Then, defining $V_2{=}\sum_{1\le i<j\le K} a_{ij}e^{\theta_i+\theta_j}$, we calculate $V_2{=}a_{12}e^{\theta_1+\theta_2}{+}a_{13}e^{\theta_1+\theta_3}{+}a_{23}e^{\theta_2+\theta_3}$, $V_3{=}b_{123}e^{\theta_1+\theta_2+\theta_3}$ (`tr3V2`, `tr3V3`). To determine b_{123}, as before, we substitute V_1, V_2, V_3 into Eq. (4.6). Evaluating the left hand side (`Eq10L`) and equate it with the right hand side (`Eq10R`), we obtain $b_{123}{=}\dfrac{(k_2{-}k_3)^2(k_1{-}k_3)^2(k_1{-}k_2)^2}{(k_1{+}k_2)^2(k_1{+}k_3)^2(k_2{+}k_3)^2}$. Finally, calculating the corrections (`Corr31`, `Corr32`, `Corr33`), we obtain the three-soliton solution (`SS3`, `SS3u`) and visualize it at different times (see Fig. 4.2).

Maple:

```
M:=3; tr33:=seq(tr3(i),i=1..M); tr3V1:=value(tr4(j,M));
LP0:=combinat[permute](3,2); LP1:=NULL: NLP0:=nops(LP0);
for i from 1 to NLP0 do
 if op(1,LP0[i])<op(2,LP0[i]) then LP1:=LP1,LP0[i]; fi; od;
LP2:=[LP1]; NLP2:=nops(LP2);
tr3V2:=subs(tr33,V[2]=add(a[op(LP2[i])]*exp(theta[op(1,LP2[i])]
 +theta[op(2,LP2[i])]),i=1..NLP2));
tr3V3:=subs(tr33,V[3]=b[1,2,3]*exp(theta[1]+theta[2]+theta[3]));
```

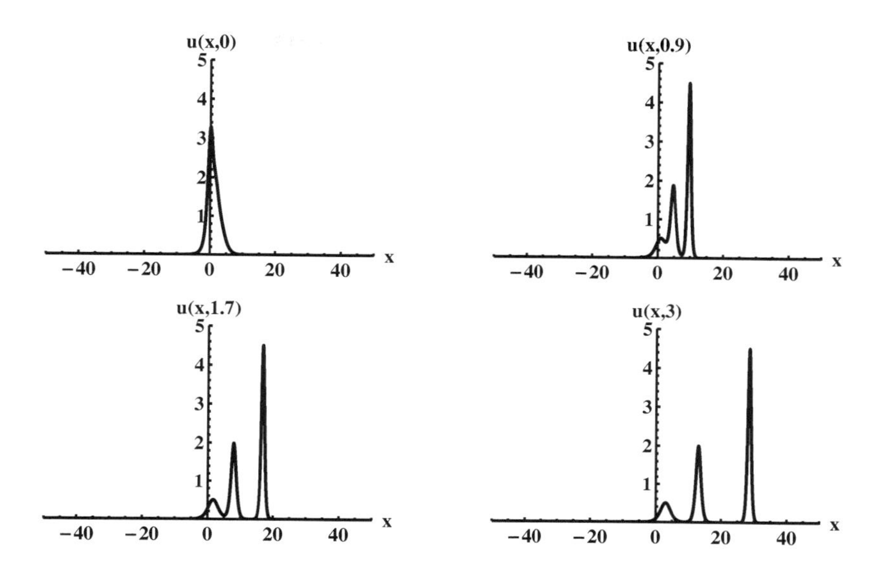

Fig. 4.2. Hirota method: three-soliton solution of the KdV equation at different times $t_k = 0, 0.9, 1.7, 3$

```
Eq10L:=algsubs(F=rhs(tr3V3),LD);
Eq51:=collect(Dxn(F1,F2,4)+DxDt(F1,F2),[F1,F2]);
Eq511:=Eq51-op(1,Eq51)-op(2,Eq51);
ND1:=algsubs(F2=rhs(tr3V2),algsubs(F1=rhs(tr3V1),Eq511)):
Eq10R:=-rhs(tr3V1)*algsubs(F=rhs(tr3V2),LD)-rhs(tr3V2)
 *algsubs(F=rhs(tr3V1),LD)-ND1:
Eq101:=simplify(Eq10L-Eq10R):
Eq102:=subs(seq(tr6(i),i=1..M),collect(Eq101,exp)); a12;
a13:=subs(a[1,2]=a[1,3],k[2]=k[3],a12);
a23:=subs(a[1,3]=a[2,3],k[1]=k[2],a13);
Eq103:=subs(a12,a13,a23,Eq102);
for i from 1 to nops(Eq103) do A||i:=simplify(op(i,Eq103)); od;
b123:=factor(isolate(subs(seq(tr6(i),i=1..3),A1),b[1,2,3]));
Corr31:=value(subs(tr6(j),tr4(j,M)));
Corr32:=subs(a12,a13,a23,subs(seq(tr6(j),j=1..M),tr3V2));
Corr33:=subs(b123,seq(tr6(i),i=1..M),tr3V3);
SS3:=subs(epsilon=1,Corr31,Corr32,Corr33,
 value(subs(SolVac,SolSer(M)))); SS3u:=subs(SS3,Eq6);
SS3uG:=subs(k[1]=1,k[2]=2,k[3]=3,rhs(SS3u));
animate(SS3uG,x=-50..50,t=-10..10,numpoints=200,frames=50);
```

Mathematica:

```
{m=3, tr33=Table[tr3[i],{i,1,m}], tr3v1=tr4[j,m]/.tr31[1]/.
 tr31[2]/.tr31[3], lP0=Permutations[Range[3],{2}],
 lP2=Select[lP0,#[[1]]<#[[2]]&], nLP2=Length[lP2]}
tr3v2=v[var][2]->Sum[a[lP2[[i,1]],lP2[[i,2]]]*Exp[
 theta[lP2[[i,1]]]+theta[lP2[[i,2]]]],{i,1,nLP2}]/.tr33
tr3v3=v[var][3]->b[1,2,3]*Exp[theta[1]+theta[2]+theta[3]]/.tr33
{trff33=f[var]->tr3v3[[2]], trff23=f[var]->tr3v2[[2]],
 trff13=f[var]->tr3v1[[2]], trf23=f2[var]->tr3v2[[2]],
 trf13=f1[var]->tr3v1[[2]]}
eq10L=lD/.trff33/.trD[trff33,x]/.D[trff33,x,t]
eq51=Collect[dxn[f1,f2,4]+dxdt[f1,f2],{f1[var],f2[var]}]
{termf1=trS1[eq51,f1[var]], termf2=trS1[eq51,f2[var]]}
eq511=eq51-termf1-termf2
nD1=eq511/.trf23/.trD[trf23,x]/.trD[trf23,t]/.D[trf23,x,t]/.
 trf13/.trD[trf13,x]/.trD[trf13,t]/.D[trf13,x,t]
eq10R=-tr3v1[[2]]*(lD/.trff23/.trD[trff23,x]/.D[trff23,x,t])-
 tr3v2[[2]]*(lD/.trff13/.trD[trff13,x]/.D[trff13,x,t])-nD1
{eq101=eq10L-eq10R//Simplify, tr63seq=Table[tr6[i],{i,1,m}]}
eq102=Collect[eq101/.tr63seq,Exp[_]]
eq1021=Map[Factor,Collect[eq102,Exp[_]]]//Simplify
{a12, a13=a12/.a[1,2]->a[1,3]/.k[2]->k[3], a23=a13/.
 a[1,3]->a[2,3]/.k[1]->k[2], eq103=eq1021/.a12/.a13/.a23}
Do[aN[i]=eq103[[i]]//Simplify; Print[aN[i]],{i,1,Length[eq103]}];
b123=Solve[(aN[3]/.tr63seq)==0,b[1,2,3]]//First//FullSimplify
corr31=tr4[j,m]/.tr6[i]/.tr31[1]/.tr31[2]/.tr31[3]
{corr32=tr3v2/.a12/.a13/.a23/.tr63seq, corr33=tr3v3/.b123/.
 tr63seq, sS3=solSer[m]/.solVac/.corr33/.corr32/.corr31/.
 epsilon->1, sS3u=eq6/.sS3/.trD[sS3,x]/.trD[sS3,t],
 trk3={k[1]->1,k[2]->2,k[3]->3}}
sS3uG[xN_,tN_]:=(sS3u[[2]]/.trk3)/.{x->xN,t->tN}; sS3uG[x,t]
Animate[Plot[Evaluate[sS3uG[x,t]],{x,-50,50},
 PlotRange->{{-50,50},{0,5}}],{t,-10,10},AnimationRate->0.9]
```

□

4.2.4 Lax Pairs

The *Lax pair*, first introduced by P. Lax in 1968 [87],* represents a set of two linear differential operators with the commutativity condition, which is identical to the nonlinear evolution equation. The existence

*One year earlier, Gardner, Greene, Kruskal, and Miura [64] proposed a method for solving exactly the Cauchy problem for the KdV equation.

of the Lax pair is a *criterium of integrability*, and it is insufficient to construct solutions. To build solutions, Lax (introducing the *Heisenberg picture*) developed a general principle for describing integrable nonlinear evolution equations that can be solved exactly by the *inverse scattering method* (which can be referred as a nonlinear analogue of the Fourier transform).

Definition 4.5 The *Lax pair* of a nonlinear PDE $\mathfrak{F}(x,t,u,\dots)=\mathfrak{F}[u]=0$ is a system of two linear differential operators $L=L(u,\lambda)$, $M=M(u,\lambda)$ (depending on a solution u of the PDE and on the spectral parameter λ) such that $[L,M]=0$ if and only if $\mathfrak{F}[u]=0$, where $[L,M]=LM-ML$ is the commutator of the operators L and M.

A Lax pair can be represented in several equivalent forms: the Lax representation, the zero-curvature representation, the projective Riccati representation, the scalar representation, the string representation (or Sato representation).

If the evolution equation $\mathfrak{F}[u]=0$ can be expressed as the *Lax equation* $L_t+[L,M]=0$ and if $L\psi=\lambda(t)\psi$ ($t\geq 0$, $x\in\mathbb{R}$) holds, then $\lambda_t=0$, and ψ satisfies the differential equation $\psi_t=M\psi$ ($t\geq 0$).

Problem 4.11 *KdV equation. Lax pairs.* Let us consider the following two *Lax pairs*:

$$L=-\frac{\partial^2}{\partial x^2}+u, \quad M=c\frac{\partial}{\partial x};$$
$$L=-\frac{\partial^2}{\partial x^2}+u, \quad M=-4\frac{\partial^3}{\partial x^3}+6u\frac{\partial}{\partial x}+3\frac{\partial u}{\partial x}.$$

Applying *Maple* predefined functions, verify that these Lax pairs are defined, respectively, for the one-dimensional wave equation $u_t-cu_x=0$ and the KdV equation $u_t-6uu_x+u_{xxx}=0$.

Maple:

```
with(PDEtools): with(Ore_algebra); declare(u(x,t));
A:=diff_algebra([Dx,x],[Dt,t],[comm,c],func={u});
L:=-Dx^2+u(x,t); M1:=c*Dx; CommLM1:=skew_product(L,M1,A)
 -skew_product(M1,L,A); LaxEq1:=diff(L,t)+CommLM1=0;
M2:=-4*Dx^3+3*(u(x,t)*Dx+skew_product(Dx,u(x,t),A));
CommLM2:=skew_product(L,M2,A)-skew_product(M2,L,A);
LaxEq2:=diff(L,t)+CommLM2=0;
```

□

It is possible to remove the restriction that the operators L and M should belong to the class of scalar operators. Let us consider the *matrix operators* L and M. This approach was proposed by V. E. Zakharov and A. B. Shabat in 1972 [174], initially known as the *Zakharov and Shabat (ZS) scheme*, for the nonlinear Schrödinger (NLS) equation. In 1974, Ablowitz, Kaup, Newell, and Segur (the AKNS group from the Clarkson University, NY) made a number of extensions of the method to other equations including the sine–Gordon equation. Then there were numerous discoveries of integrable equations, where the original Lax pair and spectral transform have been replaced by a more general approach based on a *zero-curvature representation*

$$A_t - B_x + [A, B] = 0 \tag{4.7}$$

or a *compatibility condition* for two linear problems $\phi_x = A\phi$, $\phi_t = B\phi$, where $A = A(x,t,\lambda)$, $B = B(x,t,\lambda)$ are $n \times n$ matrices (or elements of a Lie algebra) depending on the spectral parameter λ, and $[A, B] = AB - BA$. The compatibility condition allows us to discover a nonlinear evolution equation.

Problem 4.12 *Nonlinear Schrödinger equation, sine–Gordon equation, and sinh–Gordon equation. Lax pairs.* Let us consider the following three special cases of A, B_i $(i=1,2,3)$:

$$A = i\lambda \begin{pmatrix} 1 & 0 \\ 0 & -1 \end{pmatrix} + i \begin{pmatrix} 0 & q \\ r & 0 \end{pmatrix},$$

$$B_1 = 2i\lambda^2 \begin{pmatrix} 1 & 0 \\ 0 & -1 \end{pmatrix} + 2i\lambda \begin{pmatrix} 0 & q \\ r & 0 \end{pmatrix} + \begin{pmatrix} 0 & q_x \\ -r_x & 0 \end{pmatrix} - i \begin{pmatrix} rq & 0 \\ 0 & -rq \end{pmatrix},$$

$$B_2 = \frac{1}{4i\lambda} \begin{pmatrix} \cos u & -i\sin u \\ i\sin u & -\cos u \end{pmatrix}, \quad B_3 = \frac{1}{4i\lambda} \begin{pmatrix} \cosh u & -i\sinh u \\ -i\sinh u & -\cosh u \end{pmatrix},$$

where $q = q(x,t)$, $r = r(x,t)$ are complex-valued functions of x and t. Applying the compatibility condition and the matrices A and B_1, A and B_2, A and B_3, deduce, respectively, the nonlinear Schrödinger equations, the sine–Gordon equation, and the sinh–Gordon equation.

1. Substituting the matrices A and B_1 into the compatibility condition (4.7), we arrive at the equations $iq_t - q_{xx} - 2q^2 r = 0$, $ir_t + r_{xx} + 2r^2 q = 0$. Then, if we set $r = \bar{q}$ or $r = -\bar{q}$ in the above equations, we obtain the nonlinear Schrödinger equations $ir_t + r_{xx} \pm 2r|r|^2 = 0$.

2. We verify that the compatibility condition with respect to the matrices A, B_2 and A, B_3, respectively, turns into the sine–Gordon equation $u_{xt} = \sin u$ and the sinh–Gordon equation $u_{xt} = \sinh u$.

Maple:

```
with(PDEtools): with(LinearAlgebra); declare(u(x,t),r(x,t),
 q(x,t)); A0:=<<1,0>|<0,-1>>; A1:=<<0,r(x,t)>|<q(x,t),0>>;
B0:=<<0,0>|<0,0>>; A:=I*lambda*A0+I*A1; B1:=2*I*lambda^2*A0
 +2*I*lambda*A1+<<0,-diff(r(x,t),x)>|<diff(q(x,t),x),0>>
             -I*<<q(x,t)*r(x,t),0>|<0,-q(x,t)*r(x,t)>>;
CompCond:=(L,M)->simplify(map(diff,L,t)-map(diff,M,x)+(L.M-M.L));
Eqs1:=[CompCond(A,B1)[1,2]=B0[1,2],CompCond(A,B1)[2,1]=B0[2,1]];
NSEq1:=simplify(subs(q(x,t)=conjugate(r(x,t)),Eqs1))[2];
NSEq2:=simplify(subs(q(x,t)=-conjugate(r(x,t)),Eqs1))[2];
B2:=1/(4*I*lambda)*<<cos(u(x,t)),I*sin(u(x,t))>|<-I*sin(u(x,t)),
 -cos(u(x,t))>>; Eqs2:=[seq(seq(CompCond(A,B2)[i,j]=B0[i,j],
 j=1..2),i=1..2)]; SineG:=expand(subs(q(x,t)=1/2*diff(u(x,t),x),
 r(x,t)=1/2*diff(u(x,t),x),Eqs2[2]*2/I));
B3:=1/(4*I*lambda)*<<cosh(u(x,t)),-I*sinh(u(x,t))>|
 <-I*sinh(u(x,t)),-cosh(u(x,t))>>;
Eqs3:=[seq(seq(CompCond(A,B3)[i,j]=B0[i,j],j=1..2),i=1..2)];
SinhG:=expand(subs(q(x,t)=1/2*diff(u(x,t),x),r(x,t)=
 -1/2*diff(u(x,t),x),Eqs3[2]*2/I));
```

Mathematica:

```
trD[u_,var_]:=Table[D[u,{var,i}],{i,1,2}]//Flatten;
var=Sequence[x,t]; {a0={{1,0},{0,-1}},
 a1={{0,q[var]},{r[var],0}}, b0={{0,0},{0,0}}}
{a=I*lambda*a0+I*a1, b1=2*I*lambda^2*a0+2*I*lambda*a1+
 {{0,D[q[var],x]},{-D[r[var],x],0}}-I*{{q[var]*r[var],0},
 {0,-q[var]*r[var]}}}
Map[MatrixForm,{a0,a1,b0,a,b1}]
compCond[l_,m_]:=(D[l,t]-D[m,x]+(l.m-m.l))//Simplify;
eqs1={compCond[a,b1][[1,2]]==b0[[1,2]],
 compCond[a,b1][[2,1]]==b0[[2,1]]}//ComplexExpand//Expand
trq[s_]:=q[var]->s*Conjugate[r[var]];
{trr1=r[x,t]^2->r[x,t]*r1[x,t], trr2=r1[x,t]->r[x,t]}
nSEq1=(eqs1[[2]]/.trq[1]/.trr1//FullSimplify)/.trr2
nSEq2=(eqs1[[2]]/.trq[-1]/.trr1//FullSimplify)/.trr2
b2=1/(4*I*lambda)*{{Cos[u[var]],-I*Sin[u[var]]},{I*Sin[u[var]],
 -Cos[u[var]]}}
eqs2=Table[compCond[a,b2][[i,j]]==b0[[i,j]],{j,1,2},{i,1,2}]//
 Flatten//Expand
trrq={q[var]->1/2*D[u[var],x],r[var]->1/2*D[u[var],x]}
sineG=Thread[eqs2[[2]]*2/I,Equal]/.trrq/.D[trrq[[2]],t]//Expand
```

```
{b3=1/(4*I*lambda)*{{Cosh[u[var]],-I*Sinh[u[var]]},
 {-I*Sinh[u[var]],-Cosh[u[var]]}}, eqs3=Table[compCond[
 a,b3][[i,j]]==b0[[i,j]],{j,1,2},{i,1,2}]//Flatten//Expand}
sinhG=Thread[eqs3[[3]]*2/I,Equal]/.trrq/.trD[trrq,t]/.
 trD[trrq,x]//Expand
```

□

Problem 4.13 *KdV and mKdV equations. Lax pairs.* Let A_i and B_i $(i{=}1,2)$ be 2×2 matrices of the form:

$$A_1=\begin{pmatrix} i\lambda & 1 \\ u & -i\lambda \end{pmatrix},\quad A_2=\begin{pmatrix} -i\lambda & u \\ u & i\lambda \end{pmatrix},$$

$$B_1=\begin{pmatrix} 4i\lambda^3+2i\lambda u-u_x & 4\lambda^2+2u \\ 4\lambda^2u+2i\lambda u_x+2u^2-u_{xx} & -4i\lambda^3-2i\lambda u+u_x \end{pmatrix},$$

$$B_2=\begin{pmatrix} -4i\lambda^3-2i\lambda u^2 & 4\lambda^2u+2i\lambda u_x-u_{xx}+2u^3 \\ 4\lambda^2u-2i\lambda u_x-u_{xx}+2u^3 & 4i\lambda^3+2i\lambda u^2 \end{pmatrix}.$$

Verify that the compatibility condition with respect to the above defined matrices A_i and B_i $(i{=}1,2)$ turns, respectively, into the KdV equation and the modified KdV equation $u_t-6uu_x+u_{xxx}=0$, $u_t-6u^2u_x+u_{xxx}=0$*.

Maple:

```
with(PDEtools): with(LinearAlgebra); declare((u,r,q)(x,t));
A0:=<<1,0>|<0,-1>>; A11:=<<0,u(x,t)>|<1,0>>; B0:=<<0,0>|<0,0>>;
A1:=I*lambda*A0+A11; B1:=4*I*lambda^3*A0+4*lambda^2*A11
 +2*I*lambda*<<u(x,t),diff(u(x,t),x)>|<0,-u(x,t)>>
 +<<-diff(u(x,t),x),2*u(x,t)^2-diff(u(x,t),x$2)>|<2*u(x,t),
 diff(u(x,t),x)>>; CompCond:=(L,M)->simplify(map(diff,L,t)
 -map(diff,M,x)+(L.M-M.L));
Eqs1:=[seq(seq(CompCond(A1,B1)[m,n]=B0[m,n],n=1..2),m=1..2)];
KdV:=Eqs1[3]; A2:=<<-I*lambda,u(x,t)>|<u(x,t),I*lambda>>;
B2:=<<-4*I*lambda^3-2*I*lambda*u(x,t)^2,4*lambda^2*u(x,t)
 -2*I*lambda*diff(u(x,t),x)-diff(u(x,t),x$2)+2*u(x,t)^3>|
 <4*lambda^2*u(x,t)+2*I*lambda*diff(u(x,t),x)-diff(u(x,t),x$2)
 +2*u(x,t)^3,4*I*lambda^3+2*I*lambda*u(x,t)^2>>;
Eqs2:=[seq(seq(CompCond(A2,B2)[m,n]=B0[m,n],n=1..2),m=1..2)];
mKdV:=Eqs2[2];
```

*The Lax pair for the mKdV equation was first introduced by Ablowitz, Kaup, Newell, and Segur (AKNS) [3].

Mathematica:

```
trD[u_,var_]:=Table[D[u,{var,i}],{i,1,2}]//Flatten;
{var=Sequence[x,t], a0={{1,0},{0,-1}}, a11={{0,1},{u[var],0}},
 b0={{0,0},{0,0}}, a1=I*lambda*a0+a11, b1=4*I*lambda^3*a0+
 4*lambda^2*a11+2*I*lambda*{{u[var],0},{D[u[var],x],-u[var]}}+
 {{-D[u[var],x],2*u[var]},
 {2*u[var]^2-D[u[var],{x,2}],D[u[var],x]}}}
 Map[MatrixForm,{a0,a11,b0,a1,b1}]
compCond[l_,m_]:=(D[l,t]-D[m,x]+(l.m-m.l))//Simplify;
{eqs1=Table[compCond[a1,b1][[m,n]]==b0[[m,n]],{m,1,2},
 {n,1,2}]//Flatten, kdV=eqs1[[3]]}
{a2={{-I*lambda,u[var]},{u[var],I*lambda}}, b2={{-4*I*lambda^3-
 2*I*lambda*u[var]^2,4*lambda^2*u[var]+2*I*lambda*D[u[var],x]-
 D[u[var],{x,2}]+2*u[var]^3},{4*lambda^2*u[var]-
 2*I*lambda*D[u[var],x]-D[u[var],{x,2}]+2*u[var]^3,
 4*I*lambda^3+2*I*lambda*u[var]^2}}}
Map[MatrixForm,{a2,b2}]
{eqs2=Table[compCond[a2,b2][[m,n]]==b0[[m,n]],{m,1,2},
 {n,1,2}]//Flatten, mKdV=eqs2[[2]]}
```

□

4.2.5 Variational Principle

It is known that many physical systems can be described by their extremum properties of a certain physical quantity that represents as a functional in a given domain. This description is called a *variational principle* leading to the Euler–Lagrange equation (which optimizes the corresponding functional).

As we mentioned before (see Sect. 4.2), there are several criteria for complete integrability of evolution equations, one of them is the existence of an infinite number of conservation laws. For constructing conservation laws some methods require the use or existence of a variational principle. This requirement is based on the theorem, proved by E. Noether in 1918, which establish that if a variational principle can be determined for a symmetry of a PDE, a corresponding conservation law can be found.

Many nonlinear PDEs arising in applied sciences and engineering can be derived from the Euler–Lagrange variational principle, the Hamilton principle, another appropriate variational principles.

Problem 4.14 *Derivation of nonlinear PDEs. Klein–Gordon equation.* Let us derive the nonlinear Klein–Gordon equation

$$u_{tt}-u_{xx}+F'(u)=0,$$

where $\{x \in \mathbb{R}, t \geq 0\}$. Physically, $F'(u)$ is the derivative of the potential energy $F(u)$. Applying the variational principle $\delta\int\int L\,dx\,dt{=}0$, where the Lagrangian has the form $L{=}\frac{1}{2}(u_t^2{-}u_x^2){-}F(u)$, we verify that the variational principle gives the *Euler–Lagrange equation* $L_u{-}(L_{u_x})_x{-}(L_{u_t})_t{=}0$ that can be simplified to obtain the nonlinear Klein–Gordon equation.

Maple:

```
with(PDEtools): declare(u(x,t)); U:=diff_table(u(x,t));
PDE1:=U[t,t]-U[x,x]+diff(F(u),u); tr1:={U[x]=p,U[t]=q};
tr2:={p=U[x],q=U[t]}; Lag1:=1/2*(U[t]^2-U[x]^2)-F(U[]);
L:=subs(tr1,Lag1); EulerLagEq:=Diff(L,U[])-diff(subs(tr2,
   diff(L,p)),x)-diff(subs(tr2,diff(L,q)),t)=0;
value(subs(u(x,t)=u,EulerLagEq));
```

Mathematica:

```
pde1=D[u[x,t],{t,2}]-D[u[x,t],{x,2}]+D[f[u[x,t]],u]
tr1={D[u[x,t],x]->p,D[u[x,t],t]->q}
tr2={p->D[u[x,t],x],q->D[u[x,t],t]}
lag1=1/2*(D[u[x,t],t]^2-D[u[x,t],x]^2)-f[u[x,t]]
{l=lag1/.tr1, eulerLagEq=D[l,u[x,t]]-D[(D[l,p]/.tr2),x]-
 D[(D[l,q]/.tr2),t]==0}
```

□

4.3 Nonlinear Systems. Integrability Conditions

In this section, we will consider various analytical methods for solving nonlinear systems of partial differential equations. In particular, we will obtain exact solutions of overdetermined nonlinear systems and Pfaffian equations [124].

First, let us consider the overdetermined nonlinear system of the form

$$z_x{=}F(x,y,z), \quad z_y{=}G(x,y,z)$$

with respect to the unknown function $z{=}z(x,y)$. The consistency condition, $F_y{+}GF_z{=}G_x{+}FG_z$, can be derived by differentiating the two equations for obtaining z_{xy} and z_{yx}, eliminating the first derivatives z_x and z_y using the original equations, and considering the equation $z_{xy}{=}z_{yx}$.

Problem 4.15 *Nonlinear overdetermined system. Consistency condition.* For the nonlinear overdetermined system,

$$z_x{=}yz, \quad z_y{=}z^2{+}axz,$$

where $z=z(x,y)$ is the unknown function and a is a real constant, verify that the consistency condition is not satisfied identically and has the form $z(yz-1+a)=0$. Therefore, the two possible solutions for z are: $z=0$ and $z=(1-a)/y$. Verify that $z=0$ is the solution of the given system and $z=(1-a)/y$ is not.

Maple:

```
with(PDEtools): declare(z(x,y),(F,G)(x,y,z)); Uz,UF,UG:=
 diff_table(z(x,y)),diff_table(F(x,y,z)),diff_table(G(x,y,z));
F:=(x,y,z)->y*z; G:=(x,y,z)->z^2+a*x*z;
Sys1:=[Uz[x]=y*z(x,y),Uz[y]=z(x,y)^2+a*x*z(x,y)];
ConsCond:=factor(UF[y]+G(x,y,z)*UF[z]-UG[x]-F(x,y,z)*UG[z]=0);
Solz:=[solve(ConsCond,z)];
for i from 1 to 2 do simplify(algsubs(z(x,y)=Solz[i],Sys1)); od;
```

Mathematica:

```
trD[u_,var_]:=Table[D[u,{var,i}],{i,1,2}]//Flatten;
var=Sequence[x,y,z]; fF[x_,y_,z_]:=y*z; fG[x_,y_,z_]:=z^2+a*x*z;
sys1={D[z[x,y],x]==y*z[x,y],D[z[x,y],y]==z[x,y]^2+a*x*z[x,y]}
{consCond=D[fF[var],y]+fG[var]*D[fF[var],z]-D[fG[var],x]-fF[var]*
 D[fG[var],z]==0//Factor, solz=Solve[consCond,z]//Flatten}
n=Length[solz]; Do[tr[i]=z[x,y]->solz[[i,2]]; test[i]=(sys1/.
 tr[i]/.trD[tr[i],x]/.trD[tr[i],y])//Factor; Print[tr[i]];
 Print[test[i]],{i,1,2}];
```

□

Problem 4.16 *Nonlinear overdetermined system. Consistency condition.* For the nonlinear overdetermined system,

$$z_x=ae^{y-z}, \quad z_y=be^{y-z}+1,$$

where $z=z(x,y)$ is the unknown function and a, b are real constants, verify that the unique exact solution reads: $z(x,y)=y-\ln\Big(\dfrac{1}{ax+by+aC_1}\Big)$.

First, we verify that the consistency condition $F_y+GF_z=G_x+FG_z$ is satisfied identically (`ConsCond`) and therefore the system has a unique solution $z=z(x,y)$ (if the derivatives in the consistency condition are continuous). Solving the first equation of the system, we obtain the solution $z(x,y)=y-\ln\Big(\dfrac{1}{a[x+F_1(y)]}\Big)$ (`Sol1`), where $F_1(y)$ is an arbitrary function. Substituting this solution into the second equation of the system, we obtain the linear first-order ODE $F_{1y}/(x+F_1)=b/(a(x+F_1))$

(Eq1) and it's solution $F_1(y)=by/a+C_1$ (SolF1), where C_1 is an arbitrary constant. Substituting this expression into the solution $z(x,y)$ obtained above, we obtain the solution of the nonlinear overdetermined system (SolFin).

Maple:

```
with(PDEtools): declare(z(x,y),(F,G)(x,y,z)); Uz,UF,UG:=
 diff_table(z(x,y)),diff_table(F(x,y,z)),diff_table(G(x,y,z));
F:=(x,y,z)->a*exp(y-z); G:=(x,y,z)->b*exp(y-z)+1;
Sys1:=[Uz[x]=a*exp(y-z(x,y)), Uz[y]=b*exp(y-z(x,y))+1];
ConsCond:=factor(UF[y]+G(x,y,z)*UF[z]-UG[x]-F(x,y,z)*UG[z]=0);
Sol1:=simplify(pdsolve(Sys1[1]));
Eq1:=expand(algsubs(Sol1,Sys1[2])); SolF1:=dsolve(Eq1,_F1(y));
SolFin:=simplify(subs(SolF1,Sol1)); pdetest(SolFin,Sys1);
```

Mathematica:

```
Off[Solve::ifun]; trD[u_,var_]:=Table[D[u,{var,i}],{i,1,2}]//
 Flatten; var=Sequence[x,y,z]; fF[x_,y_,z_]:=a*Exp[y-z];
fG[x_,y_,z_]:=b*Exp[y-z]+1; {sys1={D[z[x,y],x]==a*Exp[y-z[x,y]],
 D[z[x,y],y]==b*Exp[y-z[x,y]]+1}, consCond=D[fF[var],y]+fG[var]*
 D[fF[var],z]-D[fG[var],x]-fF[var]*D[fG[var],z]==0//Factor}
{sol1=DSolve[sys1[[1]],z[x,y],{x,y}]//First, eq1=sys1[[2]]/.
 sol1/.trD[sol1,y]//Expand, solC1=DSolve[eq1,C[1][y],y]//First}
solFin=sol1/.solC1//FullSimplify
sys1/.solFin/.trD[solFin,x]/.trD[solFin,y]//FullSimplify
```
□

Now let us consider the *Pfaffian equation*, proposed by J. F. Pfaff in 1814, i.e., equation of the form $\omega=\alpha_1(x)\,dx_1+\cdots+\alpha_n(x)\,dx_n=0$ $(n\geq 3)$, where ω is a differential 1-form, $x\in\mathbb{R}^n$, and $\alpha_i(x)\in C^1(\mathbb{R}^n)$ are real-valued functions. A necessary and sufficient condition for the Pfaffian equation to be completely integrable,*

$$d\omega\wedge\omega=0, \tag{4.8}$$

was first obtained by Euler in 1755. We consider a particular case, three-dimensional Euclidean space $\mathbb{R}^3$. In this case, the Pfaffian equation has the form

$$P(x,y,z)dx+Q(x,y,z)dy+R(x,y,z)dz=0,$$

and always has simple solutions of the form $x=x_0$, $y=y_0$, $z=z_0$.

*$d\omega$ is the differential form of degree 2 obtained from ω by exterior differentiation, and $\wedge$ is the exterior product.

We will consider a special case and obtain the solutions of the form $z=z(x,y)$, when the variables x, y, and z are connected by the *integrability condition*. Taking into account the relations $dz=-P/R\,dx-Q/R\,dy$ and $dz=z_x\,dx+z_y\,dy$, we can show that the Pfaffian equation can be reduced to the overdetermined system $z_x=-P/R$, $z_y=-Q/R$ (considered in the previous two problems). Setting $F=-P/R$ and $G=-Q/R$, we can derive the integrability condition (4.8) for $\mathbb{R}^3$:

$$R(P_y-Q_x)+P(Q_z-R_y)+Q(R_x-P_z)=0.$$

If this condition is satisfied identically, the Pfaffian equation is completely integrable, i.e., its solution can be represented in the form of an integral $U(x,y,z)=C$, where C is an arbitrary constant.

Problem 4.17 *Pfaffian equation. Integrability condition.* For the Pfaffian equation of the form

$$y(xz+a)\,dx+x(y+b)\,dy+x^2y\,dz=0,$$

verify that the *integrability condition* is satisfied identically and therefore the Pfaffian equation is completely integrable, i.e., the integral representation of the solution has the form $xz+y+b\ln(y)+a\ln(x)=C_1$, where C_1 is an arbitrary constant.

1. Rewriting the Pfaffian equation in the equivalent overdetermined system $z_x=-(xz+a)/x^2$, $z_y=-(y+b)/(xy)$, we verify that the consistency condition is satisfied identically and apply the method considered in the previous problem for solving the nonlinear overdetermined system to obtain the solution $z(x,y)=-(y+b\ln(y)+a\ln(x)-C_1)/x$ (`solFin`).

2. Assuming that x=const, i.e., $dx=0$ and solving the resulting ODE for $z=z(y)$, we obtain the solution $z(y)=-y/x-b\ln(y)/x+F_1(x)$ (`Sol11`), where $F_1(x)$ is an arbitrary function. Substituting this solution into the original Pfaffian equation, we find the linear ODE $F_1x+a+F_{1x}x^2=0$ (`EqF1`) and it's general solution $F_1(x)=(-a\ln(x)+C_1)/x$ (`SolF1`), where C_1 is an arbitrary constant. Substituting $F_1(x)$ into the above obtained solution $z(x,y)$, we have $z(x,y)=-(y+b\ln(y)+a\ln(x)-C_1)/x$ (`SolFin` coincides to `solFin`) and the equivalent integral representation of the solution $xz+y+b\ln(y)+a\ln(x)=C_1$ (`trC1`).

Maple:

```
with(PDEtools): declare(z(x,y),phi(x),(P,Q,R,F,G)(x,y,z));
UP,UQ,UR,Uz,UF:=diff_table(P(x,y,z)),diff_table(Q(x,y,z)),
 diff_table(R(x,y,z)),diff_table(z(x,y)),diff_table(F(x,y,z));
```

```
UG:=diff_table(G(x,y,z)); P:=(x,y,z)->y*(x*z+a);
Q:=(x,y,z)->x*(y+b); R:=(x,y,z)->x^2*y; Eq1:=P(x,y,z)*dx
 +Q(x,y,z)*dy+R(x,y,z)*dz=0; IntCond:=factor(R(x,y,z)*(UP[y]
 -UQ[x])+P(x,y,z)*(UQ[z]-UR[y])+Q(x,y,z)*(UR[x]-UP[z])=0);
F:=(x,y,z)->-P(x,y,z)/R(x,y,z); G:=(x,y,z)->-Q(x,y,z)/R(x,y,z);
Sys1:=[Uz[x]=F(x,y,z),Uz[y]=G(x,y,z)];
ConsCond:=factor(UF[y]+G(x,y,z)*UF[z]-UG[x]-F(x,y,z)*UG[z]=0);
sol1:=simplify(pdsolve(Sys1[1])); eq1:=expand(algsubs(sol1,
 Sys1[2])); solF1:=dsolve(eq1,_F1(y)); solFin:=simplify(subs(
 solF1,sol1)); ODE1:=expand(subs(dx=0,Eq1)/x/dy);
ODE11:=eval(ODE1,{dz=dy*diff(z(x,y),y)});
Sol1:=dsolve(ODE11,z(x,y)); Sol11:=subs(z(x,y)=z,Sol1);
Sol12:=combine(algsubs(Sol11,Eq1)); tr1:={dz=diff(rhs(Sol11),x)
 *dx+diff(rhs(Sol11),y)*dy}; EqF1:=simplify(subs(tr1,Sol12));
SolF1:=dsolve(select(has,lhs(EqF1),_F1),_F1(x));
SolFin:=factor(eval(Sol1,SolF1)); trC1:=isolate(SolFin,_C1);
simplify(subs(trC1,simplify(algsubs(solFin,Sys1))));
```

Mathematica:

```
var=Sequence[x,y,z]; trS1[eq_,var_]:=Select[eq,MemberQ[#,var,
 Infinity]&]; trD[u_,var_]:=Table[D[u,{var,i}],{i,1,2}]//Flatten;
fP[x_,y_,z_]:=y*(x*z[x,y]+a); fQ[x_,y_,z_]:=x*(y+b);
fR[x_,y_,z_]:=x^2*y; eqP=fP[var]*dx+fQ[var]*dy+fR[var]*dz==0
fF[x_,y_,z_]:=-fP[var]/fR[var]; fG[x_,y_,z_]:=-fQ[var]/fR[var];
sys1={D[z[x,y],x]==fF[var], D[z[x,y],y]==fG[var]}
intCond=fR[var]*(D[fP[var],y]-D[fQ[var],x])+fP[var]*
 (D[fQ[var],z]-D[fR[var],y])+fQ[var]*(D[fR[var],x]-
 D[fP[var],z])==0//Factor
consCond=D[fF[var],y]+fG[var]*D[fF[var],z]-D[fG[var],x]-
 fF[var]*D[fG[var],z]==0//Factor
{sol1=DSolve[sys1[[1]],z[x,y],{x,y}]//First, eq1=sys1[[2]]/.
 sol1/.trD[sol1,y], solC1=DSolve[eq1,C[1][y],y]//First}
solFin=sol1/.solC1/.C[2]->C[1]//Factor
{ode1=Thread[(eqP/.dx->0)/x/dy,Equal]//Expand, ode11=ode1/.
 {dz->dy*D[z[x,y],y]}, sol10=DSolve[ode11,z[x,y],{x,y}]//First}
{trz=z[x,y]->z, sol11=sol10/.trz, sol12=eqP/.trz/.sol11//
 Simplify, tr1={dz->D[sol11[[1,2]],x]*dx+D[sol11[[1,2]],y]*dy}}
{eqC1=sol12/.tr1//Simplify, solC11=DSolve[trS1[eqC1[[1]],
 C[1][x]]==0,C[1][x],x]//First, solFin=sol11/.solC11/.C[2]->
 C[1]//Simplify, solFin1=solFin/.z->z[x, y]}
trC1=Solve[solFin/.Rule->Equal,C[1]]
sys1/.solFin1/.trD[solFin1,x]/.trD[solFin1,y]/.trC1//Simplify
```
□

Chapter 5
Approximate Analytical Approach

In this chapter, we will follow the approximate analytical approach for solving nonlinear PDEs. We consider the most important recently developed methods and traditional methods to find approximate analytical solutions of nonlinear PDEs and nonlinear systems. We will apply the Adomian decomposition method (ADM) and perturbation methods to solve nonlinear PDEs (e.g., the Burgers equation, the Klein–Gordon equation, the Fisher equation, etc.) and nonlinear systems.

In general, applying the ADM method (and it's improvements), we construct an approximate analytical solution as an infinite series solution, which may converge to an exact solution (if its exists). We consider exact solutions and truncated series solutions for numerical calculations, visualizations, and comparisons. As before, we perform comparative study between these methods and other approaches described in the book for the corresponding nonlinear equations and systems.

5.1 Adomian Decomposition Method

It is known that the *Adomian decomposition method* is a powerful and effective method for solving a wide class of linear and nonlinear differential (ODEs and PDEs) and integral equations. The Adomian decomposition method was first proposed by G. Adomian in 1986 (see [6], [7]). In scientific literature there is a series of research works devoted to the application of this method to a wide class of linear and nonlinear differential and integral equations (see e.g., [156]–[158], etc.).

For the nonlinear PDEs, the idea of the Adomian decomposition method consists in decomposition of the unknown function $u(x,t)$ of an equation into an infinite series

$$u(x,t)=\sum_{i=0}^{\infty} u_i(x,t), \tag{5.1}$$

where the components $u_i(x,t)$ are determined recursively. The nonlinear term $F(u)$ (e.g., uu_x, u^p, $\sin u$, e^u, $\ln u$) is represented as an infinite series of the *Adomian polynomials* A_i:

$$F(u)=\sum_{i=0}^{\infty} A_i(u_0,u_1,\ldots), \tag{5.2}$$

where the Adomian polynomials A_i can be evaluated for all types of nonlinearities. We will show this in the next section.

5.1.1 Adomian Polynomials

There exist several schemes for calculating Adomian polynomials. We consider the following scheme, introduced and justified by Adomian, for calculation of Adomian polynomials for the nonlinear term $F(u)$:

$$A_i=\frac{1}{i!}\left[\frac{d^i}{d\lambda^i}F\left(\sum_{k=0}^{\infty}\lambda^k u_k\right)\right]_{\lambda=0}, \qquad i=0,1,2,\ldots.$$

Alternative methods, based on algebraic and trigonometric identities and on Taylor series, have been developed in [158].

Problem 5.1 *Calculation of Adomian polynomials for various types of nonlinearities.* Derive Adomian polynomials for a general form of nonlinear term $F(u)$. Compute Adomian polynomials for various types of nonlinearities: u^2, u^3, u_x^2, u_x^3, $uu_x=\frac{1}{2}(u^2)_x$, $\sin u$, $\cos u$, $\sinh u$, $\cosh u$, e^u, $\ln u$.

Maple:

```
with(PDETools): declare(U(x)): N:=9; ADM1:=i->convert(subs(
 lambda=0,value(1/i!*Diff(F(Sum(lambda^k*u[k],k=0..i)),
 lambda$i))),diff);
A0:=F(u[0]);  for i from 1 to N do A||i:=ADM1(i); od;
for i from 0 to N do expand(subs(F(u[0])=u[0]^2,A||i)); od;
for i from 0 to N do expand(subs(F(u[0])=u[0]^3,A||i)); od;
for i from 0 to N do subs({seq(u[k]=u[k][x],k=0..N)},
                     expand(subs(F(u[0])=u[0]^2,A||i))); od;
for i from 0 to N do subs({seq(u[k]=u[k][x],k=0..N)},
                     expand(subs(F(u[0])=u[0]^3,A||i))); od;
for i from 0 to N do
 convert(1/2*diff(subs({seq(u[k]=U[k](x),k=0..N)},
 expand(subs(F(u[0])=u[0]^2,A||i))),x),diff); od;
```

```
for i from 0 to N do expand(subs(F(u[0])=sin(u[0]),A||i)); od;
for i from 0 to N do expand(subs(F(u[0])=cos(u[0]),A||i)); od;
for i from 0 to N do expand(subs(F(u[0])=sinh(u[0]),A||i)); od;
for i from 0 to N do expand(subs(F(u[0])=cosh(u[0]),A||i)); od;
for i from 0 to N do expand(subs(F(u[0])=exp(u[0]),A||i)); od;
for i from 0 to N do expand(subs(F(u[0])=ln(u[0]),A||i)); od;
```

Mathematica:

```
n=9; trD[u_,var_]:=Table[D[u,{var,i}],{i,1,n}]//Flatten;
fADM1[i_]:=1/i!*D[f[Sum[lambda^k*u[k],{k,0,i}]],{lambda,i}]/.
 lambda->0//Expand; a[0]=f[u[0]]
Do[a[i]=fADM1[i]; Print["a[",i,"]=",a[i]],{i,1,n}];
{trp1=f[u[0]]->u[0]^2, trp2=f[u[0]]->u[0]^3, trp3=Table[u[k]->
 D[u[x][k],x],{k,0,n}],trp5=Table[u[k]->fU[x][k],{k,0,n}]}
trp[g_]:=f[u[0]]->g[u[0]];
Do[p1[i]=a[i]/.trp1/.trD[trp1,u[0]];Print["p1[",i,"]=",p1[i]],
 {i,0,n}]; Do[p2[i]=a[i]/.trp2/.trD[trp2,u[0]];
 Print["p2[",i,"]=",p2[i]],{i,0,n}];
Do[p3[i]=(a[i]/.trp1/.trD[trp1,u[0]])/.trp3;
 Print["p3[",i,"]=",p3[i]],{i,0,n}]; Do[p4[i]=(a[i]/.trp2/.
 trD[trp2,u[0]])/.trp3;Print["p4[",i,"]=",p4[i]],{i,0,n}];
Do[p5[i]=1/2*D[(a[i]/.trp1/.trD[trp1,u[0]])/.trp5,x]//Expand;
 Print["p5[",i,"]=",p5[i]],{i,0,n}];
pFun[p_,fun_]:=Do[p[i]=a[i]/.trp[fun]/.trD[trp[fun],u[0]];
 Print[ToString[p],"[",i,"]=",p[i]],{i,0,n}];
pFun[p6,Sin]; pFun[p7,Cos]; pFun[p8,Sinh]; pFun[p9,Cosh];
pFun[p10,Exp]; pFun[p11,Log];
```

□

5.1.2 Nonlinear PDEs

It was shown [29] that if an exact solution of a given nonlinear PDE exists, then the series solution obtained converges rapidly to the exact solution. If we cannot obtain a closed form solution, we can consider a truncated series solution for numerical approximations, visualizations, and comparisons. We will show that if we evaluate few terms in the obtained series, we end up with an approximation of high degree of accuracy compared with numerical methods. Other comparisons with traditional methods (e.g., finite difference methods) have been presented in scientific literature. If we compare the ADM and the perturbation methods, we can observe the efficiency of the ADM compared to the tedious computations required by the perturbation method. In this section, applying the ADM, we will solve various nonlinear equations subject to initial and/or boundary conditions (e.g., the inviscid Burgers equation, Klein–Gordon equation, and Burgers equation).

Let us consider the nonlinear PDE (1.3) in two independent variables x, t and the initial condition $u(x,0) = g(x)$. Following the Adomian decomposition method, this equation can be rewritten in the operator form

$$\mathcal{D}_x u + \mathcal{D}_t u + \mathcal{L}(u) + F(u) = f(x,t), \tag{5.3}$$

where $\mathcal{D}_x$ and $\mathcal{D}_t$ are the highest-order differentials in x and t, $\mathcal{L}$ are linear terms of lower derivatives, $F(u)$ is a nonlinear term, and $f(x,t)$ is an inhomogeneous term. It is known that the solutions for $u(x,t)$ obtained with respect to the operators $\mathcal{D}_x u$ and $\mathcal{D}_t u$ are equivalent and converge to the exact solution [157]. For example, solving the nonlinear equation in the t-direction, we have $\mathcal{D}_t u = f(x,t) - \mathcal{D}_x u - \mathcal{L}(u) - F(u)$. Applying the inverse operator $\mathcal{D}_t^{-1}$ to both sides of this equation and using the initial condition, we obtain the solution

$$u(x,t) = \Phi - \mathcal{D}_t^{-1} f(x,t) - \mathcal{D}_t^{-1}\mathcal{D}_x u - \mathcal{D}_t^{-1}\mathcal{L}(u) - \mathcal{D}_t^{-1}F(u), \tag{5.4}$$

where the function $\Phi = u(x,0) + \sum_{i=1}^{M=n-1} \frac{1}{i!} t^i \frac{\partial^M u}{\partial t^M}(x,0)$ for $\mathcal{L}_t = \frac{\partial^n}{\partial t^n}$. Then, we represent the solution $u(x,t)$ and the nonlinear term $F(u)$ in the series forms (5.1) and (5.2), where the Adomian polynomials can be generated for all forms of nonlinearity. As a result, the components $u_i(x,t)$ $(i \geq 0)$ of the solution $u(x,t)$ can be recursively determined and hence we can obtain the solution in the series form.

Problem 5.2 *Inviscid Burgers equation. Cauchy problem.* Let us consider the Cauchy problem for the inviscid Burgers equation

$$u_t + u u_x = 0; \quad u(x,0) = x+1,$$

where $\{x \in \mathbb{R}, t \geq 0\}$. Applying the Adomian decomposition method, show that the solution of this Cauchy problem takes the following form $u(x,t) = (x+1)/(t+1)$ (`SolF`) and verify that this solution is exact solution of this problem (`Test1`). Compare this solution with the solution obtained by applying the method of characteristics (see **Problem 3.3**).

First, we rewrite the inviscid Burgers equation in the operator form $\mathcal{D}_t u = -u u_x$ (`PDE1`), where $\mathcal{D}_t = \partial/\partial t$. The inverse operator $\mathcal{D}_t^{-1}$ has the form $\mathcal{D}_t^{-1}(\cdot) = \int_0^t (\cdot)\, dt$ (`LI`). Applying the inverse operator $\mathcal{D}_t^{-1}$ to both

sides of the equation (PDE1) and using the initial condition, we obtain $u(x,t)-x-1=\mathcal{D}_t^{-1}uu_x$ (Eq2). Then, substituting the series forms of the solution (5.1) (trL) and the nonlinear term (5.2) (in which the Adomian polynomials (A[n]) are generated for the nonlinearity uu_x) (trN), into the equation Eq2, we have $\Big(\sum_{i=0}^{\infty} u_i(x,t)\Big)-x-1=\mathcal{D}_t^{-1}\Big(\sum_{i=0}^{\infty} A_i\Big)$ (Eq3).
As a result, we obtain the recursive relation (Apr[0], AprK):

$$u_0(x,t)=x+1, \quad u_{k+1}(x,t)=-\mathcal{D}_t^{-1}(A_k).$$

For example, the first two components are: $u_0=x+1$ and $u_1=-(x+1)t$ (Apr[0], Apr[1]). Therefore, the approximate analytical solution has the form $u=(x+1)(1-t+t^2-t^3+\dots)$ (Sol1), which can be written in a closed form $u(x,t)=(x+1)/(t+1)$ (SolF).

Maple:

```
with(PDEtools): declare((u,W)(x,t)); KN:=9;
ADM1:=n->convert(subs(lambda=0,value(1/n!*Diff(F(Sum(
 lambda^i*U[i],i=0..n)),lambda$n))),diff); A0[0]:=F(U[0]);
for n from 1 to KN do A0[n]:=ADM1(n); od:
for n from 0 to KN do
 A[n]:=convert(1/2*diff(subs({seq(U[i]=W[i](x,t),i=0..KN)},
       expand(subs(F(U[0])=U[0]^2,A0[n]))),x),diff); od;
L:=w->diff(w(x,t),t); NL:=w->w*diff(w(x,t),x);  L(u); NL(u);
PDE1:=w->L(w)=-NL(w); PDE1(u); IC1:=u(x,0)=x+1;
LI:=w->Int(w(x,t),t=0..t); LI(u); tr1:=u-rhs(IC1);
Eq1:=LI(lhs(PDE1(u)))=LI(rhs(PDE1(u)));
Eq2:=simplify(subs(lhs(Eq1)=tr1,Eq1));
trL:=u=add(u[j](x,t),j=0..KN);
trN:=LI(NL(u))=Int(Sum(A[i],i=0..KN),t=0..t);
Eq3:=subs(trL,lhs(Eq2))=subs(trN,rhs(Eq2));
Apr[0]:=u[0](x,t)=rhs(IC1);
AprK:=u[k+1](x,t)=-Int(AD[k],t=0..t);
for i from 0 to KN do
 Apr[i+1]:=value(subs({seq(Apr[m],m=0..i)},subs(
 {seq(W[m]=u[m],m=0..i)},subs(k=i,AD[i]=A[i],AprK)))); od;
trSol:={seq(Apr[i],i=0..KN)}; Sol:=value(subs(trSol,trL));
Sol1:=collect(combine(Sol),t); factor(Sol1);
SolF:=(x+1)*sum((-1)^j*t^j,j=0..infinity);
Test1:=subs(u=SolF,(algsubs(u(x,t)=SolF,PDE1(u))));
```

Mathematica:

```
kN=9; var=Sequence[x,t]; trD[u_,var_]:=Table[D[u,{var,i}],
 {i,1,n}]//Flatten; fADM1[n_]:=1/n!*D[f[Sum[lambda^i*fU[i],
 {i,0,n}]],{lambda,n}]/.lambda->0//Expand; a0[0]=f[fU[0]]
{trp1=f[fU[0]]->fU[0]^2, trp2=Table[fU[i]->fW[var][i],{i,0,kN}]}
Do[a0[n]=fADM1[n];Print["a0[",n,"]=",a0[n]],{n,1,kN}];
Do[a[n]=1/2*D[(a0[n]/.trp1/.trD[trp1,fU[0]])/.trp2,x]//Expand;
 Print["a[",n,"]=",a[n]],{n,0,kN}]; fL[w_]:=D[w[var],t];
fNL[w_]:=w[var]*D[w[var],x]; {fL[u], fNL[u]}
pdeT[w_]:=fL[w]==fNL[w]; pde1[w_]:=fL[w]==-fNL[w];
{pde1[u], ic1=u[x,0]->x+1}
fLI[w_]:=Integrate[w,{t,0,t}]; {fLI[u[var]], tr1=u[var]-ic1[[2]]}
{eq1=fLI[pdeT[u][[1]]]==-fLI[pdeT[u][[2]]],
 eq2=eq1/.eq1[[1]]->tr1, trL=u[var]->Sum[u[var][j],{j,0,kN}],
 trN=fLI[fNL[u]]->Hold[Integrate[Sum[a[i],{i,0,kN}],{t,0,t}]]}
{eq3=(eq2[[1]]/.trL)==(eq2[[2]]/.trN),apr[0]=u[var][0]->ic1[[2]]}
aprK=u[var][k+1]->-Hold[Integrate[ad[k],{t,0,t}]]
Do[apr[i+1]=ReleaseHold[aprK/.k->i/.ad[i]->a[i]/.
 Table[fW[var][m]->u[x,t][m],{m,0,i}]/.Table[D[fW[var][m],x]->
 D[u[x,t][m],x],{m,0,i}]/.Table[apr[m],{m,0,i}]/.
 Table[D[apr[m],x],{m,0,i}]]; Print["apr[",i+1,"]=",apr[i+1]],
{i,0,kN}]; {trSol=Table[apr[i],{i,0,kN}], sol=trL/.trSol,
 sol1=Collect[sol//Together,t], sol1//Factor}
{solF=(x+1)*Sum[(-1)^j*t^j,{j,0,Infinity}], trp3=u[x,t]->solF}
pde1[u]/.trp3/.D[trp3,x]/.D[trp3,t]
```

□

Problem 5.3 *Inviscid Burgers equation. Cauchy problem.* Let us consider the inviscid Burgers equation (as in Problem 5.2) with another initial condition:

$$u_t + u u_x = 0; \quad u(x,0) = \cos x,$$

where $\{x \in \mathbb{R}, t \geq 0\}$. Applying the Adomian decomposition method, prove that the approximate analytical solution of this Cauchy problem is represented in terms of a series of functions (`SolF`)

$$u(x,t) = \cos x + \tfrac{1}{2} t \sin(2x) - [\tfrac{3}{8}\cos(3x) + \tfrac{1}{8}\cos x] t^2 \\ - [\tfrac{1}{6}\sin(2x) + \tfrac{1}{3}\sin(4x)] t^3 + O(t^4),$$

which is more practical compared to the parametric form of the solution that can be obtained by the method of characteristics.

Maple:

```
with(PDEtools): declare((u,W)(x,t)); KN:=9;
ADM1:=n->convert(subs(lambda=0,value(1/n!*Diff(F(Sum(
 lambda^i*U[i],i=0..n)),lambda$n))),diff); A0[0]:=F(U[0]);
for n from 1 to KN do A0[n]:=ADM1(n); od:
for n from 0 to KN do
 A[n]:=convert(1/2*diff(subs({seq(U[i]=W[i](x,t),i=0..KN)},
        expand(subs(F(U[0])=U[0]^2,A0[n]))),x),diff); od;
L:=w->diff(w(x,t),t); NL:=w->w*diff(w(x,t),x);  L(u); NL(u);
PDE1:=w->L(w)=-NL(w); PDE1(u); IC1:=u(x,0)=cos(x);
LI:=w->Int(w(x,t),t=0..t); LI(u); tr1:=u-rhs(IC1);
Eq1:=LI(lhs(PDE1(u)))=LI(rhs(PDE1(u)));
Eq2:=simplify(subs(lhs(Eq1)=tr1,Eq1));
trL:=u=add(u[j](x,t),j=0..KN);
trN:=LI(NL(u))=Int(Sum(A[i],i=0..KN),t=0..t);
Eq3:=subs(trL,lhs(Eq2))=subs(trN,rhs(Eq2)); Apr[0]:=u[0](x,t)
 =rhs(IC1); AprK:=u[k+1](x,t)=-Int(AD[k],t=0..t);
for i from 0 to KN do
 Apr[i+1]:=value(subs({seq(Apr[m],m=0..i)},subs({seq(W[m]=u[m],
  m=0..i)},subs(k=i,AD[i]=A[i],AprK))));  od;
trSol:={seq(Apr[i],i=0..KN)}; Sol:=value(subs(trSol,trL));
SolF:=collect(combine(Sol),t);
```

Mathematica:

```
kN=9; var=Sequence[x,t];
trD[u_,var_]:=Table[D[u,{var,i}],{i,1,n}]//Flatten;
fADM1[n_]:=1/n!*D[f[Sum[lambda^i*fU[i],{i,0,n}]],{lambda,n}]/.
 lambda->0//Expand; a0[0]=f[fU[0]]
{trp1=f[fU[0]]->fU[0]^2, trp2=Table[fU[i]->fW[var][i],{i,0,kN}]}
Do[a0[n]=fADM1[n];Print["a0[",n,"]=",a0[n]],{n,1,kN}];
Do[a[n]=1/2*D[(a0[n]/.trp1/.trD[trp1,fU[0]])/.trp2,x]//Expand;
 Print["a[",n,"]=",a[n]],{n,0,kN}]; fL[w_]:=D[w[var],t];
fNL[w_]:=w[var]*D[w[var],x]; {fL[u], fNL[u]}
pdeT[w_]:=fL[w]==fNL[w]; pde1[w_]:=fL[w]==-fNL[w];
{pde1[u], ic1=u[x,0]->Cos[x]}
fLI[w_]:=Integrate[w,{t,0,t}]; {fLI[u[var]], tr1=u[var]-ic1[[2]]}
{eq1=fLI[pdeT[u][[1]]]==-fLI[pdeT[u][[2]]],
 eq2=eq1/.eq1[[1]]->tr1, trL=u[var]->Sum[u[var][j],{j,0,kN}],
 trN=fLI[fNL[u]]->Hold[Integrate[Sum[a[i],{i,0,kN}],{t,0,t}]]}
{eq3=(eq2[[1]]/.trL)==(eq2[[2]]/.trN),apr[0]=u[var][0]->ic1[[2]],
 aprK=u[var][k+1]->-Hold[Integrate[ad[k],{t,0,t}]]}
```

```
Do[apr[i+1]=ReleaseHold[aprK/.k->i/.ad[i]->a[i]/.
 Table[fW[var][m]->u[x,t][m],{m,0,i}]/.Table[D[fW[var][m],x]->
 D[u[x,t][m],x],{m,0,i}]/.Table[apr[m],{m,0,i}]/.
 Table[D[apr[m],x],{m,0,i}]]; Print["apr[",i+1,"]=",apr[i+1]],
{i,0,kN}]; {trSol=Table[apr[i],{i,0,kN}], sol=trL/.trSol}
solF=u[var]->Collect[Map[Together[#,Trig->True]&,sol[[2]]],t]
```

□

Problem 5.4 *Klein–Gordon equation. Cauchy problem.* Let us consider the Cauchy problem for the nonlinear inhomogeneous Klein–Gordon equation [157]

$$u_{tt}-u_{xx}+au+F(u)=f(x,t); \quad u(x,0)=g_1(x),\ u_t(x,0)=g_2(x),$$

where $\{x \in \mathbb{R}, t \geq 0\}$, $a=1$, $g_1(x)=0$, $g_2(x)=x$, $f(x,t)=xt+x^2t^2$, and $F(u)=u^2$. Following the ideas of the Adomian decomposition method and the improvement related to the noise-terms phenomenon, prove that the solution of this Cauchy problem has the form $u(x,t)=xt$ (SolF) and verify that this solution is exact solution of this problem (Test1).

The phenomenon of the self-canceling noise terms [155] may appear in inhomogeneous problems. According to the ideas of the ADM, the noise terms, defined as the identical terms (if they exist in the components $u_0(x,t)$ and $u_1(x,t)$), will provide the solution in a closed form with the two iterations. Noise terms may appear if the exact solution is part of the component $u_0(x,t)$. It is necessary to verify that the remaining terms satisfy the equation.

Following the ADM (as in the previous problems) and applying $\mathcal{D}_t^{-1}$ to both sides of the Klein–Gordon equation written in the operator form (PDE1) and using the initial conditions, we obtain

$$u(x,t)=\Phi+\mathcal{D}_t^{-1}(f(x,t))+\mathcal{D}_t^{-1}(u_{xx}-au)-\mathcal{D}_t^{-1}(F(u)),$$

where $\Phi=g_1(x)+tg_2(x)$ (in our case Eq2). Then, using the series for $u(x,t)$ and the Adomian polynomials for the nonlinear term $F(u)$, we obtain the recursive relation

$$u_0=g_1(x)+tg_2(x)+\mathcal{D}_t^{-1}(f(x,t)),\ u_{k+1}=\mathcal{D}_t^{-1}(u_{k\,xx}-au_k)-\mathcal{D}_t^{-1}(A_k).$$

In our case we have (Apr[0], Apr[1]):

$$u_0=\tfrac{1}{6}xt^3+\tfrac{1}{12}x^2t^4+xt,$$
$$u_1=\tfrac{1}{180}t^6-\tfrac{1}{12960}x^4t^{10}-\tfrac{1}{2592}x^3t^9-\tfrac{1}{2016}x^2t^8-\tfrac{1}{252}x^3t^7-\tfrac{1}{90}x^2t^6-\tfrac{1}{12}x^2t^4.$$

Canceling the noise term $\frac{1}{6}xt^3+\frac{1}{12}x^2t^4$ (termN) from the component u_0 and verifying that the remaining term satisfies the equation (Test1), we obtain exact solution $u(x,t)=xt$ (SolF).

Maple:

```
with(PDEtools): declare((u,W,U)(x,t)); KN:=9; ADM1:=n->
 convert(subs(lambda=0,value(1/n!*Diff(F(Sum(lambda^i*U[i],
 i=0..n)),lambda$n))),diff); A0[0]:=F(U[0]);
for n from 1 to KN do A0[n]:=ADM1(n); od:
for n from 0 to KN do A[n]:=subs({seq(U[i]=W[i](x,t),
 i=0..KN)},expand(subs(F(U[0])=U[0]^2,A0[n]))); od;
Dt:=w->diff(w(x,t),t$2); Dx:=w->-diff(w(x,t),x$2);
L:=w->w; NL:=w->w^2; f:=(x,t)->x*t+x^2*t^2; g1:=x->0; g2:=x->x;
PDE1:=w->Dt(w)+Dx(w)+L(w)+NL(w)=f(x,t); PDE1(u);
IC1:=[u(x,0)=g1(x),D[2](u)(x,0)=g2(x)];
LI:=w->Int(Int(w(x,t),t=0..t),t=0..t); LI(u);
tr1:=u-g1(x)-t*g2(x); KN:=1; Eq1:=LI(lhs(PDE1(u)))=LI(f);
Eq2:=simplify(subs(lhs(Eq1)=tr1,Eq1)); trL:=u=add(u[j](x,t),
 j=0..KN); trN:=LI(NL(u))=Int(Int(Sum(A[i],i=0..KN),t=0..t),
 t=0..t); Eq3:=value(subs(trL,lhs(Eq2))=subs(trN,rhs(Eq2)));
Eq31:=Eq3+g1(x)+t*g2(x); Apr[0]:=u[0](x,t)=rhs(Eq31);
AprK:=u[k+1](x,t)=Int(Int(diff(u[k](x,t),x$2),t=0..t),t=0..t)
 -Int(Int(AD[k],t=0..t),t=0..t); for i from 0 to KN do
  Apr[i+1]:=value(subs({seq(Apr[m],m=0..i)},subs(
  {seq(W[m]=u[m],m=0..i)},subs(k=i,AD[i]=A[i],AprK)))); od;
termN:=select(has,rhs(Apr[0]),[t^3,t^4]); Apr[0];
AApr[0]:=rhs(Apr[0])-termN; AApr[1]:=rhs(Apr[1])+termN;
U:=unapply(AApr[0],x,t); U(x,t); PDE1(U); SolF:=AApr[0];
Test1:=subs(u=SolF,(algsubs(u(x,t)=SolF,PDE1(u))));
```

Mathematica:

```
trS2[eq_,var1_,var2_]:=Select[eq,MemberQ[#,var1,Infinity]||
 MemberQ[#,var2,Infinity]&]; kN=9; var=Sequence[x,t];
trD[u_,var_]:=Table[D[u,{var,i}],{i,1,9}]//Flatten;
fADM1[n_]:=1/n!*D[f[Sum[lambda^i*fU[i],{i,0,n}]],{lambda,n}]/.
 lambda->0//Expand;  a0[0]=f[fU[0]]
{trp1=f[fU[0]]->fU[0]^2, trp2=Table[fU[i]->fW[var][i],{i,0,kN}]}
Do[a0[n]=fADM1[n]; Print["a0[",n,"]=",a0[n]],{n,1,kN}];
Do[a[n]=(a0[n]/.trp1/.trD[trp1,fU[0]])/.trp2//Expand;
  Print["a[",n,"]=",a[n]],{n,0,kN}]; fDt[w_]:=D[w[var],{t,2}];
fDx[w_]:=D[w[var],{x,2}]; fL[w_]:=w[var]; fNL[w_]:=w[var]^2;
f[x_, t_]:=x*t+x^2*t^2; g1[x_]:=0; g2[x_]:=x;
pde1[w_]:=fDt[w]+fDx[w]+fL[w]+fNL[w]==f[x,t];
{pde1[u], ic1={u[x,0]->g1[x],(D[u[x,t],t]/.t->0)->g2[x]}}
fLI[w_]:=Hold[Integrate[w,t,t]]; fLI[u[var]]
```

```
{tr1= u[var]-g1[x]-t*g2[x], kN=1,
 eq1=fLI[pde1[u][[1]]]==fLI[f[var]], eq2=eq1/.eq1[[1]]->tr1,
 trL=u[var]->Sum[u[var][j],{j,0,kN}], trN=fLI[fNL[u]]->
 Hold[Integrate[Sum[a[i],{i,0,1}],t,t]]}
eq3=(eq2[[1]]/.trL)==(ReleaseHold[(eq2[[2]]/.trN)])//Expand
{eq31=Thread[eq3+g1[x]+t*g2[x],Equal],
 apr[0]=u[var][0]->eq31[[2]], aprK=u[var][k+1]->Hold[Integrate[
 D[u[var][k],{x,2}],t,t]]-Hold[Integrate[ad[k],t,t]]}
Do[apr[i+1]=ReleaseHold[aprK/.k->i/.ad[i]->a[i]]/.
 Table[fW[var][m]->u[x,t][m],{m,0,i}]/.Table[D[fW[var][m],x]->
 D[u[x,t][m],x],{m,0,i}]/.Table[apr[m],{m,0,i}]/.Table[D[apr[m],
 {x,2}],{m,0,i}]; Print["apr[",i+1,"]=",apr[i+1]],{i,0,kN}];
{termN=trS2[apr[0][[2]],t^3,t^4], apr[0],
 aApr[0]=apr[0][[2]]-termN, aApr[1]=apr[1][[2]]+termN}
uN[xN_,tN_]:=aApr[0]/.x->xN/.t->tN; uN[x,t]
{pde1[uN], solF=aApr[0], trsolF=u[var]->solF,
 test1=pde1[u]/.trsolF/.trD[trsolF,x]/.trD[trsolF,t]}
```
□

Problem 5.5 *Burgers equation. Boundary value problem.* Let us consider the boundary value problem for the Burgers equation

$$u_t+uu_x=u_{xx}; \quad u(0,t)=g_1(t), \; u_x(0,t)=g_2(t),$$

where $g_1(t)=-2/(qt)$, $g_2(t)=1/t+2/(qt)^2$, and q is a parameter ($q \in \mathbb{N}$). Following the ideas of the ADM, prove that the solution of this boundary value problem has the form $u(x,t)=x/t-2/(x+qt)$ (`SolF`) and verify that this solution is exact solution of this problem (`Test1`, `Test2`).

1. As before, we generate the Adomian polynomials for the nonlinear term uu_x (`A[n]`) and rewrite the Burgers equation in the operator form $\mathcal{D}_x u=\mathcal{D}_t u+\mathcal{NL}u$ (`PDE1`), where $\mathcal{D}_x = \partial^2/\partial x^2$, $\mathcal{D}_t = \partial/\partial t$, and $\mathcal{NL}$ is the nonlinear operator. Since in this problem we have the boundary conditions, we will solve it in the x-direction, i.e., applying the inverse operator $\mathcal{D}_x^{-1}(\cdot)=\int_0^x\int_0^x(\cdot)\,dx\,dx$ (`LI`) to both sides of the equation (`PDE1`) and using the boundary conditions, we obtain the equation (`Eq2`):

$$u-\frac{2}{qt}+\frac{x}{t}+\frac{2x}{q^2t^2}=\mathcal{D}_x^{-1}u_t+\mathcal{D}_x^{-1}(uu_x).$$

Substituting the series forms of the linear term (5.1) (`trL`) and the nonlinear term (5.2) (where `A[i]` are the Adomian polynomials for the nonlinearity uu_x) (`trN`), into the equation `Eq2`, we obtain the equation `Eq3`.

Maple:

```
with(PDEtools): declare((u,W)(x,t)); KN:=9;
ADM1:=n->convert(subs(lambda=0,value(1/n!*Diff(F(Sum(
 lambda^i*U[i],i=0..n)),lambda$n))),diff); A0[0]:=F(U[0]);
for n from 1 to KN do A0[n]:=ADM1(n); od:
for n from 0 to KN do
 A[n]:=convert(1/2*diff(subs({seq(U[i]=W[i](x,t),i=0..KN)},
       expand(subs(F(U[0])=U[0]^2,A0[n]))),x),diff); od;
Dt:=w->diff(w(x,t),t); NL:=w->w*diff(w(x,t),x);
Dx:=w->diff(w(x,t),x$2); PDE1:=w->Dx(w)=Dt(w)+NL(w); PDE1(u);
BC1:=[u(0,t)=g1(t),D[1](u)(0,t)=g2(t)]; g1:=t->-2/(q*t);
g2:=t->1/t+2/(q^2*t^2); LI:=w->Int(Int(w(x,t),x=0..x),x=0..x);
LI(u); tr1:=u+(g1(t)+x*g2(t)); KN:=4; Eq1:=LI(lhs(PDE1(u)))=
 map(LI,rhs(PDE1(u))); Eq2:=expand(subs(lhs(Eq1)=tr1,Eq1));
trL:=diff(u(x,t),t)=add(diff(u[j](x,t),t),j=0..KN);
trN:=LI(NL(u))=Int(Int(Sum(A[i],i=0..KN),x=0..x),x=0..x);
Eq3:=lhs(Eq2)=value(subs(trL,trN,rhs(Eq2)));
```

Mathematica:

```
kN=9; trD[u_,var_]:=Table[D[u,{var,i}],{i,1,9}]//Flatten;
fADM1[n_]:=1/n!*D[f[Sum[lambda^i*fU[i],{i,0,n}]],
 {lambda,n}]/.lambda->0//Expand; a0[0]=f[fU[0]]
{trp1=f[fU[0]]->fU[0]^2,trp2=Table[fU[i]->fW[x,t][i],{i,0,kN}]}
Do[a0[n]=fADM1[n]; Print["a0[",n,"]=",a0[n]],{n,1,kN}];
Do[a[n]=1/2*D[(a0[n]/.trp1/.trD[trp1,fU[0]])/.trp2,x]//Expand;
  Print["a[",n,"]=",a[n]],{n,0,kN}];
fDt[w_]:=D[w[x,t],t]; fDx[w_]:=D[w[x,t],{x,2}];
fNL[w_]:=w[x,t]*D[w[x,t],x]; pde1[w_]:=fDx[w]==fDt[w]+fNL[w];
{pde1[u], bc1={u[0,t]->g1[t],(D[u[x,t],x]/.x->0)->g2[t]}}
g1[t_]:=-2/(q*t); g2[t_]:=1/t+2/(q^2*t^2); fLI[w_]:=Hold[
 Integrate[w,x,x]]; {fLI[u[x,t]], tr1=u[x,t]+(g1[t]+x*g2[t])}
{kN=4, eq1=fLI[pde1[u][[1]]]==Map[fLI, pde1[u][[2]]]}
eq2=eq1/.eq1[[1]]->tr1
trL=D[u[x,t],t]->Sum[D[u[x,t][j],t],{j,0,kN}]
trN=fLI[fNL[u]]->Hold[Integrate[Sum[a[i],{i,0,2}],x,x]]
eq3=eq2[[1]]==(eq2[[2]]/.trL/.trN)//Expand
```

2. According to the ADM, we determine the following recursive relations, $u_0 = -2/(qt) + x/t + 2x/(q^2t^2)$, $u_{k+1} = \mathcal{D}_x^{-1} u_{kt} + \mathcal{D}_x^{-1} A_k$ (`Apr[0]`, `AprK`), and obtain the approximate analytical solution (`Sol1`), which can be written in a closed form $u(x,t) = x/t - 2/(x + qt)$ (`SolF`). Finally, we verify that this solution is exact solution of this problem.

Maple:

```
Apr[0]:=u[0](x,t)=select(has, lhs(Eq3),t);
AprK:=u[k+1](x,t)=Int(Int(diff(u[k](x,t),t),x=0..x),x=0..x)
 +Int(Int(AD[k],x=0..x),x=0..x);
for i from 0 to KN do
 Apr[i+1]:=expand(value(subs({seq(Apr[m],m=0..i)},
 subs({seq(W[m]=u[m],m=0..i)},subs(k=i,AD[i]=A[i],AprK)))));
od; trSol:={seq(Apr[i],i=0..KN)}; Sol:=value(subs(trSol,
 u=add(u[j](x,t),j=0..KN)));
Sol1:=expand((rhs(Sol)-x/t)/(-2/(q*t)));
SolF:=expand(x/t-(2/(q*t))*sum((-1)^j*(1/q^j)*x^j/t^j,
 j=0..infinity)); Test1:=pdetest(u(x,t),PDE1(u));
Test2:=[subs(x=0,SolF)=g1(t),subs(x=0,diff(SolF,x))=g2(t)];
```

Mathematica:

```
trS3[eq_,var_]:=Select[eq,FreeQ[#,var]&];
apr[0]=u[x,t][0]->trS3[eq3[[1]],u[x,t]]
aprK=u[x,t][k+1]->Hold[Integrate[D[u[x,t][k],t],x,x]]+
 Hold[Integrate[ad[k],x,x]]
Do[apr[i+1]=ReleaseHold[aprK/.k->i/.ad[i]->a[i]/.
 Table[fW[x,t][m]->u[x,t][m],{m,0,i}]/.Table[D[fW[x,t][m],x]->
 D[u[x,t][m],x],{m,0,i}]/.Table[apr[m],{m,0,i}]/.
 Table[D[apr[m],t],{m,0,i}]/.Table[D[apr[m],x],{m,0,i}]]//Expand;
 Print["apr[",i+1,"]=",apr[i+1]],{i,0,kN}];
{trSol=Table[apr[i],{i,0,kN}], sol=u[x,t]->Sum[u[x,t][j],
 {j,0,kN}]/.trSol, sol1=(sol[[2]]-x/t)/(-2/(q*t))//Expand}
{solF=x/t-(2/(q*t))*Sum[(-1)^j*(1/q^j)*x^j/t^j,
 {j,0,Infinity}]//Expand, trsolF=u[x,t]->solF}
pde1[u]/.trsolF/.trD[trsolF,x]/.trD[trsolF,t]//Simplify
test2={(solF/.x->0)==g1[t],(D[solF,x]/.x->0)==g2[t]}//Simplify
```
□

5.1.3 Nonlinear Systems

In this section, applying the Adomian decomposition method, we will solve a Cauchy problem for the first-order system of nonlinear partial differential equations and compare the exact solution with the solution obtained by applying predefined functions (see Chap. 1).

Problem 5.6 *Nonlinear first-order system. Cauchy problem.* Let us consider the Cauchy problem for the nonlinear first-order system

$$u_t = vu_x + u + 1, \; v_t = -uv_x - v + 1; \quad u(x,0) = e^{-x}, \; v(x,0) = e^{x},$$

where $\{x \in \mathbb{R}, t \geq 0\}$. Following the ideas of the modified Adomian decomposition method, show that the solution of this Cauchy problem has the form $\{u(x,t)=e^{t-x}, v(x,t)=e^{x-t}\}$ (SoluF, SolvF) and verify that this solution is exact solution of this problem (Test1–Test4). Compare the results with those obtained in Problem 1.21.

1. First, let us generate the Adomian polynomials for the nonlinear terms vu_x and uv_x (A1[j], A2[j]).

Maple:

```
with(PDEtools): declare((u,v)(x,t)); KN:=9; ADM1:=n->convert(
 subs(lambda=0,value(1/n!*Diff(F(Sum(lambda^i*U[i],i=0..n)),
 lambda$n))),diff); A0[0]:=F(U[0]);
for n from 1 to KN do A0[n]:=ADM1(n); od:
for n from 0 to KN do
 A[n]:=sort(convert(1/2*diff(subs({seq(U[i]=u[i](x,t),
  i=0..KN)},expand(subs(F(U[0])=U[0]^2,A0[n]))),x),diff)); od;
A1[0]:=A[0]/u[0](x,t)*v[0](x,t); A2[0]:=A[0]/u[0,x](x,t)
 *v[0,x](x,t); for j from 1 to KN do k1:=nops(A[j]):
 A1[j]:=add(op(i,A[j])/u[k1-i](x,t)*v[k1-i](x,t),i=1..k1);
 A2[j]:=add(op(i,A[j])/u[i-1,x](x,t)*v[i-1,x](x,t),i=1..k1); od;
```

Mathematica:

```
kN=9; trD[u_,var_]:=Table[D[u,{var,i}],{i,1,9}]//Flatten;
fADM1[n_]:=1/n!*D[f[Sum[lambda^i*fU[i],{i,0,n}]],
 {lambda,n}]/.lambda->0//Expand; a0[0]=f[fU[0]]
{trp1=f[fU[0]]->fU[0]^2,trp2=Table[fU[i]->u[x,t][i],{i,0,kN}]}
Do[a0[n]=fADM1[n]; Print["a0[",n,"]=",a0[n]],{n,1,kN}];
Do[a[n]=1/2*D[(a0[n]/.trp1/.trD[trp1,fU[0]])/.trp2,x]//Expand;
 Print["a[",n,"]=",a[n]],{n,0,kN}];
{a1[0]=a[0]/u[x,t][0]*v[x,t][0],
 a2[0]=a[0]/D[u[x,t][0],x]*D[v[x,t][0],x]}
Do[k1=Length[a[j]]; a1[j]=Sum[a[j][[i]]/u[x,t][k1-i]*
 v[x,t][k1-i],{i,1,k1}]; a2[j]=Sum[a[j][[i]]/D[u[x,t][i-1],x]*
 D[v[x,t][i-1],x],{i,1,k1}]; Print["a1[",j,"]=",a1[j]];
 Print["a2[",j,"]=",a2[j]],{j,1,kN}];
```

2. As before, we rewrite the nonlinear system in the operator form (PDE1, PDE2)

$$\mathcal{D}_t u = \mathcal{NL}(v,u) + \mathcal{L}u + f(x,t), \quad \mathcal{D}_t v = -\mathcal{NL}(u,v) - \mathcal{L}v + f(x,t),$$

where $\mathcal{D}_t=\partial/\partial t$, and $\mathcal{NL}$, $\mathcal{L}$ are the corresponding nonlinear and linear operators. Applying the inverse operator $\mathcal{D}_t^{-1}(\cdot)=\int_0^t (\cdot)\,dt$ (LI) to both sides of the equations (PDE1, PDE2) and using the initial conditions, we obtain the system of equations (Eq21, Eq22):

$$u(x,t)-e^{-x}=\mathcal{D}_t^{-1}(vu_x)+\mathcal{D}_t^{-1}u+t,\ \ v(x,t)-e^{x}=-\mathcal{D}_t^{-1}(uv_x)-\mathcal{D}_t^{-1}v+t.$$

Substituting the series forms of the linear terms (5.1) (trL1, trL2) and the nonlinear terms (5.2) (where A[n] are the Adomian polynomials for the nonlinearities vu_x and uv_x) (trN1, trN2), into the equations Eq21, Eq22, we obtain the equations Eq31, Eq32.

Maple:

```
Dt:=w->diff(w(x,t),t); L:=w->w; NL:=(w1,w2)->w1*diff(w2(x,t),x);
f:=(x,t)->1; g1:=x->exp(-x); g2:=x->exp(x);
PDE1:=(w1,w2)->Dt(w1)=NL(w2,w1)+L(w1)+f(x,t); PDE1(u,v);
PDE2:=(w1,w2)->Dt(w2)=-NL(w1,w2)-L(w2)+f(x,t); PDE2(u,v);
IC1:=u(x,0)=g1(x); IC2:=v(x,0)=g2(x); LI:=w->Int(w(x,t),
 t=0..t); LI(u); tr1:=u-rhs(IC1); tr2:=v-rhs(IC2);
Eq11:=LI(lhs(PDE1(u,v)))=map(LI,rhs(PDE1(u,v)));
Eq12:=LI(lhs(PDE2(u,v)))=map(LI,rhs(PDE2(u,v)));
Eq21:=simplify(subs(lhs(Eq11)=tr1,Eq11)); Eq22:=simplify(subs(
 lhs(Eq12)=tr2,Eq12)); trL1:=u=add(u[j](x,t),j=0..KN);
trL2:=v=add(v[j](x,t),j=0..KN); trN1:=LI(NL(v,u))=Int(Sum(A[i],
 i=0..KN),t=0..t); trN2:=LI(NL(u,v))=Int(Sum(A[i],i=0..KN),
 t=0..t); Eq31:=subs(trL1,lhs(Eq21))=subs(trN1,rhs(Eq21));
Eq32:=subs(trL2,lhs(Eq22))=subs(trN2,rhs(Eq22));
```

Mathematica:

```
fDt[w_]:=D[w[x,t],t]; fL[w_]:=w[x,t]; fNL[w1_,w2_]:=w1[x,t]*
 D[w2[x,t],x]; f[x_,t_]:=1; g1[x_]:=Exp[-x]; g2[x_]:=Exp[x];
pde1[w1_,w2_]:=fDt[w1]==fNL[w2,w1]+fL[w1]+f[x,t];
pde2[w1_,w2_]:=fDt[w2]==-fNL[w1,w2]-fL[w2]+f[x,t];
{pde1[u,v],pde2[u,v], ic1=u[x,0]->g1[x], ic2=v[x,0]->g2[x]}
fLI[w_]:=Hold[Integrate[w,{t,0,t}]]; fLI[u[x,t]]
{tr1=u[x,t]-ic1[[2]], tr2=v[x,t]-ic2[[2]]}
eq11=fLI[pde1[u,v][[1]]]==Map[fLI,pde1[u,v][[2]]]
eq12=fLI[pde2[u,v][[1]]]==Map[fLI,pde2[u,v][[2]]]
{eq21=eq11/.eq11[[1]]->tr1, eq22=eq12/.eq12[[1]]->tr2}
trL1=u[x,t]->Sum[u[x,t][j],{j,0,kN}]
```

```
trL2=v[x,t]->Sum[v[x,t][j],{j,0,kN}]
trN1=fLI[fNL[v,u]]->Hold[Integrate[Sum[a[i],{i,0,kN}],{t,0,t}]]
trN2=fLI[fNL[u,v]]->Hold[Integrate[Sum[a[i],{i,0,kN}],{t,0,t}]]
eq31=(eq21[[1]]/.trL1)==(eq21[[2]]/.trN1)
eq32=(eq22[[1]]/.trL2)==(eq22[[2]]/.trN2)
```

3. Following the ideas of the modified ADM (for accelerating the convergence of the solution), we obtain the following recursive relations (`Ap1[0]`, `Ap2[0]`, `Apr1[1]`, `Apr2[1]`, `Apr1K`, `Apr2K`):

$$\begin{aligned}
&u_0=e^{-x}, \quad v_0=e^{x},\\
&u_1=t+\mathcal{D}_t^{-1}(u_{0x}v_0)+\mathcal{D}_t^{-1}(u_0), \quad v_1=t-\mathcal{D}_t^{-1}(u_0v_{0x})-\mathcal{D}_t^{-1}(v_0),\\
&u_{k+1}=\mathcal{D}_t^{-1}(A_{1k})+\mathcal{D}_t^{-1}(u_k), \qquad v_{k+1}=-\mathcal{D}_t^{-1}(A_{2k})-\mathcal{D}_t^{-1}(v_k).
\end{aligned}$$

For example, the first three components are: $u_1=e^{-x}t$, $v_1=-e^{x}t$ (`Ap1[1]`, `Ap2[1]`), $u_2=\frac{1}{2!}e^{-x}t^2$, $v_2=\frac{1}{2!}e^{x}t^2$ (`Ap1[2]`, `Ap2[2]`), $u_3=\frac{1}{3!}e^{-x}t^3$, $v_3=-\frac{1}{3!}e^{x}t^3$ (`Ap1[3]`, `Ap2[3]`), etc. As a result, we obtain the approximate analytical solution (`Solu1`, `Solv1`)

$$u=e^{-x}\Big(1+t+\frac{1}{2!}t^2+\frac{1}{3!}t^3+\dots\Big), \quad v=e^{x}\Big(1-t+\frac{1}{2!}t^2-\frac{1}{3!}t^3+\dots\Big).$$

This approximate analytical solution can be written in a closed form $u(x,t)=e^{t-x}$, $v(x,t)=e^{x-t}$ (`SoluF`, `SolvF`), which coincides with the results obtained in Problem 1.21. Finally, we verify that this solution is exact solution of this problem (`Test1`–`Test4`).

Maple:

```
Ap1[0]:=u[0](x,t)=rhs(IC1); Ap2[0]:=v[0](x,t)=rhs(IC2);
Apr1[1]:=u[1](x,t)=t+Int(A1[0],t=0..t)+Int(u[0](x,t),t=0..t);
Apr2[1]:=v[1](x,t)=t-Int(A2[0],t=0..t)-Int(v[0](x,t),t=0..t);
Apr1K:=u[k+1](x,t)=Int(A1[k],t=0..t)+Int(u[k](x,t),t=0..t);
Apr2K:=v[k+1](x,t)=-Int(A2[k],t=0..t)-Int(v[k](x,t),t=0..t);
trD[1]:={u[0,x](x,t)=-exp(-x),v[0,x](x,t)=exp(x)};
Ap1[1]:=simplify(value(subs(Ap1[0],Ap2[0],Apr1[1])));
Ap2[1]:=simplify(value(subs(trD[1],subs(Ap1[0],Ap2[0],
 Apr2[1]))));
for i from 2 to KN do
 trD[i]:={u[i-1,x](x,t)=diff(rhs(Ap1[i-1]),x),v[i-1,x](x,t)
  =diff(rhs(Ap2[i-1]),x)};
 Ap1[i]:=simplify(expand(subs({seq(Ap1[m],m=0..i-1),
  seq(Ap2[m],m=0..i-1)},value(subs(k=i-1,Apr1K)))));
```

```
 Ap2[i]:=simplify(expand(subs(seq(trD[m],m=1..i),seq(Ap1[m],
  m=0..i-1),seq(Ap2[m],m=0..i-1),value(subs(k=i-1,Apr2K)))));
od; trSol:={seq(Ap1[i],i=0..KN),seq(Ap2[i],i=0..KN)};
Solu:=value(subs(trSol,trL1)); Solv:=value(subs(trSol,trL2));
Solu1:=collect(Solu,exp); Solv1:=collect(Solv,exp);
SoluF:=combine(exp(-x)*sum((1/l!)*t^l,l=0..infinity));
SolvF:=combine(exp(x)*sum((1/l!)*(-t)^l,l=0..infinity));
Test1:=pdetest({u(x,t)=SoluF,v(x,t)=SolvF},diff(u(x,t),t)
 =v(x,t)*diff(u(x,t),x)+u(x,t)+1);
Test2:=pdetest({u(x,t)=SoluF,v(x,t)=SolvF},diff(v(x,t),t)
 =-u(x,t)*diff(v(x,t),x)-v(x,t)+1);
Test3:=subs(t=0,SoluF)=g1(x); Test4:=subs(t=0,SolvF)=g2(x);
```

Mathematica:

```
{ap1[0]=u[x,t][0]->ic1[[2]], ap2[0]=v[x,t][0]->ic2[[2]]}
{apr1[1]=u[x,t][1]->t+Hold[Integrate[a1[0],{t,0,t}]+
 Integrate[u[x,t][0],{t,0,t}]], apr2[1]=v[x,t][1]->t+
 Hold[-Integrate[a2[0],{t,0,t}]-Integrate[v[x,t][0],{t,0,t}]]}
{apr1K=u[x,t][k+1]->Hold[Integrate[a1[k],{t,0,t}]+
 Integrate[u[x,t][k],{t,0,t}]], apr2K=v[x,t][k+1]->
 Hold[-Integrate[a2[k],{t,0,t}]-Integrate[v[x,t][k],{t,0,t}]]}
ap1[1]=ReleaseHold[apr1[1]]/.ap1[0]/.ap2[0]/.D[ap1[0],x]
ap2[1]=ReleaseHold[apr2[1]]/.ap1[0]/.ap2[0]/.D[ap1[0],x]/.
 D[ap2[0],x]
Do[ap1[i]=ReleaseHold[apr1K/.k->i-1/.Flatten[Table[{D[ap2[m],x],
 D[ap1[m],x],ap2[m],ap1[m]},{m,0,i-1}]]]/.Flatten[
 Table[{D[ap2[m],x],D[ap1[m],x],ap2[m],ap1[m]},{m,0,i-1}]];
 ap2[i]=ReleaseHold[apr2K/.k->i-1/.Flatten[Table[{D[ap2[m],x],
 D[ap1[m],x],ap2[m],ap1[m]},{m,0,i-1}]]]/.Flatten[Table[
 {D[ap2[m],x],D[ap1[m],x],ap2[m],ap1[m]},{m,0,i-1}]];
 Print["ap1[",i,"]=",ap1[i]]; Print["ap2[",i,"]=",ap2[i]],
{i,2,kN}];
trSol={Table[ap1[i],{i,0,kN}],Table[ap2[i],{i,0,kN}]}//Flatten
{solu=trL1/.trSol, solv=trL2/.trSol, solu1=Collect[solu,Exp[_]],
 solv1=Collect[solv,Exp[_]], soluF=Exp[-x]*Sum[(1/l!)*t^l,
 {l,0,Infinity}], solvF=Exp[x]*Sum[(1/l!)*(-t)^l,{l,0,Infinity}]}
{trsoluF=u[x,t]->soluF, trsolvF=v[x,t]->solvF}
test1=pde1[u,v]/.trsoluF/.trsolvF/.D[trsoluF,x]/.D[trsoluF,t]
test2=pde2[u,v]/.trsoluF/.trsolvF/.D[trsolvF,x]/.D[trsolvF,t]
{test3=(soluF/.t->0)==g1[x], test4=(solvF/.t->0)==g2[x]}
```

□

5.2 Asymptotic Expansions. Perturbation Methods

Many physical systems are described by nonlinear PDEs, which include various important effects (e.g., dispersion, dissipation, nonlinearity, inhomogeneity, etc). Frequently, the governing equations can be derived from conservation laws. However, in general case, the physical systems can be too complex and the corresponding governing equations can be complicated and are not integrable by analytic methods. Therefore, it is interesting to develop mathematical methods which allow us to reduce the original problem to a more simple problem that contain all of the important features of the original problem.

The method of *asymptotic expansions* is a general mathematical method, which is very important, powerful, and well established (since the 18th century). Frequently, this method is applied based on the assumption that a solution of the given equation exists in the form of an asymptotic expansion with respect to a small parameter. In this section, we consider some applications of asymptotic methods to the solutions of nonlinear partial differential equations and nonlinear systems.

5.2.1 Nonlinear PDEs

In this section, we construct the asymptotic traveling wave solution of the initial boundary value problem for the Fisher equation and the asymptotic solutions in the near-field and far-field regions of the Cauchy problem for the nonlinear fourth-order PDE.

Problem 5.7 *Fisher equation. Asymptotic traveling wave solution. Initial boundary value problem.* Let us consider the initial boundary value problem for the Fisher equation

$$u_t{=}u_{xx}{+}u(1{-}u),\ \ u(x,t){=}u(\xi);\ \ u(0){=}\tfrac{1}{2},\ \ \lim_{\xi\to-\infty} u(\xi){=}1,\ \ \lim_{\xi\to\infty} u(\xi){=}0,$$

where $\{x \in \mathbb{R}, t \geq 0\}$. It is known that the Fisher equation admits traveling wave solutions, $u(x,t){=}u(\xi)$ $(\xi{=}x{-}ct)$, where the wave speed c is to be determined and the waveform $u(\xi)$ must satisfy the given initial and boundary conditions. Applying the regular perturbation method, construct an *asymptotic traveling wave solution* of this initial boundary value problem.

Substituting $u(x,t){=}u(\xi)$ $(\xi{=}x{-}ct)$ into the Fisher equation and the initial and boundary conditions, we obtain the nonlinear ODE with the initial and boundary conditions

$$-u_\xi c{-}u_{\xi\xi}{-}u{+}u^2{=}0;\quad u(0){=}\tfrac{1}{2},\ u(-\infty){=}1,\ u(\infty){=}0,$$

where $\{-\infty < \xi < \infty\}$. Let us apply the perturbation method to find *asymptotic solution* of this initial boundary value problem (IBVP). Introducing the new variable z and a small parameter ε according to the relations $\xi=zc$, $\xi=z\varepsilon^{-1/2}$, $c=\epsilon^{-1/2}$, we can transform the IBVP into the form

$$-u_z-\varepsilon u_{zz}-u+u^2=0; \quad u(0)=\tfrac{1}{2}, \; u(-\infty)=1, \; u(\infty)=0,$$

where $\{-\infty < z < \infty\}$. Since this problem is not a singular perturbation problem, we apply the *regular perturbation method* and look for a perturbation series expansion for $u(z)$ in powers of ε:

$$u(z;\varepsilon)=u_0(z)+u_1(z)\varepsilon+u_2(z)\varepsilon^2+\cdots,$$

where $u_i(z)$ $(i=1,\ldots,N)$ are to be determined. Substituting this form of the asymptotic solution into the differential system and setting the coefficients of like powers of ε to zero, we arrive at the more simple initial boundary value problems for each approximation. We obtain $u_0=1/(1+e^z)$, $u_1=\ln(4e^z/(1+e^z)^2)e^z/(1+e^z)^2$, and the asymptotic solution reads

$$u(z;\varepsilon)=\frac{1}{1+e^z}+\varepsilon\ln\left[\frac{4e^z}{(1+e^z)^2}\right]\frac{e^z}{(1+e^z)^2}+O(\varepsilon^2),$$

where $z=\xi/c=(x-ct)/c$ $(c\geq 2)$. This formula represents an asymptotic traveling wave solution.

Maple:

```
with(PDEtools): declare(u(xi)); tr0:=x-c*t=xi; PDE1:=v->diff(v,t)
 -diff(v,x$2)-v*(1-v)=0; Eq0:=expand(PDE1(u(lhs(tr0))));
Eq1:=collect(convert(algsubs(tr0,Eq0),diff),diff);
BC1:=(u,A)->u(-infinity)=A; BC2:=(u,B)->u(infinity)=B;
IC:=(u,C)->u(0)=C; tr1:=[xi=z*c,xi=z*epsilon^(-1/2),
 c=epsilon^(-1/2)]; Eq2:=expand(eval(dchange(tr1[2],Eq1,[z,u]),
 tr1[3])); tr2:=u(z,epsilon)=add(u||i(z)*epsilon^i,i=0..2);
Eq3:=collect(algsubs(tr2,Eq2),epsilon); Pr0:=remove(has,
 lhs(Eq3),[epsilon])=0; Sol0:=dsolve({Pr0} union {IC(u0,1/2)},
 u0(z)); limit(Sol0,z=-infinity)=BC1(u0,1);
limit(Sol0,z=infinity)=BC2(u0,0); Pr1:=remove(has,expand(
 lhs(Eq3)/epsilon),[epsilon])=0; Pr11:=collect(algsubs(Sol0,Pr1),
 [diff,u1]); Pr12:=map(factor,lhs(Pr11));
Sol1:=combine(dsolve({Pr12} union {IC(u1,0)},u1(z)));
combine(Sol1) assuming z>0; limit(Sol1,z=-infinity)=BC1(u1,0);
```

```
limit(Sol1,z=infinity)=BC2(u1,0);
Pr2:=select(has,lhs(Eq3),epsilon^2)/epsilon^2=0;
Pr21:=collect(algsubs(Sol1,algsubs(Sol0,Pr2)),[diff,u2]);
Pr22:=map(factor,lhs(Pr21)); Pr23:=op(1,Pr22)+op(2,Pr22)+
 combine(collect(factor(op(3,Pr22)),[exp,ln]));
Sol2:=combine(dsolve({Pr23} union {IC(u2,0)},u2(z)));
```

Mathematica:

```
Off[Solve::ifun]; trD[u_,var_]:=Table[D[u,{var,i}],
 {i,1,2}]//Flatten; pde1[v_]:=D[v,t]-D[v,{x,2}]-v*(1-v)==0;
{tr0=x-c*t->xi, eq0=pde1[u[tr0[[1]]]]//Expand, eq1=eq0/.tr0}
bc1[u_,a_]:=u[-Infinity]==a; bc2[u_,b_]:=u[Infinity]==b;
ic[u_,c_]:=u[z]==c; {tr1={xi->z*c,xi->z*epsilon^(-1/2),
 c->epsilon^(-1/2)}, tr11=xi->z/Sqrt[epsilon]}
eq1F[xi_]:=-u[xi]+u[xi]^2-c*u'[xi]-u''[xi]==0; eq1F[xi]
eq1FT[v_]:=First[((eq1F[xi]/.u->Function[xi,u[xi*Sqrt[
 epsilon]]])/.tr11)/.{u->v}]; eq2=(eq1FT[u]//Expand)/.tr1[[3]]
tr2=u[z]->Sum[u[i][z]*epsilon^i,{i,0,2}]
eq3=Collect[Expand[eq2/.tr2/.trD[tr2,z]],epsilon]
{pr0=Coefficient[eq3,epsilon,0]==0, sol00=DSolve[pr0,u[0][z],
 z]//First, trC10=Solve[(sol00/.z->0)[[1,2]]==ic[u,1/2][[2]],
 C[1]]//First, sol0=sol00/.trC10}
(TraditionalForm[Limit[sol0[[1,1]],z->-Infinity]]==
 Limit[sol0[[1,2]],z->-Infinity])==bc1[u[0],1]
(TraditionalForm[Limit[sol0[[1,1]],z->Infinity]]==
 Limit[sol0[[1,2]],z->Infinity])==bc2[u[0],0]
{pr1=Coefficient[eq3,epsilon,1]==0, pr11=Collect[pr1/.sol0/.
 trD[sol0,z],u[1][z]], sol10=DSolve[pr11,u[1][z],z]//First}
trC11=Solve[(sol10/.z->0)[[1,2]]==ic[u,0][[2]],C[1]]//First
sol1=sol10/.trC11//Simplify
(TraditionalForm[Limit[sol1[[1,1]],z->-Infinity]]==
 Limit[sol1[[1,2]],z->-Infinity])==bc1[u[1],0]
(TraditionalForm[Limit[sol1[[1,1]],z->Infinity]]==
 Limit[sol1[[1,2]],z->Infinity])==bc2[u[1],0]
{pr2=Coefficient[eq3,epsilon,2]==0, pr21=Collect[pr2/.sol0/.
 sol1/.trD[sol0,z]/.trD[sol1,z],u[2][z]], sol20=DSolve[
 pr21,u[2][z],z]//First, trC12=Solve[(sol20/.z->0)[[1,2]]==
 ic[u,0][[2]],C[1]]//First//FullSimplify}
sol2=sol20/.trC12//FullSimplify
```

□

Problem 5.8 *Nonlinear fourth-order PDE. Cauchy problem. Asymptotic solution. Near-field region.* Let us consider the Cauchy problem for the nonlinear partial differential equation of the fourth order

$$u_{tt}=u_{xx}+\varepsilon(u^2-u_{xx})_{xx}; \quad u(x,0)=g(x), \; u_t(x,0)=-g'(x),$$

where $\{x\in\mathbb{R}, t\geq 0\})$, and ε is a small parameter. The given initial conditions describe a wave traveling to the positive x-direction. Applying the method of asymptotic expansions, construct the asymptotic solution in the *near-field region*, where the variable $t=O(1)$ is the fast time.

We look for a perturbation series expansion in powers of ε in the form

$$u(x,t;\varepsilon)=u_0(x,t)+u_1(x,t)\varepsilon + u_2(x,t)\varepsilon^2+\cdots,$$

where $x=O(1)$, $t=O(1)$, and $u_i(x,t)$ $(i=1,\ldots,N)$ are to be determined. Substituting this asymptotic ansatz into the differential equation and the initial conditions and setting the coefficients of like powers of ε to zero, we arrive at the initial value problems for each approximation:

$$-u_{0xx}+u_{0tt}=0, \quad u_0(x,0)=g(x), \; u_{0t}(x,0)=-g'(x),$$
$$-2u_{0x}^2-2u_{0xx}u_0-u_{0xxxx}-u_{1xx}+u_{1tt}=0, \; u_1(x,0)=0, \; u_{1t}(x,0)=0.$$

Solving these initial value problems, we determine $u_0(x,t)=g(x-t)$, $u_1(x,t)=\frac{1}{4}(G(x+t)-G(x-t)-2tG'(x-t))$, where $G(x-t)=(g(x-t))^2+g''(x-t)$. Hence, the final form of the asymptotic solution has the form

$$u(x,t;\varepsilon)=g(x-t)+\frac{\varepsilon}{4}(G(t+x)-G(x-t)-2tG'(x-t))+O(\varepsilon^2).$$

This formula represents an uniform asymptotic solution of the original nonlinear equation if the function $g(x)$ is a sufficiently smooth function. In some special cases of the function $g(x)$, e.g., if $g(x)$ has a compact support or $g(x)\to 0$ sufficiently rapidly as $x\to\pm\infty$, the asymptotic solution is not uniform for $\varepsilon t=O(1)$.

Maple:

```
with(PDEtools): declare(u(x,t));
Eq1:=diff(u(x,t),t$2)-diff(u(x,t),x$2)-epsilon*diff(u(x,t)^2+
 diff(u(x,t),x$2),x$2)=0;
tr1:=u(x,t)=add(u||i(x,t)*epsilon^i,i=0..1);
Eq2:=collect(algsubs(tr1,Eq1),epsilon);
Pr0:=remove(has,lhs(Eq2),epsilon)=0;
Pr1:=remove(has,expand(lhs(Eq2)/epsilon),epsilon)=0;
```

```
ICs:=[u0(x,0)=g(x),D[2](u0)(x,0)=-diff(g(x),x)];
Sol0D:=subs(_F1(t+x)=f(t+x),_F2(t-x)=g(x-t),subs(g(t-x)=g(x-t),
 pdsolve(Pr0,u0(x,t)))); sysIP:={eval(rhs(Sol0D),t=0)=
 rhs(ICs[1]), eval(diff(rhs(Sol0D),t),t=0)=rhs(ICs[2])};
SolsysIP:=dsolve(sysIP,{f(x),g(x)}); Sol0:=subs(f(t+x)=0,Sol0D);
Pr11:=algsubs(Sol0,Pr1); tr2:={x=(xi+eta)/2,t=(eta-xi)/2};
Pr12:=dchange(tr2,Pr11,[xi,eta])/(-4); Pr13:=[selectremove(has,
 lhs(Pr12),eta)]; Pr14:=Pr13[1]=-Pr13[2];
Sol11:=map(int,Pr14,eta); Sol12:=map(int,Sol11,xi);
Sol13:=lhs(Sol12)=rhs(Sol12)+A(xi)+B(eta);
ICs1:=[eval(u1(x,t)=0,t=0),eval(D[2](u1)(x,t),t=0)=0];
sysIP1:={eval(rhs(Sol13),eta=xi)=rhs(ICs1[1]),
        eval(diff(rhs(Sol13),eta),eta=xi)=rhs(ICs1[2])};
SolsysIP:=dsolve(sysIP1,{A(xi),B(xi)});
SolsysIP1:=sort(convert(eval(SolsysIP,_C1=0),list));
A_xi:=SolsysIP1[1]; B_eta:=eval(SolsysIP1[2],xi=eta);
Sol14:=eval(Sol13,{A_xi,B_eta}); Sol15:=collect(Sol14*4,
 [xi,eta]); tr31:={g(eta)^2+diff(g(eta),eta$2)=G(eta)};
tr32:={-g(xi)^2-diff(g(xi),xi$2)=-G(xi)};
Sol16:=collect(simplify(Sol15,tr31),[xi,eta]);
Sol17:=collect(collect(simplify(Sol16,tr32),[xi,eta])/4,diff);
Sol18:=subs(diff(G(xi),xi)=Diff(G(xi),xi),Sol17);
tr4:={xi=x-t,eta=x+t}; Sol19:=u1(x,t)=subs(tr4,rhs(Sol18));
SolFin:= subs(Sol0,Sol19,tr1);
```

Mathematica:

```
trD[u_,var_]:=Table[D[u,{var,i}],{i,1,4}]//Flatten;
trS1[eq_,var_]:=Select[eq,MemberQ[#,var,Infinity]&];
trS3[eq_,var_]:=Select[eq,FreeQ[#,var]&]; {eq1=D[u[x,t],{t,2}]-
 D[u[x,t],{x,2}]-epsilon*D[u[x,t]^2+D[u[x,t],{x,2}],{x,2}]==0,
 tr1=u[x,t]->Sum[u[i][x,t]*epsilon^i,{i,0,1}]}
eq2=Collect[eq1/.tr1/.trD[tr1,x]/.trD[tr1,t],epsilon]
{pr0=Coefficient[eq2[[1]],epsilon,0]==0, pr1=Coefficient[
 eq2[[1]],epsilon,1]==0, ics={u[0][x,0]->g[x],(D[u[0][x,t],t]/.
 {t->0})->-g'[x]}, sol0D0=DSolve[pr0,u[0][x,t],{x,t}]//First}
{sol0D=sol0D0/.C[2][t+x]->f[t+x]/.C[1][t-x]->g[x-t]/.
 g[t-x]->g[x-t], sysIP={(sol0D[[1,2]]/.t->0)==ics[[1,2]],
 (D[sol0D[[1,2]],t]/.t->0)==ics[[2,2]]}, solsysIP=DSolve[sysIP,
 {f[x],g[x]},x], solfSysC=DSolve[sysIP[[2]],{f[x]},x]//First,
 trC1=Solve[sysIP[[1]]/.solfSysC,C[1]]//First,
 solfSys=solfSysC/.trC1, solgSys=ToRules[sysIP[[1]]]/.solfSys/.
 trD[solfSys,x], sol0=sol0D/.f[t+x]->0}
```

```
{pr11=pr1/.sol0/.trD[sol0,x]/.trD[sol0,t], tr2={x->(xi+eta)/2,
 t->(eta-xi)/2}, tr4=Solve[tr2/.Rule->Equal,{xi,eta}]}
pr11F[xN_,tN_]:=pr11/.x->xN/.t->tN; pr11FT[v_]:=First[((
 pr11F[x,t]/.u[1]->Function[{x,t},u[1][x-t,x+t]])/.tr2)/.
 {u->v}]//Simplify; {pr11F[x,t], pr12=Thread[(pr11FT[u]//
 Expand)/(-4),Equal]//Expand, pr14=trS1[pr12,eta]==-trS3[pr12,
 eta], sol11=Map[Integrate[#,eta]&,pr14], sol12=Map[Integrate[
 #,xi]&,sol11], sol13=sol12[[1]]==sol12[[2]]+a[xi]+b[eta]}
{ics1={u[1][x,0]->0,(D[u[1][x,t],t]/.t->0)->0},
sysIP1={(sol13[[2]]/.eta->xi)==ics1[[1,2]],(D[sol13[[2]],eta]/.
 eta->xi)==ics1[[2,2]]}, solsysIP=DSolve[sysIP1,{a[xi],b[xi]},
 xi], solsysIP1=solsysIP/.C[1]->0//First, axi=solsysIP1[[1]],
 beta=solsysIP1[[2]]/.xi->eta, sol14=sol13/.axi/.beta}
sol15=Collect[Thread[sol14*4,Equal]//Expand,{xi,eta}]
{tr31={g[eta]^2+g''[eta]->gN[eta]},tr32={-g[xi]^2-g''[xi]->
 -gN[xi]}, tr33=Map[Times[#,-1]&,tr32//First]}
{sol16=Collect[sol15/.tr31,{xi,eta}], sol17=Thread[Collect[
 sol16/.tr32/.D[tr32,xi]/.D[tr33,xi],{xi,eta}]/4,Equal]//Factor,
 sol19=Collect[u[1][x,t]->sol17[[2]]/.tr4,gN_[_]]}
solFin=tr1/.sol0/.sol19
```
□

Problem 5.9 *Nonlinear fourth-order PDE. Cauchy problem. Asymptotic solution. Far-field region.* Again (as in Problem 5.8), we consider the nonlinear PDE with the same initial conditions:

$$u_{tt}=u_{xx}+\varepsilon(u^2-u_{xx})_{xx}; \quad u(x,0)=g(x), \ u_t(x,0)=-g'(x),$$

where $\{x\in\mathbb{R}, t\geq 0\}$). Introducing the transformations, $x=\xi+\tau/\varepsilon$ and $t=\tau/\varepsilon$, and applying the method of asymptotic expansions, construct the asymptotic solution in the *far-field region*, where the variable $\tau = t\varepsilon = O(1)$ is the large time.

Applying the transformations to the nonlinear equation for $u(x,t;\varepsilon)$, we obtain the equation for $U(\xi,\tau;\varepsilon)$ (`Eq2`):

$$-2\varepsilon U_{\tau\xi}+\epsilon^2 U_{\tau\tau}-2\varepsilon U_\xi^2-2\varepsilon U U_{\xi\xi}-\varepsilon U_{\xi\xi\xi\xi}=0.$$

We look for a perturbation series expansion in powers of ε in the form

$$U(\xi,\tau;\varepsilon)=U_0(\xi,\tau)+U_1(\xi,\tau)\varepsilon+U_2(\xi,\tau)\varepsilon^2+\cdots,$$

where $\xi=O(1)$, $\tau=O(1)$ as $\varepsilon\to 0$, and $t=O(\varepsilon^{-1})$, $x=O(\varepsilon^{-1})$, i.e., the asymptotic solution in the far-field region, where the variable τ is the large time. The corrections $U_i(x,t)$ (i=1, ..., N) can be determined. Substituting this asymptotic ansatz into the differential equation and the initial conditions and setting the coefficients of like powers of ε to zero, we can arrive at the initial value problems for each approximation (Eq3). For example, we verify that the first term in the asymptotic solution, $U_0(\xi,\tau)$, satisfies the Cauchy problem for the Korteweg–de Vries equation (Eq6):

$$2(U_{0\tau}+U_0U_{0\xi})+U_{0\xi\xi\xi}=0; \quad U_0(\xi,0)=g(\xi).$$

The solution of this initial value problem exists and can be obtained by the inverse scattering transform, i.e., similar analysis can be performed for the next term, $U_1(\xi,\tau)$, in the expansion, which satisfies the equation (Pr1):

$$-2U_{1\tau\xi}+U_{0\tau\tau}-4U_{0\xi}U_{1\xi}-2U_{0\xi\xi}U_1-2U_{1\xi\xi}U_0-U_{1\xi\xi\xi\xi}=0.$$

The far-field asymptotic expansion of the solution for this problem is uniformly valid if $g(\xi)$ is sufficiently smooth and $g(\xi)\to 0$ sufficiently rapidly as $\xi\to\pm\infty$.

Maple:

```
with(PDEtools): declare(u(x,t),U(xi,tau));
FU0:=diff_table(U0(xi,tau)); Eq1:=u->diff(u,t$2)-diff(u,x$2)-
 epsilon*diff(u^2+diff(u,x$2),x$2)=0;
tr1:={x=xi+tau/epsilon,t=tau/epsilon};
Eq2:=expand(dchange(tr1,Eq1(U(x,t)),[xi,tau]));
tr2:=U(xi,tau,epsilon)=add(U||i(xi,tau)*epsilon^i,i=0..1);
Eq3:=collect(algsubs(tr2,Eq2),epsilon);
Pr0:=remove(has,expand(lhs(Eq3)/epsilon),epsilon)*(-1)=0;
Eq4:=op(1,lhs(Pr0))+Diff(U0(xi,tau)^2+FU0[xi,xi],xi,xi)=0;
Pr0-value(Eq4); Eq5:=map(int,lhs(Eq4),xi); Eq6:=value(Eq5);
Pr1:=remove(has,expand(lhs(Eq3)/epsilon^2),epsilon)=0;
```

Mathematica:

```
trD[u_,var_]:=Table[D[u,{var,i}],{i,1,4}]//Flatten;
trS1[eq_,var_]:=Select[eq,MemberQ[#,var,Infinity]&];
eq1[x_,t_]:=D[u[x,t],{t,2}]-D[u[x,t],{x,2}]-epsilon*D[
 u[x,t]^2+D[u[x,t],{x,2}],{x,2}]==0; {tr1={x->xi+tau/epsilon,
 t->tau/epsilon}, Solve[tr1/.Rule->Equal,{xi,tau}]}
eq1T[v_]:=First[((eq1[x,t]/.u->Function[{x,t},
 u[x-t,t*epsilon]])/.tr1)/.{u->v}]//Simplify; eq1[x,t]
```

```
eq2=eq1T[uN]//Expand
tr2=uN[xi,tau]->Sum[v[i][xi,tau]*epsilon^i,{i,0,1}]
eq3=Collect[eq2/.tr2/.trD[tr2,xi]/.trD[tr2,tau]/.
 D[tr2,xi,tau],epsilon]
pr0=Coefficient[eq3,epsilon,1]*(-1)==0
eq4=trS1[pr0[[1]],D[v[0][xi,tau],xi,tau]]+
 Hold[D[v[0][xi,tau]^2+D[v[0][xi,tau],{xi,2}],{xi,2}]]==0
Thread[pr0-ReleaseHold[eq4],Equal]
eq5=Map[Integrate[#,xi]&,eq4[[1]]]
{eq6=ReleaseHold[eq5], pr1=Coefficient[eq3,epsilon,2]==0}
```

□

5.2.2 Nonlinear Systems

In this section, we construct the asymptotic solution of the initial boundary value problem for the nonlinear system describing parametrically driven nonlinear fluid surface standing waves in a vertically oscillating rectangular container.

The classic example of equations describing parametric oscillations is the Mathieu and the generalized Mathieu equations. M. Faraday [42] was the first to observe experimentally parametric oscillations, when a tank containing a fluid vibrates periodically in the vertical direction, standing waves form at the free surface of the fluid when the vibration frequency is twice the frequency of the surface vibrations.

Lord Rayleigh [127] repeated most of Faraday's experiments and carried out a further series of experiments, and also developed a linear theory for these waves in terms of the Mathieu equation. Then Y. I. Sekerzh-Zenkovich [133], T. B. Benjamin and F. Ursell [17], N. N. Moiseev [104], S. V. Nesterov [111], L. N. Sretenskii [147] advanced the linear theory further.

Afterward, J. W. Miles (see [100]–[102]), J. R. Ockendon and H. Ockendon [112], S. Douady, S. Fauve, and O. Thual [41], G. A. Bordakov and S. Y. Sekerzh-Zenkovich [20] formulated a weakly nonlinear theory based on amplitude expansion.

The problem of excitation of standing nonlinear waves in a horizontally or vertically oscillating container has been extensively investigated for weak forcing analytically, numerically, and experimentally by numerous researchers (e.g., theoretical investigations of parametrically driven water waves in a vertically oscillating container [40], [20], experiments [74], [153], [78]). More recently attention has been paid to understanding the dynamics of steep and breaking standing waves in the experiments [75].

Problem 5.10 *Nonlinear standing waves in a fluid. Asymptotic solution. Initial boundary value problem.* Let us construct approximate analytical solution describing nonlinear parametrically excited standing waves in a vertically oscillating rectangular container and develop computer algebra procedures to aid in the construction of higher-order approximate analytical solutions.

In general, there are two ways for representing the fluid motion: the Eulerian approach, in which the coordinates are fixed in the reference frame of the observer, and the Lagrangian approach, in which the coordinates are fixed in the reference frame of the moving fluid. We will construct approximate analytical solutions in Lagrangian variables and will follow the analytic Lagrangian approach proposed by Y. I. Sekerzh-Zenkovich [133] for constructing approximate solutions for nonlinear waves of finite amplitude.

1. Statement of the problem. We consider two-dimensional wave motion of an inviscid fluid in a rectangular container which oscillates in the vertical direction such that the acceleration field is varied in time t as $g(t){=}g_0(1{+}\mathcal{A}\cos\Omega t)$, where $\mathcal{A}$ is the forcing amplitude, Ω is the forcing frequency of the corresponding standing mode (we consider the one-modal class of standing waves), and g_0 is the acceleration due to gravity. Let ρ be the fluid density, L the container length, and h the fluid depth. The container walls are assumed to be rigid and impermeable. We also assume that a local reference frame (x,y) is rigidly connected with the container. Applying the Lagrangian formulation to the problem with coordinates (a,b), we write out the complete system of hydrodynamic equations, i.e., two equations of motion and the continuity equation, in the form:

$$x_{tt}x_a+[y_{tt}+g_0(1+\mathcal{A}\cos\Omega t)]y_a+p_a\rho^{-1}=0,$$
$$x_{tt}x_b+[y_{tt}+g_0(1+\mathcal{A}\cos\Omega t)]y_b+p_b\rho^{-1}=0,\quad \frac{\partial(x,y)}{\partial(a,b)}=1.$$

In the moving reference frame the endwalls of the container are stationary, thus the boundary conditions at the solid walls and the free boundary are:

$$x(0,b,t)=0,\ \ x(L,b,t)=L,\ \ y(a,-h,t)=-h,\ \ p(a,0,t)=0,$$

where $x(a,b,t)$ and $y(a,b,t)$ are the coordinates of the individual moving particles, $p(a,b,t)$ is the fluid pressure. Considering a given class of standing waves, i.e., one-mode standing waves, we choose the initial conditions in the following form:

$$x_0 = a + D_0 \cosh[\kappa(b+h)] \sin(a\kappa) \cos(\psi_0),$$
$$y_0 = b + D_0 \sinh[\kappa(b+h)] \cos(a\kappa) \cos(\psi_0),$$

where $D_0 = A_0/[\kappa \sinh(h\kappa)]$, A_0 and $\psi_0 = \omega t_0 + \theta_0$ (θ_0=const) are the initial values of the amplitude and the phase. We consider the problem neglecting the surface tension effects at the free boundary of the fluid. It was shown by Andreev [10] and Yosihara [172] that if the surface tension is zero, the free boundary problem is well-posed in a finite interval of time if the following sign condition $-\nabla p \cdot \mathbf{n} > 0$ is satisfied. Here p is the fluid pressure and $\mathbf{n}$ is the outward unit normal to the free boundary. Therefore, we choose the initial value of the fluid pressure p_0 in the following form:

$$p_0 = -y\rho g_0[1 + \varepsilon \cos(\Omega t_0)] + D_0 \rho \omega^2 \kappa^{-1} \cosh[\kappa(b+h)] \cos(a\kappa) \cos(\psi_0).$$

This expression satisfies the sign condition and the condition $p > 0$ inside the fluid region.

2. Asymptotic solution. Let us consider a special case when the fluid depth is equal or close to the critical depth (the depth where the nonlinear third-order correction to the wave frequency equals zero). For unforced nonlinear standing water waves Tadjbakhsh and Keller [150] have found that the frequency-amplitude dependences are qualitatively different for fluid depths above and below a depth, called the *critical depth* by Miles and Henderson [102].

Similar to our previous results (e.g., see [139], [142]), we adopt the assumption that the small parameter ε characterizing the acceleration of the fluid motion is $\varepsilon = \frac{\mathcal{A}_m \Omega^2}{g} \ll 1$, where $\mathcal{A}_m$ is the maximum value of $\mathcal{A}$.

We introduce the dimensionless Lagrangian coordinates and time $\alpha = a\kappa$, $\beta = b\kappa$, $\psi = \omega t$ and the dimensionless variables ξ, η, σ

$$x = a + \varepsilon^{1/4} \frac{\xi}{\kappa}, \quad y = b + \varepsilon^{1/4} \frac{\eta}{\kappa}, \quad p = -\rho y g(1 + \varepsilon \cos \Omega t) + \varepsilon^{1/4} \rho \omega_0^2 \frac{\sigma}{\kappa^2},$$

where $\kappa = \pi n/L$ is the wave number (n is the number of nodes of the wave), g is the acceleration due to gravity, ω is the nonlinear frequency, $\omega_0^2 = g\lambda\kappa$ is the linear natural frequency, $\chi = h\kappa$ is the dimensionless fluid depth, and $\lambda = \tanh \chi$. We will consider weakly nonlinear standing waves or waves of small amplitude and steepness, for which the amplitude and the ratio of wave height to wavelength is assumed to be of order $\varepsilon^{1/4}$, where ε is a small parameter.

Introducing the linear differential operators

$$\mathcal{L}^1(\xi,\sigma)=\xi_{\psi\psi}+\sigma_\alpha,\ \mathcal{L}^2(\eta,\sigma)=\eta_{\psi\psi}+\sigma_\beta,\ \mathcal{L}^3(\xi,\eta)=\xi_\alpha+\eta_\beta,$$

we rewrite the problem in terms of the dimensionless coordinates and variables as follows:

$$\begin{aligned}
&\mathcal{L}^1(\xi,\sigma)=-\varepsilon^{1/4}\,(\xi_{\psi\psi}\xi_\alpha+\eta_{\psi\psi}\eta_\alpha),\ \mathcal{L}^2(\eta,\sigma)=-\varepsilon^{1/4}\,(\xi_{\psi\psi}\xi_\beta+\eta_{\psi\psi}\eta_\beta),\\
&\mathcal{L}^3(\xi,\eta)=-\varepsilon^{1/4}\,(\xi_\beta\eta_\alpha-\eta_\beta\xi_\alpha),\\
&\xi(0,\beta,\psi)=0,\ \xi(\pi n,\beta,\psi)=0,\ \eta(\alpha,-\chi,\psi)=0,\\
&\lambda\sigma(\alpha,0,\psi)-\eta(\alpha,0,\psi)=\varepsilon\eta(\alpha,0,\psi)\cos(\Omega t).
\end{aligned} \tag{5.5}$$

Applying the averaging transformations, we construct an approximate solution to the problem. The basic idea is to transform the original complicated hydrodynamic system to the dynamical system, modeling the main characteristics of the fluid motion, which is as simple as possible to investigate up to a given order in ε. We mention here the basic ideas of the method. We make an ansatz $\mathcal{F}^{111}(u)$, $u=\xi,\eta,\sigma$:

$$\begin{aligned}
&\mathcal{F}^{111}(\xi)=-C_0\sin\alpha\cosh\beta_1\cos\psi, \quad \mathcal{F}^{111}(\eta)=C_0\cos\alpha\sinh\beta_1\cos\psi,\\
&\mathcal{F}^{111}(\sigma)=C_0\cos\alpha\cosh\beta_1\cos\psi
\end{aligned}$$

in Eqs. (5.5) that corresponds to the initial conditions of a given class of standing waves. Here we denote $C_0=\mathcal{C}/\sinh\chi$ and $\beta_1=\beta+\chi$.

Following the idea of averaging method, we treat the amplitude and phase as slow functions of time, $\mathcal{C}(t)$ and $\theta(t)$, and assume that each of functions in the formal power series in the amplitude parameter ε, defining the asymptotic solution, depends on the spacial variables and on slowly varying amplitude $\mathcal{C}$, the fast phase ψ, and the slow phase θ. Therefore we look for an appropriate transformation of the dependent functions ξ,η,σ in the form

$$u=\mathcal{F}^{111}(u)+\sum_{i=2}^{\infty}\varepsilon^{\frac{i-1}{4}}u^{(i)}(\alpha,\beta,\mathcal{C}(t),\psi(t),\theta(t)), \quad u=\xi,\eta,\sigma,$$

where $\xi^{(i)}$, $\eta^{(i)}$, and $\sigma^{(i)}$ ($i=2,\dots,\infty$) are unknown 2π-periodic functions. Let $\Delta=\omega-\frac{1}{2}\Omega$ ($\omega\approx\frac{1}{2}\Omega$) be a small detuning of the order $O(\varepsilon\omega)$. Then the appropriate transformation of the main characteristics of motion, $\mathcal{C}(t)$ and $\theta(t)$, we seek in the form:

$$\frac{d\mathcal{C}}{dt}=\sum_{i=1}^{\infty}\varepsilon^{i/4}\,U_i(\mathcal{C},\theta), \qquad \frac{d\theta}{dt}=\Delta+\sum_{i=1}^{\infty}\varepsilon^{i/4}\,V_i(\mathcal{C},\theta),$$

where U_i and V_i are new unknown 2π-periodic functions, $\mathcal{C}$ is the amplitude and θ is the slow phase of the n-th wave harmonic. The fast phase of a given wave harmonic we define as $\psi(t)=\frac{1}{2}\Omega t+\theta(t)$, and according to the transformations defined above we obtain the following formula: $\frac{d\psi}{dt}=\omega+\sum_{i=1}^{\infty}\varepsilon^{i/4}V_i(\mathcal{C},\theta)$. Then, for $\varepsilon \ll 1$ and $|\Delta| \ll \omega$, we can perform averaging over the fast time $\sim 1/\omega$.

Substituting these expansions into the equations of motion and the boundary conditions (5.5) and matching the coefficients of like powers of ε, we arrive at the linear nonhomogeneous system of equations and boundary conditions for each approximation:

$$\xi_{i\psi\psi}+\sigma_\alpha=\sum_{n,m,k=0}^{\infty}\Big\{F_{1i}^{nmk}\cos(k\psi)+G_{1i}^{nmk}\sin(k\psi)\Big\}\sin(n\alpha)\cosh[m(\beta+\chi)],$$

$$\eta_{i\psi\psi}+\sigma_\beta=\sum_{n,m,k=0}^{\infty}\Big\{F_{2i}^{nmk}\cos(k\psi)+G_{2i}^{nmk}\sin(k\psi)\Big\}\cos(n\alpha)\sinh[m(\beta+\chi)],$$

$$\xi_{i\alpha}+\eta_\beta=\sum_{n,m,k=0}^{\infty}\Big\{F_{3i}^{nmk}\cos(n\alpha)\cosh[m(\beta+\chi)]\cos(k\psi)\Big\}=0,$$

$$\xi_i=0 \text{ at } \alpha=0,\pi n, \quad \eta_i=0 \text{ at } \beta=-\chi, \quad \sigma_i\tanh\chi-\eta_i=c_i \text{ at } \beta=0.$$

This system is extremely cumbersome and can be derived with the aid of computer algebra systems. In what follows, we obtain the coefficients c_i, F_{ji}^{nmk} and G_{ji}^{nmk} (j=1, 2, 3; i=2, ..., ∞) with the aid of *Maple.* We look for a solution, 2π-periodic in ψ, and θ in the form:

$$v^{(i)}=\sum_{n,m,k=0}^{i}V_i^{nmk}\mathfrak{F}^{nmk}(v), \quad v=\xi,\eta,\sigma, \quad V=\Xi,\mathrm{H},\Sigma,$$

where

$$\mathfrak{F}^{nmk}(\xi^{(i)})=\sin(n\alpha)\cosh(m\beta_1)\cos(k\psi),$$
$$\mathfrak{F}^{nmk}(\eta^{(i)})=\cos(n\alpha)\sinh(m\beta_1)\cos(k\psi),$$
$$\mathfrak{F}^{nmk}(\sigma^{(i)})=\cos(n\alpha)\cosh(m\beta_1)\cos(k\psi),$$

and Ξ_i^{nmk}, H_i^{nmk}, Σ_i^{nmk} (n,m,k=0, ..., i, i=2, ..., ∞) are unknown constants.

Substituting these series into the linear system of equations and boundary conditions for the i-th approximation and using the orthogonality conditions for periodic solutions, we obtain a family of systems of linear algebraic equations with respect to the unknown coefficients Ξ_i^{nmk}, H_i^{nmk}, and Σ_i^{nmk} $(n, m, k{=}0, \dots, i,\ i{=}2, \dots, \infty)$. Using the solvability conditions for the equations for Ξ_i^{111}, H_i^{111}, and Σ_i^{111}, we find the functions U_{i-1} and V_{i-1}.

By using formulas described above, we obtain the asymptotic solution of the order of $O(\varepsilon^{i+1})$ and the unknown corrections to the nonlinear wave frequency $\omega_{(i)}$. Setting $\beta{=}0$ in the parametric equations for κx and κy, we can obtain the profiles of surface standing waves in Lagrangian variables for the i-th approximation $(i{=}1, \dots, \infty)$.

3. Approximate analytical solution with Maple. We present the *Maple* solution for every step of the method and up to the second approximation, $i{=}2$ (`NA=2`).

(1) We find the second derivatives $\xi_{\psi\psi}$ and $\eta_{\psi\psi}$:

```
Th:=proc(Expr) local AA,Z1,Z2,j; global Z3;
 AA:=sqrt((1+th)/(1-th)); Z1:=convert(Expr, exp);
 Z2:=simplify(subs({'exp(j*chi)=AA^j' $ 'j'=-10..10},Z1));
 Z3:=combine(Z2); RETURN(Z3) end:
NA:=2: NP:=NA-1: vars1:=alpha,z,C,theta,psi: vars2:=alpha,z,C(t),
 theta(t),psi(t); Fxi:=(x->xi(alpha,z,C(x),theta(x),psi(x))):
Feta:=(x->eta(alpha,z,C(x),theta(x),psi(x))):
dxi:=diff(Fxi(t),t): deta:=diff(Feta(t),t):
FA1:=(x->A1(C(x),theta(x))): FB1:=(x->B1(C(x),theta(x))):
CT:=convert(['epsilon^i*F||A||i(t)' $ 'i'=1..NP],`+`):
psiT:=omega+convert(['epsilon^i*F||B||i(t)' $ 'i'=1..NP],`+`):
thetaT:=epsilon^4*D11+convert(['epsilon^i*F||B||i(t)'
 $ 'i'=1..NP],`+`): setsubt:={C(t)=C,psi(t)=psi,theta(t)=theta}:
setsub:={diff(C(t),t)=CT,diff(psi(t),t)=psiT,diff(theta(t),t)
 =thetaT}: sub1xi:=subs(setsub,dxi): sub2xi:=subs(setsub,
 diff(sub1xi,t)): sub1eta:=subs(setsub,deta):
sub2eta:=subs(setsub,diff(sub1eta,t)):
Xi||NP:=-sin(alpha)*cosh(I*z)/Th(sinh(chi))*C*cos(psi):
Eta||NP:=cos(alpha)*sinh(I*z)/Th(sinh(chi))*C*cos(psi):
Sigma||NP:=cos(alpha)*cosh(I*z)/Th(sinh(chi))*C*cos(psi):
xi:=(x1,x2,x3,x4,x5)->-sin(x1)*cosh(I*x2)/Th(sinh(chi))*
 x3*x4/x4*cos(x5): eta:=(x1,x2,x3,x4,x5)->cos(x1)*sinh(I*x2)
 /Th(sinh(chi))*x3*x4/x4*cos(x5): sigma:=(x1,x2,x3,x4,x5)->
 cos(x1)*cosh(I*x2)/Th(sinh(chi))*x3*x4/x4*cos(x5): xi(vars1):
eta(vars1): sigma(vars1): xi2T:=collect(sub2xi,epsilon):
xi2T:=convert(['coeff(xi2T,epsilon,i)*epsilon^(i)'$'i'=NP-1..NP],
 `+`); eta2T:=collect(sub2eta,epsilon): eta2T:=convert(
 ['coeff(eta2T,epsilon,i)*epsilon^(i)'$'i'=NP-1..NP],`+`);
```

where, according to the procedure Th, we convert the hyperbolic functions to the tanh function and denote it th.
For convenience, we exclude the symbolic expressions of linear operators $\mathcal{L}^1$, $\mathcal{L}^2$, and $\mathcal{L}^3$ in the time derivatives and the governing equations. We are working with the right-hand sides (trigonometric parts) of the expressions in (5.5).

(2) We obtain the equations of motion, the continuity equation, and the boundary conditions:

```
F1:=-omega^2*diff(sigma(vars1),alpha)-epsilon*(xi2T*diff(
 xi(vars1),alpha)+eta2T*diff(eta(vars1),alpha))-xi2T:
F2:=-omega^2*(-I)*diff(sigma(vars1),z)-epsilon*(xi2T*(-I)
 *diff(xi(vars1),z)+eta2T*(-I)*diff(eta(vars1),z))-eta2T:
F3:=-diff(xi(vars1),alpha)-(-I)*diff(eta(vars1),z)+epsilon
 *((-I)*diff(xi(vars1),z)*diff(eta(vars1),alpha)-(-I)
 *diff(eta(vars1),z)*diff(xi(vars1),alpha)):
F1:=subs({D[2](B1)(C,theta)=0,D[2](A1)(C,theta)=0},subs(
 setsubt,F1)): F1:=coeff(F1,epsilon,NP)*epsilon^NP;
F2:=subs({D[2](B1)(C,theta)=0,D[2](A1)(C,theta)=0},subs(
 setsubt,F2)): F2:=coeff(F2,epsilon,NP)*epsilon^NP;
F3:=subs(setsubt,F3): F3:=coeff(F3,epsilon,NP)*epsilon^NP;
setAB:={A1(C,theta)=A1,B1(C,theta)=B1}: F1S:=expand(subs(
 setAB,F1)); F2S:=expand(subs(setAB,F2)); F3S:=expand(F3);
Bc11:=eval(coeff(xi(vars2),epsilon,NP-1),alpha=0);
Bc12:=simplify(eval(coeff(xi(vars2),epsilon,NP-1),alpha=Pi*n))
 assuming n::natural; Bc2:=eval(coeff(eta(vars2),epsilon,NP-1),
 z=0); bc3:=(x,y)->lambda*x-y-epsilon^4*y*cos(2*psi-2*theta):
Bc3:=bc3(sigma(vars2),eta(vars2)): lambda:=Th(tanh(chi));
Bc3:=subs(setsubt,eval(Bc3,z=-I*chi)):
Bc3:=coeff(Bc3,epsilon,NP)*epsilon^NP;
```

(3) We rewrite the governing equations describing the standing wave motion and the fourth boundary condition at the free surface in the form of *Maple* functions:

```
Eq1:=(N,M,K)->-K^2*Xi||N||M||K-N*Sigma||N||M||K=EqC(F1S,N,M,K)
 /epsilon^NP; Eq2:=(N,M,K)->-K^2*Eta||N||M||K+M*Sigma||N||M||K
 =EqC(F2S,N,M,K)/epsilon^NP; Eq3:=(N,M,K)->N*Xi||N||M||K+
 M*Eta||N||M||K=EqC(F3S,N,M,K)/epsilon^NP; Eq4:=(N,M,K)->lambda
 *Sigma||N||M||K*Th(cosh(M*chi))-Eta||N||M||K*Th(sinh(M*chi));
Eq1s:=(N,M,K)->-K^2*Xi||N||M||K-N*Sigma||N||M||K=EqCA(F1S,N,M,K)
 /epsilon^NP;
```

```
Eq2s:=(N,M,K)->-K^2*Eta||N||M||K+M*Sigma||N||M||K
 =EqCA(F2S,N,M,K)/epsilon^NP; Eq3s:=(N,M,K)->N*Xi||N||M||K+
 M*Eta||N||M||K=EqCA(F3S,N,M,K)/epsilon^NP; Eq4s:=(N,M,K)->lambda
 *Sigma||N||M||K*Th(cosh(M*chi))-Eta||N||M||K*Th(sinh(M*chi));
```

(4) The coefficients F_{ji}^{nmk} and G_{ji}^{nmk} (j=1, 2, 3, n, m, k=0, ..., i) can be calculated according to the orthogonality conditions. We create the procedures `S1_NMK`–`S3_NMK`, `S1A_NMK`–`S3A_NMK`, which are the functions in the procedures `EqC` and `EqCA`:

```
S1_NMK:=proc(x,N,M,K) local I_N,I_M,I_K; global I_T;
 I_N:=1/Pi*int(x*sin(N*alpha),alpha=0..2*Pi);
 if M=0 then I_M:=1/(2*Pi)*int(I_N*cos(M*z),z=0..2*Pi);
 else I_M:=1/Pi*int(I_N*cos(M*z),z=0..2*Pi); fi:
 if K=0 then I_K:=1/(2*Pi)*int(I_M*cos(K*psi),psi=0..2*Pi);
 else I_K:=1/Pi*int(I_M*cos(K*psi),psi=0..2*Pi); fi:
 I_T:=I_K/omega^2; RETURN(I_T); end:
S1A_NMK:=proc(x,N,M,K) local I_N,I_M,I_K; global I_T;
 I_N:=1/Pi*int(x*sin(N*alpha),alpha=0..2*Pi);
 if M=0 then I_M:=1/(2*Pi)*int(I_N*cos(M*z),z=0..2*Pi);
 else I_M:=1/Pi*int(I_N*cos(M*z),z=0..2*Pi); fi:
 I_K:=1/Pi*int(I_M*sin(K*psi),psi=0..2*Pi);
 I_T:=I_K/omega^2; RETURN(I_T); end:
S2_NMK:=proc(x,N,M,K) local I_N,I_M,I_K; global I_T;
 if N=0 then I_N:=1/(2*Pi)*int(x*cos(N*alpha),alpha=0..2*Pi);
 else I_N:=1/Pi*int(x*cos(N*alpha),alpha=0..2*Pi); fi:
 I_M:=1/(Pi*I)*int(I_N*sin(M*z),z=0..2*Pi);
 if K=0 then I_K:=1/(2*Pi)*int(I_M*cos(K*psi),psi=0..2*Pi);
 else I_K:=1/Pi*int(I_M*cos(K*psi),psi=0..2*Pi); fi:
 I_T:=I_K/omega^2; RETURN(I_T); end:
S2A_NMK:=proc(x,N,M,K) local I_N,I_M,I_K; global I_T;
 if N=0 then I_N:=1/(2*Pi)*int(x*cos(N*alpha),alpha=0..2*Pi);
 else I_N:=1/Pi*int(x*cos(N*alpha),alpha=0..2*Pi); fi:
 I_M:=1/(Pi*I)*int(I_N*sin(M*z),z=0..2*Pi);
 I_K:=1/Pi*int(I_M*sin(K*psi),psi=0..2*Pi);
 I_T:=I_K/omega^2; RETURN(I_T); end:

S3_NMK:=proc(x,N,M,K) local I_N,I_M,I_K; global I_T;
 if N=0 then I_N:=1/(2*Pi)*int(x*cos(N*alpha),alpha=0..2*Pi);
 else I_N:=1/Pi*int(x*cos(N*alpha),alpha=0..2*Pi); fi:
 if M=0 then I_M:=1/(2*Pi)*int(I_N*cos(M*z),z=0..2*Pi);
 else I_M:=1/(Pi)*int(I_N*cos(M*z),z=0..2*Pi); fi:
```

```
 if K=0 then I_K:=1/(2*Pi)*int(I_M*cos(K*psi),psi=0..2*Pi);
 else I_K:=1/Pi*int(I_M*cos(K*psi),psi=0..2*Pi); fi:
 I_T:=I_K; RETURN(I_T); end:
S3A_NMK:=proc(x,N,M,K) local I_N,I_M,I_K; global I_T;
 if N=0 then I_N:=1/(2*Pi)*int(x*cos(N*alpha),alpha=0..2*Pi);
 else I_N:=1/Pi*int(x*cos(N*alpha),alpha=0..2*Pi); fi:
 if M=0 then I_M:=1/(2*Pi)*int(I_N*cos(M*z),z=0..2*Pi);
 else I_M:=1/(Pi)*int(I_N*cos(M*z),z=0..2*Pi); fi:
 I_K:=1/Pi*int(I_M*sin(K*psi),psi=0..2*Pi);
 I_T:=I_K; RETURN(I_T); end:
```

(5) To find the coefficients in the governing equations, we create the procedures EqC, EqCA, Bc3_NK, Bc3A_NK, and FORS.

```
EqC:=proc(Eq,N,M,K) local NEq,i; global SS; NEq:=nops(Eq); SS:=0:
 for i from 1 to NEq do
  if Eq=F1S then SS:=SS+S1_NMK(factor(op(i,Eq)),N,M,K):
  elif Eq=F2S then SS:=SS+S2_NMK(factor(op(i,Eq)),N,M,K):
  elif Eq=F3S then SS:=SS+S3_NMK(factor(op(i, Eq)),N,M,K): fi:
 od: RETURN(SS) end:
EqCA:=proc(Eq,N,M,K) local NEq,i; global SS; NEq:=nops(Eq);SS:=0:
  for i from 1 to NEq do
   if Eq=F1S then SS:=SS+S1A_NMK(op(i,Eq),N,M,K):
   elif Eq=F2S then SS:=SS+S2A_NMK(op(i,Eq),N,M,K):
   elif Eq=F3S then SS:=SS+S3A_NMK(op(i,Eq),N,M,K): fi:
  od: RETURN(SS) end:
Bc3_NK:=proc(x,N,K) local I_N,I_K; global I_T;
 if N=0 then I_N:=1/(2*Pi)*int(x*cos(N*alpha),alpha=0..2*Pi);
 else I_N:=1/Pi*int(x*cos(N*alpha),alpha=0..2*Pi); fi:
 if K=0 then I_K:=1/(2*Pi)*int(I_N*cos(K*psi),psi=0..2*Pi);
 else I_K:=1/Pi*int(I_N*cos(K*psi),psi=0..2*Pi); fi:
 I_T:=I_K; RETURN(I_T); end:
Bc3A_NK:=proc(x,N,K) local I_N,I_K; global I_T;
 if N=0 then I_N:=1/(2*Pi)*int(x*cos(N*alpha),alpha=0..2*Pi);
 else I_N:=1/Pi*int(x*cos(N*alpha),alpha=0..2*Pi); fi:
 I_K:=1/Pi*int(I_N*sin(K*psi),psi=0..2*Pi);
 I_T:=I_K; RETURN(I_T); end:
FORS:=proc(Syst) local i,nm; global SYST;
 nm:=nops(Syst): SYST:={}:
 for i from 1 to nm do SYST:=SYST union
  {cat(lhs(op(i,Syst)),9)=rhs(op(i,Syst))} od: RETURN(SYST) end:
```

(6) We solve these systems and obtain the asymptotic solution of the order of $O(\varepsilon^{i+1})$, using the equations described above:

```
SetSol:={}:
for N from 0 to NA do for M from 0 to NA do for K from 0 to NA do
 SetSol:=SetSol union {Xi||N||M||K=0} union {Eta||N||M||K=0}
        union {Sigma||N||M||K=0}: od: od: od:
sys1:=[Eq1(2,0,0),Eq2(2,0,0),Eq3(2,0,0),Eq1(2,2,0),Eq2(2,2,0),
 Eq3(2,2,0),Eq4(2,0,0)+Eq4(2,2,0)=-Bc3_NK(Bc3,2,0)/epsilon^NP];
sys1:=convert(sys1,set) union {Eta200=0}: var1:={Xi200,Eta200,
 Sigma200,Xi220,Eta220,Sigma220}: DF1:=solve(sys1,var1);
sys2:=[Eq1(2,0,2),Eq2(2,0,2),Eq3(2,0,2),Eq1(2,2,2),Eq2(2,2,2),
 Eq3(2,2,2),Eq4(2,0,2)+Eq4(2,2,2)=-Bc3_NK(Bc3,2,2)/epsilon^NP];
sys2:=convert(sys2,set) union {Eta202=0}: var2:={Xi202,Eta202,
 Sigma202,Xi222,Eta222,Sigma222}: DF2:=solve(sys2,var2);
sys3:=[Eq1(0,2,0),Eq2(0,2,0),Eq3(0,2,0),Eq1(0,0,0),Eq2(0,0,0),
 Eq3(0,0,0),Eq4(0,2,0)+Eq4(0,0,0)=-Bc3_NK(Bc3,0,0)/epsilon^NP];
sys3:=convert(sys3,set) union {Eta000=0,Xi020=0,Xi000=0}:
var3:={Xi020,Eta020,Sigma020,Xi000,Eta000,Sigma000}: DF3:=solve(
 sys3,var3); sys4:=[Eq1(0,2,2),Eq2(0,2,2),Eq3(0,2,2),Eq1(0,0,2),
 Eq2(0,0,2),Eq3(0,0,2),Eq4(0,2,2)+Eq4(0,0,2)=-Bc3_NK(Bc3,0,2)/
 epsilon^NP]; sys4:=convert(sys4,set) union {Eta002=0}:
var4:={Xi022,Eta022,Sigma022,Xi002,Eta002,Sigma002}:
DF4:=solve(sys4,var4); C_2:=0: S_2:=0:
sys5:=[Eq1(1,1,1),Eq2(1,1,1),Eq3(1,1,1),Eq4(1,1,1)=
 -Bc3_NK(Bc3,1,1)/epsilon^N]; sys5:=convert(sys5,set) union
 {Sigma111=C_2}: var5:={Xi111,Eta111,Sigma111,B||NP}: DF5:=
 solve(sys5,var5); sys6:=[Eq1s(1,1,1),Eq2s(1,1,1),Eq3s(1,1,1),
 Eq4s(1,1,1)=-Bc3A_NK(Bc3,1,1)/epsilon^N]; sys6:=convert(sys6,
 set) union {Sigma111=S_2}: var6:={Xi111,Eta111,Sigma111,A||NP}:
DF6S:=solve(sys6,var6); for i from 1 to nops(DF6S) do
 if has(op(i,DF6S),A||NP) then ZZ:=i fi: od: DF6s:=subsop(
 ZZ=NULL,DF6S): DF6:=FORS(DF6s); sys7:=[Eq1s(2,0,0),Eq2s(2,0,0),
 Eq3s(2,0,0),Eq1s(2,2,0),Eq2s(2,2,0),Eq3s(2,2,0),Eq4s(2,0,0)+
 Eq4s(2,2,0)=-Bc3A_NK(Bc3,2,0)/epsilon^NP]; sys7:=convert(sys7,
 set): var7:={Xi200,Eta200,Sigma200,Xi220,Eta220,Sigma220}:
DF7S:=solve(sys7,var7); DF7:=FORS(DF7S);
sys8:=[Eq1s(2,0,2),Eq2s(2,0,2),Eq3s(2,0,2),Eq1s(2,2,2),
 Eq2s(2,2,2),Eq3s(2,2,2),Eq4s(2,0,2)+Eq4s(2,2,2)=
 -Bc3A_NK(Bc3,2,2)/epsilon^NP]; sys8:=convert(sys8,set)
 union {Eta202=0}: var8:={Xi202,Eta202,Sigma202,Xi222,Eta222,
 Sigma222}: DF8S:=solve(sys8,var8); DF8:=FORS(DF8S);
```

(7) We generate the approximate analytical solution for the second approximation:

```
FF:=proc(x::list) global Xi_S,Eta_S,Sigma_S,NA; local i,Nx,a,b,
 c,s,N,M,K; Nx:=nops(x);  Xi_S:={}: Eta_S:={}: Sigma_S:={}:
 for i from 1 to Nx do a:=lhs(x[i]); for N from 0 to NA do
 for M from 0 to NA do for K from 0 to NA do
 if a=Xi||N||M||K then XiS||N||M||K:=rhs(x[i])*sin(N*alpha)
 *cosh(M*(beta+chi))*cos(K*psi): Xi_S:=Xi_S union {XiS||N||M||K};
 elif a=Eta||N||M||K then EtaS||N||M||K:=rhs(x[i])*cos(N*alpha)
 *sinh(M*(beta+chi))*cos(K*psi): Eta_S:=Eta_S union
 {EtaS||N||M||K};
 elif a=Sigma||N||M||K then SigmaS||N||M||K:=rhs(x[i])
 *cos(N*alpha)*cosh(M*(beta+chi))*cos(K*psi):
 Sigma_S:=Sigma_S union {SigmaS||N||M||K}; fi:
 if a=Xi||N||M||K||9 then XiS||N||M||K:=rhs(x[i])*sin(N*alpha)
 *cosh(M*(beta+chi))*sin(K*psi): Xi_S:=Xi_S union {XiS||N||M||K};
 elif a=Eta||N||M||K||9 then EtaS||N||M||K:=rhs(x[i])
 *cos(N*alpha)*sinh(M*(beta+chi))*sin(K*psi):
 Eta_S:=Eta_S union {EtaS||N||M||K};
 elif a=Sigma||N||M||K||9 then SigmaS||N||M||K:=rhs(x[i])
 *cos(N*alpha)*cosh(M*(beta+chi))*sin(K*psi):
 Sigma_S:=Sigma_S union {SigmaS||N||M||K}; fi: od: od: od: od:
RETURN(Xi_S, Eta_S, Sigma_S); end:
SetSol:=convert({op(SetSol)} union {'op(DF||i)'$'i'=1..8},list):
Xi||NA:=convert(FF(SetSol)[1],`+`); Eta||NA:=convert(
 FF(SetSol)[2],`+`); Sigma||NA:=convert(FF(SetSol)[3],`+`);
 Xi||NP:=-sin(alpha)*cosh(beta+chi)/Th(sinh(chi))*C*cos(psi);
Eta||NP:=cos(alpha)*sinh(beta+chi)/Th(sinh(chi))*C*cos(psi);
Sigma||NP:=cos(alpha)*cosh(beta+chi)/Th(sinh(chi))*C*cos(psi);
XiSer:=Xi||NP+epsilon^NP*(Xi||NA); EtaSer:=Eta||NP+epsilon^NP
 *(Eta||NA); SigmaSer:=Sigma||NP+epsilon^NP*(Sigma||NA);
AN:=nops(DF6S): for j from 1 to AN do if lhs(op(j,DF6S))=A||NP
 then AA||NP:=rhs(op(j,DF6S)): fi: od: BN:=nops(DF5):
for j from 1 to BN do if lhs(op(j,DF5))=B||NP then BB||NP:=
 rhs(op(j,DF5)): fi: od: combine(AA||NP), combine(BB||NP);
```

4. Hydrodynamic-type system. Following the described method and applying the averaging transformations, we can obtain the system of ODEs or hydrodynamic-type system, which models the main characteristics of the original complicated system and is more simple to investigate (e.g., by performing the qualitative analysis of differential equations).

Therefore performing symbolic calculations according to the averaging transformations described above, we obtain the first values $A_1=0$ and $B_1=0$ (`AA||NP`, `BB||NP`) in the second approximation.

Then doing so in a similar manner, we can obtain the following values for A_i and B_i up to the fifth approximation: $A_2=A_3=0$, $A_4=\frac{1}{4}\omega\mathcal{C}\sin(2\theta)$, $B_2=-\omega\phi_2\mathcal{C}^2$, $B_3=0$, $B_4=\frac{1}{4}\omega[\cos(2\theta)-\phi_4\mathcal{C}^4]$. The second- and fourth-order corrections to the wave frequency, ϕ_2 and ϕ_4, take the form:

$$\phi_2=\frac{1}{64}z^{-4}(2z^6+3z^4+12z^2-9),$$
$$\phi_4=\frac{1}{16384}z^{-10}(36z^{16}-64z^{14}-883z^{12}+691z^{10}-1611z^8+6138z^6 -4077z^4+1323z^2-81),$$

where $z=\tanh\chi$. Hence we arrive at the following hydrodynamic-type system:

$$\frac{d\mathcal{C}}{dt}=\frac{1}{4}\varepsilon\omega\mathcal{C}\sin(2\theta), \quad \frac{d\theta}{dt}=\Delta-\varepsilon^{1/2}\omega\phi_2\mathcal{C}^2+\varepsilon\omega\left[\frac{1}{4}\cos(2\theta)-\phi_4\mathcal{C}^4\right]. \quad (5.6)$$

This system describes evolution of the amplitude $\mathcal{C}$ and the slow phase θ. The terms containing the ϕ_{2i} $(i=1,2)$ describe the nonlinear frequency shift of the standing wave. We perform the qualitative analysis of this hydrodynamic-type system in Problem 3.22 and solve this problem following analytical-numerical approach in Problem 7.3.

5. Surface profiles. We construct the standing wave profiles, using the approximate analytical solution obtained up to the second approximation and the nonlinear second-order correction ϕ_2 to the wave frequency (obtained in the third approximation).

According to the Lagrangian formulation, the free surface $y=\eta(x,t)$ is defined by the parametric curve $\{x(a,0,t), y(a,0,t)\}$, $b=0$. Setting $b=0$ in the solutions obtained for x and y and choosing the parameters that correspond to the experimental data performed by W. W. Schultz et. al [145], we can observe the standing wave motion. The theoretical surface profile is shown in Fig. 5.1 at different moments of time.

We can compare the theoretical and experimental results [145]. We can see that the theoretical surface wave profiles are in good agreement with the experimental profiles.

```
with(plots): Digits:=30: tr1:=th=tanh(chi);
xiS:=subs(tr1,op(1,XiSer)+epsilon^(1/4)*op(2,op(2,XiSer)));
etaS:=subs(tr1,op(1,EtaSer)+epsilon^(1/4)*op(2,op(2,EtaSer)));
beta:=0: alpha:=a*kappa: rho:=1: psi:=OmegaT/2+Pi: g:=981.7:
n:=2: L:=10.5: w:=1.7: h:=15: s:=0.11: Omega:=44.647;
```

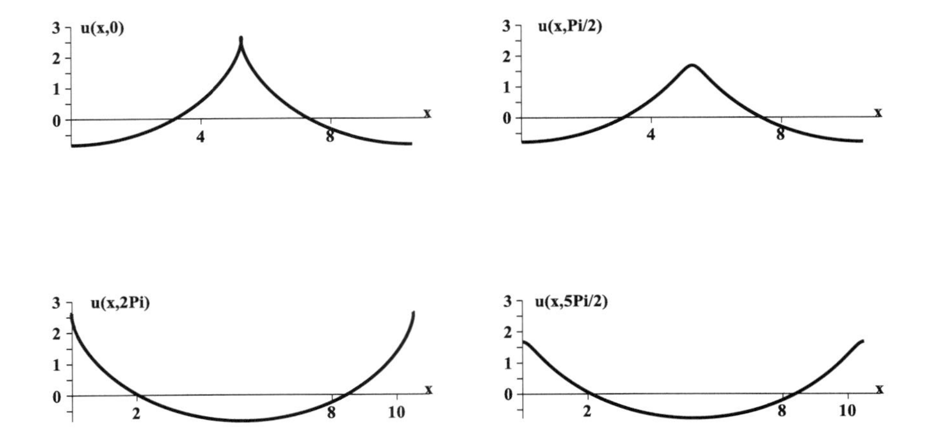

Fig. 5.1. Theoretical surface profile at different times $t_k = 0, \frac{1}{2}\pi, 2\pi, \frac{5}{2}\pi$

```
kappa:=evalf((Pi*n)/L): chi:=evalf(h*kappa):
lambda:=evalf(tanh(chi)): omega:=evalf(sqrt(lambda*g*kappa)):
Lambda:=evalf((2*Pi)/kappa): epsilon:=evalf(s*Omega^2/g):
phi2:=subs(z=lambda,(2*z^6+3*z^4+12*z^2-9)/(64*z^4));
BB2:=-omega*phi2*C^2; Delta:=evalf(omega-Omega/2);
EqC:=Delta+BB2*epsilon^(1/2)+(1/4)*epsilon*omega;
C:=fsolve(EqC,C=0..100); C2:=epsilon^(1/4)*C;
y2:=epsilon^(1/4)*etaS/kappa; x2:=a+epsilon^(1/4)*xiS/kappa;
X2:=unapply(x2,[a,OmegaT]); Y2:=unapply(y2,[a,OmegaT]);
animate([X2(a,t),Y2(a,t),a=0..L],t=0..40,color=blue,
 thickness=4,scaling=constrained,frames=50);
Y2Max:=Y2(L/2,0); Y2Min:=Y2(L/2,2*Pi); DeltaY:=(Y2Max-Y2Min)/2;
HL:=(Y2Max-Y2Min)/Lambda; lprint(`Omega=`,evalf(Omega/(2*Pi)),
 `2Pi Omega=`,Omega,`h=`,h,`s=`,s); lprint(`C2=`,C2,`HL=`,HL,
 `Y2Max=`,Y2Max); lprint(`DeltaY=`,DeltaY);
X2mx:=subs(OmegaT=0,x2): Y2mx:=subs(OmegaT=0,y2):
Mmx:=plot([X2mx,Y2mx,a=0..L],color=blue,thickness=2):
display(Mmx);
```

□

Chapter 6
Numerical Approach

It is known that nonlinear partial differential equations represent important mathematical models of real-world phenomena (e.g., physical, chemical, biological, economic, social, etc). Frequently it is not possible to find exact or approximate analytical solutions of a complicated nonlinear initial boundary value problem, therefore it is necessary to investigate the given problem numerically and analyze some simple special cases.

In this chapter, first, we will consider the construction of numerical and graphical solutions of various initial boundary value problems using predefined functions and embedded methods in both systems (*Maple* and *Mathematica*), e.g., for the second-order PDEs (linear heat equation and Burgers equation, linear and nonlinear wave equations, the Klein–Gordon equation). In some cases we compare the numerical solutions with the corresponding analytical solutions and obtain the corresponding error function. Additionally, we will show how to find numerical and graphical solutions by specifying a particular numerical method and numerical boundary conditions.

Then, we will concentrate on finite difference methods and solve various important initial boundary value problems, e.g., for the Burgers and the inviscid Burgers equations, for the nonlinear wave equation, for the KdV equation (investigating N-soliton interaction), for the sine–Gordon equation (investigating antikink-kink interaction), for the nonlinear Poisson equation. In some cases we obtain numerical solutions using the predefined functions in both systems and compare two numerical solutions, also we compare the numerical solutions and the corresponding analytical solutions (obtained in the previous chapters of the book), the corresponding linear and nonlinear problems.

6.1 Embedded Numerical Methods

First, let us consider the predefined functions in both systems with the aid of which we can obtain the approximate numerical solutions solving various linear and nonlinear PDE problems.

In *Maple*, with the aid of the predefined function `pdsolve` (with option `numeric`), we can solve numerically initial boundary value problems for a single PDE (of higher order), PDE systems by using the embedded methods, or by specifying a particular method for solving a single PDE. It is possible to impose Dirichlet, Neumann, Robin, or periodic boundary conditions.

In *Mathematica*, various initial boundary value problems can be solved numerically with the aid of the function `NDSolve` (with various options).

6.1.1 Nonlinear PDEs

In *Maple*, numerical solutions can be obtained automatically (without specifying a numerical method) by using embedded θ-methods (see [89], [106]). The θ-method is a generalization of the known finite difference approximations (explicit and implicit) by introducing a parameter θ ($0\leq \theta \leq 1$) and taking a weighted average of the two formulas, where the special case $\theta = \frac{1}{2}$ corresponds to the Crank–Nicolson method [34]* and $\theta=0$, $\theta=1$ are the explicit and implicit methods, respectively. In *Maple* (Ver. ≤ 14) it is possible to solve numerically only evolution equations via the predefined functions.

Maple:

```
infolevel[all]:=5;    Sol:=pdsolve(PDEs,IBCs,numeric,funcs,ops);
Num_vals:=Sol:-value();          Sol:-plot3d(func,t=t0..t1,ops);
Num_vals(num1,num2);  Sol:-animate(func,t=t0..t1,x=x0..x1,ops);
pdsolve(PDE,IBC,numeric,numericalbcs=val,method=M1,startup=M2);
```

`pdsolve,numeric`, finding numerical solutions to a partial differential equation `PDE` or a system of `PDEs`;

`plot3d`, `animate`, visualizing the numerical solution `Sol` obtained by applying `pdsolve,numeric`;

`value`, displaying numerical values of the numerical solution `Sol`.

*We note that the name of the numerical method, `CrankNicolson`, known in scientific literature is slightly different from Maple's name `CrankNicholson`.

The solution obtained is represented as a *module* (similar to a procedure, with the operator `:-`), which can be used for obtaining visualizations (`plot`, `plot3d`, `animate`, `animate3d`) and numerical values (`value`), in more detail, see `?pdsolve[numeric]`.

In *Mathematica*, with the aid of the predefined function `NDSolve`, it is possible to obtain approximate numerical solutions of various linear and nonlinear PDE problems (initial boundary value problems). This can be done by the *Mathematica* system applying the *method of lines*. We note that in *Mathematica* (Ver. ≤ 8) it is possible to solve numerically only *evolution equations* using `NDSolve`. Additionally, it is possible to specify explicitly the method of lines for solving PDEs and the proper suboptions for the method of lines.

Mathematica:

```
NDSolve[{pde,ic,bc}, depVars, indVars, ops]
NDSolve[{pde,ic,bc},  u, {x,x1,x2}, {t,t1,t2},...]
NDSolve[{pde,ic,bc}, {u1,...,un}, {x,x1,x2}, {t,t1,t2},...]
NDSolve[{pde,ic,bc},u,{x,x1,x2},{t,t1,t2},Method->{m,subOps}]
Options[NDSolve`MethodOfLines]
```

`NDSolve`, finding numerical solutions to PDEs problems (initial boundary value problems), where `depVars` and `indVars` are the dependent and independent variables, respectively;

`NDSolve,Method`, finding numerical solutions to PDEs problems by the method of lines with some specific suboptions `subOps`;

the option `Method` and the most important suboptions:

```
Method->{"MethodOfLines","SpatialDiscretization"->{
 "TensorProductGrid","MinPoints"->val,"MaxPoints"->val,
 "MaxStepSize"->val,"PrecisionGoal"->val,"DifferenceOrder"->val}}
```

Now we write out schematically numerical solution `sol` with the aid of `NDSolve` and then we use it to obtain various visualizations (e.g., `Plot`, `Plot3D`, `Animate`) and numerical values (`numVals`).

```
sol=NDSolve[{pde,ic,bc},u,{x,x1,x2},{t,t1,t2},ops]
Plot[Evaluate[u[x,tk]/.sol],{x,x1,x2},ops]
numVals=Evaluate[u[xk,tk]/.sol]
Plot3D[Evaluate[u[x,t]/.sol],{x,x1,x2},{t,t1,t2},ops]
Animate[Plot[Evaluate[u[x,t]/.sol],{x,x1,x2},ops],{t,t1,t2},ops]
```

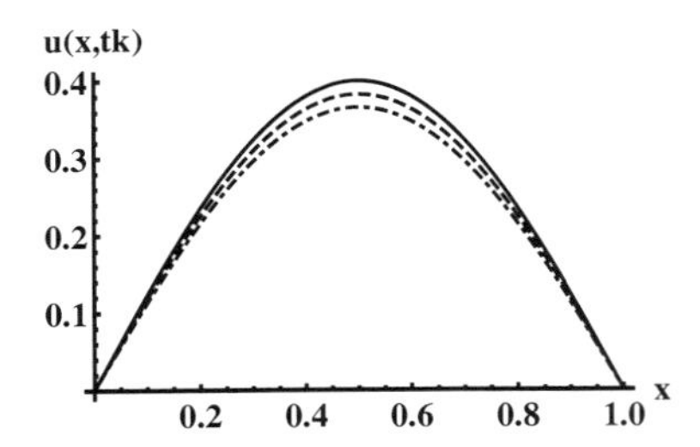

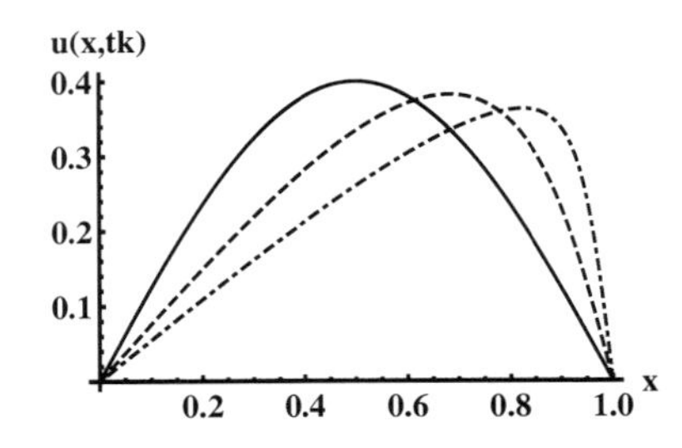

Fig. 6.1. Plots of numerical solutions of the initial boundary value problems for the linear heat equation and the Burgers equation at different times $t_k = 0$ (thick lines), $\frac{1}{2}$ (dashed lines), 1 (dot-dashed lines)

Problem 6.1 *Linear heat equation and Burgers equation. Numerical and graphical solutions.* Let us consider the initial boundary value problems for the linear heat equation and the Burgers equation:

$$u_t{=}\nu u_{xx};\ u(x,0){=}f(x),\ u(0,t){=}0,\ u(L,t){=}0,$$
$$u_t{=}\nu u_{xx} - uu_x;\ u(x,0){=}f(x),\ u(0,t){=}0,\ u(L,t){=}0,$$

defined in the domain $\mathcal{D} = \{0 < x < L,\ 0 < t < T\}$, where $L{=}1$, $T{=}40$, $f(x){=}A\sin(\pi x/L)$, the initial amplitude $A{=}0.4$, and the kinematic viscosity $\nu{=}0.009$. Find numerical and graphical solutions of the given initial boundary value problems, compare the numerical solutions of these initial boundary value problems with the corresponding analytical solutions, obtain an error function at $t{=}t_k$ and plot it, and compare the corresponding numerical values of the numerical and analytical solutions at $t{=}t_k$.

1. First, let us find numerical and graphical solutions (see Fig. 6.1) of the given initial boundary value problems for the linear heat equation and the Burgers equation applying predefined functions.

Maple:

```
with(plots): with(PDEtools): declare(u(x,t)); nu:=0.009; A:=0.4:
S:=1/100; tR:=0..40; xR:=0..1; NF:=30; NP:=100; N:=3; L:=1;
L1:=[red,blue,green]; L2:=[0,1/2,1]; Ops:=spacestep=S,timestep=S;
Op1:=frames=NF,numpoints=NP; f:=x->A*sin(Pi*x/L);
PDE1:=diff(u(x,t),t)-nu*diff(u(x,t),x$2)=0;
PDE2:=diff(u(x,t),t)-nu*diff(u(x,t),x$2)+u(x,t)*diff(u(x,t),x)=0;
IC:={u(x,0)=f(x)}; BC:={u(0,t)=0,u(L,t)=0};
```

```
for i from 1 to 2 do
 Sol||i:=pdsolve(PDE||i,IC union BC,numeric,u(x,t),Ops); od;
for i from 1 to N do for j from 1 to 2 do
 G||j||i:=Sol||j:-plot(t=L2[i],color=L1[i],numpoints=200): od:od:
display({seq(G1||i,i=1..N)}); display({seq(G2||i,i=1..N)});
for i from 1 to 2 do
 Num_vals||i:=Sol||i:-value(); Num_vals||i(1/2,Pi);
 Sol||i:-plot3d(u(x,t),t=tR,shading=zhue,axes=boxed);
 Sol||i:-animate(u(x,t),x=xR,t=tR,Op1,thickness=3); od;
```

Mathematica:

```
f[x_]:=a*Sin[Pi*x/l]; {nu=0.009,a=0.4,s=0.01,nP=100,nN=3,l=1,
 l1={Red,Blue,Green},l2={0,1/2,1}}
SetOptions[Plot,ImageSize->500,PlotRange->All,PlotPoints->nP*2,
 PlotStyle->{Blue,Thickness[0.01]}];
{pde[1]=D[u[x,t],t]-nu*D[u[x,t],{x,2}]==0,
 pde[2]=D[u[x,t],t]-nu*D[u[x,t],{x,2}]+u[x,t]*D[u[x,t],x]==0,
 ic={u[x,0]==f[x]},bc={u[0,t]==0,u[l,t]==0}}
Do[sol[i]=NDSolve[{pde[i],ic,bc},u,{x,0,1},{t,0,40},
 MaxStepSize->s],{i,1,2}]; Do[g[j,i]=Plot[Evaluate[u[x,l2[[i]]]/.
 sol[j]],{x,0,1},PlotStyle->{l1[[i]],Thickness[0.01]}],
 {j,1,2},{i,1,nN}]; GraphicsRow[{Show[Table[g[1,i],{i,1,nN}]],
 Show[Table[g[2,i],{i,1,nN}]]}]
Do[numVals[i]=Evaluate[u[1/2,Pi]/.sol[i]]; Print[numVals[i]];
 g3D[i]=Plot3D[Evaluate[u[x,t]/.sol[i]],{x,0,1},{t,0,40},
 ColorFunction->Function[{x,y},Hue[x]],BoxRatios->1,ViewPoint->
 {-1,2,1},ImageSize->900]; gCP[i]=ContourPlot[Evaluate[
  u[x,t]/.sol[i]],{x,0,1},{t,0,40},ColorFunction->Hue,
  ImageSize->300],{i,1,2}]; GraphicsRow[{g3D[1],g3D[2]}]
GraphicsRow[{gCP[1],gCP[2]}]
Animate[Row[{Plot[Evaluate[u[x,t]/.sol[1],{x,0,1}],PlotRange->
 {0,0.4}], Plot[Evaluate[u[x,t]/.sol[2],{x,0,1}],PlotRange->
 {0,0.4}]}],{t,0,40},AnimationRate->0.5]
```

2. We compare the numerical solutions of these initial boundary value problems with the corresponding analytical solutions. The solution of the first initial boundary value problem for the linear heat equation (e.g., see [125], p. 1269) has the form: $u(x,t)=\int_0^L f(\xi)G(x,\xi,t)\,d\xi$, where $G(x,\xi,t)=\frac{2}{L}\sum_{n=1}^{\infty}\sin\left(\frac{n\pi x}{L}\right)\sin\left(\frac{n\pi\xi}{L}\right)e^{P}$, $P=-\frac{\nu n^2\pi^2 t}{L^2}$.

Applying the Hopf–Cole transformation (see Problem 2.14), we obtain the solution of the initial boundary value problem for the Burgers equation $u(x,t) = -2\nu\phi_x/\phi$, where $\phi(x,t) = A_0 + \sum_{n=1}^{\infty} A_n e^P \cos(n\pi x/L)$ is the solution of the initial boundary problem for the linear heat equation, and the coefficients A_0 and A_n take the form:

$$A_0 = \frac{1}{L}\int_0^L e^{-X(1-\cos q)}\,dx, \quad A_n = \frac{2}{L}\int_0^L e^{-X(1-\cos q)}\cos(nq)\,dx.$$

Here $X{=}AL/(2\pi\nu)$ and $q = \pi x/L$. Moreover, this solution can be written in terms of the modified Bessel function of the first kind, $I_n(x)$, as follows:

$$u(x,t){=}\frac{4\nu\pi}{L}\left(\frac{\sum_{n=1}^{\infty} nI_n(X)e^P \sin(n\pi x/L)}{I_0(X){+}2\sum_{n=1}^{\infty} I_n(X)e^P\cos(n\pi x/L)}\right).$$

We plot the exact and numerical solutions for $(x_k, t_k) \in \mathcal{D}$.

It should be noted that the solution of the linear heat equation consists of only one fundamental harmonic $u(x,t) = A\exp(P_1)\sin(\pi x/L)$ (where $P_1{=}-\nu\pi^2 t/L^2$), whereas the solution of the Burgers equation consists of an infinite set of harmonics. The linear solution depends on the initial amplitude A, whereas the nonlinear solution depends on the *Reynolds number* $R{=}AL/\nu$. If $R{\to}\infty$, the series solution converges rapidly for all values of $B{=}\nu t/L^2$. If $B{\gg}1$, then we obtain the only one first harmonic $u(x,t){=}(4\nu\pi/L)\exp(P_1)\sin(\pi x/L)$ (as in the linear case).

Maple:

```
N1:=1; N2:=50; tk:=1: X:=(A*L)/(2*Pi*nu); q:=Pi*x/L;
p:=nu*Pi^2*t/L^2; GrF:=(x,xi,t,N)->2/L*add(sin(n*q)
 *sin(n*Pi*xi/L)*exp(-p*n^2),n=1..N1);
SolAn1:=unapply(int(expand(f(xi)*GrF(x,xi,t,N)),xi=0..L),x,t);
A0:=evalf(1/L*int(exp(-X*(1-cos(q))),x=0..L));
An:=2/L*int(exp(-X*(1-cos(q)))*cos(n*q),x=0..L);
phi:=A0+add(evalf(An)*exp(-n^2*p)*cos(n*q),n=1..N2);
SolAn2:=unapply(-2*nu*diff(phi,x)/phi,x,t);
for i from 1 to 2 do
 G1||i:=Sol||i:-plot(t=tk,color=red,numpoints=2*NP):
 G2||i:=plot(SolAn||i(x,tk),x=xR,color=blue): od:
display({G11,G21}); display({G12,G22});
```

Table 6.1. Comparison of numerical and analytical solutions of Burgers equation

x	Numerical sol.	Analytical sol.	Absolute error
0.000	0.000	0.000	0.000
0.100	0.055	0.055	6.293×10^{-6}
0.200	0.109	0.109	0.000
0.300	0.161	0.161	0.000
0.400	0.212	0.212	0.000
0.500	0.260	0.260	0.000
0.600	0.304	0.304	4.955×10^{-6}
0.700	0.340	0.340	8.184×10^{-6}
0.800	0.362	0.362	0.000
0.900	0.338	0.338	0.000
1.000	1.802×10^{-17}	2.334×10^{-16}	2.154×10^{-16}

Mathematica:

```
{n1=1,n2=50,tk=1,xN=(a*l)/(2*Pi*nu),q=Pi*x/l,p=nu*Pi^2*t/l^2}
grF[x_,xi_,t_,nN]:=2/l*Sum[Sin[n*q]*Sin[n*Pi*xi/l]*
 Exp[-p*n^2],{n,1,nN}];
solAn[1][x1_,t1_]:=Integrate[Expand[f[xi]*grF[x,xi,t,nN]],
 {xi,0,l}]/.{x->x1,t->t1}; solAn[1][x,t]
a0=1/l*NIntegrate[Exp[-xN*(1-Cos[q])],{x,0,l}]
an=2/l*Hold[NIntegrate[Exp[-xN*(1-Cos[q])]*Cos[n*q],{x,0,l},
 AccuracyGoal->6]]
phi=a0+Sum[Re[ReleaseHold[an]]*Exp[-n^2*p]*Cos[n*q],{n,1,n2}]
solAn[2][x1_,t1_]:=-2*nu*D[phi,x]/phi/.x->x1/.t->t1;
solAn[2][x,t]
Do[g1[i]=Plot[Evaluate[u[x,tk]/.sol[i]],{x,0,l},PlotStyle->
 {Red,Thickness[0.01]},PlotPoints->nP*2];
   g2[i]=Plot[Evaluate[solAn[i][x,tk]],{x,0,l},PlotStyle->
 {Blue,Thickness[0.005]}],{i,1,2}]; Show[{g1[1],g2[1]}]
Show[{g1[2],g2[2]}]
```

3. Finally, for one of these initial boundary value problems, e.g., for the second problem, we obtain an error function at $t=t_k$ and plot it. We can compare the numerical values of the numerical and analytical solutions at $t=t_k$, e.g., $t_k=1$ (see Tab. 6.1).

Maple:

```
ErFun2:=rhs(Sol2:-value(abs(u(x,t)-SolAn2(x,t)),t=tk,output=
 listprocedure)[3]); seq(ErFun2(n/10),n=0..10);
utk:=Sol2:-value(t=tk,output=listprocedure);
uVal:=rhs(op(3,utk)); plot(uVal(x),x=xR);
for x from 0 to L by 0.1 do print(uVal(x),evalf(SolAn2(x,tk)),
      abs(uVal(x)-evalf(SolAn2(x,tk)))); od;
```

Mathematica:

```
uVal[x1_]:=Evaluate[u[x,tk]/.sol[2]]/.{x->x1}; Plot[Evaluate[
 u[x,tk]/.sol[2]],{x,0,1},PlotStyle->{Red,Thickness[0.01]},
 PlotPoints->nP*2]
PaddedForm[Table[{First[uVal[x1]],solAn[2][x1,tk]//N,Abs[First[
 uVal[x1]]-solAn[2][x1,tk]//N]},{x1,0,1,0.1}]//TableForm,
 {12,5}]
```
□

Problem 6.2 *Linear and nonlinear wave equations. Numerical and graphical solutions.* Let us consider the initial boundary value problems for the linear and nonlinear wave equations:

$$u_{tt}{=}c^2\,u_{xx};\;\; u(x,0){=}0,\;\; u_t(x,0){=}\sin(4\pi x),\;\; u(0,t){=}u(L,t){=}0,$$
$$u_{tt}{=}c^2\,u_{xx}+F(u);\;\; u(x,0){=}0,\;\; u_t(x,0){=}\sin(4\pi x),\;\; u(0,t){=}u(L,t){=}0,$$

defined in $\mathcal{D}=\{0<x<L,\ 0<t<T\}$, where $c=1/(4\pi)$, $L{=}0.5$, $T{=}1.5$, $\lambda{=}1$, and $F(u){=}e^{\lambda u}$. We find numerical and graphical solutions (see Fig. 6.2) of the linear and nonlinear wave equations subject to the same initial and boundary value conditions.

Maple:

```
with(VectorCalculus): with(plots): with(PDEtools):
declare(u(x,t)); c:=evalf(1/(4*Pi)): lambda:=1: L:=0.5; T:=1.5;
S:=1/100; tR:=0..T; xR:=0..L; NF:=30; NP:=100; Ops:=spacestep=S,
 timestep=S; Op1:=frames=NF,numpoints=NP,thickness=3; N:=3;
L1:=[red,blue,green];L2:=[0.3,0.7,1.5]; F:=u->exp(lambda*u(x,t));
PDE1:=diff(u(x,t),t$2)-c^2*Laplacian(u(x,t),'cartesian'[x])=0;
PDE2:=diff(u(x,t),t$2)=c^2*diff(u(x,t),x$2)+F(u);
Ics:={u(x,0)=0,D[2](u)(x,0)=sin(4*Pi*x)};
Bcs:={u(0,t)=0,u(L,t)=0};
for i from 1 to 2 do
 Sol||i:=pdsolve(PDE||i,Ics union Bcs,numeric,u(x,t),Ops); od;
```

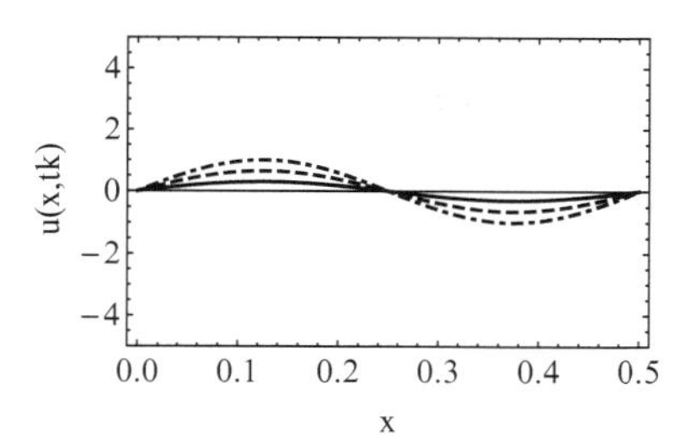

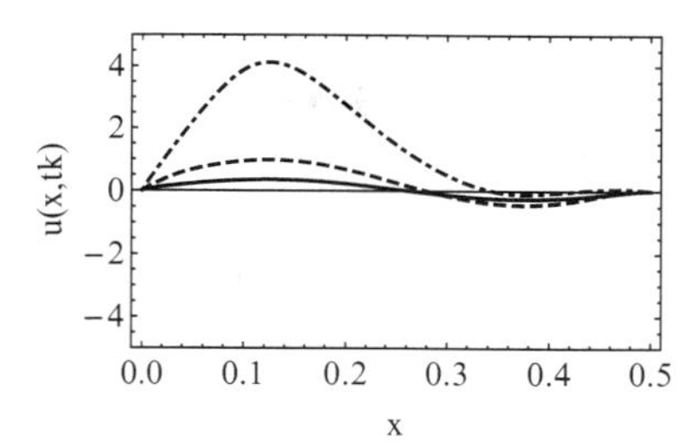

Fig. 6.2. Plots of numerical solutions of the initial boundary value problems for the linear and nonlinear wave equations at different times $t_k = 0.3$ (thick lines), 0.7 (dashed lines), 1.5 (dot-dashed lines)

```
for i from 1 to N do for j from 1 to 2 do
 G||j||i:=Sol||j:-plot(t=L2[i],color=L1[i],numpoints=NP*2): od:
od: display({seq(G1||i,i=1..N)}); display({seq(G2||i,i=1..N)});
for i from 1 to 2 do
 Num_vals||i:=Sol||i:-value(); Num_vals||i(0.1,1/2);
 Sol||i:-plot3d(u(x,t),t=tR,shading=zhue,axes=boxed);
 Sol||i:-animate(u(x,t),x=xR,t=tR,Op1); od;
```

Mathematica:

```
<<DifferentialEquations`InterpolatingFunctionAnatomy`
SetOptions[Plot,ImageSize->500,PlotRange->{All,{-5,5}},
 PlotPoints->nP*2,PlotStyle->{Blue,Thickness[0.01]}];
SetOptions[Plot3D,ImageSize->500,PlotRange->All]; c=N[1/(4*Pi)];
lambda=1; f[u_]:=Exp[lambda*u]; {l=0.5,tF=1.5,s=1/100,
 nP=100,nN=3,l1={Red,Blue,Green},l2={0.3,0.7,1.5}}
pde[1]=D[u[x,t],{t,2}]-c^2*D[u[x,t],{x,2}]==0
pde[2]=D[u[x,t],{t,2}]-c^2*D[u[x,t],{x,2}]-f[u[x,t]]==0
ic={u[x,0]==0,(D[u[x,t],t]/.t->0)==Sin[4*Pi*x]}
bc={u[0,t]==0,u[l,t]==0}
Do[sol[j]=NDSolve[Flatten[{pde[j],ic,bc}],u,{x,0,l},{t,0,tF},
 MaxStepSize->s,PrecisionGoal->2];
 f[j]=u/.First[sol[j]],{j,1,2}];
Map[Length,InterpolatingFunctionCoordinates[f[2]]]
Do[g[j,i]=Plot[Evaluate[u[x,l2[[i]]]/.sol[j]],{x,0,l},
 PlotStyle->{l1[[i]],Thickness[0.01]}],{j,1,2},{i,1,nN}];
GraphicsRow[{Show[Table[g[1,i],{i,1,nN}]],
 Show[Table[g[2,i],{i,1,nN}]]}]
```

```
Do[numVals[i]=Evaluate[u[0.1,1/2]/.sol[i]]; Print[numVals[i]];
 g3D[i]=Plot3D[Evaluate[u[x,t]/.sol[i]],{x,0,1},{t,0,tF},
 ColorFunction->Function[{x,y},Hue[x]],BoxRatios->1,
 ViewPoint->{-1,2,1},ImageSize->500]; gCP[i]=ContourPlot[
 Evaluate[u[x,t]/.sol[i]],{x,0,1},{t,0,tF},ColorFunction->Hue,
 ImageSize->300],{i,1,2}]; GraphicsRow[{g3D[1],g3D[2]}]
GraphicsRow[{gCP[1],gCP[2]}]
Animate[Row[{Plot[Evaluate[u[x,t]/.sol[1],{x,0,1}],PlotRange->
 {-1,5}],Plot[Evaluate[u[x,t]/.sol[2],{x,0,1}],PlotRange->
 {-1,5}]}],{t,0,tF,0.001},AnimationRate->0.1]
```

□

Problem 6.3 *Klein–Gordon equation. Numerical and graphical solutions.* Let us consider the initial boundary value problem for the Klein–Gordon equation

$$u_{tt}-a^2\,u_{xx}+F(u)=0;$$
$$u(x,0)=f_1(x),\;\; u_t(x,0)=f_2(x),\;\; u(\alpha,t)=f_3(t),\;\; u(\beta,t)=f_4(t),$$

where $F(u)=bu-cu^3$, $\mathcal{D}=\{x_I\leq x\leq x_F,\;\; 0\leq t\leq t_F\}$, and a, b, c, α, β, $f_1(x)$, $f_2(x)$, $f_3(t)$, $f_4(t)$ are the given parameters and functions corresponding to various types of solutions (one-soliton and two-soliton solutions, periodic solution) of the Klein–Gordon equation (see Problems 2.21–2.23, 2.32 of the present book and Examples 39.28., 40.26. in [144]).

We find numerical and graphical solutions of the given Klein–Gordon equation subject to the corresponding initial and boundary conditions.

1. Let us define the initial boundary value problem (`KG`, `IC`, `BC`).

Maple:

```
with(plots): with(PDEtools): declare(u(x,t));
S:=1/100; NF:=30; NP:=100; N:=3; L:=1; L1:=[red,blue,green];
L2:=[0,1/2,1]; Ops:=spacestep=S,timestep=S; Ops1:=frames=NF,
 numpoints=NP; Ops2:=grid=[20,20],shading=zhue,axes=boxed;
F:=u->b*u(x,t)-c*u(x,t)^3; F(u);
KG:=diff(u(x,t),t$2)-a^2*diff(u(x,t),x$2)+F(u)=0;
IC:=(f1,f2)->{u(x,0)=f1,D[2](u)(x,0)=f2};
BC:=(c,f1,d,f2)->{D[1](u)(c,t)=f1,D[1](u)(d,t)=f2};
```

Mathematica:

```
SetOptions[Plot,ImageSize->500,PlotRange->All,PlotPoints->
 nP*2,PlotStyle->{Blue,Thickness[0.01]}]; {s=1/100,nP=100}
{nN=3,l=1,l1={Red,Blue,Green},l2={0,1/2,1}}
f[u_]:=b*u[x,t]-c*u[x,t]^3; f[u]
eKG=D[u[x,t],{t,2}]-a^2*D[u[x,t],{x,2}]+f[u]==0
fIC1[f1_]:=u[x,0]==f1; fIC2[f2_]:=(D[u[x,t],t]/.t->0)==f2;
fBC1[c_,f1_]:=(D[u[x,t],x]/.x->c)==f1;
fBC2[d_,f2_]:=(D[u[x,t],x]/.x->d)==f2;
{fIC1[f1],fIC2[f2],fBC1[c,f1],fBC2[d,f2]}
```

2. We consider the Klein–Gordon equation in $\mathcal{D}$ with the initial and boundary conditions, where (f12,f22,f32,f42)

$$f_1(x){=}q\tanh(kx), \quad f_2(x){=}-\lambda kq(1-\tanh(kx)^2),$$
$$f_3(t){=}kq(1-\tanh(k(-5{-}\lambda t))^2), \quad f_4(t){=}kq(1{-}\tanh(k(5{-}\lambda t))^2),$$

and $k{=}\sqrt{a/(2\lambda^2{-}2a^2)}$, $q{=}\sqrt{a/c}$. As a result, we obtain the one-soliton solution (kink solution) and visualize it.

Maple:

```
params2:={a=0.1,b=0.1,c=1,lambda=0.3}; c2:=-5; d2:=5; tR2:=0..4;
xR2:=c2..d2; k:=sqrt(a/(2*(lambda^2-a^2))); SolEx2:=sqrt(a/c)*
 tanh(k*(x-lambda*t)); f12:=eval(SolEx2,t=0); f22:=eval(diff(
 SolEx2,t),t=0); f32:=eval(diff(SolEx2,x),x=c2); f42:=eval(
 diff(SolEx2,x),x=d2); KG2:=evalf(subs(params2,KG));
IC2:=evalf(subs(params2,IC(f12,f22)));
BC2:=evalf(subs(params2,BC(c2,f32,d2,f42)));
Sol2:=pdsolve(KG2,IC2 union BC2,numeric,u(x,t),Ops);
for i from 1 to N do G||i:=Sol2:-plot(t=L2[i],color=L1[i],
 numpoints=NP*2): od: display({seq(G||i,i=1..N)});
Sol2:-plot3d(u(x,t),t=tR2,Ops2,orientation=[95,55]);
Sol2:-animate(u(x,t),x=xR2,t=tR2,Ops1,thickness=3);
```

Mathematica:

```
{params2={a->0.1,b->0.1,c->1,lambda->0.3},c2=-5,d2=5,tF2=4,
 xI2=c2,xF2=d2,k=Sqrt[a/(2*(lambda^2-a^2))]}
{solEx2=Sqrt[a/c]*Tanh[k*(x-lambda*t)], f12=solEx2/.t->0,
 f22=D[solEx2,t]/.t->0, f32=D[solEx2,x]/.x->c2,
 f42=D[solEx2,x]/.x->d2}
```

```
{eKG2=N[eKG/.params2], ic2=N[{fIC1[f12],fIC2[f22]}/.
 params2], bc2=N[{fBC1[c2,f32],fBC2[d2,f42]}/.params2]}
sol2=NDSolve[Flatten[{eKG2,ic2,bc2}],u,{x,xI2,xF2},{t,0,tF2},
 MaxStepSize->s,PrecisionGoal->2]
Do[g[i]=Plot[Evaluate[u[x,l2[[i]]]/.sol2],{x,xI2,xF2},
 PlotStyle->{l1[[i]],Thickness[0.01]}],{i,1,nN}];
Show[Table[g[i],{i,1,nN}]]
Plot3D[Evaluate[u[x,t]/.sol2],{x,xI2,xF2},{t,0,tF2},
 ColorFunction->Function[{x,y},Hue[x]],BoxRatios->1,ViewPoint->
 {-1,2,1},ImageSize->500]
Animate[Plot[Evaluate[u[x,t]/.sol2,{x,xI2,xF2}],PlotRange->
 {-0.5,0.5}],{t,0,tF2},AnimationRate->0.5]
```

3. We consider the given Klein–Gordon equation in $\mathcal{D}$ with the initial and boundary conditions, where (f13,f23,f33,f43)

$$f_1(x)=q\,\mathrm{sech}\,(kx), \quad f_2(x)=\lambda kq\,\mathrm{sech}\,(kx)\tanh(kx),$$
$$f_3(t)=-\,kq\,\mathrm{sech}\,[k(-5-\lambda t)]\tanh[k(-5-\lambda t)],$$
$$f_4(t)=-\,kq\,\mathrm{sech}\,[k(5-\lambda t)]\tanh[k(5-\lambda t)],$$

and $k=\sqrt{a/(a^2-\lambda^2)}$, $q=\sqrt{2a/c}$. As a result, we obtain the one-soliton solution (hump-shaped soliton) and visualize it.

Maple:

```
params3:={a=0.3,b=0.3,c=1,lambda=0.25}; c3:=-5; d3:=5;
tR3:=0..4; xR3:=c3..d3; k:=sqrt(a/(a^2-lambda^2));
SolEx3:=sqrt(2*a/c)*sech(k*(x-lambda*t));
f13:=eval(SolEx3,t=0); f23:=eval(diff(SolEx3,t),t=0);
f33:=eval(diff(SolEx3,x),x=c3); f43:=eval(diff(SolEx3,x),x=d3);
KG3:=evalf(subs(params3,KG));
IC3:=evalf(subs(params3,IC(f13,f23)));
BC3:=evalf(subs(params3,BC(c3,f33,d3,f43)));
Sol3:=pdsolve(KG3,IC3 union BC3,numeric,u(x,t),Ops);
for i from 1 to N do
 G||i:=Sol3:-plot(t=L2[i],color=L1[i],numpoints=NP*2): od:
display({seq(G||i,i=1..N)}); Sol3:-plot3d(u(x,t),t=tR3,Ops2);
Sol3:-animate(u(x,t),x=xR3,t=tR3,Ops1,thickness=3);
```

Mathematica:

```
{params3={a->0.3,b->0.3,c->1,lambda->0.25},c3=-5,d3=5,tF3=4,
 xI3=c3,xF3=d3,k=Sqrt[a/(a^2-lambda^2)]}
```

```
{solEx3=Sqrt[2*a/c]*Sech[k*(x-lambda*t)],f13=solEx3/.t->0,
 f23=D[solEx3,t]/.t->0,f33=D[solEx3,x]/.x->c3, f43=D[solEx3,x]/.
 x->d3, eKG3=N[eKG/.params3], ic3=N[{fIC1[f13],fIC2[f23]}/.
 params3], bc3=N[{fBC1[c3,f33],fBC2[d3,f43]}/.params3]}
sol3=NDSolve[Flatten[{eKG3,ic3,bc3}],u,{x,xI3,xF3},{t,0,tF3},
 MaxStepSize->s,PrecisionGoal->2]
Do[g[i]=Plot[Evaluate[u[x,l2[[i]]]/.sol3],{x,xI3,xF3},PlotStyle->
 {l1[[i]],Thickness[0.01]}],{i,1,nN}]; Show[Table[g[i],{i,1,nN}]]
Plot3D[Evaluate[u[x,t]/.sol3],{x,xI3,xF3},{t,0,tF3},
 ColorFunction->Function[{x,y},Hue[x]],BoxRatios->1,ViewPoint->
 {1,2,1},PlotRange->All,PlotPoints->{20,20},ImageSize->500]
Animate[Plot[Evaluate[u[x,t]/.sol3,{x,xI3,xF3}],PlotRange->
 {0,1}],{t,0,tF3},AnimationRate->0.5]
```

4. We consider the Klein–Gordon equation in $\mathcal{D}$ with the initial and boundary conditions, where (f14,f24,f34,f44)

$$f_1(x)=q\sum_{i=1}^{2}\operatorname{sech}\left[k_i(x-\tilde{x}_i)\right],\ f_2(x)=q\sum_{i=1}^{2}k_i\lambda_i\operatorname{sech}\left[k_i(x-\tilde{x}_i)\right]\tanh[k_i(x-\tilde{x}_i)],$$

$$f_3(t)=q\sum_{i=1}^{2}-k_i\operatorname{sech}\left[k_i(-5-\lambda_i t-\tilde{x}_i)\right]\tanh[k_i(-5-\lambda_i t-\tilde{x}_i)],$$

$$f_4(t)=q\sum_{i=1}^{2}-k_i\operatorname{sech}\left[k_i(5-\lambda_i t-\tilde{x}_i)\right]\tanh[k_i(5-\lambda_i t-\tilde{x}_i)],$$

and $k_i=\sqrt{a/(a^2-\lambda_i^2)}$, $q=\sqrt{2a/c}$. As a result, we obtain the two-soliton solution (the interaction of two equal hump-shaped solitons) and visualize it.

Maple:

```
params4:={a=0.3,b=0.3,c=1,lambda[1]=0.25,lambda[2]=-0.25,x0[1]=
 -2,x0[2]=2}; for i to 2 do k[i]:=sqrt(a/(a^2-lambda[i]^2)); od:
c4:=-5; d4:=5; tR4:=0..7; xR4:=c4..d4;
SolEx4:=sqrt(2*a/c)*add(sech(k[i]*(x-lambda[i]*t-x0[i])),i=1..2);
f14:=eval(SolEx4,t=0); f24:=eval(diff(SolEx4,t),t=0);
f34:=eval(diff(SolEx4,x),x=c4); f44:=eval(diff(SolEx4,x),x=d4);
KG4:=evalf(subs(params4,KG)); IC4:=evalf(subs(params4,
 IC(f14,f24))); BC4:=evalf(subs(params4,BC(c4,f34,d4,f44)));
Sol4:=pdsolve(KG4,IC4 union BC4,numeric,u(x,t),Ops);
for i from 1 to N do
 G||i:=Sol4:-plot(t=L2[i],color=L1[i],numpoints=NP*2): od:
```

```
display({seq(G||i,i=1..N)}); Sol4:-plot3d(u(x,t),t=tR4,Ops2);
Sol4:-animate(u(x,t),x=xR4,t=tR4,Ops1,thickness=3);
```

Mathematica:

```
params4={a->0.3,b->0.3,c->1,lambdaN[1]->0.25,lambdaN[2]->-0.25,
 x0[1]->-2,x0[2]->2}; Do[kN[i]=Sqrt[a/(a^2-lambdaN[i]^2)],
 {i,1,2}]; {c4=-5,d4=5,tF4=7,xI4=c4,xF4=d4}
{solEx4=Sqrt[2*a/c]*Sum[Sech[kN[i]*(x-lambdaN[i]*t-x0[i])],
 {i,1,2}], f14=solEx4/.t->0,f24=D[solEx4,t]/.t->0,
 f34=D[solEx4,x]/.x->c4, f44=D[solEx4,x]/.x->d4,
 eKG4=N[eKG/.params4],ic4=N[{fIC1[f14], fIC2[f24]}/.params4],
 bc4=N[{fBC1[c4,f34],fBC2[d4,f44]}/.params4]}
sol4=NDSolve[Flatten[{eKG4,ic4,bc4}],u,{x,xI4,xF4},{t,0,tF4},
 MaxStepSize->s,PrecisionGoal->2]
Do[g[i]=Plot[Evaluate[u[x,l2[[i]]]/.sol4],{x,xI4,xF4},
 PlotStyle->{l1[[i]],Thickness[0.01]}],{i,1,nN}];
Show[Table[g[i],{i,1,nN}]]
Plot3D[Evaluate[u[x,t]/.sol4],{x,xI4,xF4},{t,0,tF4},
 ColorFunction->Function[{x,y},Hue[x]],BoxRatios->1,ViewPoint->
 {1,2,1},PlotRange->All,PlotPoints->{20,20},ImageSize->500]
Animate[Plot[Evaluate[u[x,t]/.sol4,{x,xI4,xF4}],PlotRange->
 {0,1.8}],{t,0,tF4},AnimationRate->0.5]
```

5. We consider the Klein–Gordon equation in $\mathcal{D}$ with the initial and boundary conditions, where $f_1(x)=A(1+\cos(\omega x))$, $f_2(x)=0$, $f_3(t)=0$, $f_4(t)=0$, and $\omega=\frac{2}{5}\pi$. As a result, we obtain the periodic solution and visualize it.

Maple:

```
params5:={a=1,b=1,c=-1,A=1.5}; c5:=-5; d5:=5; tR5:=0..4;
xR5:=c5..d5; f15:=A*(1+cos(2*Pi*x/d5)); f25:=0; f35:=0;
f45:=0; KG5:=evalf(subs(params5,KG)):
IC5:=evalf(subs(params5,IC(f15,f25))):
BC5:=evalf(subs(params5,BC(c5,f35,d5,f45)));
Sol5:=pdsolve(KG5,IC5 union BC5,numeric,u(x,t),Ops);
for i from 1 to N do
 G||i:=Sol5:-plot(t=L2[i],color=L1[i],numpoints=NP): od:
 display({seq(G||i,i=1..N)}); Sol5:-plot3d(u(x,t),t=tR5,Ops2);
Sol5:-animate(u(x,t),x=xR5,t=tR5,Ops1,thickness=3);
```

Mathematica:

```
params5={a->1,b->1,c->-1,aN->1.5}; {c5=-5,d5=5,tF5=4,
 xI5=c5,xF5=d5,f15=aN*(1+Cos[2*Pi*x/d5]),f25=0,f35=0,f45=0,
 eKG5=N[eKG/.params5], ic5=N[{fIC1[f15],fIC2[f25]}/.params5],
 bc5=N[{fBC1[c5,f35],fBC2[d5,f45]}/.params5]}
sol5=NDSolve[Flatten[{eKG5,ic5,bc5}],u,{x,xI5,xF5},{t,0,tF5},
 MaxStepSize->s,PrecisionGoal->2]
Do[g[i]=Plot[Evaluate[u[x,l2[[i]]]/.sol5],{x,xI5,xF5},PlotStyle->
 {l1[[i]],Thickness[0.01]}],{i,1,nN}]; Show[Table[g[i],{i,1,nN}]]
Plot3D[Evaluate[u[x,t]/.sol5],{x,xI5,xF5},{t,0,tF5},
 ColorFunction->Function[{x,y},Hue[x]],BoxRatios->1,ViewPoint->
 {1,2,1},PlotRange->All,PlotPoints->{20,20},ImageSize->500]
Animate[Plot[Evaluate[u[x,t]/.sol5,{x,xI5,xF5}],PlotRange->
 {-3,3}],{t,0,tF5},AnimationRate->0.5]
```

□

6.1.2 Specifying Classical Numerical Methods

As we mentioned above, in *Maple* it is possible to find numerical solutions specifying one of the 11 *classical methods**, specifying numerical boundary conditions and finite difference schemes for two-stage methods (the description table of these methods see in [144]). But there is some restriction with respect to these classical methods: a single PDE must be parabolic or hyperbolic and of the first-order in time. PDEs that are greater than first-order in time can be solved by converting to an equivalent first-order system.

Problem 6.4 *Inviscid Burgers equation. Numerical and graphical solutions. Single-stage numerical method.* For the inviscid Burgers equation with the initial and boundary conditions,

$$u_t+u\,u_x=0;\quad u(x,0)=f(x),\quad u(0,t)=\tfrac{1}{4}e^{-10},$$

where $f(x)=\frac{1}{4}e^{-10(4x-2)^2}$, obtain the numerical and graphical solutions of this initial boundary value problem applying the *single stage explicit* `ForwardTime1Space[backward]` method.

*The classical numerical methods embedded in *Maple* are: the forward time forward/backward space method, the centered time forward/backward space method, the backward time forward/backward space method, the forward time centered space (or Euler) method, the centered time centered space (or Crank–Nicolson) method, the backward time centered space method (or backward Euler), the box method, the Lax–Friedrichs method, the Lax–Wendroff method, the Leapfrog method, the DuFort–Frankel method.

Since the boundary condition is given on the left (according to the method), we consider the domain $\mathcal{D}=\{0 \le x \le 1,\ t \ge 0\}$, We find the numerical and graphical solutions of this initial boundary value problem and visualize it, e.g., at times $t=0, 0.15, 0.3$.

Maple:

```
with(PDEtools): with(plots): declare(u(x,t)); NF:=30; NP:=100;
xR:=0..1; tR:=0..1; S:=1/100; Ops:=timestep=S,spacestep=S;
N:=3; L1:=[red,blue,magenta]; L2:=[0,0.15,0.3];
PDE1:=diff(u(x,t),t)+diff(u(x,t),x)*u(x,t)=0;
f:=x->exp(-10*(4*x-2)^2)/4; IBC:={u(x,0)=f(x),u(0,t)=exp(-10)/4};
M1:=ForwardTime1Space[backward];
Sol1:=pdsolve(PDE1,IBC,numeric,time=t,range=xR,method=M1,Ops);
Num_vals1:=Sol1:-value(); Num_vals1(0,0.5);
for i from 1 to N do
 G||i:=Sol1:-plot(t=L2[i],color=L1[i],numpoints=NP*2): od:
display({seq(G||i,i=1..N)}); Ops1:=frames=NF,numpoints=NP;
Sol1:-animate(u(x,t),x=xR,t=tR,Ops1,thickness=3);
```

If $f'(x_{cr})=\frac{1}{4}(-320x+160)e^{-10(4x-2)^2}<0$ for $x_{cr}\in(\frac{1}{2},\infty)$, solutions of this initial value problem break down at time t_{cr} by generating a vertical line in the wave profile which corresponds to a discontinuity in the solution. For $t>t_{cr}$, the solution must be continued as a *weak solution*, one of the examples of which is a *shock wave*. □

Problem 6.5 *Nonlinear first-order PDE. Numerical and graphical solutions. Single-stage numerical method. Numerical boundary condition (NBC).* For the nonlinear first-order equation with the initial and boundary conditions,

$$u_t=\nu\, u_x\, u^3; \quad u(x,0)=\cos(\tfrac{1}{2}\pi x), \quad u(-1,t)=u(1,t),$$

defined in $\mathcal{D}=\{-1\le x\le 1,\ 0\le t\le 5\}$, obtain the numerical and graphical solutions of this initial boundary value problem applying the well-known and popular Crank–Nicolson method [34].

Since this PDE is odd-order in space, it requires a numerical boundary condition, which we can choose as Box discretization of the PDE at the right boundary, `[Box,n]`. Then we find the numerical solution of this initial boundary value problem and visualize it, e.g., at times $t=0, \frac{1}{2}$.

Maple:

```
with(PDEtools): with(plots): declare(u(x,t)); NF:=30; NP:=100;
xR:=-1..1; tR:=0..5; S:=1/100; Ops:=timestep=S,spacestep=S;
N:=2; L1:=[red,blue]; L2:=[0,1/2]; nu:=0.1; NBCs:=[Box,n];
PDE1:=diff(u(x,t),t)=nu*diff(u(x,t),x)*u(x,t)^3;
IBC:={u(x,0)=cos(Pi*x/2),u(-1,t)=u(1,t)}; M1:=CrankNicholson;
Sol1:=pdsolve(PDE1,IBC,numeric,numericalbcs=NBCs,method=M1,Ops);
for i from 1 to N do
 G||i:=Sol1:-plot(t=L2[i],color=L1[i],numpoints=NP): od:
display({seq(G||i,i=1..N)}); Ops1:=frames=NF,numpoints=NP;
Sol1:-animate(u(x,t),x=xR,t=tR,Ops1,thickness=3);
```

□

Problem 6.6 *Inviscid Burgers equation. Numerical and graphical solutions. Two-stage numerical method. Startup method. NBC.* Let us consider the initial boundary value problem for the inviscid Burgers equation,

$$u_t+uu_x=0; \quad u(x,0)=f(x), \quad u_x(-10,t)=0,$$

defined in $\mathcal{D}=\{-10 \le x \le 10,\ t \ge 0\}$, where $f(x)=\begin{cases} 0, & x<-1, \\ x+1, & -1 \le x \le 0, \\ 1, & x>0. \end{cases}$

Obtain the numerical and graphical solutions of this problem applying the two-stage explicit DuFort–Frankel method.

We have to indicate how to compute the additional stage required for two-stage methods, i.e., the option `startup`. Since the inviscid Burgers equation is odd-order in space, we have to indicate a numerical boundary condition, which we can choose in the form `u[1, n]-u[1,n-1]`. Then, we find the numerical solution of this initial boundary value problem and visualize it at times $t=0, 0.4, 0.5$.

Maple:

```
with(PDEtools): with(plots): declare(u(x,t));
NF:=30; NP:=100; xR:=-10..10; tR:=0..5; S:=1/50;
Ops:=timestep=S,spacestep=S;
N:=3; L1:=[red,blue,magenta]; L2:=[0,0.4,0.5];
PDE1:=diff(u(x,t),t)+u(x,t)*diff(u(x,t),x)=0;
f:=x->piecewise(x<-1,0,x>=-1 and x<=0,x+1,1);
IBC:={u(x,0)=f(x),(D[1](u))(-10,t)=0}; M1:=DuFortFrankel;
Sol1:=pdsolve(PDE1,IBC,type=numeric,time=t,range=xR,
      numericalbcs=u[1,n]-u[1,n-1],method=M1,startup=Euler,Ops);
```

```
for i from 1 to N do
 G||i:=Sol1:-plot(t=L2[i],color=L1[i],numpoints=NP,thickness=2):
od: display({seq(G||i,i=1..N)}); Ops1:=frames=NF,numpoints=NP;
Sol1:-animate(u(x,t),x=xR,t=tR,Ops1,thickness=3);
```
□

6.1.3 Nonlinear Systems

In this section, we will show how to obtain numerical and graphical solutions of nonlinear systems in *Maple* and *Mathematica* with the aid of the predefined functions (`pdsolve` and `NDSolve`).

Problem 6.7 *Nonlinear first-order system. Numerical and graphical solutions.* Let us consider the initial boundary value problem for the nonlinear first-order system (discussed in Problems 1.21, 5.6),

$$u_t = vu_x + u + 1, \quad v_t = -uv_x - v + 1;$$
$$u(x,0) = e^{-x}, \; v(x,0) = e^{x}, \; u(1,t) = e^{t-1}, \; v(0,t) = e^{-t},$$

defined in $\mathcal{D} = \{0 \le x \le 1, 0 \le t \le 0.5\}$. Obtain numerical solution of the given initial boundary value problem, visualize it at different times and in the parametric form.

Maple:

```
Digits:=30: with(PDEtools):  with(plots): Ops1:=numpoints=100:
Ops2:=color=magenta: Ops3:=color=blue: Ops4:=color="BlueViolet":
Ops5:=axes=boxed,shading=zhue,orientation=[40,50]; a:=0:
b:=1: Tf:=0.5; U,V:=diff_table(u(x,t)),diff_table(v(x,t)):
sys1:={U[t]=V[]*U[x]+U[]+1,V[t]=-U[]*V[x]-V[]+1};
IBC1:={u(x,0)=exp(-x),v(x,0)=exp(x),u(1,t)=exp(t-1),
 v(0,t)=exp(-t)}; S:=1/100; Ops:=spacestep=S,timestep=S;
Sol1:=pdsolve(sys1,IBC1,[u,v],numeric,time=t,range=a..b,Ops);
L1:=[0.1,0.2,0.5]; NL1:=nops(L1); for i from 1 to NL1 do
 G||i:=Sol1:-plot(t=L1[i],Ops1,Ops||(i+1)); od:
display({G1,G2,G3}); GU:=Sol1:-plot(u(x,t),t=Tf,Ops1,Ops2):
GV:=Sol1:-plot(v(x,t),t=Tf,Ops1,Ops3): display({GU,GV});
PU:=r->eval(u,(Sol1:-value(u,t=Tf))(r)):
PV:=r->eval(v,(Sol1:-value(v,t=Tf))(r)):
plot([PU,PV,a..b],Ops2,labels=[u,v],Ops1); Sol1:-animate(
 t=Tf,Ops1,Ops2); GU3D:=Sol1:-plot3d(u(x,t),t=0..Tf):
GV3D:=Sol1:-plot3d(v(x,t),t=0..Tf): display(GU3D,Ops5);
display(GV3D,Ops5); P1:=Sol1:-value(u,t=0); P1(0.1,0);
```

Mathematica:

```
{nP=100, iS=500, l1={0.1,0.2,0.5}, l2={Magenta,Blue,Hue[0.8]}}
nl1=Length[l1]; SetOptions[Plot,ImageSize->iS,PlotRange->All,
 PlotPoints->nP*2]; SetOptions[Plot3D,ImageSize->500,
 ColorFunction->Function[{x,y},Hue[x]],PlotRange->All,
 BoxRatios->1,ViewPoint->{1,2,1}]; {s=1/100,a=0,b=1,tF=0.5}
sys1={D[u[x,t],t]==v[x,t]*D[u[x,t],x]+u[x,t]+1,
      D[v[x,t],t]==-u[x,t]*D[v[x,t],x]-v[x,t]+1}
{ibc1={u[x,0]==Exp[-x],v[x,0]==Exp[x],u[1,t]==Exp[t-1],
 v[0,t]==Exp[-t]}, sol1=NDSolve[Flatten[{sys1,ibc1}],{u,v},
 {x,a,b},{t,0,tF},MaxStepSize->s]}
Do[g[i]=Plot[Evaluate[u[x,l1[[i]]]/.sol1],{x,a,b},PlotStyle->
 {l2[[i]],Thickness[0.01]}],{i,1,nl1}];
Show[Table[g[i],{i,1,nl1}]]
gu=Plot[Evaluate[u[x,tF]/.sol1],{x,a,b},PlotStyle->{l2[[1]],
 Thickness[0.01]}]; gv=Plot[Evaluate[v[x,tF]/.sol1],{x,a,b},
 PlotStyle->{l2[[2]],Thickness[0.01]}]; Show[{gu,gv}]
ParametricPlot[{Evaluate[{u[x,tF],v[x,tF]}/.sol1]},{x,a,b},
 PlotRange->All,PlotStyle->{l2[[1]],Thickness[0.01]}]
Animate[Plot[{Evaluate[{u[x,t]}/.sol1]},{x,a,b},PlotRange->
 {0.3,1.6},PlotStyle->{l2[[1]],Thickness[0.01]}],{t,0,tF},
 AnimationRate->0.1]
gu3D=Plot3D[Evaluate[u[x,t]/.sol1],{x,a,b},{t,0,tF}];
gv3D=Plot3D[Evaluate[v[x,t]/.sol1],{x,a,b},{t,0,tF}];
GraphicsRow[{gu3D,gv3D}]
p1=Evaluate[u[0.1,0]/.sol1]
```

□

Problem 6.8 *FitzHugh–Nagumo equations. Numerical and graphical solutions.* Let us consider the FitzHugh–Nagumo type equations [107] arising in mathematical biology and model the nerve impulse propagation along an axon. One of the examples is the nonlinear reaction-diffusion system of the form*

$$u_t=u_{xx}+u(\alpha-u)(u-1)-v, \quad v_t=\beta u,$$

defined in $\mathcal{D}=\{x \in \mathbb{R},\ t \geq 0\}$, where function $u(x,t)$ is a membrane potential, $v(x,t)$ is a phenomenological recovery variable, $0<\alpha<1$ and $\beta>0$.

*The numerical solution of this nonlinear system has been obtained with MATLAB in Example 41.3.in [144].

We find numerical and graphical solutions of this nonlinear system subject to the following boundary and initial conditions:

$$u_x(a,t)=0,\ u_x(b,t)=0,\ v_x(a,t)=0,\ v_x(b,t)=0,$$
$$u(x,0)=\begin{cases}1, & x<0,\\ 0, & x>0,\end{cases} \quad v(x,0)=\begin{cases}0, & x<0,\\ 1, & x>0.\end{cases}$$

Setting $\mathcal{D}=\{-50\le x\le 50,\ 0\le t\le 50\}$, $\alpha=0.1$, $\beta=0.01$, we obtain numerical solution of the given initial boundary value problem and visualize it at different times.

Maple:

```
with(PDEtools): with(plots): Ops1:=numpoints=100:
Ops2:=color=magenta: Ops3:=color=blue: Ops4:=color="BlueViolet":
alpha:=0.1: beta:=0.01: a:=-50: b:=50:
U,V:=diff_table(u(x,t)),diff_table(v(x,t)):
sys1:=[U[t]=U[x,x]+U[]*(alpha-U[])*(U[]-1)-V[],V[t]=beta*U[]];
IBC1:={u(x,0)=piecewise(x<0,1,0),v(x,0)=piecewise(x<0,0,1),
 D[1](u)(a,t)=0, D[1](u)(b,t)=0}; L1:=[0.5,5,7]; NL1:=nops(L1);
Sol1:=pdsolve(sys1,IBC1,[u,v],numeric);
for i from 1 to NL1 do
 G||i:=Sol1:-plot(t=L1[i],Ops1,Ops||(i+1)); od:
display({G1,G2,G3}); GU:=Sol1:-plot(u(x,t),t=5,Ops1,Ops2):
GV:=Sol1:-plot(v(x,t),t=5,Ops1,Ops3): display({GU,GV});
Sol1:-animate(t=0.5,Ops1,Ops2);
```

Mathematica:

```
{nP=100,iS=500,l1={0.5,5,7},l2={Magenta,Blue,Hue[0.8]}}
nl1=Length[l1]; SetOptions[Plot,ImageSize->iS,PlotRange->All,
 PlotPoints->nP*2]; SetOptions[Plot3D,ImageSize->iS,
 ColorFunction->Function[{x,y},Hue[x]], PlotRange->All,
 BoxRatios->1, ViewPoint->{1,2,1}]; {alpha=0.1,beta=0.01,
 a=-50,b=50,tF=7, sys1={D[u[x,t],t]==D[u[x,t],{x,2}]+
 u[x,t]*(alpha-u[x,t])*(u[x,t]-1)-v[x,t],D[v[x,t],t]==
 beta*u[x,t]}, ibc1={u[x,0]==Piecewise[{{1,x<0},{0,x>0}}],
 v[x,0]==Piecewise[{{0,x<0},{1,x>0}}],(D[u[x,t],x]/.x->a)==0,
 (D[u[x,t],x]/.x->b)==0}}
sol1=NDSolve[Flatten[{sys1,ibc1}],{u,v},{x,a,b},{t,0,7},
 AccuracyGoal->1,PrecisionGoal->1,MaxSteps->{15,Infinity},
 Method->{"MethodOfLines","SpatialDiscretization"->
 {"TensorProductGrid"}}]
```

```
Do[g[i]=Plot[Evaluate[u[x,l1[[i]]]/.sol1],{x,a,b},PlotStyle->
 {l2[[i]],Thickness[0.01]}],{i,1,nl1}];
Show[Table[g[i],{i,1,nl1}]]
gu=Plot[Evaluate[u[x,5]/.sol1],{x,a,b},PlotStyle->{l2[[1]],
 Thickness[0.01]}]; gv=Plot[Evaluate[v[x,5]/.sol1],{x,a,b},
 PlotStyle->{l2[[2]],Thickness[0.01]}]; Show[{gu,gv}]
Animate[Plot[{Evaluate[{u[x,t]}/.sol1]},{x,a,b},PlotRange->
 {-1,1.3},PlotStyle->{l2[[1]],Thickness[0.01]}],{t,0,tF},
 AnimationRate->0.5]
```

□

6.2 Finite Difference Methods

There are the four big classes of methods for constructing numerical solutions of nonlinear PDEs: finite difference methods (FDM), finite element methods (FEM) (with the related finite volume methods), spectral methods (SM) (with the related spectral element methods, discontinuous spectral element methods), and boundary element methods (BEM). In this section, we will show a helpful role of computer algebra systems for generating and applying various finite difference methods in order to construct numerical solutions of nonlinear PDEs. Finite difference methods approximate a given PDE in differential form and are based on Taylor expansions to approximate local derivatives (at grid points).

6.2.1 Evolution Equations

In order to approximate the linear and nonlinear PDEs by finite differences, we have to generate a *mesh* (or grid) in a domain $\mathcal{D}$, e.g., $\mathcal{D}=\{a\leq x\leq b, c\leq t\leq d\}$. The mesh can be of various types, e.g., rectangular, along the characteristics, polar, etc. We assume (for simplicity) that the sets of lines of the mesh are equally spaced and that $u(x,t)$ is the dependent variable in a given PDE. We write h and k for the line spacings and define the *mesh points* $X_i=a+ih$, $T_j=c+jk$ $(i=0,\dots,NX, j=0,\dots,NT)$ and $h=(b-a)/NX$, $k=(d-c)/NT$. We calculate approximations of the solution at these mesh points, these approximate points will be denoted by $U_{i,j}\approx u(X_i,T_j)$. We approximate the derivatives in a given equation by finite differences (of various types) and then solve resulting difference equations.

Problem 6.9 *Burgers equation. Explicit method.* Let us consider the initial boundary value problem for the Burgers equation

$$u_t=\nu\, u_{xx}-uu_x;\quad u(x,0)=f(x),\quad u(a,t)=0,\quad u(b,t)=0,$$

defined in the domain $\mathcal{D}=\{a\leq x\leq b,\ 0\leq t\leq T\}$, where $f(x)=\sin(\pi x/L)$, $a=0$, $b=1$, $L=1$, $T=0.4$, $\nu=0.009$. Applying the forward difference method, construct the approximate numerical solution of the given initial boundary value problem.

Let us generate the rectangular mesh: $X=a+ih$, $T=jk$, where $i=0,\dots,NX$, $j=0,\dots,NT$, $h=(b-a)/NX$, $k=T/NT$. We denote the approximate solution of $u(x,t)$ at the mesh point (i,j) as $U_{i,j}$. In forward difference method, the second-order derivative u_{xx} is replaced by a central difference approximation (CDA) and the first-order derivatives u_t and u_x, by a forward difference approximation (FWDA). The FD scheme for the nonlinear Burgers equation has the form:

$$U_{i,j+1}=U_{i,j}+r(U_{i+1,j}-2U_{i,j}+U_{i-1,j})-(k/h)U_{i,j}(U_{i+1,j}+U_{i,j}),$$

where $r=\nu k/h^2$. In this explicit FD scheme, the unknown value $U_{i,j+1}$ (at $(j+1)$-th step) is determined from the 3 known values $U_{i-1,j}$, $U_{i,j}$, and $U_{i+1,j}$ (at j-th step).

Maple:

```
with(plots): nu:=0.009: NX:=100: NT:=100: a:=0: b:=1: L:=1.:
T:=0.4; h:=evalf((b-a)/NX); k:=evalf(T/NT); r:=nu*k/h^2;
f:=x->evalf(sin(Pi*x/L)); for i from 0 to NX do X[i]:=a+i*h od:
Ops1:=thickness=3,labels=["X","U"]; IC:={seq(U(i,0)=f(X[i]),
 i=0..NX)}; BC:={seq(U(a,j)=0,j=0..NT),seq(U(NX,j)=0,j=0..NT)}:
IBC:=IC union BC: FD:=(i,j)->U(i,j)+r*(U(i+1,j)-2*U(i,j)
 +U(i-1,j))-k/h*U(i,j)*(U(i+1,j)-U(i,j));
for j from 0 to NT do for i from 1 to NX-1 do
 U(i,j+1):=subs(IBC,FD(i,j)); od: od:
G:=j->plot([seq([X[i],subs(IBC,U(i,j))],i=0..NX)],color=blue):
display([seq(G(j),j=0..NT)],insequence=true,Ops1);
```

Mathematica:

```
SetOptions[ListPlot,ImageSize->500,PlotRange->{{0,1},{0,1.05}},
 Joined->True]; {nu=0.009,nX=100,nT=100,a=0,b=1,l=1.,tF=0.4,
 h=(b-a)/nX//N, k=tF/nT//N, r=nu*k/h^2}
f[x_]:=Sin[Pi*x/l]//N; Table[xN[i]=a+i*h,{i,0,nX}];
ic=Table[uN[i,0]->f[xN[i]],{i,0,nX}];
bc={Table[uN[a,j]->0,{j,0,nT}], Table[uN[nX,j]->0,{j,0,nT}]};
ibc=Flatten[{ic,bc}]
fd[i_,j_]:=uN[i,j]+r*(uN[i+1,j]-2*uN[i,j]+uN[i-1,j])-k/h*uN[i,j]*
 (uN[i+1,j]-uN[i,j]);
```

```
Do[uN[i,j+1]=fd[i,j]/.ibc,{j,0,nT},{i,1,nX-1}];
g[j_]:=ListPlot[Table[{xN[i],uN[i,j]/.ibc},{i,0,nX}],PlotStyle->
 {Blue,Thickness[0.01]},AxesLabel->{"X","U"}];
grs=Evaluate[Table[g[j],{j,0,nT}]]; ListAnimate[grs]
```

The numerical and graphical solutions obtained in this problem can be compared with the results obtained in Problem 6.1 (by applying the predefined functions `pdsolve` and `NDSolve`) and in Problem 7.1 (by applying the numerical method of lines). □

Problem 6.10 *Inviscid Burgers equation. Explicit method.* Let us consider the initial boundary value problem for the inviscid Burgers equation

$$u_t + uu_x = 0; \quad u(x,0) = \arctan(4x) + 2, \quad u(a,t) = 0,$$

defined in the domain $\{a \le x \le b,\ 0 \le t \le T\}$, where $a{=}-4$, $b{=}6$, $T{=}0.4$. We note that this equation is of hyperbolic type and can be rewritten in the conservation form, i.e., as the physical conservation law of the form $u_t{+}(F(u))_x{=}0$, where $F(u){=}\frac{1}{2}u^2$. For the discretization of the problems we will use this form of the Burgers equation.

We generate the rectangular mesh: $X{=}a{+}ih$, $T{=}jk$ ($i{=}0,\dots,NX$, $j{=}0,\dots,NT$, $h{=}(b{-}a)/NX$, $k{=}T/NT$). We denote the approximate solution of $u(x,t)$ at the mesh point (i,j) as $U_{i,j}$. In the *Lax method*, the derivative u_t is replaced by a forward difference approximation (FWDA) and the derivative $(F(u))_x$, by a central difference approximation (CDA) (together with the corresponding substitution). The FD scheme for the inviscid Burgers equation has the form:

$$U_{i,j+1}{=}\tfrac{1}{2}\left(U_{i+1,j}{+}U_{i-1,j}\right) - \tfrac{1}{4}(k/h)\left(U_{i+1,j}^2{-}U_{i-1,j}^2\right).$$

In this explicit FD scheme, the unknown value $U_{i,j+1}$ (at $(j{+}1)$-th step) is determined from the 3 known values $U_{i-1,j}$, $U_{i,j}$, and $U_{i+1,j}$ (at j-th step). This FD scheme is stable for $k/h < 1/|u|$, i.e., the stability condition depends on the solution. We construct the approximate numerical solution of the initial boundary value problem by applying the Lax method.

Maple:

```
with(plots): NX:=50: NT:=9: a:=-4; b:=6; T:=0.4;
h:=evalf((b-a)/NX); k:=evalf(T/NT); f:=x->arctan(4*x)+2;
for i from 0 to NX do X[i]:=a+i*h od: L1:=[1,3,5,7,9];
IC:={seq(U[i,0]=f(X[i]),i=0..NX)}; BC:={seq(U[a,j]=0,j=0..NT)}:
```

```
IBC:=IC union BC:
FD:=(i,j)->1/2*(U[i+1,j]+U[i-1,j])-k/(2*h)*(U[i+1,j]^2/2-
 U[i-1,j]^2/2);
for j from 0 to NT do for i from 1 to NX-1 do
 U[i,j+1]:=subs(IBC,FD(i,j)); od: od: NL1:=nops(L1);
for i from 1 to NL1 do
 G||(L1[i]):=[seq([X[j],U[j,L1[i]]],j=1..NX+1)]; od:
plot([seq(G||(L1[i]),i=1..NL1)]);
```

Mathematica:

```
SetOptions[ListPlot,ImageSize->300,PlotRange->{All,All},Joined->
 True]; {nX=50,nT=9,a=-4,b=6,tF=0.4,h=(b-a)/nX//N,k=tF/nT//N}
f[x_]:=ArcTan[4*x]+2//N; Table[xN[i]=a+i*h,{i,0,nX}];
ic=Table[uN[i,0]->f[xN[i]],{i,0,nX}]; bc={Table[uN[a,j]->0,
 {j,0,nT}]}; ibc=Flatten[{ic,bc}]
fd[i_,j_]:=0.5*(uN[i+1,j]+uN[i-1,j])-k/(2*h)*(uN[i+1,j]^2/2-
 uN[i-1,j]^2/2); Do[uN[i,j+1]=fd[i,j]/.ibc,{j,0,nT},
 {i,1,nX-1}]; {l1={1,3,5,7,9},nl1=Length[l1]}
Do[g[j]=ListPlot[Table[{xN[i],uN[i,l1[[j]]]/.ibc},{i,0,nX}],
 PlotStyle->{Hue[0.7+i/10],Thickness[0.01]},AxesLabel->
 {"X","U"}],{j,1,nl1}];
Show[Table[g[i],{i,1,nl1}],Frame->True,Axes->False]
```
□

Problem 6.11 *Linear heat equation and inviscid Burgers equation. Forward/backward FD methods. Crank–Nicolson and Lax methods.* Let us consider the initial boundary value problems for the linear heat equation and the inviscid Burgers equation,

$$u_t = \nu\, u_{xx}; \;\; u(x,0)=f(x), \;\; u(0,t)=0, \;\; u(L,t)=0,$$
$$u_t + u u_x = 0; \;\; u(x,0)=f(x), \;\; u(a,t)=0,$$

defined in the domain $\mathcal{D}=\{0 \le x \le L, \; 0 \le t \le T\}$, where $f(x)=\sin(\pi x/L)$, $L=1$, $T=0.2$, $\nu=1$. Applying the forward/backward difference methods, the Crank–Nicolson and the Lax methods, find the approximate numerical solution of the given initial boundary value problems.

1. Linear heat equation. We generate the rectangular mesh: $X=ih$, $T=jk$ $(i=0,\ldots,NX,\; j=0,\ldots,NT,\; h=L/NX,\; k=T/NT)$. We denote the approximate solution of $u(x,t)$ at the mesh point (i,j) as $U_{i,j}$. In *forward difference method*, the second-order derivative u_{xx} is replaced by a central difference approximation (CDA) and the first-order derivative u_t, by a forward difference approximation (FWDA). The final FD scheme for the linear heat equation is

$$U_{i,j+1}=(1-2r)U_{i,j}+r(U_{i+1,j}+U_{i-1,j}),$$

where $r=\nu k/h^2$. In this explicit FD scheme, the unknown value $U_{i,j+1}$ (at $(j+1)$-th step) is determined from the 3 known values $U_{i-1,j}$, $U_{i,j}$, and $U_{i+1,j}$ (at j-th step). This FD scheme is unstable for $r>0.5$.

Maple:

```
with(plots): nu:=1: NX:=15: NT:=100: L:=1.: T:=0.2; h:=L/NX;
k:=T/NT; r:=nu*k/h^2; f:=x->evalf(sin(Pi*x));
for i from 0 to NX do X[i]:=i*h od:
IC:={seq(U(i,0)=f(X[i]),i=0..NX)}; BC:={seq(U(0,j)=0,j=0..NT),
 seq(U(NX,j)=0,j=0..NT)}: IBC:=IC union BC:
FD:=(i,j)->(1-2*r)*U(i,j)+r*(U(i+1,j)+U(i-1,j));
for j from 0 to NT do for i from 1 to NX-1 do
 U(i,j+1):=subs(IBC,FD(i,j)); od: od:
G:=j->plot([seq([X[i],subs(IBC,U(i,j))],i=0..NX)],color=blue):
Ops1:=thickness=3,labels=["X","U"];
display([seq(G(j),j=0..NT)],insequence=true,Ops1);
```

Mathematica:

```
SetOptions[ListPlot,ImageSize->300,PlotRange->{{0,1},{0,1}},
 Joined->True]; f[x_]:=N[Sin[Pi*x]]; {nu=1,nX=15,nT=100,l=1,
 tF=0.2,h=l/nX,k=tF/nT,r=nu*k/h^2}
Table[xN[i]=i*h,{i,0,nX}]; ic=Table[uN[i,0]->f[xN[i]],{i,0,nX}];
bc={Table[uN[0,j]->0,{j,0,nT}],Table[uN[nX,j]->0,{j,0,nT}]};
ibc=Flatten[{ic,bc}]
fd[i_,j_]:=(1-2*r)*uN[i,j]+r*(uN[i+1,j]+uN[i-1,j]);
Do[uN[i,j+1]=fd[i,j]/.ibc,{j,0,nT},{i,1,nX-1}];
g[j_]:=ListPlot[Table[{xN[i],uN[i,j]/.ibc},{i,0,nX}],
 PlotStyle->{Blue,Thickness[0.01]},AxesLabel->{"X","U"}];
grs=Evaluate[Table[g[j],{j,0,nT}]]; ListAnimate[grs]
```

2. We note that this FD scheme can be represented in the matrix form: $U_i=MU_{i-1}$, where $U_0=(f(X_1),\dots,f(X_{NX-1}))$, and M is the $NX\times NX$ tridiagonal band matrix (with $1-2r$ along the main diagonal, r along the first subdiagonals, and zeros everywhere else). We obtain the numerical solution using this matrix representation of the FD scheme.

Maple:

```
with(plots): with(LinearAlgebra): nu:=1: NX:=40: NT:=800:
L:=1.: T:=0.2; h:=L/NX; k:=T/NT; r:=nu*k/h^2; NG:=90;
interface(rtablesize=NX): M:=BandMatrix([r,1-2*r,r],1,NX-1);
f:=x->evalf(sin(Pi*x)); U0:=Vector([[seq(f(i*h),i=1..NX-1)]]);
for k from 1 to NG do  U||k:=M.U||(k-1);
 G||k:=plot([[0,0],seq([i/NX,U||k[i]],i=1..NX-1),[L,0]]); od:
display([seq(G||i,i=1..NG)],insequence=true,thickness=3);
```

Mathematica:

```
SetOptions[ListPlot,ImageSize->500,PlotStyle->{Blue,
 Thickness[0.01]},PlotRange->{{0,1},{0,1}},Joined->True];
{nu=1,nX=40,nT=800,l=1,tF=0.2,h=l/nX,k=tF/nT,r=nu*k/h^2,nG=90}
f[x_]:=N[Sin[Pi*x]]; mat=SparseArray[{Band[{2,1}]->r,
 Band[{1,1}]->1-2*r,Band[{1,2}]->r},{nX-1,nX-1}];
Print[MatrixForm[mat]]; uN[0]=Table[f[i*h],{i,1,nX-1}]
Do[uN[k]=mat.uN[k-1]; gr[k]={{0,0}};
 Do[gr[k]=Append[gr[k],{i/nX,uN[k][[i]]}],{i,1,nX-1}];
 gr[k]=Append[gr[k],{1,0}]; g[k]=ListPlot[gr[k]],{k,1,nG}];
grs=Evaluate[Table[g[j],{j,1,nG}]]; ListAnimate[grs]
```

3. In *backward difference method*, the second-order derivative u_{xx} is replaced by a central difference approximation (CDA) and the first-order derivative u_t, by a backward difference approximation (BWDA). The final FD scheme for the linear diffusion equation is

$$(1+2r)U_{i,j}-r(U_{i+1,j}+U_{i-1,j})-U_{i,j-1}=0,$$

where $r=\nu k/h^2$. In this *implicit FD scheme*, we have to solve numerically these difference equations at each of the internal mesh points at each j-th step (where $j=1,\dots,NT$) with the initial and boundary conditions. This FD scheme is unconditionally stable. We calculate the approximate numerical solution of the initial boundary value problem by applying the backward difference method.

Maple:

```
with(plots): nu:=1: NX:=50: NT:=50: L:=1.: T:=0.2; h:=L/NX;
 k:=T/NT; r:=nu*k/h^2; for i from 0 to NX do X[i]:=i*h; od:
f:=i->evalf(sin(Pi*X[i])): IBC:={seq(U[i,0]=f(i),i=0..NX),
 seq(U[0,j]=0,j=0..NT),seq(U[NX,j]=0,j=0..NT)}: Sol0:=IBC;
```

```
FD:=(i,j)->(1+2*r)*U[i,j]-r*(U[i+1,j]+U[i-1,j])-U[i,j-1];
for j from 1 to NT do
 Eqs||j:={seq(FD(i,j)=0,i=1..NX-1)}; Eqs1||j:=subs(
  Sol||(j-1),IBC,Eqs||j); vars||j:={seq(U[i,j],i=1..NX-1)};
  Sol||j:=fsolve(Eqs1||j,vars||j); od:
G:=j->plot([seq([X[i],subs(Sol||j,IBC,U[i,j])],i=0..NX)],
            color=blue,thickness=3):
display([seq(G(j),j=0..NT)],insequence=true,labels=["X","U"]);
```

Mathematica:

```
SetOptions[ListPlot,PlotRange->{{0,1},{0,1}},Joined->True];
{nu=1,nX=50,nT=50,l=1,tF=0.2,h=l/nX,k=tF/nT,r=nu*k/h^2}
Table[xN[i]=i*h,{i,0,nX}]; f[i_]:=N[Sin[Pi*xN[i]]];
ibc={Table[uN1[i,0]->f[i],{i,0,nX}],Table[uN1[0,j]->0,
 {j,0,nT}], Table[uN1[nX,j]->0,{j,0,nT}]}//Flatten
sol[0]=ibc; fd[i_,j_]:=(1+2*r)*uN1[i,j]-
 r*(uN1[i+1,j]+uN1[i-1,j])-uN1[i,j-1];
Do[eqs[j]=Table[Expand[fd[i,j]]==0,{i,1,nX-1}];
 eqs1[j]=eqs[j]/.sol[j-1]/.ibc; vars[j]=Table[uN1[i,j],
 {i,1,nX-1}]; sol[j]=NSolve[eqs1[j],vars[j]],{j,1,nT}];
g[j_]:=ListPlot[Table[{xN[i],uN1[i,j]/.Flatten[sol[j]]/.ibc},
 {i,0,nX}],PlotStyle->{Blue,Thickness[0.01]},AxesLabel->
 {"X","U"}]; grs=Evaluate[Table[g[j],{j,1,nT}]];
ListAnimate[grs]
```

4. The *Crank–Nicolson method* consists in averaging the forward FD scheme at j-th time step and the backward FD scheme at $(j+1)$-th time step. The final FD scheme for the linear heat equation is

$$-rU_{i-1,j+1}+2(1+r)U_{i,j+1}-rU_{i+1,j+1}=rU_{i-1,j}+2(1-r)U_{i,j}+rU_{i+1,j},$$

where $r=k/h^2$. In this FD scheme we have 3 unknown values of U at the $(j+1)$-th time step and 3 known values at the j-th time step. This FD scheme is unconditionally stable. We calculate the approximate numerical solution of the initial boundary value problem by applying the Crank–Nicolson method [34].

We compare the approximate numerical solution with the exact solution of this problem $u(x,t)=\exp(-\pi^2 t)\sin(\pi x)$ at $(x_k,t_k)\in\mathcal{D}$. Additionally, we obtain the numerical solution using the *Maple* function `pdsolve` (with the option `method=CrankNicholson`) and compare the resulting two numerical solutions with the exact solution.

Maple:

```
with(PDEtools): with(plots): declare(v(x1,t1)); L:=1; T:=0.2:
nu:=1: NX:=20: NT:=20: NX1:=NX-1: NX2:=NX-2: h:=L/NX; k:=T/NT;
r:=nu*k/(h^2); SX:=h; ST:=k; tR:=0..T; xR:=0..L; NF:=30:
NP:=100: tk:=0.2: Ops1:=spacestep=SX,timestep=ST:
F:=i->sin(Pi*i); IC:={v(x1,0)=F(x1)}; BC:={v(0,t1)=0,v(L,t1)=0};
U[NX-1]:=0: PDE1:=diff(v(x1,t1),t1)-nu*diff(v(x1,t1),x1$2)=0;
for i from 1 to NX1 do U[i-1]:=evalf(F(i*h)); od:
LM[0]:=1+r: UM[0]:=-r/(2*LM[0]):
for i from 2 to NX2 do LM[i-1]:=1+r+r*UM[i-2]/2;
 UM[i-1]:=-r/(2*LM[i-1]); od: LM[NX1-1]:=1+r+0.5*r*UM[NX2-1]:
for j from 1 to NT do t:=j*k; Z[0]:=((1-r)*U[0]+r*U[1]/2)/LM[0];
 for i from 2 to NX1 do Z[i-1]:=((1-r)*U[i-1]+0.5*r*(U[i]+U[i-2]+
  Z[i-2]))/LM[i-1]; od:
 U[NX1-1]:=Z[NX1-1]:
 for i1 to NX2 do i:=NX2-i1+1; U[i-1]:=Z[i-1]-UM[i-1]*U[i]; od:
od: ExSol:=(x,t)->exp(-Pi^2*t)*sin(Pi*x):
printf(`Crank-Nicolson Method\n`); for i from 1 to NX1 do X:=i*h;
 printf(`%3d %11.8f %13.8f %13.8f,%13.8f\n`,i,X,U[i-1],
 evalf(ExSol(X,tk)),U[i-1]-evalf(ExSol(X,tk))); od:
NSol1:=pdsolve(PDE1,IC union BC,numeric,v(x1,t1),Ops1,time=t1,
 range=0..L,method=CrankNicholson):
printf(`Crank-Nicolson Method\n`); vtk:=NSol1:-value(t1=tk,
 output=listprocedure): vVal:=rhs(op(3,vtk)):
for i from 1 to NX1 do X1:=i*h:
 printf(`%3d %11.8f %13.8f %13.8f,%13.8f\n`,i,evalf(X1),vVal(X1),
  evalf(ExSol(X1,tk)),abs(vVal(X1)-evalf(ExSol(X1,tk)))); od:
```

We note that to obtain the coincidence between the numerical solution (using `pdsolve`, `numeric`, `method`) and our solution (with FD scheme), it is necessary to establish the coincidence between the parameters of the two solutions: `SX:=h`, `ST:=k`, `spacestep=SX`, `timestep=ST`.

Mathematica:

```
fF[i_] := Sin[Pi*i]; {l=1,tF=0.2,nu=1,nX=20,nT=20,nX1=nX-1,
 nX2=nX-2,h=l/nX,k=tF/nT,r=nu*k/(h^2),tk=0.2}
{ic={v[x1,0]==fF[x1]}, bc={v[0,t1]==0,v[l,t1]==0}}
{lM=Table[0,{i,0,nX}],uN=Table[0,{i,0,nX}],uM=Table[0,{i,0,nX}],
 z=Table[0,{i,0,nX}], uN[[nX-1]]=0}
pde1=D[v[x1,t1],t1]-nu*D[v[x1,t1],{x1,2}]==0 Do[uN[[i-1]]=
 N[fF[i*h]],{i,1,nX1}]; {lM[[0]]=1+r, uM[[0]]=-r/(2*lM[[0]])}
```

```
Do[lM[[i-1]]=1+r+r*uM[[i-2]]/2; uM[[i-1]]=-r/(2*lM[[i-1]]),
 {i,2,nX2}]; lM[[nX1-1]]=1+r+0.5*r*uM[[nX2-1]]
Do[t=j*k; z[[0]]=((1-r)*uN[[0]]+r*uN[[1]]/2)/lM[[0]];
 Do[z[[i-1]]=((1-r)*uN[[i-1]]+0.5*r*(uN[[i]]+uN[[i-2]]+
  z[[i-2]]))/lM[[i-1]],{i,2,nX1}]; uN[[nX1-1]]=z[[nX1-1]];
 Do[i=nX2-i1+1; uN[[i-1]]=z[[i-1]]-uM[[i-1]]*uN[[i]],{i1,1,nX2}],
{j,1,nT}]; nD=10; extSol[x1_,t1_]:=Exp[-Pi^2*t]*Sin[Pi*x]/.
 {x->x1,t->t1}; Print["Crank-Nicolson Method"];
Do[xN=i*h; Print[i," ",PaddedForm[N[xN,nD],{12,10}]," ",
 PaddedForm[uN[[i-1]],{12,10}]," ", PaddedForm[N[extSol[xN,tk],
 nD],{12,10}]," ", PaddedForm[uN[[i-1]]-N[extSol[xN,tk],nD],
 {12,10}]],{i,1,nX1}];
```

5. The inviscid Burgers equation can be rewritten in the *conservation form*, i.e., as the physical conservation law of the form $u_t+(F(u)=_x=0$, where $F(u)=\frac{1}{2}u^2$. For the discretization of the problem we will use this form of the inviscid Burgers equation. We generate the rectangular mesh: $X=a+ih$, $T=jk$ ($i=0,\ldots,NX$, $j=0,\ldots,NT$, $h=(b-a)/NX$, $k=T/NT$). We denote the approximate solution of $u(x,t)$ at the mesh point (i,j) as $U_{i,j}$. In the *Lax method*, the derivative u_t is replaced by a forward difference approximation (FWDA) and the derivative $(F(u))_x$, by a central difference approximation (CDA) (together with the corresponding substitution). The final FD scheme for the linear diffusion equation reads

$$U_{i,j+1}=\frac{1}{2}\Big(U_{i+1,j}+U_{i-1,j}\Big)-\frac{k}{2h}\left(\frac{U_{i+1,j}^2}{2}-\frac{U_{i-1,j}^2}{2}\right).$$

In this explicit FD scheme, the unknown value $U_{i,j+1}$ (at $(j+1)$-th step) is determined from the 3 known values $U_{i-1,j}$, $U_{i,j}$, and $U_{i+1,j}$ (at j-th step). This FD scheme is stable for $k/h<1/|u|$, i.e., the stability condition depends on the solution.

We find the approximate numerical solution of the initial boundary value problem by applying the Lax method and introducing a computational boundary condition ($u(b,t)=0$).

Maple:

```
restart: with(plots): NX:=20: NT:=300: A:=0.4; L:=1; a:=0; b:=1;
T:=40.; h:=evalf((b-a)/NX); k:=evalf(T/NT); f:=x->evalf(
 A*sin(Pi*x/L)); for i from 0 to NX do X[i]:=a+i*h od:
IC:={seq(U(i,0)=f(X[i]),i=0..NX)};
BC:={seq(U(a,j)=0,j=0..NT),seq(U(NX,j)=0,j=0..NT)}:
IBC:=IC union BC: FD:=(i,j)->0.5*(U(i+1,j)+U(i-1,j))-
 k/(2*h)*(U(i+1,j)^2/2-U(i-1,j)^2/2);
```

```
for j from 0 to NT do for i from 1 to NX-1 do
 U(i,j+1):=subs(IBC,FD(i,j)); od: od:
G:=j->plot([seq([X[i],subs(IBC,U(i,j))],i=0..NX)],color=blue,
 thickness=3,numpoints=100):
display([seq(G(j),j=0..NT)],insequence=true,labels=["X","U"]);
```

Mathematica:

```
ClearAll["Global`*"]; SetOptions[ListPlot,PlotRange->{{0,1},
 {0,0.4}},Joined->True]; {nX=20,nT=300,am=0.4,l=1,a=0,b=1,
 tF=40,h=N[(b-a)/nX],k=N[tF/nT]}
f[x_]:=N[am*Sin[Pi*x/l]]; Table[xN[i]=a+i*h,{i,0,nX}];
ic=Table[uN[i,0]->f[xN[i]],{i,0,nX}]; bc={Table[uN[a,j]->0,
 {j,0,nT}],Table[uN[nX,j]->0,{j,0,nT}]}; ibc=Flatten[{ic,bc}]
fd[i_,j_]:=0.5*(uN[i+1,j]+uN[i-1,j])-k/(2*h)*(uN[i+1,j]^2/2-
 uN[i-1,j]^2/2); Do[uN[i,j+1]=fd[i,j]/.ibc,{j,0,nT},{i,1,nX-1}];
g[j_]:=ListPlot[Table[{xN[i],uN[i,j]/.ibc},{i,0,nX}],
 PlotStyle->{Blue,Thickness[0.01]},AxesLabel->{"X","U"}];
grs=Evaluate[Table[g[j],{j,0,nT}]]; ListAnimate[grs]
```
□

Problem 6.12 *Linear and nonlinear wave equations. Explicit difference methods.* Let us consider the initial boundary value problems for the linear and nonlinear wave equations describing the motion of a fixed string,

$$u_{tt}=c^2\, u_{xx};\ u(x,0)=f(x),\ u_t(x,0)=g(x),\ u(0,t)=u(L,t)=0,$$
$$u_{tt}=c^2\, u_{xx}+F(u);\ u(x,0)=f(x),\ u_t(x,0)=g(x),\ u(0,t)=u(L,t)=0,$$

defined in the domain $\mathcal{D}=\{0\le x\le L,\ 0\le t\le T\}$, where $F(u)=e^{\lambda u}$, $f(x)=0$, $g(x)=\sin(4\pi x)$, $L=0.5$, $T=1.5$, $c=1/(4\pi)$. Applying the explicit central finite difference method, find the approximate numerical solution of the initial boundary value problems.

1. Linear wave equation. In *explicit central difference method*, each second derivative is replaced by a central difference approximation. The FD scheme for the linear wave equation is

$$U_{i,j+1}=2(1-r)U_{i,j}+r(U_{i+1,j}+U_{i-1,j})-U_{i,j-1},$$

where $r=(ck/h)^2$. In this FD scheme we have one unknown value $U_{i,j+1}$ that depends explicitly on the four known values $U_{i,j}$, $U_{i+1,j}$, $U_{i-1,j}$, $U_{i,j-1}$ at the previous time steps (j and $j-1$). To start the process we have to know the values of U at the time steps j=0 and j=1. So we can define the initial conditions at these time steps: $U_{i,0}=f(X_i)$ and $U(X_i,0)_t \approx (U_{i,1} - U_{i,0})/k = g(X_i)$, $U_{i,1}=f(X_i)+kg(X_i)$. This FD scheme is stable for $r \leq 1$. We find the approximate numerical solution of the initial boundary value problem by applying the explicit central finite difference method.

Maple:

```
c:=1/(4*Pi); L:=0.5; T:=1.5; NX:=40; NT:=40; NX1:=NX+1;
NX2:=NX-1; NT1:=NT+1; NT2:=NT-1; h:=L/NX; k:=T/NT;
r:=evalf(c*k/h); F:=i->0; G:=i->sin(4*Pi*i);
for j from 2 to NT1 do U[0,j-1]:=0; U[NX1-1,j-1]:=0; od:
U[0,0]:=evalf(F(0)); U[NX1-1,0]:=evalf(F(L));
for i from 2 to NX do
 U[i-1,0]:=F(h*(i-1)); U[i-1,1]:=(1-r^2)*F(h*(i-1))
  +r^2*(F(i*h)+F(h*(i-2)))/2+k*G(h*(i-1)); od:
for j from 2 to NT do for i from 2 to NX do
 U[i-1,j]:=evalf(2*(1-r^2)*U[i-1,j-1]+r^2*(U[i,j-1]
  +U[i-2,j-1])-U[i-1,j-2]); od; od;
printf(` i X(i) U(X(i),NT)\n`);
for i from 1 to NX1 do  X[i-1]:=(i-1)*h:
 printf(`%3d %11.8f %13.8f\n`,i,X[i-1],U[i-1,NT1-1]); od:
Points:=[seq([X[i-1],U[i-1,NT1-1]],i=1..NX1)];
plot(Points,style=point,color=blue,symbol=circle);
```

Mathematica:

```
SetOptions[ListPlot,PlotRange->All,Joined->False];  f[x_]:=0;
g[x_]:=Sin[4*Pi*x]; {c=N[1/(4*Pi)],l=0.5,tF=1.5,nX=40,nX1=nX+1,
 nT=40,nT1=nT+1,h=l/nX,k=tF/nT,r=N[c*k/h]}
fi[i_]:=f[h*(i-1)]; gi[i_]:=g[h*(i-1)]; uN=Table[0,{nT1},{nX1}];
For[i=1,i<=nT1,i++,uN[[i,1]]=fi[i]]; For[i=2,i<=nT,i++,
 uN[[i,2]]=(1-r^2)*fi[i]+r^2*(fi[i+1]+fi[i-1])/2+k*gi[i]];
For[j=3,j<=nX1,j++, For[i=2,i<=nT,i++, uN[[i,j]]=2*(1-r^2)*
 uN[[i,j-1]]+r^2*(uN[[i+1,j-1]]+uN[[i-1,j-1]])-uN[[i,j-2]]//N];];
Print["  i       xN[i]        uN[xN[i],nT1]", "\n"];
For[i=1,i<=nT1,i++,Print[PaddedForm[i,2],PaddedForm[h*(i-1),7],
   "   ",PaddedForm[uN[[i,nT1]],10]]];
points=Table[{h*(i-1),uN[[i,nT1]]},{i,1,nT1}]
ListPlot[points,PlotStyle->{Blue,PointSize[0.02]}]
Print[NumberForm[TableForm[Transpose[Chop[uN]]],3]];
ListPlot3D[uN,ViewPoint->{3,1,3},ColorFunction->Hue]
```

2. We find the same approximate numerical solution of the initial boundary value problem by applying the explicit central finite difference method and following the other style of programming.

We plot the numerical solution for $(x_k, t_k) \in \mathcal{D}$ and compare the results with the numerical solution obtained in Problem 6.2.

Maple:

```
with(plots): c:=evalf(1/(4*Pi)); L:=0.5: T:=1.5: NX:=40:
NT:=40: h:=L/NX; k:=T/NT; r:=(c*k/h)^2; f:=x->0:
g:=x->evalf(sin(4*Pi*x)): IC:={seq(U1(i,0)=f(i*h),i=1..NX-1),
 seq(U1(i,1)=f(i*h)+k*g(i*h),i=1..NX-1)}:
BC:={seq(U1(0,j)=0,j=0..NT),seq(U1(NX,j)=0,j=0..NT)}:
IBC:=IC union BC: FD:=(i,j)->2*(1-r)*U1(i,j)+r*(U1(i+1,j)
 +U1(i-1,j))-U1(i,j-1);
for j from 1 to NT-1 do for i from 1 to NX-1 do
 U1(i,j+1):=subs(IBC,FD(i,j)); od: od:
G:=j->plot([seq([i*h,subs(IBC,U1(i,j))],i=0..NX)],color=blue):
display([seq(G(j),j=0..NT)],insequence=true,thickness=3,
 labels=["X","U"]);
```

Mathematica:

```
SetOptions[ListPlot,ImageSize->500,PlotRange->{{0,0.5},{-1,1}},
 Joined->True]; f[x_]:=0; g[x_]:=N[Sin[4*Pi*x]]; {c=N[1/(4*Pi)],
 l=0.5,tF=1.5,nX=40,nT=40,h=l/nX,k=tF/nT,r=(c*k/h)^2}
ic={Table[uN1[i,0]->f[i*h],{i,1,nX-1}],Table[uN1[i,1]->f[i*h]+
 k*g[i*h],{i,1,nX-1}]}; bc={Table[uN1[0,j]->0,{j,0,nT}],
 Table[uN1[nX,j]->0,{j,0,nT}]}; ibc=Flatten[{ic,bc}]
fd[i_,j_]:=2*(1-r)*uN1[i,j]+r*(uN1[i+1,j]+uN1[i-1,j])-uN1[i,j-1];
Do[uN1[i,j+1]=fd[i,j]/.ibc,{j,1,nT-1},{i,1,nX-1}];
g[j_]:=ListPlot[Table[{i*h,uN1[i,j]/.ibc},{i,0,nX}],PlotStyle->
 {Blue,Thickness[0.01]},AxesLabel->{"X","U"}];
grs=N[Table[g[j],{j,0,nT}]]; ListAnimate[grs]
```

3. Nonlinear wave equation. Applying the *explicit central difference method*, we generate the FD scheme for the nonlinear wave equation:

$$U_{i,j+1} = 2(1-r)U_{i,j} + r(U_{i+1,j} + U_{i-1,j}) - U_{i,j-1} + \exp(U(i,j))k^2,$$

where $r = (ck/h)^2$. In this FD scheme we have one unknown value $U_{i,j+1}$ that depends explicitly on the four known values $U_{i,j}$, $U_{i+1,j}$, $U_{i-1,j}$, $U_{i,j-1}$ at the previous time steps (j and $j-1$).

To start the process we have to know the values of U at the time steps j=0 and j=1. So we can define the initial conditions at these times steps: $U_{i,0}=f(X_i)$ and $U(X_i,0)_t \approx (U_{i,1}-U_{i,0})/k=g(X_i)$, $U_{i,1}=f(X_i)+kg(X_i)$.

We construct the approximate numerical solution of the given initial boundary value problem by applying the explicit finite difference method, plot the numerical solution for $(x_k,t_k) \in \mathcal{D}$, and compare the results with the approximate solution obtained in Problem 6.2 by applying the predefined functions (pdsolve and NDSolve).

Maple:

```
restart: with(plots): L:=0.5: T:=1.5: NX:=25: NT:=20:
c:=evalf(1/(4*Pi)): h:=L/NX; lambda:=1: k:=T/NT; r:=(c*k/h)^2;
f:=x->0: g:=x->evalf(sin(4*Pi*x)): IC:={seq(U(i,0)=f(i*h),
 i=1..NX-1),seq(U(i,1)=f(i*h)+k*g(i*h),i=1..NX-1)}:
BC:={seq(U(0,j)=0,j=0..NT),seq(U(NX,j)=0,j=0..NT)}:
IBC:=IC union BC: FD:=(i,j)->evalf(2*(1-r)*U(i,j)+r*(U(i+1,j)
 +U(i-1,j))-U(i,j-1)+exp(lambda*U(i,j))*k^2);
for j from 1 to NT-1 do for i from 1 to NX-1 do
  U(i,j+1):=subs(IBC,FD(i,j)); od: od:
G:=j->plot([seq([i*h,subs(IBC,U(i,j))],i=0..NX)],color=blue):
display([seq(G(j),j=0..NT)],insequence=true,thickness=3,
  labels=["X","U"]);
```

Mathematica:

```
ClearAll["Global`*"]; SetOptions[ListPlot,ImageSize->500,
 PlotRange->{{0,0.5},{-1,5}},Joined->True]; {l=0.5,tF=1.5,
 nX=20,nT=20,c=N[1/(4*Pi)],h=l/nX,k=tF/nT,r=(c*k/h)^2}
f[x_]:=0; g[x_]:=Sin[4*Pi*x]//N; ic={Table[uN[i,0]->f[i*h],
 {i,1,nX-1}],Table[uN[i,1]->f[i*h]+k*g[i*h],{i,1,nX-1}]};
bc={Table[uN[0,j]->0,{j,0,nT}],Table[uN[nX,j]->0,{j,0,nT}]};
ibc=Flatten[{ic,bc}]
fd[i_,j_]:=2*(1-r)*uN[i,j]+r*(uN[i+1,j]+uN[i-1,j])-uN[i,j-1]+
 Exp[uN[i,j]]*k^2;
Do[uN[i,j+1]=fd[i,j]/.ibc,{j,1,nT-1},{i,1,nX-1}];
g[j_]:=ListPlot[Table[{i*h,uN[i,j]/.ibc},{i,0,nX}],
  PlotStyle->{Blue,Thickness[0.01]},AxesLabel->{"X","U"}];
grs=Evaluate[Table[g[j],{j,0,nT}]]; ListAnimate[grs]
```

□

6.2.2 Interaction of Solitons

Let us consider the Korteweg–de Vries equation

$$u_t + auu_x + bu_{xxx} = 0,$$

defined in $\mathcal{D} = \{x \in \mathbb{R},\ t \geq 0\}$, where a, b are real constants. We investigate numerically *soliton interactions* and confirm that solitary wave solutions of the KdV equation are solitons, i.e., we verify an ability of a soliton solution to survive an interaction with another soliton solutions of the KdV equation. For this, we have to consider the special boundary conditions for a soliton solution, $\lim_{x\to\pm\infty} u(x,t) = 0$ in $\mathcal{D}$, and special initial conditions that can generate a family of solitons with different speeds, pulse centers and traveling in the (x,t) plane.

We choose a solitary wave solution of the KdV equation in the form $f(x) = -(3c/a)(\tanh[\sqrt{cb}\,(-z+C_1)/(2b)]^2 - 1)$, where $z = x - ct$, $C_1 = 0$. This solution has been obtained analytically in Problem 2.18. Then we extend this solution to generate a family of solitons with different speeds c, where the pulses are localized at time $t = t_k$ and space $x = x_k$.

We follow the approach proposed by N. Zabusky and M. Kruskal [173] in 1965 for numerical investigation of solitary wave collisions using finite difference (FD) approximations. The Zabusky–Kruskal FD scheme has the form

$$\begin{aligned} U_{i,j+1} = U_{i,j-1} - \tfrac{1}{3} arh^2 (U_{i+1,j} + U_{i,j} + U_{i-1,j})(U_{i+1,j} - U_{i-1,j}) \\ - rb(U_{i+2,j} - 2U_{i+1,j} + 2U_{i-1,j} - U_{i-2,j}), \end{aligned}$$

where h is the spatial step size, k is the time step size, and the ratio $r = k/h^3$ satisfies the stability condition of the FD scheme $r < 0.38$. We note that in this FD scheme the first derivatives are replaced by central difference approximation (CDA), the nonlinear term uu_x by the average value of three u terms at the grid points $(i+1,j)$, (i,j), and $(i-1,j)$, and the third derivative by $(U_{i+2,j} - 2U_{i+1,j} + 2U_{i-1,j} - U_{i-2,j})/(2h^3)$.

Problem 6.13 *Korteweg–de Vries equation. Zabusky–Kruskal finite difference scheme. N-soliton* interaction.* Following the Zabusky–Kruskal FD scheme, find the approximate numerical solution of the initial boundary value problem

$$u_t + auu_x + bu_{xxx} = 0;\ u(x,0) = f(x),\ \lim_{x\pm\infty} u(x,t) = 0,$$

*In the present problem we consider $N=2$ solitary wave solutions.

where $f(x)=-(3c/a)(\tanh[\sqrt{cb}\,(-z+C_1)/(2b)]^2-1)$, $a=2$, $b=1$, $C_1=0$. Confirm that the solitary waves are stable and survive the interaction process without changing the shape.

Maple:

```
with(plots): Digits:=50: L:=100: T:=70: NX:=100: NT:=360:
h:=evalf(L/NX); k:=evalf(T/NT); r:=evalf(k/h^3); a:=2: b:=1:
xR:=0..NX; tR:=0..NT; tk:=0; NP:=1000: L1:=[20,40];
L2:=[0.7,0.2]; K:=nops(L1); Ops1:=numpoints=NP,thickness=3:
S:=(xk,c)->3*c/a*(sech(sqrt(c*b)/(2*b)*(-(x-xk)+c*t))^2):
F1:=unapply(add(S(L1[i],L2[i]),i=1..K),x,t);
F2:=unapply(diff(F1(x,t),t),x,t);  F1(x,t); F2(x,t);
for i from 0 to NX do for j from 0 to NT do U(i,j):=0: od: od:
for i from 0 to NX do U(i,0):=evalf(F1(i,tk)): od:
for i from 0 to NX do U(i,1):=evalf(F1(i,tk)+k*F2(i,tk)): od:
FD:=(i,j)->U(i,j-1)-a*r*h^2*(U(i+1,j)+U(i,j)+U(i-1,j))*
 (U(i+1,j)-U(i-1,j))/3-r*b*(U(i+2,j)-2*U(i+1,j)+2*U(i-1,j)-
 U(i-2,j));
for j from 1 to NT do for i from 2 to NX-2 do
 U(i,j+1):=evalf(FD(i,j)): od: od:
G:=j->plot([seq([i,U(i,j)],i=2..NX-2)],Ops1,color=blue,
 view=[0..100,0..1.2]): LG:=[seq(G(4*j),j=0..NT/4)]:
display(LG,insequence=true);
```

Mathematica:

```
{l=100,tF=70,nX=100,nT=360,h=N[l/nX],k=N[tF/nT],r=N[k/h^3],
 a=2,b=1,tk=0,l1={20,40},l2={0.7,0.2},kl1=Length[l1]}
SetOptions[ListPlot,ImageSize->500,PlotRange->{{0,100},{0,1.2}},
 PlotStyle->{Blue,Thickness[0.01]},Joined->True];
fS[xk_,c_]:=3*c/a*(Sech[Sqrt[c*b]/(2.*b)*(-(x-xk)+c*t)]^2);
f1[x1_,t1_]:=Sum[fS[l1[[i]],l2[[i]]],{i,1,kl1}]/.x->x1/.t->t1;
f2[x1_,t1_]:=D[f1[x,t],t]/.x->x1/.t->t1; {f1[x,t],f2[x,t]}
Do[uN[i,j]=0,{j,0,nT},{i,0,nX}]; Do[uN[i,0]=f1[i,tk],{i,0,nX}];
Do[uN[i,1]=f1[i,tk]+k*f2[i,tk],{i,0,nX}];
fd[i_,j_]:=uN[i,j-1]-a*r*h^2*(uN[i+1,j]+uN[i,j]+uN[i-1,j])*
 (uN[i+1,j]-uN[i-1,j])/3-r*b*(uN[i+2,j]-2*uN[i+1,j]+2*uN[i-1,j]-
 uN[i-2,j]); Do[uN[i,j+1]=fd[i,j],{j,1,nT},{i,2,nX-2}];
g[j_]:=ListPlot[Table[{i,uN[i,j]},{i,2,nX-2}],PlotStyle->
 {Blue,Thickness[0.01]},AxesLabel->{"X","U"}];
grs=Evaluate[Table[g[4*j],{j,0,nT/4}]]; ListAnimate[grs]
```

□

Problem 6.14 *Sine–Gordon equation. Finite difference scheme with space-time mesh along the characteristics. Kink-kink interaction.* Let us consider the initial boundary problem for the sine–Gordon equation

$$u_{xx}-u_{tt}=\sin u;\ u(x,0)=S(x,0),\ p(x,0)=S_x(x,0),\ q(x,0)=S_t(x,0),$$
$$u(-L,t)=-2\pi,\ p(-L,t)=q(-L,t)=0,\ u(L,t)=2\pi,\ p(L,t)=q(L,t)=0,$$

defined in the domain $\mathcal{D}=\{-L\le x\le L, 0\le t\le T\}$, where $p=u_x$, $q=u_t$, and $S(x,t)=-4\arctan\left(\frac{\sqrt{2}}{2}\frac{(1-e^{2x\sqrt{2}})e^{-x+t}}{1+e^{2t}}\right)$.

Applying the FD scheme with the space-time mesh along the characteristics, we find the approximate numerical solution of the initial boundary value problem describing the kink-kink interaction, visualize the interaction process, and compare it with the same interaction process (with the same parameters) described in Problem 2.32.

Let us obtain the numerical solution of this problem describing the kink-kink interaction. Since this problem we have studied analytically applying the generalized method of separation of variables in Problem 2.32, we can introduce the initial profile $S(x,t)$ according to the solution (2.2) of Problem 2.32. Setting the same parameters as in the analytic solution, i.e., $A_2=1$, $A_1=2$, $C_1=0$, we obtain the above presented initial profile $S(x,t)$ for the kin-kink interaction.

Let us generate (instead of a rectangular mesh) a space-time mesh along the characteristics. For this, we solve the sine–Gordon equation by the generalized method of characteristics (see Problem 3.15) and obtain that this equation has the two characteristic directions $dt/dx=\pm 1$ with the slopes $\pm 45°$, respectively. Therefore, in the FD scheme we will have the 3 points at the intersection of the characteristics: the point $(i+1,j+1)$ (in which the value of the solution we have to determined), the left and right points at the previous time step, (i,j) and $(i+2,j)$, respectively.

Then, following R. H. Enns and G. C. McGuire [44], we construct the FD scheme and obtain the value of the solution at the $(i+1,j+1)$-th grid point $U_{i+1,j+1}$:

$$2U_{i+1,j+1}=U_L+U_R,\ U_L=U_{i,j}+\tfrac{h}{2}(Up_{i,j}+Up_{i+1,j+1}+Uq_{i,j}+Uq_{i+1,j+1}),$$
$$U_R=U_{i+2,j}+\tfrac{h}{2}(-Up_{i+2,j}-Up_{i+1,j+1}+Uq_{i+2,j}+Uq_{i+1,j+1}),$$
$$2Up_{i+1,j+1}=Up_{i+2,j}+Up_{i,j}+Uq_{i+2,j}-Uq_{i,j}+h[-\sin(U_{i+2,j})+\sin(U_{i,j})],$$
$$2Uq_{i+1,j+1}=Up_{i+2,j}-Up_{i,j}+Uq_{i+2,j}+Uq_{i,j}+h[-\sin(U_{i+2,j})-\sin(U_{i,j})],$$

where $U_{i,j}$, $Up_{i,j}$, $Uq_{i,j}$ are, respectively, the approximation values of $u(x,t)$, $p(x,t)$, and $q(x,t)$.

Maple:

```
with(plots): NX:=100: NT:=100: xR:=0..NX; tR:=0..NT; NP:=100;
A2:=1.; A1:=2; L:=10; L1:=[-L,L]; K:=nops(L1); Ops1:=
 numpoints=NP,thickness=3,color=blue: S:=(x,t)->-4*arctan(
 (A2-exp(2*x*sqrt(A1)))*exp(-x*sqrt(A1)+t*sqrt(A1-1))
 *sqrt(A1-1)/(sqrt(A1)*(A2+exp(2*t*sqrt(A1-1)))));
h:=evalf((L1[2]-L1[1])/NX);
Sx:=unapply(diff(S(x,t),x),x,t):St:=unapply(diff(S(x,t),t),x,t):
for i from 0 to NX by 2 do
 X:=L1[1]+i*h; U[i,0]:=evalf(S(X,-10)); Up[i,0]:=evalf(
 Sx(X,-10)); Uq[i,0]:=evalf(St(X,-10)); od:
for j from 2 to NT by 2 do
 U[0,j]:=evalf(-2*Pi): Up[0,j]:=0: Uq[0,j]:=0;
 U[NX,j]:=evalf(2*Pi): Up[NX,j]:=0: Uq[NX,j]:=0; od:
for j from 0 to NT-1 do
 if modp(j,2)=0 then k:=0: else k:=1: fi;
 for i from k to NX-2 by 2 do
  Up[i+1,j+1]:=0.5*(Up[i+2,j]+Up[i,j]+Uq[i+2,j]-Uq[i,j]
   +(-sin(U[i+2,j])+sin(U[i,j]))*h);
  Uq[i+1,j+1]:=0.5*(Up[i+2,j]-Up[i,j]+Uq[i+2,j]+Uq[i,j]
   +(-sin(U[i+2,j])-sin(U[i,j]))*h);
  UL:=U[i,j]+0.5*h*(Up[i,j]+Up[i+1,j+1]+Uq[i,j]+Uq[i+1,j+1]);
  UR:=U[i+2,j]+0.5*h*(-Up[i+2,j]-Up[i+1,j+1]+Uq[i+2,j]
   +Uq[i+1,j+1]); U[i+1,j+1]:=0.5*(UL+UR); od: od:
for j from 0 to NT by 2 do
 G:=j->plot([seq([L1[1]+2*i*h,U[2*i,j]],i=0..NX/2)],Ops1): od:
LG:=[seq(G(2*j),j=0..NT/2)]: display(LG,insequence=true);
```

Mathematica:

```
{nX=100,nT=100,a2=1,a1=2,l=10,l1={-l,l},ll1=Length[l1]}
SetOptions[ListPlot,ImageSize->500,PlotRange->{{-l,l},{-7,7}},
 PlotStyle->{Blue,Thickness[0.01]},Joined->True];
fS[x_,t_]:=-4*ArcTan[(a2-Exp[2*x*Sqrt[a1]])*Exp[-x*Sqrt[a1]+
 t*Sqrt[a1-1]]*Sqrt[a1-1]/(Sqrt[a1]*(a2+Exp[2*t*Sqrt[a1-1]]))];
h=N[(l1[[2]]-l1[[1]])/nX];
sx[x1_,t1_]:=D[fS[x,t],x]/.x->x1/.t->t1;
st[x1_,t1_]:=D[fS[x,t],t]/.x->x1/.t->t1;
Do[xN=l1[[1]]+i*h; uN[i,0]=N[fS[xN,-10]];
   up[i,0]=N[sx[xN,-10]]; uq[i,0]=N[st[xN,-10]],{i,0,nX,2}];
Do[uN[0,j]=N[-2*Pi]; up[0,j]=0; uq[0,j]=0;
   uN[nX,j]=N[2*Pi]; up[nX,j]=0; uq[nX,j]=0,{j,2,nT,2}];
```

```
For[j=0,j<=nT-1,j++, If[Mod[j,2]==0,k=0,k=1];
 For[i=k,i<=nX-2,i=i+2, up[i+1,j+1]=0.5*(up[i+2,j]+up[i,j]+
 uq[i+2,j]-uq[i,j]+(-Sin[uN[i+2,j]]+Sin[uN[i,j]])*h);
  uq[i+1,j+1]=0.5*(up[i+2,j]-up[i,j]+uq[i+2,j]+uq[i,j]+
   (-Sin[uN[i+2,j]]-Sin[uN[i,j]])*h);
  uL=uN[i,j]+0.5*h*(up[i,j]+up[i+1,j+1]+uq[i,j]+uq[i+1,j+1]);
  uR=uN[i+2,j]+0.5*h*(-up[i+2,j]-up[i+1,j+1]+uq[i+2,j]+
   uq[i+1,j+1]); uN[i+1,j+1]=0.5*(uL+uR)]];
Do[g[j_]:=ListPlot[Table[{l1[[1]]+2*i*h,uN[2*i,j]},{i,0,nX/2}],
 PlotStyle->{Blue,Thickness[0.01]},AxesLabel->{"X","U"}],
 {j,0,nT,2}]; grs=Evaluate[Table[g[2*j],{j,0,nT/2}]];
ListAnimate[grs]
```
□

6.2.3 Elliptic Equations

Since in *Maple* (Ver. ≤ 14) and *Mathematica* (Ver. ≤ 8), it is possible to solve numerically (via predefined functions) only evolution equations, we apply finite difference methods and the method of lines (see Sect. 7.1) for constructing numerical and graphical solutions of initial boundary value problems for elliptic equations, e.g., the linear and nonlinear Poisson equations. It should be noted that in MATLAB, it is possible to solve numerically various initial boundary value problems for nonlinear elliptic equations applying predefined functions and embedded methods, e.g., to solve scalar nonlinear elliptic PDEs and their systems in two space dimensions, nonlinear problems defined on a more complicated geometry (for details, see [144]).

Problem 6.15 *Linear and nonlinear Poisson equations. Central difference scheme.* Let us consider the two-dimensional linear and nonlinear Poisson equations

$$u_{xx}+u_{yy}=f(x,y), \quad u_{xx}+u_{yy}=F(u),$$

defined in the domain $\mathcal{D}=\{a\leq x\leq b,\ c\leq y\leq d\}$ and subject to the same boundary conditions:

$$u(x,c)=f_1(x),\ u(x,d)=f_2(x),\ u(a,y)=f_3(y),\ u(b,y)=f_4(y).$$

Such boundary value problems describe a steady-state process $u(x,y)$ in a bounded rectangular object. Let us choose $f(x,y)=\sin x\cos y$, $F(u)=\sin(u)$, $f_1(x)=f_2(x)=-\frac{1}{2}\sin x$, $f_3(y)=f_4(y)=0$, $a=0$, $b=\pi$, $c=0$, $d=2\pi$. For the linear Poisson equation, applying the explicit finite difference scheme, obtain the approximate numerical solution of the boundary value problem, visualize it in $\mathcal{D}$, and compare with the exact solution (which can be obtained via predefined functions). For the nonlinear Poisson equation, modifying the FD scheme, construct the approximate numerical solution of the boundary value problem and visualize it in $\mathcal{D}$.

1. Exact solution. Applying *Maple* predefined functions, we find the exact solution of the boundary value problem for the linear Poisson equation, $u(x,y){=}-\frac{1}{2}\sin x\cos y$ (`Sol32`), and visualize it in $\mathcal{D}$.

Maple:

```
with(linalg): with(PDEtools): f:=(x,y)->sin(x)*cos(y);
PDE1:=laplacian(u(x,y),[x,y])-f(x,y)=0;
Sol1:=pdsolve(PDE1,build); Test1:=pdetest(Sol1,PDE1);
Sol11:=unapply(subsop(1=0,2=0,rhs(Sol1)),x,y);
Sol12:=expand(Sol11(x,y));
plot3d(Sol12,x=0..Pi,y=0..2*Pi,shading=zhue);
Sol11(x,0);  Sol11(x,2*Pi); Sol11(0,y); Sol11(Pi,y);
```

2. Numerical and graphical solutions. Let us generate the rectangular mesh: $x{=}a{+}ih$, $y{=}c{+}jk$ ($i{=}0,\ldots,NX$, $j{=}0,\ldots,NY$, $h{=}\dfrac{b-a}{NX}$, $k{=}\dfrac{d-c}{NY}$). We denote the approximate solution of $u(x,y)$ at the mesh point (i,j) as $U_{i,j}$. The second derivatives in Poisson equation are replaced by a central difference approximation (CDA). The FD scheme takes the form:

$$2(1{+}r)U_{i,j}{-}U_{i+1,j}{-}U_{i-1,j}{-}rU_{i,j+1}{-}rU_{i,j-1}{=}\sin(ih)\cos(jk),$$

where $r{=}(h/k)^2$.

Maple:

```
with(plots): a:=0; b:=Pi; c:=0; d:=2*Pi; NX:=20; NY:=20;
h:=(b-a)/NX; k:=(d-c)/NY; r:=(h/k)^2; XY:=seq(x[i]=a+i*h,
 i=0..NX),seq(y[j]=c+j*k,j=0..NY); FD:=(i,j)->2*(1+r)*U[i,j]
 -U[i+1,j]-U[i-1,j]-r*U[i,j+1]-r*U[i,j-1]-cos(j*k)*sin(i*h)=0;
F1:=i->-1/2*sin(i*h); F2:=i->-1/2*sin(i*h); F3:=j->0; F4:=j->0;
BC:=seq(U[i,0]=F1(i),i=0..NX),seq(U[i,NY]=F2(i),i=0..NX),
    seq(U[0,j]=F3(j),j=0..NY),seq(U[NX,j]=F4(j),j=0..NY);
Eqs:={seq(seq(FD(i,j),i=1..NX-1),j=1..NY-1)}: Eqs1:=subs(BC,Eqs):
vars:={seq(seq(U[i,j],i=1..NX-1),j=1..NY-1)}: Sol:=evalf(
 fsolve(Eqs1,vars)); Points:=[seq(seq([x[i],y[j],U[i,j]],
 i=0..NX),j=0..NY)]: Points1:=subs({XY,BC,op(Sol)},Points):
pointplot3d(Points1,symbol=solidsphere,shading=z,
 orientation=[50,70],axes=frame);
```

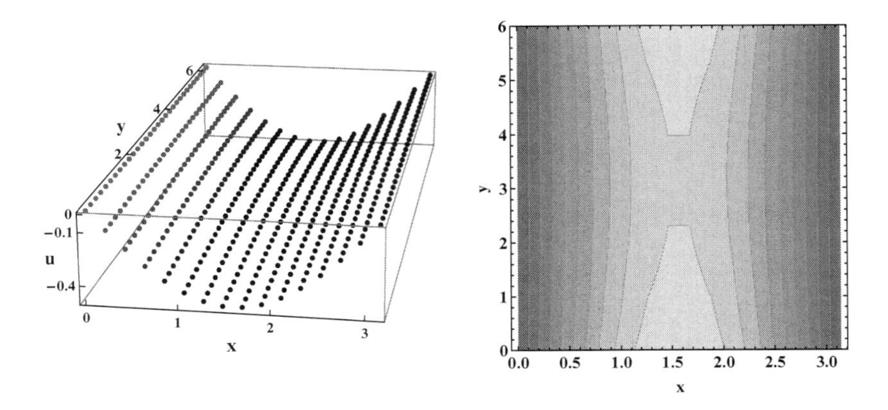

Fig. 6.3. The 3D and contour plots of numerical solution of the boundary value problem for the nonlinear Poisson equation in $\mathcal{D}$

Mathematica:

```
SetAttributes[{x,y},NHoldAll]; {nD=10,a=0,b=Pi,c=0,d=2*Pi,nX=20,
 nY=20,h=N[(b-a)/nX],k=N[(d-c)/nY],r=N[(h/k)^2]}
xN=Table[x[i]->a+i*h,{i,0,nX}]//N; yN=Table[y[j]->c+j*k,
 {j,0,nY}]//N; fd[i_,j_]:=2*(1+r)*uN[i,j]-uN[i+1,j]-uN[i-1,j]-
 r*uN[i,j+1]-r*uN[i,j-1]-Cos[j*k]*Sin[i*h]; f1[i_]:=
 -1/2*Sin[i*h]; f2[i_]:=-1/2*Sin[i*h]; f3[j_]:=0; f4[j_]:=0;
bc=Flatten[{Table[uN[i,0]->f1[i],{i,0,nX}],Table[uN[i,nY]->
 f2[i],{i,0,nX}], Table[uN[0,j]->f3[j],{j,0,nY}],
 Table[uN[nX,j]->f4[j],{j,0,nY}]}]
eqs=Flatten[Table[fd[i,j]==0,{i,1,nX-1},{j,1,nY-1}]];
eqs1=Flatten[eqs/.bc]; vars=Flatten[Table[uN[i,j],{i,1,nX-1},
 {j,1,nY-1}]]; sol=NSolve[eqs1,vars]
points=Table[{x[i],y[j],uN[i,j]},{i,0,nX},{j,0,nY}];
points1=Flatten[N[points/.xN/.yN/.bc/.sol[[1]],nD],1];
g3D=ListPointPlot3D[points1,BoxRatios->{1,1,1},PlotStyle->
 PointSize[0.01],PlotRange->All]; gCP=ListContourPlot[points1,
 PlotRange->All]; GraphicsRow[{g3D,gCP},ImageSize->800]
```

3. Nonlinear Poisson equation. Modifying the above FD scheme, we construct the numerical solution of the boundary value problem for the nonlinear Poisson equation and visualize it in $\mathcal{D}$ (see Fig. 6.3).

Maple:

```
with(plots): a:=0; b:=Pi; c:=0; d:=2*Pi; NX:=15; NY:=15;
h:=(b-a)/NX; k:=(d-c)/NY; r:=(h/k)^2;
XY:=seq(x[i]=a+i*h,i=0..NX),seq(y[j]=c+j*k,j=0..NY);
FD:=(i,j)->evalf(2*(1+r)*U[i,j]-U[i+1,j]-U[i-1,j]-r*U[i,j+1]
          -r*U[i,j-1]-h^2*sin(U[i,j]))=0;
F1:=i->-1/2*sin(i*h); F2:=i->-1/2*sin(i*h);
F3:=j->0; F4:=j->0;
BC:=seq(U[i,0]=F1(i),i=0..NX),seq(U[i,NY]=F2(i),i=0..NX),
    seq(U[0,j]=F3(j),j=0..NY),seq(U[NX,j]=F4(j),j=0..NY);
Eqs:={seq(seq(FD(i,j),i=1..NX-1),j=1..NY-1)}:
Eqs1:=subs(BC,Eqs);
vars:={seq(seq(U[i,j]=0.,i=1..NX-1),j=1..NY-1)};
Sol:=fsolve(Eqs1,vars,real);
Points:=[seq(seq([x[i],y[j],U[i,j]],i=0..NX),j=0..NY)];
Points1:=evalf(subs(XY,BC,op(Sol),Points));
pointplot3d(Points1,symbol=solidsphere,shading=z,
 orientation=[50,60],axes=frame);
Points2:=[seq([seq([x[i],y[j],U[i,j]],i=0..NX)],j=0..NY)]:
Points3:=evalf(subs(XY,BC,op(Sol),Points2)):
surfdata(Points3,axes=boxed,labels=[x,y,u], shading=zhue);
```

Mathematica:

```
{nD=30,a=0,b=Pi,c=0,d=2*Pi,nX=15,nY=30,h=(b-a)/nX,k=(d-c)/nY}
r=(h/k)^2; xN=Table[x[i]->a+i*h,{i,0,nX}]; yN=Table[y[j]->c+j*k,
 {j,0,nY}]; fd[i_,j_]:=2*(1+r)*uN[i,j]-uN[i+1,j]-uN[i-1,j]-
 r*uN[i,j+1]-r*uN[i,j-1]-h^2*Sin[uN[i,j]]; f1[i_]:=
 -1/2*Sin[i*h]; f2[i_]:=-1/2*Sin[i*h]; f3[j_]:=0; f4[j_]:=0;
bc=Flatten[{Table[uN[i,0]->f1[i],{i,0,nX}],Table[uN[i,nY]->
 f2[i],{i,0,nX}], Table[uN[0,j]->f3[j],{j,0,nY}],
 Table[uN[nX,j]->f4[j],{j,0,nY}]}]
eqs=Flatten[Table[fd[i,j]==0,{i,1,nX-1},
 {j,1,nY-1}]]; eqs1=Flatten[eqs/.bc];
vars=Flatten[Table[{uN[i,j],0},{i,1,nX-1},{j,1,nY-1}],1];
sol=FindRoot[eqs1,vars]
points=Table[{x[i],y[j],uN[i,j]},{i,0,nX},{j,0,nY}];
points1=Flatten[N[points/.xN/.yN/.bc/.sol,nD],1];
g3D=ListPointPlot3D[points1,BoxRatios->Automatic,PlotStyle->
 PointSize[0.01],PlotRange->All]; gCP=ListContourPlot[points1,
 PlotRange->All]; GraphicsRow[{g3D,gCP},ImageSize->800]
```

It should be noted that if we slightly modify the boundary conditions, the results of this problem can be compared with the corresponding numerical solution obtained with MATLAB and described in Example 41.4 (see [144]).

Generally speaking, in the case of nonlinear elliptic PDEs, the resulting finite difference system is nonlinear and if this system is large it can be more difficult to find an appropriate numerical solution (with the aid of the predefined functions `fsolve` and `FindRoot`).

For this purpose, it is recommended to add various options to these predefined functions, e.g., specifying an appropriate initial guess for each variable (see this problem and Problem 7.3) in order to enter to the corresponding neighborhood of a point in the n-dimensional space.

Since in *Maple* (Ver. ≤ 14) and *Mathematica* (Ver. ≤ 8) it is not possible to solve numerically elliptic partial differential equations via the predefined functions (`pdsolve` and `NDSolve`), we now reformulate our problem following the approach proposed by C. A. Silebi and W. E. Schiesser [136]. According to this idea, we add a time derivative to the steady state elliptic partial differential equation and apply the method of lines (see Sect. 7.1), i.e., constructing finite differences in both x and y directions (applying the predefined function `NDSolve`). Finally, we have to obtain the steady state solution by increasing the time interval, e.g., `tF=100`).

Mathematica:

```
f1[x_]:=-1/2*Sin[x]; f2[x_]:=-1/2*Sin[x]; f3[y_]:=0;
f4[y_]:=0; {a=0,b=Pi,c=0,d=2*Pi,tF=100}
{pde1=D[u[x,y,t],t]==D[u[x,y,t],{x,2}]+D[u[x,y,t],{y,2}]+
 Sin[u[x,y,t]], ic={u[x,y,0]==-1/2*Sin[x]},
 bc={u[x,c,t]==f1[x],u[x,d,t]==f2[x],u[a,y,t]==f3[y],
 u[b,y,t]==f4[y]}}
sol1=NDSolve[{pde1,ic,bc},u,{x,a,b},{y,c,d},{t,0,tF},
 Method->{"MethodOfLines","SpatialDiscretization"->
 {"TensorProductGrid","DifferenceOrder"->"Pseudospectral"}}]
Plot3D[Evaluate[u[x,y,tF]/.sol1],{x,a,b},{y,c,d},
 BoxRatios->Automatic,PlotRange->All]
```

We can see that this numerical solution coincides with the numerical solution obtained by applying the finite difference method. □

Chapter 7
Analytical-Numerical Approach

One of the main points (related to computer algebra systems) is based on the implementation of a whole solution process, e.g., starting from an analytical derivation of exact governing equations, constructing discretizations and analytical formulas of a numerical method, performing numerical procedure, obtaining various visualizations, and analyzing and comparing the numerical solution obtained with other types of solutions.

In this chapter, we will show a very helpful role of computer algebra systems in the implementation of a whole solution process. In particular, following the analytical-numerical approach, we will construct analytical formulas of numerical methods (method of lines, spectral collocation method), obtain numerical and graphical solutions of nonlinear problems (the initial boundary value problem for the Burgers equation and the nonlinear first-order system, the initial boundary value problem describing nonlinear parametrically excited standing waves in a vertically oscillating rectangular container in Eulerian coordinates), and compare the numerical solutions with other types of solutions obtained in the book (e.g., numerical and asymptotic solutions obtained in Eulerian and Lagrangian coordinates).

7.1 Method of Lines

The method of lines, proposed by E. N. Sarmin and L. A. Chudov [131] in 1963, allows us to solve initial boundary value problems for linear and nonlinear PDEs. The first step in the solution process consists in the construction and analysis of a numerical method for the given PDEs by applying finite differences in all but one dimension and leaving the other variable continuous. Then we can solve the resulting system of ODEs (by applying the classical numerical methods), visualize the solutions obtained, analyze them, and compare with other types of solutions or experimental data. There exist the semi-analytical method of lines ([149],

[164]) for finding semi-analytical solutions of linear evolution equations (where the discretization process results in a system of ODEs, which is solved by applying the matrix exponential method) and the numerical method of lines for finding numerical solutions of linear and nonlinear evolution equations. In scientific literature, there are many papers discussing various aspects of the method of lines for various types of PDEs, e.g., see important works in this field written by W. E. Schiesser ([134], [135], [137], [138]).

In this section, following the analytical-numerical approach, we describe a whole solution process based on the numerical method of lines for evolution equations (in particular, the Burgers equation and the nonlinear first-order system). It should be noted that in some cases the method of lines can be applied to elliptic equations (e.g., see [136], Problem 6.15) by adding a time derivative to an elliptic PDE, applying finite differences in both directions (x, y), and obtaining the steady-state solution (by waiting for the process may reach a steady state).

7.1.1 Nonlinear PDEs

In this section, considering the Burgers equation and the nonlinear first-order system and applying the numerical method of lines, we will construct the corresponding numerical methods with *Maple*, obtain numerical and graphical solutions, and compare them with other types of solutions (considered in the book).

Since the predefined *Mathematica* function `NDSolve` is based on the method of lines, we will obtain (automatically) the numerical solutions for the given nonlinear equation and system by specifying various options that can serve to avoid various warning messages and to control the discretization and integration processes. We compare the results obtained in both computer algebra systems.

Problem 7.1 *Burgers equation. Initial boundary value problem.* Let us consider the initial boundary value problem for the Burgers equation (considered in Problem 6.9)

$$u_t = \nu\, u_{xx} - u u_x; \quad u(x,0) = f(x), \quad u(a,t) = 0, \quad u(b,t) = 0,$$

defined in the domain $\mathcal{D} = \{a \le x \le b,\ 0 \le t \le T_f\}$, where $f(x) = \sin(\pi x/L)$, $a=0$, $b=1$, $L=1$, $T_f=0.4$, $\nu=0.009$. Following the numerical method of lines, for the given initial boundary value problem, construct analytically the numerical method, obtain the corresponding numerical solution, visualize it, and compare with those obtained in Problem 6.9.

1. Analytical derivation of a numerical method. Following the numerical method of lines and applying finite difference approximations (see Chap. 6) for the spatial derivatives (`FWD1`, `BWD1`, `CD1`, `CD2`), we convert the Burgers equation, the initial and boundary conditions to the following system of nonlinear ODEs in time (`Eqs`, `Eqs1`) with the corresponding initial conditions (`ICs`):

$$\frac{du_i}{dt}=\nu\Big(\frac{u_{i+1}-2u_i+u_{i-1}}{h^2}\Big)-u_i\Big(\frac{u_{i+1}-u_{i-1}}{2\,h}\Big), \quad u_i(0)=f(x_i),$$

where i is the node number, h is the node spacing (`tr1`), $x_i=ih$ is the value of x at the node point i (`tr2`), the variable u_i corresponds to the dependent variable at node point i. We denote the number of interior node points in the x-direction as `NX`. Since the values u_0 and u_{NX+1} are eliminated using the boundary conditions (`BC1D`, `BC2D`), we obtain `NX` nonlinear ODEs for `NX` dependent variables u_i ($i=1,\ldots,$ `NX`).

Maple:

```
with(LinearAlgebra):  with(PDEtools): declare(u(x,t)):
with(plots): Digits:=15; nu:=0.009; L:=1; a:=0; b:=1; Tf:=0.4;
NX:=20; tr1:=h=evalf((b-a)/(NX+1)); tr2:=x=i*h; NT:=20;
EAbs:=0.1e-6; Guxt:=Matrix(NX+2,NT+1);
PDE1:=diff(u(x,t),t)=nu*diff(u(x,t),x$2)-u(x,t)*diff(u(x,t),x);
BC1:=u(x,t)=0; BC2:=u(x,t)=0; IC1:=u(x,0)=sin(Pi*x/L);
FWD1:=(-u[k+2](t)-3*u[k](t)+4*u[k+1](t))/(2*h);
BWD1:=(u[k-2](t)+3*u[k](t)-4*u[k-1](t))/(2*h);
CD1:=(u[k+1](t)-u[k-1](t))/(2*h);
CD2:=(u[k-1](t)-2*u[k](t)+u[k+1](t))/h^2;
BC1D:=subs(lhs(BC1)=u[0](t),x=a,BC1);
BC2D:=subs(lhs(BC2)=u[NX+1](t),x=b,BC2);
tr3:={diff(u(x,t),x$2)=subs(k=i,CD2),diff(u(x,t),x)=
 subs(k=i,CD1),u(x,t)=u[i](t)}; EqD[0]:=BC1D; EqD[NX+1]:=BC2D;
for i from 1 to NX do
 EqD[i]:=diff(u[i](t),t)=expand(subs(tr3,tr2,rhs(PDE1))); od;
for i from 1 to NX do EqD[i]:=eval(EqD[i],tr1); od;
Eqs:=seq(EqD[i],i=1..NX); Eqs1:=subs(EqD[0],EqD[NX+1],[Eqs]);
vars:=seq(u[i](t),i=1..NX);
ICs:=seq(u[i](0)=evalf(subs(tr1,subs(tr2,rhs(IC1)))),i=1..NX);
```

2. Numerical and graphical solutions. Then, integrating numerically in time the resulting nonlinear system of ODEs with the initial conditions by applying the predefined functions `dsolve` and `DSolve`, we obtain the numerical solution of the problem (`SolN`, `U[i]`). Finally, we perform various visualizations of the numerical solution obtained (`Gut[i]`, `Gux[i]`, `Guxt[i,j]`, `Points`, `Points1`) and can compare them with the corresponding solution of Problem 6.9.

Maple:

```
SolN:=dsolve({op(Eqs1),ICs},{vars},type=numeric,output=
 listprocedure,abserr=EAbs);
for i from 1 to NX do U[i]:=subs(SolN,u[i](t)); od;
U[0]:=subs(u[1](t)=U[1],u[2](t)=U[2],u[0](t));
U[NX+1]:=subs(u[NX](t)=U[NX],u[NX-1](t)=U[NX-1],u[NX+1](t));
for i from 1 to NX do
 Gut[i]:=plot(U[i](t),t=0..Tf,thickness=2,color=blue): od:
display({seq(Gut[i],i=1..NX)},labels=[t,u],axes=boxed);
X1:=[seq(subs(tr1,(i-1)*h),i=1..NX+2)]; T1:=[seq(i*Tf/NT,
 i=0..NT)]; for j from 1 to NT+1 do
 Gux[j]:=plot([seq([subs(tr1,i*h),U[i](T1[j])],i=0..NX+1)],
 style=line,thickness=2,color=blue,axes=boxed,numpoints=100):
od: display({seq(Gux[j],j=1..NT+1)},labels=[x,u]);
for i from 1 to NX+2 do
 Guxt[i,1]:=evalf(subs(tr1,subs(x=(i-1)*h,rhs(IC1)))); od;
for i from 1 to NX+2 do for j from 2 to NT+1 do
  Guxt[i,j]:=eval(U[i-1](t),t=T1[j]): od: od:
Points:=[evalf(seq(seq([X1[i],T1[j],Guxt[i,j]],i=2..NX-1),
 j=1..NT-1))]; pointplot3d(Points,symbol=solidsphere,
 shading=z,labels=[x,t,u],orientation=[-50,60],axes=frame);
Points1:=[seq([seq([X1[i],T1[j],Guxt[i,j]],i=2..NX-1)],
 j=1..NT-1)]: surfdata(Points1,axes=boxed,labels=[x,t,u],
 shading=zhue,grid=[20,20]);
```

In *Mathematica*, defining the given initial boundary value problem for the Burgers equation, we obtain the numerical solution represented as a function that uses interpolation and visualize it. To avoid various warning messages and to control the solution process, we specify the option `Method`, that can be written in the general form as follows: `Method->{MethodName,MethodOptions}`. In our case, we specify `MethodOfLines` and other options (e.g., `MinPoints`, for a smaller grid spacing).

Mathematica:

```
{l=1,a=0,b=1,tF=0.4, pde1=D[u[x,t],t]==nu*D[u[x,t],{x,2}]-u[x,t]*
 D[u[x,t],x],bc1=u[a,t]==0,bc2=u[b,t]==0,ic1=u[x,0]==Sin[Pi*x/l]}
```

```
sol1=Block[{nu=0.009}, NDSolve[{pde1,ic1,bc1,bc2},u,{x,0,1},
 {t,0,tF},Method->{"MethodOfLines","SpatialDiscretization"->
 {"TensorProductGrid","MinPoints"->500,"DifferenceOrder"->
 "Pseudospectral"}}]][[1]]
Plot3D[u[x,t]/.sol1,{x,0,1},{t,0,tF},BoxRatios->{1,1,1},
 AxesLabel->{x,t,u},Mesh->{70,0},ImageSize->500,
 ViewPoint->{-2,-2,3}]
Plot[Table[u[x,i]/.sol1,{i,0,tF,0.01}],{x,0,1},ImageSize->500]
```

The numerical and graphical solutions obtained in this problem can be compared with the results obtained in Problem 6.1 (by applying the predefined functions pdsolve and NDSolve) and in Problem 6.9 (by applying the finite difference method). □

7.1.2 Nonlinear Systems

In this section, following the numerical method of lines, we generalize the *Maple* procedure (developed for solving nonlinear PDEs) to solve systems of nonlinear PDEs. As in Sect. 7.1.1, we obtain (automatically) the numerical solution of the given nonlinear system with *Mathematica*, visualize it, and compare the results obtained in both computer algebra systems.

Problem 7.2 *Nonlinear first-order system. Initial boundary value problem.* Let us consider the initial boundary value problem for the nonlinear first-order system (considered in Problems 1.21, 5.6, 6.7)

$$u_t = vu_x + u + 1, \quad v_t = -uv_x - v + 1;$$
$$u(x,0) = e^{-x}, \; v(x,0) = e^{x}, \; u(1,t) = e^{t-1}, \; v(0,t) = e^{-t},$$

defined in the domain $\mathcal{D} = \{a \le x \le b, 0 \le t \le T_f\}$, where $a=0$, $b=1$, $T_f=0.4$. Following the ideas of the numerical method of lines, for the given initial boundary value problem, construct analytically the numerical method, obtain the corresponding numerical solution, visualize it, and compare with the corresponding solution of Problem 6.7.

1. Analytical derivation of a numerical method. Following the numerical method of lines and applying finite difference approximations (see Chap. 6) for the spatial derivatives (FWD11, FWD12, BWD11, BWD12, CD11, CD12, CD21, CD22), we convert the given nonlinear system, the initial and boundary conditions to the following system of nonlinear ODEs in time

(`Eqs`, `Eqs1`) with the corresponding initial conditions (`ICs`):

$$\frac{du_{i,1}}{dt}=u_{i,2}\Big(\frac{u_{i+1,1}-u_{i-1,1}}{2\,h}\Big)+u_{i,1}+1, \quad u_{i,1}(0)=e^{-x_i},$$
$$\frac{du_{i,2}}{dt}=-u_{i,1}\Big(\frac{u_{i+1,2}-u_{i-1,2}}{2\,h}\Big)-u_{i,2}+1, \quad u_{i,2}(0)=e^{x_i},$$

where i is the node number, h is the node spacing (`tr1`), $x_i=ih$ is the value of x at the node point i (`tr2`), the variables $u_{i,1}$ and $u_{i,2}$ correspond to the dependent variables $u(x,t)$ and $v(x,t)$ at node point i. We denote the number of interior node points in the x-direction as `NX` and the number of equations as `N` ($N=2$). In order to eliminate the values $u_{0,1}$, $u_{0,2}$, $u_{NX+1,1}$, $u_{NX+1,2}$, we use the boundary conditions (`BC1D`, `BC2D`) and introduce the computational boundary conditions (`BC3D`, `BC4D`). Then, we obtain 2×`NX` nonlinear ODEs for 2×`NX` dependent variables $u_{i,1}$, $u_{i,2}$ ($i=1,\ldots,$ `NX`).

Maple:

```
with(LinearAlgebra):  with(PDEtools): declare(u(x,t)):
with(plots): Digits:=15; L:=1; a:=0; b:=1; Tf:=0.5;
N:=2; NX:=10; tr1:=h=evalf((b-a)/(NX+1)); tr2:=x=i*h; NT:=40;
EAbs:=0.1e-6; Guxt:=Matrix(NX+2,NT+1); Gvxt:=Matrix(NX+2,NT+1);
PDE1:=diff(u[1](x,t),t)=u[2](x,t)*diff(u[1](x,t),x)+u[1](x,t)+1;
PDE2:=diff(u[2](x,t),t)=-u[1](x,t)*diff(u[2](x,t),x)-u[2](x,t)+1;
IC1:=u[1](x,0)=exp(-x); IC2:=u[2](x,0)=exp(x); BC1:=u[1](x,t)
 =exp(t-1); BC2:=u[2](x,t)=exp(-t); BC3:=u[1](x,t)=exp(t);
BC4:=u[2](x,t)=exp(1-t); for i from 1 to N do
 FWD1||i:=1/2*(-u[k+2,i](t)-3*u[k,i](t)+4*u[k+1,i](t))/h;
 BWD1||i=1/2*(u[k-2,i](t)+3*u[k,i](t)-4*u[k-1,i](t))/h;
 CD1||i:=1/2*(u[k+1,i](t)-u[k-1,i](t))/h;
 CD2||i:=(u[k-1,i](t)-2*u[k,i](t)+u[k+1,i](t))/h^2; od;
BC1D:=subs(lhs(BC1)=u[NX+1,1](t),x=b,BC1); BC2D:=subs(lhs(BC2)=
 u[0,2](t),x=a,BC2); BC3D:=subs(lhs(BC3)=u[0,1](t),x=a,BC3);
BC4D:=subs(lhs(BC4)=u[NX+1,2](t),x=b,BC4);
tr3:={diff(u[1](x,t),x$2)=subs(k='i',CD21), diff(u[2](x,t),x$2)
 =subs(k='i',CD22), diff(u[1](x,t),x)=subs(k='i',CD11),
 diff(u[2](x,t),x)=subs(k='i',CD12), u[1](x,t)=u['i',1](t),
 u[2](x,t)=u['i',2](t)}; EqD[NX+1,1]:=BC1D; EqD[0,2]:=BC2D;
EqD[0,1]:=BC3D; EqD[NX+1,2]:=BC4D; for i from 1 to NX do
 EqD[i,1]:=diff(u[i,1](t),t)=expand(subs(tr3,tr2, rhs(PDE1)));
 EqD[i,2]:=diff(u[i,2](t),t)=expand(subs(tr3,tr2, rhs(PDE2)));od;
for i from 1 to NX do EqD[i,1]:=eval(EqD[i,1],tr1); EqD[i,2]:=
 eval(EqD[i,2],tr1); od; Eqs:=seq(seq(EqD[i,j],i=1..NX),j=1..N);
```

```
Eqs1:=subs(EqD[NX+1,1],EqD[0,2],EqD[0,1],EqD[NX+1,2],[Eqs]);
vars:=seq(seq(u[i,j](t),i=1..NX),j=1..N);
ICs:=seq(u[i,1](0)=evalf(subs(tr1,subs(tr2,rhs(IC1)))),i=1..NX),
 seq(u[i,2](0)=evalf(subs(tr1,subs(tr2,rhs(IC2)))),i=1..NX);
```

2. Numerical and graphical solutions. Then, integrating numerically in time the resulting nonlinear system of ODEs with the initial conditions by applying the predefined functions `dsolve` and `DSolve`, we obtain the numerical solution of the problem (`SolN`, `U[i,j]`). Finally, we perform various visualizations of the numerical solution obtained (`Guxt[i,j]`, `Gvxt[i,j]`, `PsU`, `PsU1`, `PsV`, `PsV1`) and can compare them with the corresponding solution of Problem 6.7.

Maple:

```
SolN:=dsolve({op(Eqs1),ICs},{vars},type=numeric,
 output=listprocedure,abserr=EAbs);
for j from 1 to N do for i from 1 to NX do U[i,j]:=subs(SolN,
 u[i,j](t)); od; od; X1:=[seq(subs(tr1,(i-1)*h),i=1..NX+2)];
T1:=[seq(i*Tf/NT,i=0..NT)]; for i from 1 to NX+2 do
 Guxt[i,1]:=evalf(subs(tr1,subs(x=(i-1)*h,rhs(IC1)))); od;
for i from 1 to NX+2 do for j from 2 to NT+1 do
  Guxt[i,j]:=eval(U[i-1,1](t),t=T1[j]): od: od:
PsU:=[evalf(seq(seq([X1[i],T1[j],Guxt[i,j]],i=2..NX+1),
 j=1..NT+1))]; pointplot3d(PsU,symbol=solidsphere,shading=z,
 labels=[x,t,u],orientation=[40,60],axes=frame);
PsU1:=[seq([seq([X1[i],T1[j],Guxt[i,j]],i=2..NX)],j=1..NT+1)]:
surfdata(PsU1,axes=boxed,labels=[x,t,u],shading=zhue);
for i from 1 to NX+2 do
 Gvxt[i,1]:=evalf(subs(tr1,subs(x=(i-1)*h,rhs(IC2)))); od;
for i from 1 to NX+2 do for j from 2 to NT+1 do
  Gvxt[i,j]:=eval(U[i-1,2](t),t=T1[j]): od: od:
PsV:=[evalf(seq(seq([X1[i],T1[j],Gvxt[i,j]],i=2..NX+1),
 j=1..NT+1))]; pointplot3d(PsV,symbol=solidsphere,shading=z,
 labels=[x,t,v],orientation=[40,60],axes=frame);
PsV1:=[seq([seq([X1[i],T1[j],Gvxt[i,j]],i=2..NX)],j=1..NT+1)]:
surfdata(PsV1,axes=boxed,labels=[x,t,v], shading=zhue);
eval(U[1,1](t),t=0);
```

In *Mathematica*, defining the given initial boundary value problem for the nonlinear first-order system, we obtain the numerical solution and visualize it.

Mathematica:

```
SetOptions[Plot3D,BoxRatios->{1,1,1},AxesLabel->{x,t,u},
 Mesh->{70,0},ImageSize->500,ViewPoint->{-2,-2,3},
 PlotRange->All]; {l=1,a=0,b=1,tF=0.5, pde1=D[u[x,t],t]==
 v[x,t]*D[u[x,t],x]+u[x,t]+1, pde2=D[v[x,t],t]==-u[x,t]*
 D[v[x,t],x]-v[x,t]+1, ic1=u[x,0]==Exp[-x],ic2=v[x,0]==Exp[x],
 bc1=u[1,t]==Exp[t-1], bc2=v[0,t]==Exp[-t]}
sol1=NDSolve[{pde1,pde2,ic1,ic2,bc1,bc2},{u,v},
 {x,a,b},{t,0,tF},Method->{"MethodOfLines",
 "SpatialDiscretization"->{"TensorProductGrid"}}]
Plot3D[u[x,t]/.sol1,{x,a,b},{t,0,tF}]
Plot3D[v[x,t]/.sol1,{x,a,b},{t,0,tF}]
p1=Evaluate[u[0.1,0]/.sol1]
```

The numerical and graphical solutions obtained in this problem can be compared with the analytical solutions obtained in Problem 1.21 (by applying the *Maple* predefined function `pdsolve`) and Problem 5.6 (by applying the modified Adomian decomposition method), and with numerical results obtained in Problem 6.7 (by applying the predefined functions `pdsolve` and `NDSolve`). □

7.2 Spectral Collocation Method

In this section, considering one of the large class of methods for constructing numerical solutions of nonlinear PDEs, i.e., spectral methods (SM), we will show a helpful role of computer algebra systems for implementing a whole solution process (generation of analytical formulas of a numerical method, construction of numerical and graphical solutions, and comparison of numerical solutions with other types of solutions).

The Galerkin methods, introduced by B. G. Galerkin [63] in 1915, represent a wide class of methods for constructing approximate solutions of differential (or integral) equations by finite linear combinations of *basis functions* with some desirable properties. In spectral methods, we apply global polynomials as the basis functions (instead of piecewise polynomials in finite element methods) for constructing numerical solutions of nonlinear PDEs (for details see [116], [26], [68]). For a rectangular domain $\mathcal{D}$, we can choose a family of orthogonal polynomials (as the basis functions), which can be considered as an extension of the concept of eigenfunctions.

Moreover, spectral methods have the two main approaches: we can choose an exact-fitting approach (as in FDM), *called spectral collocation methods* (where we define collocation points), and we can choose a best fitting approach (as in FEM), called the *discontinuous Galerkin method.*

The main advantages of spectral methods over finite element methods and finite difference methods are:

a) accuracy (spectral methods give very accurate approximations for smooth solutions with few degrees of polynomials);

b) order of convergence (in finite element methods, the order of convergence is limited by the degree of the polynomials, and in spectral methods is limited by the regularity of the solution, e.g., is exponential for analytical solutions);

c) flexibility (spectral methods are more flexible than finite difference methods, e.g., spectral methods can be developed with an arbitrary mesh, with some nodes of an element located between two nodes of another element).

7.2.1 Nonlinear Systems

We will illustrate the spectral method on an initial boundary value problem, describing parametrically driven nonlinear standing waves of finite amplitude in a vertically oscillating rectangular container. We develop a spectral collocation method to solve this problem. It should be noted that for this nonlinear problem, the high-order asymptotic solution has been obtained in Lagrangian coordinates and the qualitative analysis has been performed with the aid of computer algebra systems [139] (see Problems 5.2.2, 3.22).

As we noted before, there are two ways for representing the fluid motion: the Eulerian approach (in which the coordinates are fixed in the reference frame of the observer) and the Lagrangian approach (in which the coordinates are fixed in the reference frame of the moving fluid). In this section, we will follow the Eulerian approach for constructing the numerical solution and compare the results with the asymptotic solution obtained in Lagrangian coordinates.

Problem 7.3 *Nonlinear standing waves in a fluid. Numerical solution. Initial boundary value problem.* Let us construct the numerical and graphical solutions describing nonlinear parametrically excited standing waves in a vertically oscillating rectangular container and develop computer algebra procedures to aid in a whole solution process.

1. Statement of the problem. Following the Eulerian approach, let us consider the two-dimensional motion of an ideal incompressible fluid in a rectangular container oscillating in the vertical direction with amplitude* $\tilde{A}_c$ and frequency $\tilde{\Omega}$ such that $\tilde{y}(\tilde{t}){=}{-}\tilde{A}_c\cos(\tilde{\Omega}\tilde{t})$. Let $\tilde{\rho}{=}1$ be the fluid density, $\tilde{L}=\frac{1}{2}\tilde{\lambda}$ the length of the container, $\tilde{h}$ the fluid depth, and $\tilde{\lambda}$ the wave length. The container walls are assumed to be rigid and impermeable. We assume that the fluid motion is periodic in time $\tilde{t}$ (with the period $\tilde{T}{=}2\pi/\tilde{\omega}$) and in the horizontal direction $\tilde{x}$ with the period equal to the wave length $\tilde{\lambda}$.

We define Cartesian coordinates according to the requirements: the coordinate system is rigidly connected to the container; the fluid motion is symmetric with respect to the vertical planes $\tilde{x}{=}0$ and $\tilde{x}{=}\frac{1}{2}\tilde{\lambda}$; the free surface is $\tilde{y}{=}\tilde{\eta}(\tilde{x},\tilde{t})$; the line $\tilde{y}{=}0$ corresponds to the mean water level; at $\tilde{t}{=}0$ the fluid is at rest; and the finite depth is defined as $\tilde{y}{=}{-}\tilde{h}$.

Introducing the dimensionless parameters, the fluid depth, $h{=}\tilde{h}\kappa$, the frequency of the nonlinear wave, $\omega{=}\tilde{\omega}/\sqrt{\kappa g}$, the spatial coordinates and time, $x{=}\tilde{x}\kappa$, $y{=}\tilde{y}\kappa$, $t{=}\tilde{t}\omega\sqrt{\kappa g}$, the wave amplitude, $C{=}\tilde{C}\kappa$, the amplitude of the container oscillations $A_c{=}\tilde{A}_c\kappa$, the wave profile and the velocity potential, $\eta(x,t){=}\kappa\tilde{\eta}(\tilde{x},\tilde{t})/C$ and $\phi(x,y,t){=}\tilde{\phi}(\tilde{x},\tilde{y},\tilde{t})\sqrt{\kappa^3}/(C\sqrt{g})$, we investigate this two-dimensional nonlinear standing wave problem and assume that the flow is irrotational. We have to solve the following dimensionless equations for a fluid with initial and boundary conditions in the domain $\mathfrak{D}=\{0\le x\le\pi, -h\le y\le C\eta(x,t)\}$ for the wave profile $\eta(x,t)$, the velocity potential $\phi(x,y,t)$, and the angular frequency ω of the nonlinear standing wave:

$$
\begin{aligned}
&\phi_{xx}+\phi_{yy}=0 \quad \text{in } \mathfrak{D}=\{0\le x\le\pi, -h\le y\le C\eta(x,t)\};\\
&\eta[1{+}4A_c\omega^2\cos(2t)]{+}\omega\phi_t{+}\tfrac{1}{2}C[(\phi_x)^2{+}(\phi_y)^2]{=}0 \quad \text{at } y{=}C\eta(x,t);\\
&\phi_y{-}\omega\eta_t{-}C\phi_x\eta_x{=}0 \quad \text{at } y{=}C\eta(x,t);\\
&\phi_x{=}0 \quad \text{at } x{=}0,\pi; \qquad \phi_y{=}0 \quad \text{at } y{=}{-}h;\\
&\int_0^\pi \eta(x,t)\,dx{=}0, \quad \nabla\phi(x,y,\,t{+}2\pi){=}\nabla\phi(x,y,t), \quad \eta(0,0){-}\eta(\pi,0){=}2.
\end{aligned}
\tag{7.1}
$$

In this problem we consider the subharmonic excitation of surface standing nonlinear water waves in a vertically oscillating container, that is $\omega\approx\frac{1}{2}\Omega$. The desired solution will depend upon the following dimensionless parameters: the fluid depth h, the wave amplitude C, and the amplitude of the container oscillations A_c.

2. Analytical derivation of the spectral collocation method. Let us describe the method and perform analytical derivation of formulas of the method.

*We denote the dimensional variables with the symbol ~.

(a) We will study periodic solutions (in x and t) of the standing wave problem. In order to solve the nonlinear problem numerically let us represent the unknown functions $\eta(x,t)$ and $\phi(x,y,t)$ in the form of infinite series:

$$\begin{aligned}\eta(x,t)=&B_{11}\cos x+B_{12}\cos x\cos t+B_{13}\cos x\cos(2t)+\dots\\&+B_{21}\cos(2x)+B_{22}\cos(2x)\cos t+B_{23}\cos(2x)\cos(2t)+\dots\\&+B_{31}\cos(3x)+B_{32}\cos(3x)\cos t+B_{33}\cos(3x)\cos(2t)+\dots\end{aligned}$$

and

$$\begin{aligned}\phi(x,y,t)=&A_0t+A_{11}\sin t+A_{12}\sin(2t)+A_{13}\sin(3t)+\dots\\&+A_{21}\sin t\cos x\cosh(y+h)+A_{22}\sin(2t)\cos x\cosh(y+h)+\dots\\&+A_{31}\sin t\cos(2x)\cosh[2(y+h)]+\dots,\end{aligned}$$

where A_0, $A_{11},\dots$, $B_{11},\dots$, are the unknown coefficients.

We note that these forms of the functions $\eta(x,t)$ and $\phi(x,y,t)$ allow us to convert the first, fourth, fifth, sixth, seventh equations of the governing system (7.1) to the identities. Additionally, it can be shown that according to the symmetry and periodicity of the problem, some unknown coefficients equal to zero, i.e., $A_{nm}=0$, $B_{nm}=0$ if $n+m$ is even.

Therefore, we have simplified our problem, i.e., instead of solving the eight equations, we have to find (for a given value of the amplitude C and the amplitude of the container oscillations A_c) the unknown nonzero coefficients A_0, A_{nm}, B_{nm} and the nonlinear frequency ω satisfying the following three equations:

$$\begin{aligned}&\eta[1+4A_c\omega^2\cos(2t)]+\omega\phi_t+\tfrac{1}{2}C[(\phi_x)^2+(\phi_y)^2]=0 \quad \text{at } y=C\eta(x,t);\\&\phi_y-\omega\eta_t-C\phi_x\eta_x=0 \quad \text{at } y=C\eta(x,t);\\&\eta(0,0)-\eta(\pi,0)=2.\end{aligned}\tag{7.2}$$

We will perform this approximately by replacing the infinite series by finite sums of the form:

$$\begin{aligned}\eta(x,t)=&\sum_{n=1}^{N-1}\sum_{m=1}^{N}B_{nk}\cos(nx)\cos[(m-1)t],\\\phi(x,y,t)=&A_0t+\sum_{n=1}^{N}\sum_{m=1}^{N-1}A_{nm}\sin(mt)\cos[(n-1)x]\cosh[(n-1)(y+h)].\end{aligned}\tag{7.3}$$

If N and $n+m$ are even, then $A_{nm}=0$, $B_{nm}=0$ and we can determine that the number of nonzero coefficients of the A_{nm} and of the B_{nm}, are equal to $N(N-1)$.

Therefore, for a given value of the amplitude C and the amplitude of the container oscillations A_c, the total number of the unknown coefficients (A_{nm}, B_{nm}, A_0, ω) is $N(N-1)+2$.

These unknown coefficients have to satisfy the nonlinear system (7.2), which we derive analytically as follows:

Maple:

```
N:=4: S:=0; tr1:={y=C*eta}; tr2:={x=0,t=0}; tr3:={x=0,t=Pi};
for n from 1 to N-1 do for m from 1 to N do
 if type(n+m,odd) then S:=S+B||n||m*cos(n*x)*cos((m-1)*t) fi:
od: od: eta:=S; S:=0:
for n from 1 to N do for m from 1 to N-1 do
 if type(n+m,odd) then
  S:=S+A||n||m*cos((n-1)*x)*cosh((n-1)*(y+h))*sin(m*t)
 fi: od: od: phi:=A0*t+S;
Eq1:=eval(eta*(1+eps*omega^2*cos(2*t))+omega*diff(phi,t)
 +1/2*C*((diff(phi,x))^2+(diff(phi,y))^2),tr1);
Eq2:=eval(diff(phi,y)-omega*diff(eta,t)-C*diff(phi,x)
 *diff(eta,x),tr1);
Eq3:=2-eval(eta,tr2)+eval(eta,tr3); Eq4:=eval(Eq1,tr3);
```

Mathematica:

```
{nN=4, sN=0, tr1=y->cN*eta, tr2={x->0,t->0}, tr3={x->0,t->Pi}}
Do[If[OddQ[n+m],sN=sN+bN[n,m]*Cos[n*x]*Cos[(m-1)*t]],
 {m,1,nN},{n,1,nN-1}]; {eta=sN, sN=0}
Do[If[OddQ[n+m], sN=sN+aN[n,m]*Cos[(n-1)*x]*Cosh[(n-1)*(y+h)]*
 Sin[m*t]],{m,1,nN-1},{n,1,nN}]; phi=a0*t+sN
eq1=(eta*(1+eps*omega^2*Cos[2*t])+omega*D[phi,t]+1/2*cN*
 (D[phi,x]^2+D[phi,y]^2))/.tr1
eq2=(D[phi,y]-omega*D[eta,t]-cN*D[phi,x]*D[eta,x])/.tr1
{eq3=2-(eta/.tr2)+(eta/.tr3), eq4=eq1/.tr3}
```

(b) Now, let us find these unknown coefficients by applying the collocation approach, i.e., satisfying the governing system (7.2). Therefore, we define the mesh (X_j, T_i) $(j=1,\dots,N-1,\ i=1,\dots,N/2)$ in $\mathfrak{D}$, in which we choose the mesh points as follows:

$$X_j=\begin{cases}\pi(j-1)/(N-2) & \text{if } N\neq 4,\\ \pi(j-\frac{1}{2})/(N-1) & \text{if } N=4,\end{cases}\qquad T_i=\pi(i-\tfrac{1}{2})/N,$$

where N is even (e.g., $N=4, 6, 8, \ldots$). Substituting the truncated series (7.3) into the governing system (7.2) at the mesh points, we obtain $N(N-1)+1$ nonlinear transcendental equations, the last equation we obtain substituting the initial values $x=0$, $t=0$ into the first equation of the governing system (7.2). Therefore we construct analytically the system of $N(N-1)+2$ nonlinear transcendental equations with respect to $N(N-1)+2$ unknown coefficients A_{nm}, B_{nm}, A_0, and ω as follows:

Maple:

```
tr4:=(j,i)->[x=X||j,t=T||i];
for i from 1 to N/2 do T||i:=Pi/N*(i-1/2):
for j to N-1 do
 if N=4 then X||j:=Pi/3*(j-1/2) else X||j:=Pi/(N-2)*(j-1) fi;
 Eq1||i||j:=eval(subs(tr4(j,i),Eq1));
 Eq2||i||j:=eval(subs(tr4(j,i),Eq2));
 print("Eq1"||i||j,Eq1||i||j); print("Eq2"||i||j,Eq2||i||j);
od; od; L1:=NULL: L2:=NULL:
for n to N do for m to N-1 do if type(n+m,odd) then
 L1:=L1,A||n||m fi: od: od:
for n to N-1 do for m to N do if type(n+m,odd) then
 L2:=L2,B||n||m fi: od: od: LA:=[L1]; LB:=[L2]; k:=1:
for i from 1 to N/2 do for j from 1 to N-1 do
 F||k:=simplify(Eq1||i||j); F||(k+N/2*(N-1)):=
 simplify(Eq2||i||j); k:=k+1: od; od; LF:=NULL:
F||(N*(N-1)+1):=simplify(Eq3); F||(N*(N-1)+2):=simplify(Eq4);
for i from 1 to N*(N-1)+2 do LF:=LF,F||i; od: LFforNum:={LF}:
```

Mathematica:

```
tr4[j_,i_]:={x->xN[j],t->tN[i]};
Do[tN[i]=Pi/nN*(i-1/2); Print["T","[",i,"]=",tN[i]];
 Do[If[nN==4, xN[j]=Pi/3*(j-1/2), xN[j]=Pi/(nN-2)*(j-1)];
  eqD1[i,j]=eq1/.tr4[j,i]; eqD2[i,j]=eq2/.tr4[j,i];
  Print["Eq1","[",i,",",j,"]=",eqD1[i,j]];
  Print["Eq2","[",i,",",j,"]=",eqD2[i,j]],
{j,1,nN-1}],{i,1,nN/2}]; {l1={}, l2={},lF={}}
Do[If[OddQ[n+m],l1=Append[l1,aN[n,m]]],{m,1,nN-1},{n,1,nN}];
Do[If[OddQ[n+m],l2=Append[l2,bN[n,m]]],{m,1,nN},{n,1,nN-1}];
{lA=l1, lB=l2, k=1}
Do[fF[k]=eqD1[i,j]; fF[k+nN/2*(nN-1)]=eqD2[i,j]; k=k+1,
 {j,1,nN-1},{i,1,nN/2}];
{fF[nN*(nN-1)+1]=eq3,fF[nN*(nN-1)+2]=eq4}
Do[lF=Append[lF,fF[i]],{i,1,nN*(nN-1)+2}]; lFforNum=lF;
```

We note that the analytical derivation of the spectral collocation method can be performed for various even values of N, e.g., $N = 4, 6, 8$, that correspond, respectively, to $14, 32, 58$ equations.

3. Construction of the numerical solution. The system of nonlinear equations obtained above can be solved numerically, e.g., we solve it by Newton's iteration method. Thus, we find standing waves for various values of the amplitude C and the fixed values of the amplitude of the container oscillations A_c and the fluid depth h. In particular, we determine the nonlinear frequency and wave profiles of standing waves.

(a) The solution process. The process of constructing the numerical solution, based on the procedure developed in [144] (for unforced nonlinear standing waves), is follows: for the fixed values of A_c and h, the Newton's iterations are performed starting with a small value of C that corresponds to the solution of the given linearized problem. It is known that the convergence of Newton's method depends on an appropriate initial guess. As follows from [23], there exist infinite set of solutions of this problem. But we restrict ourselves to the construction of approximate numerical *one-mode solution* (see Problem 5.10), in which the only main harmonic dominates, $\cos x \cos t$, $\cos x \sin t \cosh(y{+}h)$.

(b) Initial guess. Therefore, in order to specify an appropriate initial guess, i.e., to enter to the neighborhood of a point in the space $\mathbb{R}^{N(N-1)+2}$, that corresponds to the neighborhood of the point of the linear approximation, it is important to specify the coefficients A_{21} of the main harmonic $\phi(x, y, t)$, B_{12} of the main harmonic $\eta(x, t)$, and ω for all values of N ($N{=}4, 6, 8, \ldots$) according to the following formulas: $A_{21}{=}-\omega/\sinh h$, $B_{12}{=}1$, and $\omega{=}\sqrt{\tanh h}$. These formulas can be obtained as a particular one-mode solution, that is, $\eta(x, t){=}\cos x \cos t$, $\phi(x, y, t){=}-(\omega/\sinh h)\sin t \cos x \cosh(y + h)$, $\omega^2{=}\tanh h$, of the corresponding linearized problem (for $A_c{=}0$) $\eta{+}\omega\phi_t{=}0$, $\phi_y{-}\omega\eta_t{=}0$ at $y{=}0$.

The other coefficients, A_{nm}, B_{nm}, i.e., the coordinates of a point in the space $\mathbb{R}^{N(N-1)+2}$, can be specified as small values (or zeros) with respect to A_{21} and B_{12}.

(c) The phase of oscillations. Since the last equation of the system (7.2) is fixing the phase of fluid oscillations, let us describe the qualitative construction of numerical solutions. In the asymptotic theory and in the numerical approximations, the wave profile in the linear approximation is described, respectively, by the formulas

$$\tilde{\eta}^a{=}\tilde{\zeta}\cos\kappa\tilde{x}\cos\tilde{\Psi}{=}\tilde{\zeta}\cos\kappa\tilde{x}\cos\left(\tfrac{1}{2}\tilde{\Omega}\,\tilde{t}^a{+}\theta\right), \quad \tilde{\eta}^n{=}A_{21}\cos\tilde{x}\cos\tilde{t}.$$

Also, as a result of the qualitative analysis of the solutions we have established (see Problem 3.22) that there exist the two different cases of the slope of the resonance curves (that correspond to the two classes of nonzero fixed points):

The positive slope, where the phase is equal to $\theta = \frac{1}{2}\pi,\ \frac{3}{2}\pi$.
The negative slope, where the phase is equal to $\theta = 0,\ \pi$.

Now let us find the transformation of the variable $\tilde{t}^a$ such that $\tilde{\eta}^a = \tilde{\eta}^n$.

Case 1. If $\theta = \frac{1}{2}\pi,\ \frac{3}{2}\pi,\ \tilde{\omega} = \frac{1}{2}\tilde{\Omega}$, we obtain

$$\tilde{\eta}^a = \cos\left(\tfrac{1}{2}\tilde{\Omega}\tilde{t}^a + \theta\right) = \cos\left(\tilde{\omega}\tilde{t}^a + \tfrac{\pi}{2}\right), \quad \tilde{\eta}^n = \cos\tilde{t},$$

hence we find $\tilde{t}^a = \dfrac{\tilde{t}}{\tilde{\omega}} - \dfrac{\pi}{2\tilde{\omega}}$.

In this case, the oscillations of the container are described by the formula: $\tilde{y} = -\tilde{A}_c\cos(\tilde{\Omega}\tilde{t}^a) = -\tilde{A}_c\cos(2\tilde{\omega}\tilde{t}^a) = -\tilde{A}_c\cos(2\tilde{t} - \pi) = \tilde{A}_c\cos(2\tilde{t})$.

Therefore the first equation of the system (7.2) in the dimensionless variables can be rewritten in the form

$$\eta[1 - 4A_c\omega^2\cos(2t)] + \omega\phi_t + \tfrac{1}{2}C[(\phi_x)^2 + (\phi_y)^2] = 0 \quad \text{at } y = C\eta(x,t). \qquad (7.4)$$

Case 2. If $\theta = 0,\ \pi$, we can derive the relations in a similar manner and find that

$$\tilde{t}^a = \frac{\tilde{t}}{\tilde{\omega}}, \qquad \tilde{y} = -\tilde{A}_c\cos(\tilde{\Omega}\tilde{t}^a) = -\tilde{A}_c\cos(2\tilde{\omega}\tilde{t}^a) = -\tilde{A}_c\cos(2\tilde{t}).$$

Therefore the first equation of the system (7.2) in the dimensionless variables can be rewritten in the form

$$\eta[1 + 4A_c\omega^2\cos(2t)] + \omega\phi_t + \tfrac{1}{2}C[(\phi_x)^2 + (\phi_y)^2] = 0 \quad \text{at } y = C\eta(x,t). \qquad (7.5)$$

Thus, to construct the numerical solutions that will correspond to different phases, we have to solve the two separate systems of equations (7.2), in which the first equations are represented, respectively, in the form (7.4) and (7.5). However, it is more simple the following procedure: first, to construct the numerical solution for the fixed value of the fluid depth h and the amplitude of the container oscillations $+A_c$, then to repeat all this procedure for the same values of the parameters except for the amplitude of the container oscillations, $-A_c$.

(d) Solution procedures. We define the solution procedure `SolOmega0` for the first calculation with specifying an appropriate initial guess as follows:

Maple:

```
SolOmega0:=proc(E,H,c,LEqs::set,A::list,B::list)
 local param,param1,LF1,IVA,IVB,IVA1,IVB1,IVA0,IVals,Sol0,
  OmegaNum0; param:=[C=c,h=H,eps=E];
 param1:=omega0=sqrt(tanh(H));
 LF1:=evalf(eval(LEqs,param)); IVA:={op(A)} minus {A21};
 IVB:={op(B)} minus {B12}; IVA:=[op({op(A)} minus {A21})];
 IVB:=[op({op(B)} minus {B12})];
 IVA1:={seq(IVA[i]=0.,i=1..nops(IVA))};
 IVB1:={seq(IVB[i]=0.,i=1..nops(IVB))}; IVA0:={A0=0.};
 IVals:=eval(eval({A21=-omega0/sinh(H),B12=1.,omega=omega0},
  param1),param) union IVA1 union IVB1 union IVA0;
 Sol0:=fsolve(LF1,IVals);
 OmegaNum0:=eval(rhs(op(select(has,Sol0,omega))),param1);
RETURN(OmegaNum0,Sol0); end proc;
```

Mathematica:

```
solOmega0[eN_,hN_,c_,lEqs_,lA_,lB_]:=Module[
 {param,param1,lF1,ivA,ivB,ivA1,ivB1,ivA0,iVals,sol0,
  omegaNum0},param={cN->c,h->hN,eps->eN}; param1=omega0->
  Sqrt[Tanh[hN]]; lF1=lEqs/.param;
 ivA=Complement[lA,{aN[2,1]}]; ivB=Complement[lB,{bN[1,2]}];
 ivA1=Table[{ivA[[i]],0.},{i,1,Length[ivA]}];
 ivB1=Table[{ivB[[i]],0.},{i,1,Length[ivB]}];
 ivA0={{a0,0.}}; iVals=Union[{{aN[2,1],-omega0/Sinh[hN]},
  {bN[1,2],1.},{omega,omega0}}/.param1/.param,ivA1,ivB1,ivA0];
 sol0=FindRoot[lF1,iVals,PrecisionGoal->Infinity];
 omegaNum0=Select[sol0,MemberQ[#,omega]&][[1,2]]/.param1/.param;
 Return[{omegaNum0,sol0}]];
```

Also we define the solution procedure `SolOmega` for the subsequent calculations using solutions obtained in the previous step as an appropriate initial guess as follows:

Maple:

```
SolOmega:=proc(E,H,c,LEqs::set,IVals::set)
 local param,param1,LF1,Sol,OmegaNum;
 param:=[C=c,h=H,eps=E]; param1:=omega0=sqrt(tanh(H));
 LF1:=evalf(eval(LEqs,param));
 Sol:=fsolve(LF1,IVals);
 OmegaNum:=eval(rhs(op(select(has,Sol,omega))),param1);
 RETURN(OmegaNum,Sol); end proc;
```

Mathematica:

```
solOmega[eN_,hN_,c_,lEqs_,iVals_]:=Module[
 {param,param1,lF1,sol,omegaNum},
 param={cN->c,h->hN,eps->eN};
 param1=omega0->Sqrt[Tanh[hN]]; lF1=lEqs/.param;
 sol=FindRoot[lF1,iVals,PrecisionGoal->Infinity];
 omegaNum=Select[sol,MemberQ[#,omega]&][[1,2]]/.param1/.param;
 Return[{omegaNum,sol}]];
```

(e) General Procedure. Now we describe the general procedure of constructing the subsequent numerical solutions as follows:

We construct a solution applying Newton's method (the procedure `SolOmega0`) for a small value of C after a few Newton's iterations with a given accuracy (it can be changed varying the environment variable `Digits`).

Assuming that the nonlinear frequency is a continuous function of the amplitude, we choose the solution obtained in the previous step as an initial guess, increase gradually the wave amplitude, and apply Newton's method (the procedure `SolOmega`).

We repeat the second step until the wave amplitude will have the maximal value, that we choose according to the experimental data performed by W. W. Schultz et. al [145] on the parametrical excitation of nonlinear standing waves on the surface of a fluid of finite depth.

We visualize the resonance curve (frequency-amplitude dependence) of a standing wave (see Fig. 7.1).

Maple:

```
Digits:=30: with(plots): Ac:=0.11; M:=2; L:=10.5;
kappa:=evalf(Pi*M/L); Hd:=3.*kappa; E:=evalf(-4*Ac*kappa);
H:=3.; g:=981.7;  Cn:=20.; C0:=0.03; Cf:=0.6; Ch:=(Cf-C0)/Cn;
Ampl:=C0; LAmpl:=[seq(C0+i*Ch,i=0..Cn)];
LNum:=NULL: for i to nops(LAmpl) do
 if i=1 then Sol||i:=SolOmega0(E,H,LAmpl[i],LFforNum,LA,LB);
    LNum:=LNum,Sol||i[1];
 else Sol||i:=SolOmega(E,H,LAmpl[i],LFforNum,Sol||(i-1)[2]);
    LNum:=LNum,Sol||i[1]; fi; od: LOmegaNum:=[LNum];
PNum:=evalf([seq([LOmegaNum[i]*sqrt(kappa*g)*2,LAmpl[i]/kappa],
 i=1..nops(LOmegaNum))]);
GNum:=plot(PNum,color=blue,thickness=3): display(GNum);
```

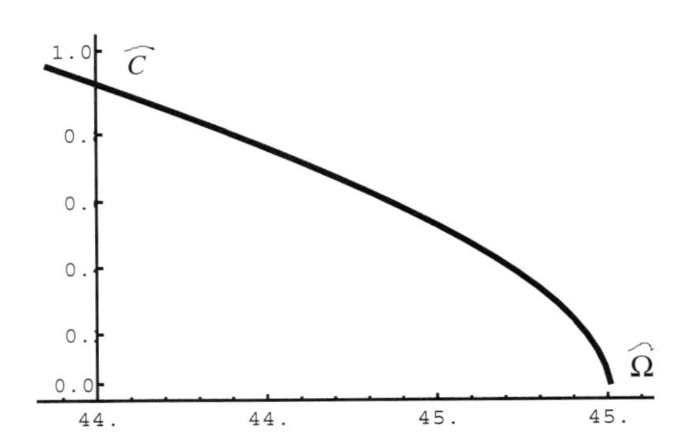

Fig. 7.1. The resonance curve of nonlinear parametrically excited standing waves in a vertically oscillating rectangular container ($\widetilde{C}$ is the amplitude, $\widetilde{\Omega}$ is the frequency)

Mathematica:

```
SetOptions[ListPlot,ImageSize->500,Joined->True,PlotRange->
 {All,{0.,1.1}},TicksStyle->Directive[Blue,9],Ticks->
 {{44.5,45.,45.5},Automatic}]; {nD=30, ac=0.11, mN=2, lN=10.5,
 kappa=Pi*mN/lN, hd=3.*kappa, eN=-4*ac*kappa, hN=3., g=981.7,
 cn=20., c0=0.03, cf=0.6, ch=(cf-c0)/cn, ampl=c0}
{lAmpl=Table[c0+i*ch,{i,0,cn}], lNum={}, nlAmpl=Length[lAmpl]}
Do[If[i==1, {sol[i]=solOmega0[eN,hN,lAmpl[[i]],lFforNum,lA,lB],
 lNum=Append[lNum,sol[i][[1]]]},
 {solP[i]=Table[{sol[i-1][[2]][[k,1]],sol[i-1][[2]][[k,2]]},
 {k,1,Length[sol[1][[2]]]}];
 sol[i]=solOmega[eN,hN,lAmpl[[i]],lFforNum,solP[i]],
 lNum=Append[lNum,sol[i][[1]]]}],
{i,1,nlAmpl}]; lOmegaNum=lNum
pNum=Table[{lOmegaNum[[i]]*Sqrt[kappa*g]*2,lAmpl[[i]]/kappa},
 {i,1,Length[lOmegaNum]}]
gNum=ListPlot[pNum,PlotStyle->{Blue,Thickness[0.01]}]; Show[gNum]
```

5. Surface profiles. Finally, we construct the *surface profiles* in the dimensional variables, $\tilde{\eta}(\tilde{x},\tilde{t})$, using the approximate numerical solutions obtained up to the third approximation (see Fig. 7.2).

It is interesting to observe the standing wave motions described in Eulerian variables and to compare them with the corresponding standing wave motions described in Lagrangian variables (see Problem 5.10, Fig. 5.1) and with the corresponding experimental data [145].

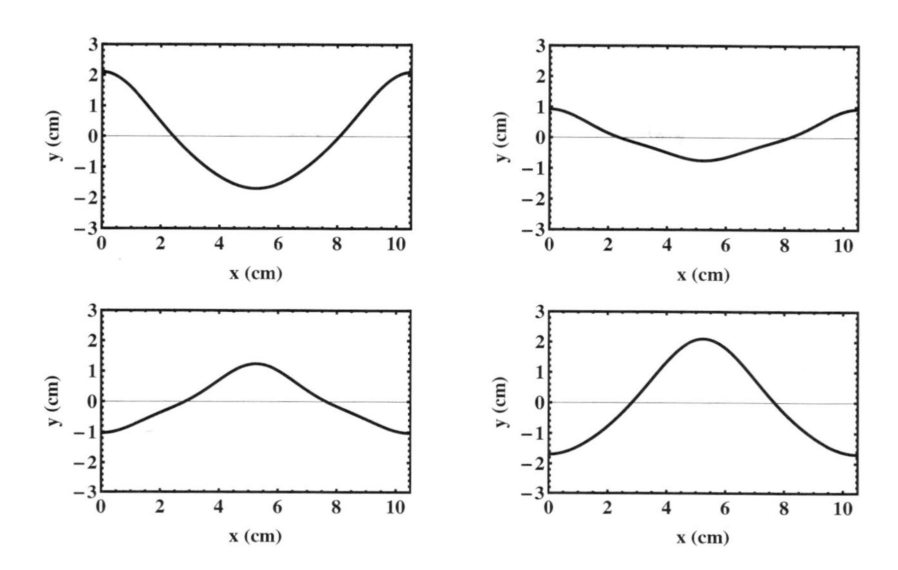

Fig. 7.2. Surface wave profiles at different times $t_k = 0, \frac{1}{4}\pi, \frac{1}{17}\pi, 2\pi$

Maple:

```
LT:=[0,Pi/2,Pi]; LC:=[blue,magenta,"BlueViolet"]; eta; Sol11;
C11:=LAmpl[11]; Prof11:=eval(eta,Sol11[2]); Sol11[1];
Prof12:=subs(x=xD*kappa,t=tD*Sol11[1]*sqrt(kappa*g),Prof11);
Prof13:=Prof12*kappa/C11;
animate(Prof13,xD=0..L,tD=0..1,color=blue,frames=50,
 numpoints=100,thickness=3,scaling=constrained);
for j from 1 to 3 do Gr||j:=plot(subs(tD=LT[j],Prof13),
 xD=0..L,scaling=constrained,color=LC[j],thickness=2); od:
display({Gr1, Gr2, Gr3});
plot3d(Prof13,xD=0..L,tD=0..1,shading=z,style=patchnogrid,
 transparency=0.9,axes=boxed);
contourplot(Prof13,xD=0..L,tD=0..1,grid=[50,50],contours=10,
 filled=true,coloring=["BlueViolet",magenta]);
```

Mathematica:

```
lTime={0,Pi/2,Pi}; lAspRat={1/2,1/2,0.1}; nP=100;
SetOptions[Plot,ImageSize->300,PlotRange->All,PlotStyle->
 {Blue,Thickness[0.01]},PlotPoints->nP]; {eta,sol[11]}
{cN11=lAmpl[[11]], sol[11][[1]], prof11=eta/.sol[11][[2]]}
```

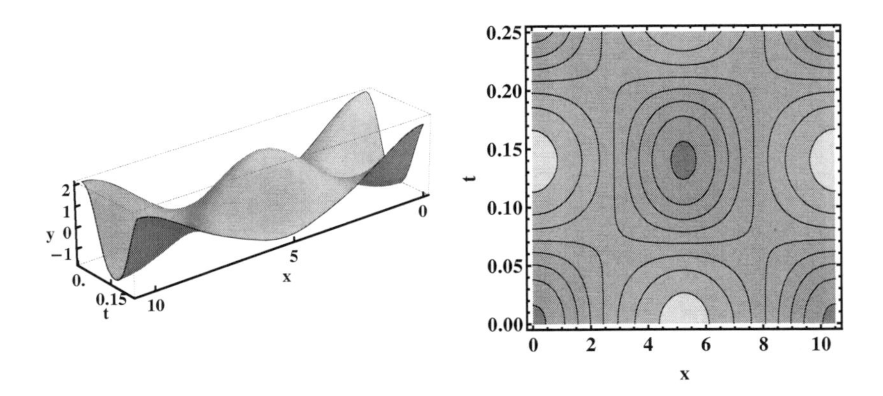

Fig. 7.3. Numerical solution of the nonlinear standing waves problem: the 3D and contour plots for $x \in [0, L]$ and $t \in [0, \frac{1}{4}]$

```
prof12=prof11/.x->xD*kappa/.t->tD*sol[11][[1]]*Sqrt[kappa*g]
prof13[x1_,t1_]:=(prof12*kappa/cN11)/.xD->x1/.tD->t1;
Animate[Plot[prof13[xD,tD],{xD,0,lN},AspectRatio->Automatic,
 ImageSize->500,PlotRange->{All,{-2,3}}],{tD,0,1},
 AnimationRate->0.1]
Do[gr[j]=Plot[prof13[x,lTime[[j]]],{x,0,lN},AspectRatio->
 lAspRat[[j]]],{j,1,3}];
GraphicsRow[{gr[1],gr[2],gr[3]},ImageSize->1000]
gr3D=Plot3D[prof13[x,t],{x,0,lN},{t,0,1},ImageSize->500,
 PlotRange->All,PlotStyle->{Hue[0.47],Thickness[0.001]},
 Mesh->None]; grCP=ContourPlot[prof13[x,t],{x,0,lN},{t,0,1},
 ImageSize->500,PlotRange->All]; GraphicsRow[{gr3D,grCP}]
```

The 3D and contour plots of the numerical solution obtained in Eulerian variables are shown in Fig. 7.3.

Moreover, it is interesting to compare the resonance curves constructed for different cases of the fluid depth (that is, the infinity depth, critical depth, and small depth) in Eulerian variables (this problem) with the corresponding resonance curves obtained according to the asymptotic theory (see Problem 3.22). □

Appendix A
Brief Description of Maple

A.1 Introduction

Maple is a general purpose computer algebra system (CAS), in which symbolic computation can be easily combined with exact, approximate (floating-point) numerical computation, and with arbitrary-precision numerical computation. *Maple* provides powerful scientific graphics capabilities, for details, see [85], [33], [48], [128], [1], [97], [143], etc.

The first concept of *Maple* and initial versions were developed by the Symbolic Computation Group at the University of Waterloo in the early 1980s. The company Maplesoft was created in 1988. The development of *Maple* was done mainly in research labs at Waterloo University and at the University of Western Ontario (see [27], [65]), with important contributions from worldwide research groups in other universities.

The most important features of *Maple* are: fast symbolic, numerical computation, and interactive visualization; easy to use; easy to incorporate new user defined capabilities; understandable, open-source software development path; available for almost all operating systems; powerful programming language, intuitive syntax, easy debugging; extensive library of mathematical functions and specialized packages; free resources, collaborative character of development (e.g. see Maple Web Site `www.maplesoft.com` (`MWS`), Maple Application Center `MWS/applications`, Teacher Resource Center `MWS/TeacherResource`, Student Help Center `MWS/studentcenter`, Maple Community `MWS/community`).

Maple consists of three parts: the interface, the kernel (basic computational engine), and the library. The interface and the kernel (written in the C programming language) form a smaller part of the system (they are loaded when a *Maple* session is started). The interface handles the input of mathematical expressions, output display, plotting of functions, and support of other user communication with the system. The interface medium is the *Maple* worksheet. The kernel interprets the user input and carries out the basic algebraic operations, and deals with storage management. The library consists of two parts: the main library and a collection of packages. The main library (written in the *Maple* programming language) includes many functions in which resides most of the common mathematical knowledge of *Maple*.

A.2 Basic Concepts

The *prompt symbol* (>) serves for typing the *Maple* function; the semicolon (;) or colon (:) symbols* with pressing `Enter` at the end of the function serves for evaluating the *Maple* function, displaying the result, and inserting a new prompt.

Maple contains a complete *online help system*, the command line help system, e.g. by typing `?NameOfFunction` or `help(NameOfFunction);` or `?help`; reference information can be accessed by using the `Help` menu, by highlighting a function and then pressing `Ctrl-F1` or `F1` or `F2` (for Ver. ≥ 9), by pressing `Ctrl-F2`.

Maple worksheets are files that keep track of a working process and organize it as a collection of expandable groups (for details, see `?worksheet`, `?shortcut`). The new worksheet (or the new problem) it is best to begin with the statement `restart` for cleaning *Maple'* s memory. All problems in the book assume that they begin with `restart`.

Previous results (during a session) can be referred with symbols `%` (the last result), `%%` (the next-to-last result), `%%...%`, k times (the k-th previous result).

Comments can be included with the sharp sign `#` and all characters following it up to the right end of a line. Also the text can be inserted with `Insert` → `Text`.

Incorrect response. If you get no response or an incorrect response you may have entered or executed the function incorrectly. Do correct the function or interrupt the computation (the stop button in the Tool Bar menu).

Maple source code can be viewed for most of the functions, general and specialized (package functions), e.g. `interface(verboseproc=2); print(factor);`

More detailed information about how the problem is being solved can be printed with the function `infolevel`, e.g. `infolevel[pdsolve]:=5; pdsolve(diff(u(x,t),t)=diff(u(x,t),x)^2*u(x,t),u(x,t));`

Palettes can be used for building or editing mathematical expressions without the need of remembering the *Maple* syntax.

---

*In earlier versions of *Maple* and in *Classic Worksheet Maple*, we have to end a function with a colon or semicolon. In this book we follow this tradition in every problem.

The Maplet User Interface (for Ver. ≥ 8) consists of *Maplet applications* that are collections of windows, dialogs, actions (see `?Maplets`).

A number of specialized functions are available in various specialized *packages (subpackages)* (see `?index[package]`, `with`):

```
with(package); func(args);     package[subpackage][func](args);
```

Numerical approximations: numerical approximation of an expression `expr` to 10 significant digits, `evalf(expr)`; global changing a precision, `Digits:=n` (for details, see `?environment`); local changing a precision, `evalf(expr,n)`; numerical approximation of `expr` using a binary hardware floating-point system, `evalhf(expr)`; performing numerical approximations using the hardware or software floating-point systems, `UseHardwareFloats:=value` (see `?UseHardwareFloats`, `?environment`).

A.3 Maple Language

Maple language is a high-level programming language, well-structured, comprehensible. It supports a large collection of data structures or *Maple* objects (functions, sequences, sets, lists, arrays, tables, matrices, vectors, etc.) and operations among these objects (type-testing, selection, composition, etc.). The *Maple* procedures in the library are available in readable form. The library can be complemented with locally user developed programs and packages.

Arithmetic operators: `+ - * / ^`.

Logic operators: `and`, `or`, `xor`, `implies`, `not`.

Relation operators: `<`, `<=`, `>`, `>=`, `=`, `<>`.

A variable name is a combination of letters, digits, or the underline symbol (_), beginning with a letter, e.g. `a12_new`.

Abbreviations for the longer *Maple* functions or expressions: `alias`, e.g. `alias(H=Heaviside); diff(H(t),t);` to remove this abbreviation, `alias(H=H);`

Maple is case sensitive, there is a difference between lowercase and uppercase letters, e.g. `evalf(Pi)` and `evalf(pi)`.

Various reserved keywords, symbols, names, and functions, these words cannot be used as variable names, e.g. operator keywords, additional language keywords, global names that start with (_) (for details, see `?reserved`, `?ininames`, `?inifncs`, `?names`).

The assignment/unassignment operators: a variable can be "free" (with no assigned value) or can be assigned any value (symbolic, numeric) by the assignment operators `a:=b` or `assign(a=b)`. To unassign (clear) an assigned variable (see `?:=` and `?'`), e.g. `x:='x'`, `evaln(x)`, or `unassign('x')`.

The difference between the *operators (:=) and (=)*: the operator `var:=expr` is used to assign `expr` to the variable `var`, and the operator `A=B` — to indicate equality (not assignment) between the left- and the right-hand sides (see `?rhs`), e.g. `Equation:=A=B; Equation; rhs(Equation); lhs(Equation);`

The range operator `(..)`, an expression of type range `expr1..expr2`, for example `a[i]$ i=1..9; plot(sin(x),x=-Pi..Pi);`

Statements are input instructions from the keyboard that are executed by *Maple* (e.g. `break`, `by`, `do`, `end`, `for`, `function`, `if`, `proc`, `restart`, `return`, `save`, `while`).

The statement separators semicolon (;) and colon (:). The result of a statement followed with a semicolon (;) will be displayed, and it will not be displayed if it is followed by a colon (:), e.g. `plot(sin(x),x=0..Pi); plot(sin(x),x=0..Pi):`

An expression is a valid statement, and is formed as a combination of constants, variables, operators and functions. Every expression is represented as a tree structure in which each node (and leaf) has a particular data type. For the analysis of any node and branch, the functions `type`, `whattype`, `nops`, `op` can be used. *A boolean expression* is formed with the *logical operators* and the relation operators.

An equation is represented using the binary operator `(=)`, and has two operands, the left-hand side, `lhs`, and the right-hand side, `rhs`.

Inequalities are represented using the relation operators and have two operands, the left-hand side, `lhs`, and the right-hand side, `rhs`.

A string is a sequence of characters having no value other than itself, cannot be assigned to, and will always evaluate to itself. For instance, `x:="string";` and `sqrt(x);` is an invalid function. Names and strings can be used with the `convert` and `printf` functions.

Maple is sensitive to types of brackets and quotes.

Types of brackets: parentheses for grouping expressions, `(x+9)*3`, for delimiting the arguments of functions, `sin(x)`; square brackets for constructing lists, `[a,b,c]`, vectors, matrices, arrays; curly brackets for constructing sets, `{a,b,c}`.

Types of quotes: forward-quotes to delay evaluation of expression, `'x+9+1'`, to clear variables, `x:='x'`; back-quotes to form a symbol or a name, `` `the name:=7`; k:=5; print(`the value of k is`,k); `` double quotes to create strings, and a single double quote ", to delimit strings.

Types of numbers: integer, rational, real, complex, roots of equations, for example `-55`, `5/6`, `3.4`, `-2.3e4`, `Float(23,-45)`, `3-4*I`, `Complex(2/3,3); RootOf(_Z^3-2,index=1);`

Predefined constants: symbols for definitions of commonly used mathematical constants, `true`, `false`, `gamma`, `Pi`, `I`, `infinity`, `FAIL`, `exp(1)` (for details see `?ininames`, `?constants`).

Functions or function expressions have the form `f(x)` or `expr(args)` and represent a function call, or application of a function (or procedure) to arguments (`args`). *Active functions* (beginning with a lowercase letter) are used for computing, e.g. `diff`, `int`, `limit`. *Inert functions* (beginning with a capital letter) are used for showing steps in the problem-solving process; e.g. `Diff`, `Int`, `Limit`.

Library functions (or predefined functions) and user-defined functions.

Predefined functions: most of the well known functions are predefined by *Maple* and they are known to some *Maple* functions (e.g., `diff`, `evalc`, `evalf`, `expand`, `series`, `simplify`). Numerous special functions are defined (see `?FunctionAdvisor`).

User-defined functions: the functional operator (`->`) (see `?->`):

```
f:=x->expr;   f:=(x1,...,xn)->expr;      f:=t-><x1(t),...,xn(t)>;
```

Alternative definitions of functions: `unapply` converts an expression to a function, and a procedure is defined with `proc`:

```
f:=proc(x) option operator,arrow; expr end proc;
f:=proc(x1,...,xn) expr end;                  f:=unapply(expr,x);
```

Evaluation of function $f(x)$ at $x = a$, $\{x = a, y = b\}$:

```
f(a,b);          subs(x=a,y=b,f(x,y));    eval(f(x,y),{x=a,y=b});
```

In *Maple* language there are two forms of modularity: *procedures* and *modules*.

A procedure (see `?procedure`) is a block of statements which one needs to use repeatedly. A procedure can be used to define a function (if the function is too complicated to write by using the arrow operator), to create a matrix, graph, logical value.

```
proc(args) local v1; global v2; options ops; stats; end proc;
```

where `args` is a sequence of arguments, `v1` and `v2` are the names of local and global variables, `ops` are special options (see `?options`), and `stats` are statements that are realized inside the procedure.

A module (see `?module`) is a generalization of the procedure concept. Since the procedure groups a sequence of statements into a single statement (block of statements), the module groups related functions and data.

```
module() export v1; local v2; global v3; option ops; stats;
end module;
```

where v1, v2, and v3 are the names of export, local, and global variables, respectively, ops are special options (see ?module[options]), and stats are statements that are realized inside the module.

In *Maple* language there are essentially *two control structures*: the selection structure if and the repetition structure for.

```
if cond1 then expr1 else expr2  end if;
if cond1 then expr1 elif cond2 then expr2 else expr3 end if;
for i from i1 by step to i2 do stats end do;
for i from i1 by step to i2 while cond1 do stats end do;
for i in expr1 do stats end do;  for i in expr1 do stats od;
```

where cond1 and cond2 are conditions, expr1, expr2 are expressions, stats are statements, i, i1, i2 are, respectively, the loop variable, the initial and the last values of i. These operators can be nested. The operators break, next, while inside the loops are used for breaking out of a loop, to proceed directly to the next iteration, or for imposing an additional condition. The operators end if and fi, end do and od are equivalent.

Maple objects, sequences, lists, sets, tables, arrays, vectors, matrices, are used for representing more complicated data.

```
Sequence1:=expr1,...,exprn;          Sequence2:=seq(f(i),i=a..b);
List1:=[Sequence1];   Range1:=a..n;            Set1:={Sequence1};
Table1:=table([expr1=A1,...,exprN=AN]); Arr1:=Array(a..n,a,,m);
Vec1:=Vector(<a1,...,an>);         Vec2:=Vector(1..n,[a1,...,an]);
Matrix(<<a11,a21>|<a12,a22>>);   Matrix([[a11,a12],[a21,a22]]);
```

Sequences a1,a2,a3, *lists* [a1,a2,a3], *sets* $\{a1, a2, a3\}$ are groups of expressions. *Maple* preserves the order and repetition in sequences and lists and does not preserves it in sets. The order in sets can change during a *Maple* session. *A table* is a group of expressions represented in tabular form. Each entry has an index (an integer or any arbitrary expression) and a value (see ?table). *An array* is a table with integer range of indices (see ?Array). In *Maple* arrays can be of any dimension (depending of computer memory). *A vector* is a one-dimensional array with positive range integer of indices (see ?vector, ?Vector). *A matrix* is a two-dimensional array with positive range integer of indices (see ?matrix, ?Matrix).

Appendix B
Brief Description of Mathematica

B.1 Introduction

Mathematica is a general purpose computer algebra system, in which symbolic computation can be easily combined with exact, approximate (floating-point), and with arbitrary-precision numerical computation. *Mathematica* provides powerful scientific graphics capabilities, for details, see [14], [66], [70], [71], [72], [129], [143], [154], [175], etc.

The first concept of *Mathematica* and first versions were developed by Stephen Wolfram in 1979–1988. The company Wolfram Research was founded in 1987, which continues to develop *Mathematica* ([168], [169]).

The most important features of *Mathematica* are: fast symbolic, numerical, acoustic, parallel computation; static and dynamic computation, and interactive visualization; it is possible to incorporate new user defined capabilities; available for almost all operating systems; powerful and logical programming language; extensive library of mathematical functions and specialized packages; interactive mathematical typesetting system; free resources (e.g. see the Mathematica Learning Center `www.wolfram.com/support/learn`, Wolfram Demonstrations Project `demonstrations.wolfram.com`, Wolfram Information Center `library.wolfram.com`).

Mathematica consists of two basic parts: the *kernel*, computational engine and the *interface*, *front end*. These two parts are separate, but communicate with each other via the *MathLink* protocol. The kernel interprets the user input and performs all computations. The kernel assigns the labels `In[number]` to the input expression and `Out[number]` to the output. These labels can be used for keeping the computation order. In this book, we will not include these labels in the solutions of problems. The result of kernel's work can be viewed with the function `InputForm`. The interface between the user and the kernel is called *front end* and is used to display the input and the output generated by the kernel. The medium of the front end is the *Mathematica notebook*.

There are significant changes to numerous *Mathematica* functions incorporated to the new versions of the system. The description of important differences for Ver. < 6 and Ver. ≥ 6 is reported in the literature (e.g., see [143])*.

B.2 Basic Concepts

If we type a *Mathematica* command and press the `RightEnter` key or `Shift+Enter` (or `Enter` to continue the command on the next line), *Mathematica* evaluates the command, displays the result, and inserts a horizontal line (for the next input).

Mathematica contains many sources of online help, e.g. Wolfram Documentation Center, Wolfram Demonstrations Project (for Ver. ≥ 6), *Mathematica Virtual Book* (for Ver. ≥ 7); `Help` menu; it is possible to mark a function and to press F1; to type `?func`, `??func`, `Options[func]`; to use the symbols (`?`) and (`*`), e.g., `?Inv*`, `?*Plot`, `?*our*`.

Mathematica notebooks are electronic documents that may contain *Mathematica* output, text, graphics (see `?Notebook`). It is possible to work simultaneously with many notebooks. A *Mathematica* notebook consists of a list of cells. Cells are indicated along the right edge of the notebook by brackets. Cells can contain subcells, and so on. The kernel evaluates a notebook cell by cell. There are *different types of cells:* input cells (for evaluation), text cells (for comments), Title, Subtitle, Section, Subsection, etc., can be found in the menu `Format` → `Style`.

Previous results (during a session) can be referred with symbols `%` (the last result), `%%` (the next-to-last result), and so on.

Comments can be included within the characters `(*comments*)`.

Incorrect response: if some functions take an "infinite" computation time, you may have entered or executed the command incorrectly. To terminate a computation, you can use: `Evaluation` → `Quit Kernel` → `Local`.

Palettes can be used for building or editing mathematical expressions, texts, graphics and allows one to access by clicking the mouse to the most common mathematical symbols.

In *Mathematica*, there exist many specialized functions and modules which are not loaded initially. They must be loaded separately from files in the *Mathematica* directory. These files are of the form `filename.m`. The full name of a package consists of a `context` and a `short` name, and it is written as `` context`short ``. To load a package corresponding to a context: ``<<context` ``. To get a list of the functions in a package: ``Names["context`*"]``.

Numerical approximations: numerical approximation of `expr` to 6 significant digits, `N[expr]`, `expr//N`, numerical approximation of the expression to n significant digits, `N[expr,n]`, `NumberForm[expr,n]`, scientific notation of

*A complete list of all changes can be found in the Documentation Center and on the Wolfram Web Site `www.wolfram.com`.

numerical approximation of an expression `expr` to n significant digits, `ScientificForm[expr,n]`.

B.3 Mathematica Language

Mathematica language is a very powerful programming language based on systems of transformation rules, functional, procedural, and object-oriented programming techniques (see [93]). This distinguishes it from traditional programming languages. It supports a large collection of data structures or *Mathematica objects* (functions, sequences, sets, lists, arrays, tables, matrices, vectors, etc.) and operations on these objects (type-testing, selection, composition, etc.). The library can be enlarged with custom programs and packages.

Symbol refers to a symbol with the specified name, e.g., expressions, functions, objects, optional values, results, argument names. *A name of symbol*, is a combination of letters, digits, or certain special characters, not beginning with a digit, e.g., `a12new`. Once defined, a symbol retains its value until it is changed or removed.

Expression is a symbol that represents an ordinary *Mathematica* expression `expr` in readable form. The head of `expr` can be obtained with `Head[expr]`. The structure and various forms of `expr` can be analyzed with `TreeForm`, `FullForm[expr]`, `InputForm[expr]`. *A boolean expression* is formed with the *logical operators* and the relation operators.

Basic arithmetic operators and the corresponding functions:
`+ - * / ^`, `Plus`, `Subtract`, `Minus`, `Times`, `Divide`, `Power`.

Logic operators and their equivalent functions: `&&`, `||`, `!`, `=>`, `And`, `Or`, `Xor`, `Not`, `Implies`.

Relation operators and their equivalent functions: `==`, `!=` `<`, `>`, `<=` `>=`, `Equal`, `Unequal`, `Less`, `Greater`, `LessEqual`, `GreaterEqual`.

Mathematica is case sensitive, there is a difference between lowercase and uppercase letters, e.g., `Sin[Pi]` and `sin[Pi]` are different. All *Mathematica* functions begin with a capital letter. Some functions (e.g., `PlotPoints`) use more than one capital. To avoid conflicts, it is best to begin with a lower-case letter for all user-defined symbols.

The result of each calculation is displayed, but it can be suppressed by using a semicolon (;), e.g., `Plot[Sin[x],x,0,2*Pi]; a=9; b=3; c=a*b`.

Patterns: *Mathematica* language is based on pattern matching. A pattern is an expression which contains an underscore character (_). The pattern can stand for any expression. Patterns can be constructed from the templates, e.g., `x_`, `x_/;cond`, `pattern?test`, `x_:IniValue`, `x^n_`, `x_^n_`, `f[x_]`, `f_[x_]`.

Basic transformation rules: `->`, `:>`, `=`, `:=`, `^:=`, `^=`.

The rule `lhs->rhs` transforms `lhs` to `rhs`. *Mathematica* regards the left-hand side as a pattern.

The rule `lhs:>rhs` transforms `lhs` to `rhs`, evaluating `rhs` only after the rule is actually used.

The assignment `lhs=rhs` (or `Set`) specifies that the rule `lhs->rhs` should be used whenever it applies.

The assignment `lhs:=rhs` (or `SetDelayed`) specifies that `lhs:>rhs` should be used whenever it applies, i.e., `lhs:=rhs` does not evaluate `rhs` immediately but leaves it unevaluated until the rule is actually called.

The rule `lhs^:=rhs` assigns `rhs` to be the delayed value of `lhs`, and associates the assignment with symbols that occur at level one in `lhs`.

The rule `lhs^=rhs` assigns `rhs` to be the value of `lhs`, and associates the assignment with symbols that occur at level one in `lhs`.

Transformation rules are useful for making substitutions without making the definitions permanent and are applied to an expression using the operator `/.` (`ReplaceAll`) or `//.` (`ReplaceRepeated`).

The difference between the operators (=) and (==): the operator `lhs=rhs` is used to assign `rhs` to `lhs`, and the equality operator `lhs==rhs` indicates equality (not assignment) between `lhs` and `rhs`.

Unassignment of definitions:
`Clear[symb]`, `ClearAll[symb]`, `Remove[symb]`, `symb=.`;
`Clear["Global`*"]; ClearAll["Global`*"]; Remove["`*"];`
(to clear all global symbols defined in a *Mathematica* session),
`?symb`, `?`*` (to recall a symbol's definition).

`ClearAll["Global'*"]; Remove["Global'*"];` is a useful initialization for starting working with a problem.

An equation is represented using the binary operator `==`, and has two operands, the left-hand side `lhs` and the right-hand side `rhs`.

Inequalities are represented using the relational operators and have two operands, the left-hand side `lhs` and the right-hand side `rhs`.

A string is a sequence of characters having no value other than itself and can be used as labels for graphs, tables, and other displays. The strings are enclosed within double-quotes, e.g., `"abc"`.

Data types: every expression is represented as a tree structure in which each node (and leaf) has a particular data type. For the analysis of any node and branch can be used a variety number of functions, e.g., `Length`, `Part`, a group of functions ending in the letter `Q` (`DigitQ`, `IntegerQ`, etc.).

Types of brackets: parentheses for grouping, `(x+9)*3`; square brackets for function arguments, `Sin[x]`; curly brackets for lists, `{a,b,c}`.

Types of quotes: back-quotes for context mark, format string character, number mark, precision mark, accuracy mark; double-quotes for strings.

Types of numbers: integer, rational, real, complex, root, e.g., `-5, 5/6, -2.3^-4, ScientificForm[-2.3^-4], 3-4*I, Root[#^2+#+1&,2]`.

Mathematical constants: symbols for definitions of selected mathematical constants, for example, `I`, `Pi`, `Catalan`, `Degree`, `E`, `EulerGamma`, `Infinity`, `GoldenRatio`.

Two classes of functions: *pure functions* and functions defined in terms of a variable (*predefined* and *user-defined* functions).

Pure functions are defined without a reference to any specific variable. The arguments are labeled `#1,#2,...`, and an ampersand `&` is used at the end of definition. Predefined functions. Most of the mathematical functions are predefined. *Mathematica* includes all the common *special functions* of mathematical physics.

The names of mathematical functions are complete English words or the traditional abbreviations (for a few very common functions), for example, `Conjugate`, `Mod`. Person's name mathematical functions have names of the form `PersonSymbol`, e.g. the Legendre polynomials $P_n(x)$, `LegendreP[n,x]`.

User-defined functions are defined using the pattern `x_`:

```
f[x_]:=expr;      f=Function[x,expr];      f[x1_,...,xn_]:=expr;
f[t_]:={x1[t],...,xn[t]};      f=Function[t,{x1[t],...,xn[t]}];
```

Evaluation of a function or an expression without assigning a value can be performed using the replacement operator `/.`, e.g. `f[a]`, `expr/.x->a`.

```
f[a]          f[a,b]          expr /. x->a          expr /.{x->a,x->b}
```

Function application: `expr//func` is equivalent to `fun[expr]`.

A module is a local object that consists of several functions which one needs to use repeatedly (see `?Module`). A module can be used to define a function (if the function is too complicated to write by using the notation `f[x_]:=expr`), to create a matrix, a graph, a logical value, etc.

Block is similar to `Module`, the main difference between them is that `Block` treats the values assigned to symbols as local, but the names as global, whereas `Module` treats the names of local variables as local.

With is similar to `Module`, the principal difference between them is that `With` uses local constants that are evaluated only once, but `Module` uses local variables whose values may change many times.

```
Module[{var1,...},body];            Module[{var1=val1,...},body];
Block[{var1,...},expr];              Block[{var1=val1,...},expr];
With[{var1=val1,var2=val2,...},expr];
```

where `var1,...` are local variables, `val1,...` are initial values of local variables, `body` is the body of the module (as a sequence of statements separated by semicolons).

The final result of the module is the result of the last statement (without a semicolon). Also `Return[expr]` can be used to return an expression.

In *Mathematica* language there are the following *two control structures*: the selection structures `If`, `Which`, `Switch` and the repetition structures `Do`, `While`, `For`.

```
If[cond,exprTrue]                         If[cond,exprTrue,exprFalse]
                                If[cond,exprTrue,exprFalse,exprNeither]
Which[cond1,expr1,...]       Switch[expr,patt1,val1,patt2,val2,...]
Do[expr,{i,i1,i2,iStep}] Do[expr,{i,i1,i2,iS},{j,j1,j2,jS},...]
While[cond,expr]                          For[i=i1,cond,iStep,expr]
```

where `exprTrue, exprFalse, exprNeither` are expressions that execute, respectively, if the condition `cond` is `True`, `False`, and is neither `True` or `False`; `i,i1,i2` (`j,j1,j2`) are the loop variable and the initial and the last values of `i` (`j`).

Mathematica objects: *lists* are the fundamental objects in *Mathematica*.

The other objects (e.g., sets, matrices, tables, vectors, arrays, tensors, objects containing data of mixed type) are represented as lists. A list is an ordered set of objects separated by commas and enclosed in curly braces, `{elements}`, or defined with the function `List[elements]`.

Nested lists are lists that contain other lists. There are many functions which manipulate lists.

Sets are represented as lists.

Vectors are represented as lists, vectors are simple lists. Vectors can be expressed as single columns with `ColumnForm[list,horiz,vert]`.

Tables, matrices, and tensors are represented as nested lists. There is no difference between the way they are stored: they can be generated using the functions `MatrixForm[list]`, `TableForm[list]`, or using the nested list functions. Matrices and tables can also be conveniently generated using menu *Palettes* or `Insert`.

A matrix is a list of vectors.

A tensor is a list of matrices with the same dimensionality.

In *Mathematica* (for Ver. ≥ 6), the new kind of output, the *dynamic output* has been introduced allowing to create dynamic interfaces of different types. Numerous new functions for creating various dynamic interfaces have been developed.

References

[1] Abel, M. L. and Braselton, J. P.: Maple by Example, 3rd edition. AP Professional, Boston, MA 2005

[2] Ablowitz, M. J., Ramani, A., and Segur, H.: A Connection between Nonlinear Evolution Equations and Ordinary Differential Equations of P-type I and II. *J. Math. Phys.* 21, 715–721; 1006–1015 (1980)

[3] Ablowitz, M. J., Kaup, D. J., Newell, A. C., and Segur, H.: The Inverse Scattering Transform — Fourier Analysis for Nonlinear Problems. *Studies in Appl. Math.* 53(4), 249–315 (1974)

[4] Ablowitz, M. J. and Clarkson, P. A.: Solitons, Nonlinear Evolution Equations and Inverse Scattering. Cambridge University Press, Cambridge 1991

[5] Abramowitz, M. and Stegun, I. A. (Eds.): Handbook of Mathematical Functions with Formulas, Graphs and Mathematical Tables. National Bureau of Standards Applied Mathematics, Washington, D.C. 1964

[6] Adomian, G.: Nonlinear Stochastic Operator Equations. Academic Press, San Diego 1986

[7] Adomian, G.: Solving Frontier Problems of Physics: The Decomposition Method. Kluwer, Boston 1994

[8] Akritas, A. G.: Elements of Computer Algebra with Applications. Wiley, New York 1989

[9] Anco, S. C. and Bluman, G. W.: Direct Construction Method for Conservation Laws of Partial Differential Equations. Part I: Examples of Conservation Law Classifications. *Eur. J. Appl. Math.* 13, 545–566 (2002)

[10] Andreev, V. K.: Stability of Unsteady Motions of a Fluid with a Free Boundary. Nauka, Novosibirsk 1992

[11] Andrew, A. D. and Morley, T. D.: Linear Algebra Projects Using Mathematica. McGraw-Hill, New York 1993

[12] Aoki, H.: Higher-order Calculation of Finite Periodic Standing Waves by Means of the Computer. *J. Phys. Soc. Jpn.* 49, 1598–1606 (1980)

[13] Arnold, V. I.: Geometric Methods in the Theory of Ordinary Differential Equations. Springer, New York 1988

[14] Bahder, T. B.: Mathematica for Scientists and Engineers. Addison-Wesley, Redwood City, CA 1995

[15] Baldwin D., Hereman W., and Sayers J.: Symbolic Algorithms for the Painlevé Test, Special Solutions, and Recursion Operators of Nonlinear PDEs. *In: CRM Proceedings and Lecture Series 39* (Winternitz P. and Gomez-Ullate D, Eds.) 17–32, American Mathematical Society, Providence, RI 2004

[16] Benjamin, T., Bona, J., and Mahony, J.: Model Equations for Long Waves in Nonlinear Dispersive Systems. *Philos. Trans. R. Soc. London Ser. A* 272, 47–78 (1972)

[17] Benjamin, T. B. and Ursell, F.: The Stability of the Plane Free Surface of a Liquid in Vertical Periodic Motion. *Proc. Roy. Soc. Lond. A* 225, 505–515 (1954)

[18] Birkhoff, G.: Hydrodynamics. Princeton University Press, Princeton, NJ 1950

[19] Bluman, G. W.: Applications of the General Similarity Solution of the Heat Equation to Boundary Value Problems. *Quart. Appl. Math.* 31, 403–415 (1974)

[20] Bordakov, G. A. and Sekerzh-Zenkovich, S. Ya.: Nonlinear Faraday Resonance in Two-layer Fluid of Finite Depth. *Preprint of IPM RAS.* 475, Moscow 1990

[21] Boussinesq, J.: Théorie des Ondes et des Remous qui Se Propagent le Long d'un Canal Rectangulaire Horizontal, en Communiquant au Liquide Continu Dans ce Canal des Vitesses Sensiblement Pareilles de la Surface au Fond. *J. Math. Pures Appl.* 17(2), 55–108 (1872)

[22] Bronshtein, I. N., Semendyayev, K. A., Musiol, G., and Muehlig, H.: Handbook of Mathematics, 5th edition. Springer, Berlin 2007

[23] Bryant, P.J., Stiassnie, M.: Different Forms for Nonlinear Standing Waves in Deep Water. *J. Fluid Mech.* 272, 135–156 (1994)

[24] Calmet, J. and van Hulzen, J. A.: Computer Algebra Systems. Computer Algebra: Symbolic and Algebraic Computations (Buchberger, B., Collins, G. E., and Loos, R., Eds.), 2nd edition. Springer, New York 1983

[25] Camassa, R. and Holm, D.: An Integrable Shallow Water Equation with Peaked Solitons. *Phys. Rev. Lett.* 71(11), 1661–1664 (1993)

[26] Canuto, C., Hussaini, M. Y., Quarteroni, A., and Zang, T. A.: Spectral Methods in Fluid Dynamics. Springer Series in Computational Physics. Springer, Berlin Heidelberg New York 1989

[27] Char, B. W., Geddes, K. O., Gonnet, G. H., Monagan, M. B., and Watt, S. M.: Maple Reference Manual, 5th edition. Waterloo Maple Publishing, Waterloo, Ontario, Canada 1990

[28] Cheb-Terrab, E. S. and von Bulow, K.: A Computational Approach for the Analytical Solving of Partial Differential Equations. *Comput. Phys. Commun.* 90, 102–116 (1995)

[29] Cherruault, Y. and Adomian, G.: Decomposition Methods: A New Proof of Convergence. *Math. Comput. Modelling* 18(12), 103–106 (1993)

[30] Concus, P.: Standing Capillary-gravity Waves of Finite Amplitude. *J. Fluid Mech.* 14, 568–576 (1962)

[31] Conte, R.: Invariant Painlevé Analysis of Partial Differential Equations. *Phys. Lett. A* 140, 383–389 (1989)

[32] Conte R., Fordy A. P., and Pickering, A.: A Perturbation Painlevé Approach to Nonlinear Differential Equations. *Phys. D* 69, 33–58 (1993)

[33] Corless, R. M.: Essential Maple. Springer, Berlin 1995

[34] Crank, J. and Nicolson, P.: A Practical Method for Numerical Evaluation of Solutions of Partial Differential Equations of the Heat-conduction Type. *Proc. Camb. Philos. Soc.* 43, 50–67 (1947)

[35] Darboux, G.: Sur les Surfaces dont la Courbure Totale est Constante. *C. R. Acad. Sc. Paris* 97, 848–850 (1883)

[36] Davenport, J. H., Siret, Y., and Tournier, E.: Computer Algebra Systems and Algorithms for Algebraic Computation. Academic Press, London 1993

[37] Debnath, L.: Nonlinear Partial Differential Equations for Scientists and Engineers, 2nd edition. Birkhäuser, Boston, MA 2005

[38] Ding, X.-X. and Liu, T.-P. (Eds.): Nonlinear Evolutionary Partial Differential Equations. Studies in Advanced Mathematics (Yau, S.-T., Ed.). American Mathematical Society. International Press, Providence, RI 1997

[39] d'Alembert, J. L. R.: Investigation of the Curve Formed by a Vibrating String. *In: Acoustics: Historical and Philosophical Development* (R. B. Lindsay, Ed.) 119–123, Dowden, Hutchinson and Ross, Stroudsburg 1973

[40] Dodge, F. T., Kana, D. D., and Abramson, N.: Liquid Surface Oscillations in Longitudinally Excited Rigid Cylindrical Containers. *AIAA J.* 3, 685–695 (1965)

[41] Douady, S., Fauve, S., and Thual, O.: Oscillatory Phase Modulation of Parametrically Forced Surface Waves. *Europhys. Lett.* 10(4), 309–315 (1989)

[42] Faraday, M.: On a Peculiar Class of Acoustical Figures, and on Certain Forms Assumed by Groups of Particles upon Vibrating Elastic Surfaces. *Phil. Trans. Roy. Soc. London.* 121, 299–340 (1831)

[43] FitzHugh, R.: Impulses and Physiological States in Theoretical Models of Nerve Membrane. *Biophys. J.* 1, 445–466 (1961)

[44] Enns, R. H. and McGuire, G. C.: Computer Algebra Recipes: An Advanced Guide to Scientific Modeling. Springer, New York 2007

[45] He, J. H.: A Variational Iteration Method — a Kind of Nonlinear Analytical Technique: Some Examples. *Int. J. Nonlinear Mech.* 34(4), 699–708 (1999)

[46] He, J. H. and Wu, X. H.: Exp-function Method for Nonlinear Wave Equations. *Chaos Solitons and Fractals* 30(3), 700–708 (2006)

[47] He, J. H. and Abdou, M. A.: New Periodic Solutions for Nonlinear Evolution Equation Using Exp-method. *Chaos Solitons and Fractals* 34, 1421–1429 (2007)

[48] Heck, A.: Introduction to Maple, 3rd edition. Springer, New York 2003

[49] Hereman, W. and Zhaung, W.: Symbolic Software for Soliton Theory. Acta Applicandae Mathematicae. *Phys. Lett. A* 76, 95–96 (1980)

[50] Hereman, W. and Angenent, S.: The Painlevé Test for Nonlinear Ordinary and Partial Differential Equations. *MACSYMA Newsletter* 6, 11–18 (1989)

[51] Hereman, W. and Nuseir, A.: Symbolic Methods to Construct Exact Solutions of Nonlinear Partial Differential Equations. *Math. Comput. Simulation* 43, 13–27 (1997)

[52] Hereman, W.: Shallow Water Waves and Solitary Waves. *In: Encyclopedia of Complexity and Systems Science* 8112–8125, Springer, Heibelberg 2009

[53] Hietarinta, J.: Hirotas Bilinear Method and Its Connection with Integrability. *In: Lect. Notes Phys.* 279–314, Springer, Berlin Heidelberg 2009

[54] Higham, N. J.: Functions of Matrices. Theory and Computation. SIAM, Philadelphia 2008

[55] Hirota, R.: Exact Solution of the Korteweg–de Vries Equation for Multiple Collisions of Solitons. *Phys. Rev. Lett.* 27, 1192–1194 (1971)

[56] Hirota, R. and Satsuma, J.: N-Soliton Solutions of Model Equations for Shallow Water Waves. *J. Phys. Soc. Jpn.* 40, 611–612 (1976)

[57] Hirota, R.: Direct Methods in Soliton Theory. *In: Topics in Current Physics* (R. Bullough and P. Caudrey, Eds.) 157–175, Springer, New York 1980

[58] Hirota, R. and Ito, M.: Resonance of Solitons in One Dimension. *J. Phys. Soc. Jpn.* 52, 744–748 (1983)

[59] Hirota, R.: The Direct Method in Soliton Theory. Cambridge University Press, Cambridge 2004

[60] Galaktionov, V. A. and Posashkov, S. A.: On New Exact Solutions of Parabolic Equations with Quadratic Nonlinearities. *Zh. Vych. Matem. i Mat. Fiziki* 29(4), 497–506 (1989)

[61] Galaktionov, V. A.: On New Exact Blow-up Solutions for Nonlinear Heat Conduction Equations. *Differential and Integral Equations* 3, 863–874 (1990)

[62] Galaktionov, V. A.: Invariant Subspaces and New Explicit Solutions to Evolution Equations with Quadratic Nonlinearities. *Proc. Royal. Soc. Edinburgh Sect. A* 125(2), 225–246 (1995)

[63] Galerkin, B. G.: Rods and Plates. Series in Some Problems of Elastic Equilibrium of Rods and Plates. *Vestn. Inzh. Tch. (USSR)* 19, 897–908 (1915)

[64] Gardner, C. S., Greene, J. M., Kruskal, M. D., and Miura, R. M.: Method for Solving the Korteweg–de Vries Equation. *Phys. Rev. Lett.* 19, 1095–1097 (1967)

[65] Geddes, K. O., Czapor, S. R., and Labahn, G.: Algorithms for Computer Algebra. Kluwer Academic Publishers, Boston 1992

[66] Getz, C. and Helmstedt, J.: Graphics with Mathematica: Fractals, Julia Sets, Patterns and Natural Forms. Elsevier Science and Technology Book, Amsterdam Boston 2004

[67] Göktaş, Ü. and Hereman, W.: Symbolic Computation of Conserved Densities for Systems of Nonlinear Evolution Equations. *J. Symb. Comput.* 24, 591–621 (1997)

[68] Gottlieb, D. and Orszag, S.A.: Numerical Analysis of Spectral Methods: Theory and Applications. CBMS–NSF Regional Conference Series in Applied Mathematics 26, SIAM, Philadelphia 1993

[69] Gradshteyn, I. S. and Ryzhik, I. M.: Tables of Integrals, Series, and Products. Academic Press, New York 1980

[70] Gray, J. W.: Mastering Mathematica: Programming Methods and Applications. Academic Press, San Diego 1994

[71] Gray, T. and Glynn, J.: Exploring Mathematics with Mathematica: Dialogs Concerning Computers and Mathematics. Addison-Wesley, Reading, MA 1991

[72] Green, E., Evans, B., and Johnson, J.: Exploring Calculus with Mathematica. Wiley, New York 1994

[73] Grosheva, M. V. and Efimov, G. B.: On Systems of Symbolic Computations (in Russian). *In: Applied Program Packages. Analytic Transformations* (Samarskii, A. A., Ed.) 30–38, Nauka, Moscow 1988

[74] Gu, X. M., Sethna, P. R., and Narain, A.: On Three-dimensional Nonlinear Subharmonic Resonant Surface Waves in a Fluid: Part I–Theory. *J. Appl. Mech.* 55, 213–219 (1988)

[75] Jiang, L., Perlin, M., and Schultz, W. W.: Period Tripling and Energy Dissipation of Breaking Standing Waves. *J. Fluid Mech.* 369, 273–299 (1998)

[76] Jimbo, M. and Miwa, T.: Solitons and Infinite Dimensional Lie Algebras. *RIMS, Kyoto Univ.* 19, 943–1001 (1983)

[77] Kadomtsev, B. B. and Petviashvili, V. I.: On the Stability of Solitary Waves in Weakly Dispersive Media. *Sov. Phys. Dokl.* 15, 539–541 (1970)

[78] Kalinitchenko, V. A., Nesterov, S. V., Sekerzh-Zenkovich, S. Ya., and Chaykovskii, A. A.: Experimental Investigation of Standing Waves Excited under the Faraday Resonance. *J. Fluid Dynamics* 30(1), 101–106 (1995)

[79] Kawahara, T.: Oscillatory Solitary Waves in Dispersive Media. *J. Phys. Soc. Japan* 33, 260–264 (1972)

[80] Jimbo M., Kruskal M. D., and Miwa T.: The Painlevé Test for the Self-dual Yang–Mills Equations. *Phys. Lett. A.* 92(2), 59–60 (1982)

[81] Krasnov, M., Kiselev, A., and Makarenko, G.: Problems and Exercises in Integral Equations. Mir Publishers, Moscow 1971

[82] Klerer, M. and Grossman, F.: A New Table of Indefinite Integrals Computer Processed. Dover, New York 1971

[83] Korn, G. A. and Korn, T. M.: Mathematical Handbook for Scientists and Engineers: Definitions, Theorems, and Formulas for Reference and Review, 2nd edition. Dover Publications, New York 2000

[84] Kowalevski, S.: Sur le Problème de la Rotation dun Corps Solide Autour dun Point Fixe. *Acta Mathematica* 12, 177–232 (1889) (Reprinted in: Kovalevskaya, S. V.: Scientific Works, AS USSR Publ. House, Moscow 1948)

[85] Kreyszig, E.: Maple Computer Guide for Advanced Engineering Mathematics, 8th edition. Wiley, New York 2000

[86] Kudryashov, N. A.: Seven Common Errors in Finding Exact Solutions of Nonlinear Differential Equations. *Commun. Nonlinear Sci. Numer. Simulat.* 14, 3507–3529 (2009)

[87] Lax, P. D.: Integrals of Nonlinear Equations of Evolution and Solitary Waves. *Comm. Pure Appl. Math.* 21, 467–490 (1968)

[88] Lapidus, L. and Pinder, G. F.: Numerical Solution of Partial Differential Equations in Science and Engineering. Wiley-Interscience, New York 1999

[89] Larsson, S. and Thomee, V.: Partial Differential Equations with Numerical Methods. Springer, New York 2008

[90] LeVeque, R. J.: Finite Difference Methods for Ordinary and Partial Differential Equations: Steady-state and Time-dependent Problems. SIAM, Philadelphia 2007

[91] Lie, S.: Theorie der Transformationsgruppen. Vol. III. Teubner, Leipzig 1893

[92] Lynch, S.: Dynamical Systems with Applications using Maple, 2nd edition. Birkhäuser, Boston 2009

[93] Maeder, R. E.: Programming in Mathematica, 3rd edition. Addison-Wesley, Reading, MA 1996

[94] Malfliet, W.: Solitary Wave Solutions of Nonlinear Wave Equations. *Am. J. Phys.* 60(7), 650–654 (1992)

[95] Matsuno, Y.: Bilinearization of Nonlinear Evolution Equations. II. Higher-order Modified Korteweg–de Vries Equations. *J. Phys. Soc. Jpn.* 49, 787–794 (1980)

[96] Matsuno, Y.: Bilinear Transformation Method. Academic Press, London 1984

[97] Meade, D. B., May, S. J. M., Cheung, C-K., and Keough, G. E.: Getting Statrted with Maple, 3rd edition. Wiley, Hoboken, NJ 2009

[98] Mickens, R. E.: Exact Solutions to a Population Model: The Logistic Equation with Advection. *SIAM Rev.* 30, 629–633 (1988)

[99] Mikhailov, A. V., Shabat, A. B. and Yamilov, R. I.: The Symmetry Approach to the Classification of Non-linear Equations. Complete Lists of Integrable Systems. *Russian Math. Surveys* 42(4), 1–63 (1987)

[100] Miles, J. W.: Nonlinear Surface Waves in Closed Basins. *J. Fluid Mech.* 75, 419–448 (1976)

[101] Miles, J. W.: Nonlinear Faraday Resonance. *J. Fluid Mech.* 146, 285–302 (1984)

[102] Miles, J. W. and Henderson, D.: Parametrically Forced Surface Waves. *Annu. Rev. Fluid Mech.* 22, 143–165 (1990)

[103] Miura, R. M., Gardner, C. S., and Kruskal, M. D.: Korteweg–de Vries Equation and Generalizations. II. Existence of Conservation Laws and Constants of Motion. *J. Math. Phys.* 9, 1204–1209 (1968)

[104] Moiseyev, N. N.: On the Theory of Nonlinear Vibrations of a Liquid. *Prikl. Mat. Mekh.* 22, 612–621 (1958)

[105] Morawetz, C. S.: Potential Theory for Regular and Mach Reflection of a Shock at a Wedge. *Comm. Pure Appl. Math.* 47(5), 593–624 (1994)

[106] Morton, K. W. and Mayers, D. F.: Numerical Solution of Partial Differential Equations: An Introduction. Cambridge University Press, Cambridge 1995

[107] Murray, J. D.: Mathematical Biology. Springer, Berlin Heidelberg 1993

[108] Nagumo, J. S., Arimoto, S., and Yoshizawa, S.: An Active Pulse Transmission Line Simulating Nerve Axon. *Proc. Inst. Radio Engineers* 20, 2061–2071 (1962)

[109] Nayfeh, A.: Perturbation Methods. Wiley, New York 1973

[110] Naz, R.: Symmetry Solutions and Conservation Laws for Some Partial Differential Equations in Fluid Mechanics, Doctor of Philosophy Thesis. University of the Witwatersrand, Johannesburg 2008

[111] Nesterov, S. V.: Resonant Interactions of Surface and Internal Waves. *Izv. Atmos. Ocean. Phys.* 8, 252–254 (1972)

[112] Ockendon, J. R. and Ockendon, H.: Resonant Surface Waves. *J. Fluid Mech.* 59, 397–413 (1973)

[113] Okamura, M.: Resonant Standing Waves on Water of Uniform Depth. *J. Phys. Soc. Jpn.* 66, 3801–3808 (1997)

[114] Olver, P. J.: Evolution Equations Possessing Infinitely Many Symmetries. *J. Math. Phys.* 18, 1212–1215 (1977)

[115] Olver, P. J.: Euler Operators and Conservation Laws of the BBM Equation. *Math. Proc. Camb. Phil. Soc.* 85, 143–160 (1979)

[116] Orszag, S. A.: Spectral Methods for Problems in Complex Geometries. *J. Comput. Phys.* 37, 70–92 (1980)

[117] Painlevé, P.: Leçons sur la Théorie Analytique des Équations Différentielles, Hermann, Paris 1897

Online version: The Cornell Library Historical Mathematics Monographs, *http://historical.library.cornell.edu/*

[118] Parkes, E. J. and Duffy, B. R.: An Automated Tanh-function Method for Finding Solitary Wave Solutions to Nonlinear Evolution Equations. *Comput. Phys. Commun.* 98, 288–300 (1996)

[119] Penney, W. G. and Price, A. T.: Finite Periodic Stationary Gravity Waves in a Perfect Liquid, Part 2. *Phil. Trans. R. Soc. Lond.* A 224, 254–284 (1952)

[120] Perring, J. K. and Skyrme, T. R.: A Model Unified Field Equation. *Nuclear Physics* 31, 550–555 (1962)

[121] Poincarè, H.: Sur les Équations aux Dérivés Partielles de la Physique Mathématique, *Amer. J. Math.* 12, 211–294 (1890)

[122] Polyanin, A. D. and Manzhirov, A. V.: Handbook of Integral Equations, 2nd edition. CRC Press, Boca Raton 2008

[123] Polyanin, A. D.: Handbook of Linear Partial Differential Equations for Engineers and Scientists. Chapman and Hall/CRC Press, Boca Raton 2002

[124] Polyanin, A. D. and Zaitsev, V. F.: Handbook of Nonlinear Partial Differential Equations, 2nd edition. Chapman and Hall/CRC Press, Boca Raton 2011

[125] Polyanin, A. D. and Manzhirov, A. V.: Handbook of Mathematics for Engineers and Scientists. Chapman and Hall/CRC Press, Boca Raton, London 2006

[126] Polyanin, A. D. and Zhurov, A. I.: Exact Solutions to Nonlinear Equations of Mechanics and Mathematical Physics. *Doklady Physics* 43(6), 381–385 (1998)

[127] Rayleigh, Lord: Deep Water Waves, Progressive or Stationary, to the Third Order of Approximation. *Phil. Trans. R. Soc. Lond.* A 91, 345–353 (1915)

[128] Richards, D.: Advanced Mathematical Methods with Maple. Cambridge University Press, Cambridge 2002

[129] Ross, C. C.: Differential Equations: An Introduction with Mathematica. Springer, New York 1995

[130] Sanders, J. A. and Roelofs, M.: An Algorithmic Approach to Conservation Laws Using the 3-dimensional Heisenberg Algebra. Tech. Rep. 2. RIACA, Amsterdam 1994

[131] Sarmin, E. N. and Chudov, L. A.: On the Stability of the Numerical Integration of Systems of Ordinary Differential Equations Arising in the Use of the Straight Line Method. *USSR Computational Mathematics and Mathematical Physics* 3(6), 1537–1543 (1963)

[132] Sedov, L. I.: Similarity and Dimensional Methods in Mechanics. Chapters 4, 5. CRC Press, Boca Raton 1993

[133] Sekerzh-Zenkovich, Y. I.: On the Theory of Standing Waves of Finite Amplitude. *Doklady Akad. Nauk. USSR* 58, 551–554 (1947)

[134] Schiesser, W. E.: The Numerical Methods of Lines. Academic Press, New York 1991

[135] Schiesser, W. E.: Computational Mathematics in Engineering and Applied Science: ODEs, DAEs, and PDEs. CRC Press, Boca Raton 1994

[136] Silebi, C. A. and Schiesser, W. E.: Dynamic Modeling of Transport Process Systems. Academic Press, San Diego 1992

[137] Lie, H. J. and Schiesser, W. E.: Ordinary and Partial Differential Equation Routines in C, C++, Fortran, Java, Maple, and MATLAB. Chapman and Hall/CRC Press, Boca Raton 2004

[138] Schiesser, W. E. and Griffiths, G. W.: A Compendium of Partial Differential Equation Models: Method of Lines Analysis with MATLAB. Cambridge University Press, Cambridge 2009

[139] Shingareva, I.: Investigation of Standing Surface Waves in a Fluid of Finite Depth by Computer Algebra Methods. PhD thesis, Institute for Problems in Mechanics, Russian Academy of Sciences, Moscow 1995

[140] Shingareva, I. and Lizárraga-Celaya, C.: High-order Asymptotic Solutions to Free Standing Water Waves by Computer Algebra. Proc. Maple Summer Workshop (Lopez, R. J., Ed.), 1–28, Waterloo, Ontario, Canada 2004

[141] Shingareva, I., Lizárraga Celaya, C., and Ochoa Ruiz, A. D.: Maple y Ondas Estacionarias. Problemas y Soluciones. Editorial Unison, Universidad de Sonora, Hermosillo, México 2006

[142] Shingareva, I. and Lizárraga-Celaya, C.: On Frequency-amplitude Dependences for Surface and Internal Standing Waves. *J. Comp. Appl. Math.* 200, 459–470 (2007)

[143] Shingareva, I. and Lizárraga-Celaya, C.: Maple and Mathematica. A Problem Solving Approach for Mathematics, 2nd edition. Springer, Wien New York 2009

[144] Shingareva, I. and Lizárraga-Celaya, C: Symbolic and Numerical Solutions of Nonlinear Partial Differential Equations with Maple, Mathematica, and MATLAB, Part III. Chapters 39–41. *In: Handbook of Nonlinear Partial Differential Equations, 2nd edition.* Chapman and Hall/CRC Press, Boca Raton 2011

[145] Schultz, W. W., Vanden-Broeck, J. M., Jiang, L., and Perlin, M.: Highly Nonlinear Standing Water Waves with Small Capillary Effect. *J. Fluid Mech.* 369, 253–272 (1998)

[146] Sokolov, V. V. and Shabat, A. B.: Classification of Integrable Evolution Equations. *Soviet Scientific Rev., Section C, Math. Phys. Rev.* 4, 221–280 (1984)

[147] Sretenskii, L. N.: Theory of Wave Motions of Fluids. Nauka, Moscow, 1977

[148] Strikwerda, J.: Finite Difference Schemes and Partial Differential Equations, 2nd edition. SIAM, Philadelphia 2004

[149] Subramanian, V. R. and White, R. E.: Semianalytical Method of Lines for Solving Elliptic Partial Differential Equations. *Chemical Engineering Science* 59, 781–788 (2004)

[150] Tadjbakhsh, I. and Keller, J. B.: Standing Surface Waves of Finite Amplitude. *J. Fluid Mech.* 8, 442–451 (1960)

[151] Taylor, G. I.: The Formation of a Blast Wave by a Very Intense Explosion. I. Theoretical Discussion. *Proc. Roy. Soc.* A 201, 159–174 (1950)

[152] Thomas, W.: Numerical Partial Differential Equations: Finite Difference Methods. Springer, New York 1995

[153] Virnig, J. C., Berman, A. S., and Sethna, P. R.: On Three-dimensional Nonlinear Subharmonic Resonant Surface Waves in a Fluid: Part II — Experiment. *J. Appl. Mech.* 55, 220–224 (1988)

[154] Vvedensky, D. D.: Partial Differential Equations with Mathematica. Addison-Wesley, Wokingham 1993

[155] Wazwaz, A. M.: Necessary Conditions for the Appearance of Noise Terms in Decomposition Solution Series. *Appl. Math. Comput.* 81, 199–204 (1997)

[156] Wazwaz, A. M.: A Reliable Modification of Adomians Decomposition Method. *Appl. Math. Comput.* 92(1), 1–7 (1998)

[157] Wazwaz, A. M.: Partial Differential Equations: Methods and Applications. Balkema Publishers, Leiden 2002

[158] Wazwaz, A. M.: The Modified Decomposition Method for Analytic Treatment of Differential Equations. *Appl. Math. Comput.* 173(1), 165–176 (2006)

[159] Wazwaz, A. M.: Peakons, Kinks, Compactons and Solitary Patterns Solutions for a Family of Camassa–Holm Equations by Using New Hyperbolic Schemes. *Appl. Math. Comput.* 182(1), 412–424 (2006)

[160] Weiss J., Tabor M., and Carnevale G.: The Painlevé Property for Partial Differential Equations. *J. Math. Phys.* 24, 522–526 (1983)

[161] Weiss J.: The Painlevè Property for Partial Differential Equations. II. Bäcklund Transformation, Lax Pairs, and the Schwarzian Derivative. *J. Math. Phys.* 24(6), 1405–1413 (1983)

[162] Weisstein, E. W.: CRC Concise Encyclopedia of Mathematics, 2nd edition. CRC Press, Boca Raton 2003

[163] Wester, M. J.: Computer Algebra Systems: A Practical Guide. Wiley, Chichester, UK 1999

[164] White, R. E. and Subramanian, V. R.: Computational Methods in Chemical Engineering with Maple. Springer, Berlin Heidelberg 2010

[165] Whittaker, E. T.: A Treatise on Analytical Dynamics of Particles and Rigid Bodies. Dover Publications, New York 1944

[166] Wickam-Jones, T.: Mathematica Graphics: Techniques and Applications. Springer, New York 1994

[167] Wolf, T.: A Comparison of Four Approaches for the Calculation of Conservation Laws. *Europ. J. Appl. Math.* 13, 129–152 (2002)

[168] Wolfram, S.: A New Kind of Science. Wolfram Media, Champaign, IL 2002

[169] Wolfram, S.: The Mathematica Book, 5th edition. Wolfram Media, Champaign, IL 2003

[170] Xie, F. D. and Chen, Y.: An Algorithmic Method in Painlevé Analysis of PDE. *Comput. Phys. Commun.* 154, 197–204 (2003)

[171] Xu, G. Q. and Li, Z. B.: Symbolic Computation of the Painlevé Test for Nonlinear Partial Differential Equations using Maple. *Comput. Phys. Commun.* 161, 65–75 (2004)

[172] Yosihara, H.: Gravity Waves on the Free Surface of an Incompressible Perfect Fluid of Finite Depth. *Kyoto Univ. Math.* 18, 49–96 (1982)

[173] Zabusky, N. J. and Kruskal, M. D.: Interaction of Solitons in a Collisionless Plasma and the Recurrence of Initial States. *Phys. Rev. Lett.* 15, 240–243 (1965)

[174] Zakharov, V. E. and Shabat, A. B.: Exact Theory of Two-Dimensional Self-focusing and One-dimensional Self-modulation of Waves in Nonlinear Media. *Soviet Phys. JETP* 34, 62–69 (1972)

[175] Zimmerman, R. L. and Olness, F.: Mathematica for Physicists. Addison-Wesley, Reading, MA 1995

Index